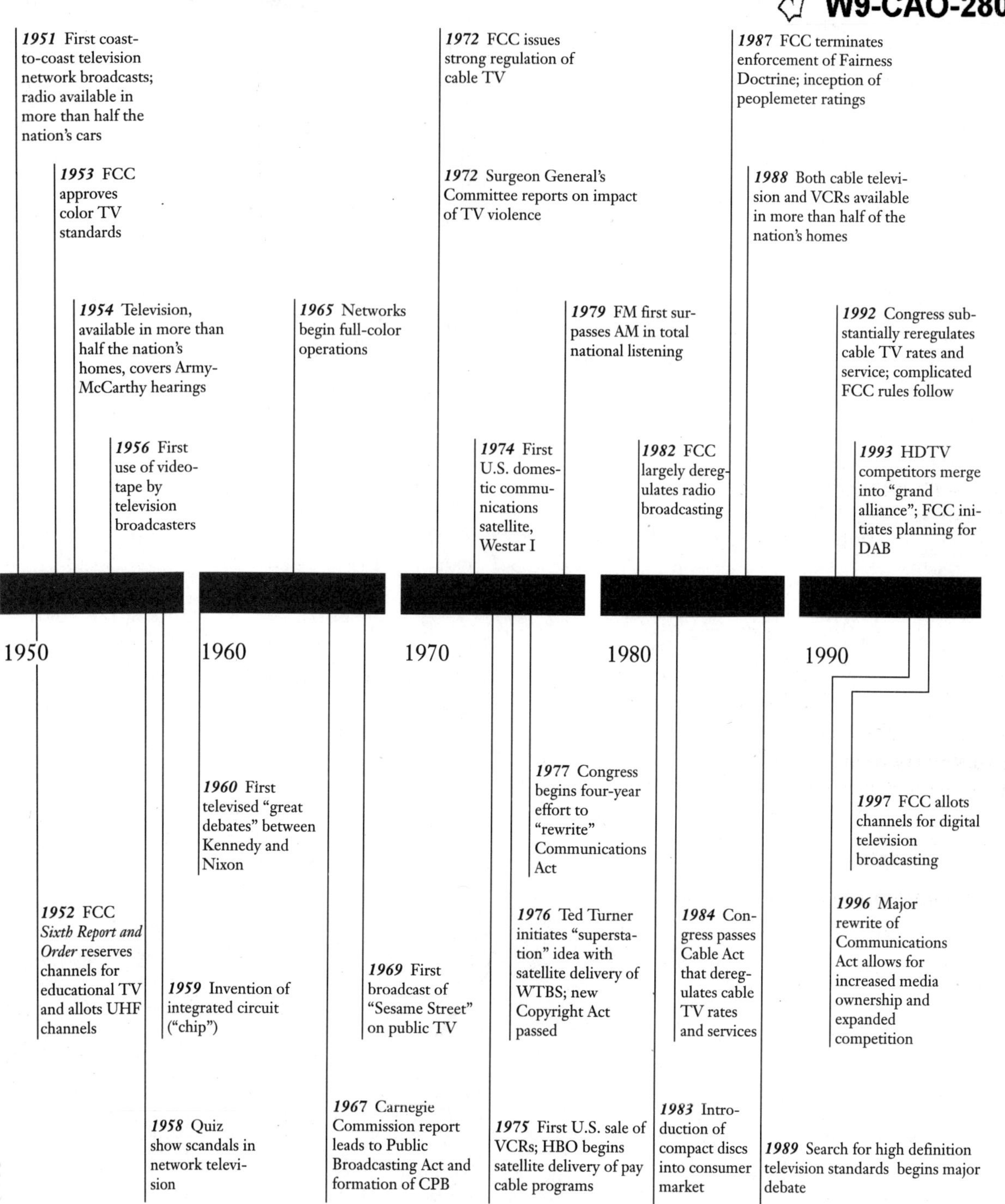
1951 First coast-to-coast television network broadcasts; radio available in more than half the nation's cars
1953 FCC approves color TV standards
1954 Television, available in more than half the nation's homes, covers Army-McCarthy hearings
1956 First use of video-tape by television broadcasters
1965 Networks begin full-color operations
1972 FCC issues strong regulation of cable TV
1972 Surgeon General's Committee reports on impact of TV violence
1974 First U.S. domestic communications satellite, Westar I
1979 FM first surpasses AM in total national listening
1982 FCC largely deregulates radio broadcasting
1987 FCC terminates enforcement of Fairness Doctrine; inception of peoplemeter ratings
1988 Both cable television and VCRs available in more than half of the nation's homes
1992 Congress substantially reregulates cable TV rates and service; complicated FCC rules follow
1993 HDTV competitors merge into "grand alliance"; FCC initiates planning for DAB
1950
1960
1970
1980
1990
1952 FCC *Sixth Report and Order* reserves channels for educational TV and allots UHF channels
1958 Quiz show scandals in network television
1959 Invention of integrated circuit ("chip")
1960 First televised "great debates" between Kennedy and Nixon
1967 Carnegie Commission report leads to Public Broadcasting Act and formation of CPB
1969 First broadcast of "Sesame Street" on public TV
1975 First U.S. sale of VCRs; HBO begins satellite delivery of pay cable programs
1976 Ted Turner initiates "superstation" idea with satellite delivery of WTBS; new Copyright Act passed
1977 Congress begins four-year effort to "rewrite" Communications Act
1983 Introduction of compact discs into consumer market
1984 Congress passes Cable Act that deregulates cable TV rates and services
1989 Search for high definition television standards begins major debate
1996 Major rewrite of Communications Act allows for increased media ownership and expanded competition
1997 FCC allots channels for digital television broadcasting

Broadcasting in America

A Survey of Electronic Media

Eighth Edition

SYDNEY W. HEAD
Late of the University of Miami

CHRISTOPHER H. STERLING
The George Washington University

LEMUEL B. SCHOFIELD
University of Miami

THOMAS SPANN
University of Nebraska

MICHAEL A. McGREGOR
Indiana University

Houghton Mifflin Company Boston New York

Sponsoring Editor: George T. Hoffman
Editorial Assistant: Kara Maltzahn
Packaging Services Supervisor: Charline Lake
Senior Production/Design Coordinator: Sarah Ambrose
Senior Manufacturing Coordinator: Priscilla J. Abreu
Marketing Manager: Pamela J. Laskey

Cover design: Linda Wade, Wade Design
Cover illustration: Linda Wade, Wade Design

Chapter Opening Photo Credits
2 Eric Sander / The Gamma Liaison; **18** UPI/Corbis-Bettmann; **54** AP/Wide World; **82** © Harry Heleotis / Liaison International; **118** Courtesy NASA; **150** The Gamma Liaison; **192** Photofest; **214** Photofest; **238** Photofest; **276** Nielsen Media Research; **302** Michael Siluk / The Image Works; **328** © Dunn Photographic Associates, Inc.; **360** AP Photo / Susan Walsh; **394** Photo Researchers

Printed in the U.S.A.

Library of Congress Catalog Card Number: 97-72479

ISBN: 0-395-87371-1

456789-DH-01 00 99

Brief Contents

Contents

List of Exhibits

Preface

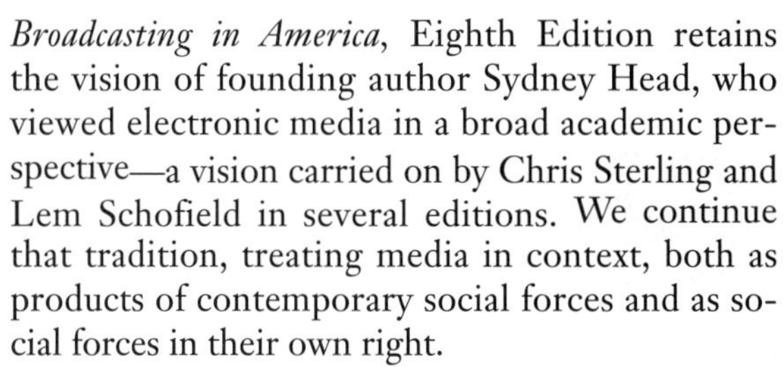

Broadcasting in America, Eighth Edition retains the vision of founding author Sydney Head, who viewed electronic media in a broad academic perspective—a vision carried on by Chris Sterling and Lem Schofield in several editions. We continue that tradition, treating media in context, both as products of contemporary social forces and as social forces in their own right.

This eighth edition of *Broadcasting in America*, however, differs from previous editions in several ways. In response to a changing market and concerns expressed about the size of this text, we have attempted in this edition to "split the difference" between previous editions of the full text and the streamlined brief versions. We have reduced the text to fourteen chapters, which better suits classes taught on the quarter system and those in which supplementary readings are assigned. However, the breadth of content more closely parallels previous full-size editions than the brief versions.

Coverage of Recent Developments

This edition continues the ongoing process of updating the reader about the never-ending changes in electronic media. Such changes seemingly occur with greater rapidity every year: Congress enacts new legislation, the FCC is charged with implementing it, industries react and reconfigure themselves, and consumers face a constant need to reassess their telecommunications needs in an ever-changing environment. We have covered developments that occurred in the four years between publication of the seventh edition and the production of this text, including the following:

- Passage and implementation of the Telecommunications Act of 1996, which fundamentally changed the electronic media landscape
- Mergers and acquisitions of electronic media companies, including the Cap Cities/ABC merger with Disney, the Westinghouse acquisition of CBS, and the increased consolidation of media ownership, especially in radio
- The demise of the 25-year-old Prime Time Access Rule and the Syndication and Financial Interest rules
- The establishment of an advanced digital television standard and the FCC's efforts to implement the transition to digital broadcasting
- The roll out of DBS and its effects on existing broadband delivery services such as cable television
- The O.J. Simpson trial, whose many facets and ramifications dictate its inclusion in several sections of the book

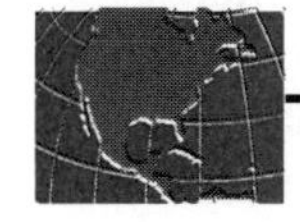

Changes in the Eighth Edition

Broadcasting in America, Eighth Edition continues coverage of all electronic media, some of which have only recently emerged as *mass* media. Many chapters, for example, contain expanded consideration of the Internet and telephone networks. We have also provided updates on broadband delivery systems, such as DBS and MMDS, plus covered newer systems, such as DARS. Specific changes in this edition include the following:

- A new Chapter 1, modified from parts of Chapter 16 of the seventh edition and Chapter 1 of the Brief Version, examines the emerging information infrastructure and the accelerating convergence of various electronic media forms. The chapter also introduces students to some of the vocabulary of electronic media in the United States.
- Chapter 2 combines Chapters 1 and 2 of the seventh edition into one chapter that covers the history of traditional broadcasting. Chapter 3 addresses the history and increasing importance of other electronic media, including telephony and the Internet.
- New technical material in Chapters 4 and 5 explores the developing digital revolution, high-definition television, and the complexities of sophisticated networking and switching, while retaining traditional coverage of electronic media basics.
- A revised Chapter 6 blends information from the seventh edition's Chapters 6 and 7 into one chapter on the basics of commercial electronic media organization and economics.
- An updated Chapter 7, "Noncommercial Services," which corresponds to Chapter 8 in the seventh edition, integrates program and audience material previously scattered throughout the book.
- Revised Chapters 8 and 9 describe how media mergers and acquisitions as well as technological developments such as DBS and the Internet affect programming and program production. These chapters place special emphasis on the emergence of cable television as a major player in electronic media programming.
- New information in Chapter 10 treats the increasingly complex problem of measuring electronic media audiences; and an updated Chapter 11 discusses our understanding of how the electronic media affect us as individuals and as a society.
- Extensive revision of information is provided in chapters dealing with the legal arena. Much of this results from major modifications in the law brought about by the Telecommunications Act of 1996. These changes and their effects are detailed in Chapter 12, which deals primarily with licensing and noncontent regulation, and Chapter 13, which treats constitutional and content-related issues.
- A concluding Chapter 14, "Global View," reviews the rapid and radical changes in electronic media in other countries.

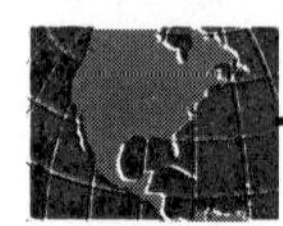

Ancillary Support

We fortunately retained the services of Dr. Louise Benjamin of the University of Georgia to prepare a new *Instructor's Resource Manual* for this eighth edition. Professor Benjamin prepared similar materials for earlier editions of the full text and the brief versions and brings with her invaluable experience. The *Manual* presents chapter analyses and summaries, reviews learning objectives and key concepts, provides lecture and project sections, and includes a test bank. As always, the publisher makes these materials available to adopters and provides the test items to instructors on computer disk in IBM or Macintosh formats. Also, frequent "newsletter" updates will be available on the Houghton Mifflin Web site (http://www.hmco.com/), offering the latest developments in an ever-changing field.

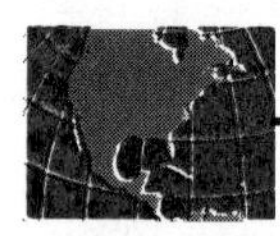

Acknowledgments

As always in a work like this, many people deserve our thanks and appreciation. Walter Gantz, Susan Eastman, Herb Terry, Ron Osgood, and Rod Rightmire, all colleagues of Mike McGregor at Indiana University, as well as Pete Seel at Colorado State and Ed Fink at California State–Fullerton, contributed support, information, and advice throughout. Hailing Li, a graduate student at Indiana, provided valuable research assistance, especially with Chapter 14. Also many thanks to Lisa Lindsey for her sense of humor and for doing all those "little things" that make an author's life easier.

Likewise, Tom Spann wishes to thank his Nebraska colleagues: Richard Alloway, Laurie Lee, Peter Mayeux, Jerry Renaud, and Larry Walklin. They provided continuous support and encouragement, without which his participation in this project would have been impossible.

Jack Loftus and Chris Carrico at Nielsen Media Research and Shelly Cagner at Arbitron offered valuable insights and updates on audience measurement. Scott MacDonald at *Advertising Age*, Serena Wyckoff at CPB, and Rick Ducey and Brian Savoie at the National Association of Broadcasters pointed us to important information on the current status of the electronic media industries. Many members of the Federal Communications Commission staff, including Mary Izzard, Roger Holberg, Sandra Watson, Brian Carter, Rebecca Willman, Nikki Shears, Son Nguyen, Debra Gonzales, and Gary Kalagian, provided crucial information on a wide variety of topics. Many people also helped with information about programming policies and practices, including Richard Bates (CableVision), Donald Browers (KMTV, Omaha), Brad Gonzales (KPTM, Omaha), and Nancy Woulfe (Home Box Office). Mark Potts provided helpful information about New Century Network and the World Wide Web. Paula Henderson of Scientific Atlanta helped with updates related to cable technology and Nebraska Educational Telecommunications Commission's Paul Sautter answered many questions dealing with technical issues. Thanks also to Marty Bender and Dean Metcalf at WFBQ, Indianapolis.

Kara Maltzahn at Houghton Mifflin deserves special thanks for her encouragement and for keeping us on track when the task seemed a bit overwhelming. Thanks also to Robine Andrau at PrePress Company, Inc. who provided encouragement and discipline as the deadline approached ever closer. Mike McGregor also thanks his sons, Geoff and Dan, for being supportive of this project and for thinking of so many ways to spend whatever royalties accrue.

Finally, we would like to thank the following reviewers for their comments and suggestions:

David Martin
California State University–Sacramento

Gary C. Dreibelbis
Solano College

Rich Houlberg
San Francisco State University

Gerald Weaver
Mesa State College

Charles Clift
Ohio University

John Sanchez
American University

Edward Galizia
Art Institute of Fort Lauderdale

Tom Spann
Mike McGregor

CNN CNN

Chapter 1

Introducing Electronic Media

CHAPTER CONTENTS

You already know something about American electronic mass media even before you pick up this book. Like most of your friends, you probably use radio, television, cable, videocassette recorders, CDs, and perhaps networked computers daily—to the point where they take up a substantial amount of your time. So why, then, read a textbook on something readily available in your life? What can you possibly learn that you don't already know? Well, ask yourself these questions:

- Where did today's media giants come from? How, for example, did CBS, MTV, and Ted Turner get their start?
- How are programs made, how are they paid for, who chooses them, and why?
- Do we watch the same kinds of programs our parents did? Is anything really "new"?
- Just how accurate are those ratings we hear about? ("Nobody ever asks me what I watch!")
- How much does government control what we see and hear? Should it control more? Or less?
- And why do we have a system of electronic media so different from those in most other countries?

Studying electronic media as the year 2000 approaches is an increasingly exciting thing to do. Rarely has such dramatic and rapid change occurred in the ways people communicate. Reports of new developments appear almost daily. Yet many of us take all this as matter-of-fact evolution. After all, most young people in America have grown up with computers, video games, and multichannel television, and they take these artifacts quite for granted. But staying on top of what some regard as a media revolution, with all of its twists and turns, its subtleties and its implications, is a challenge for any serious student of communication.

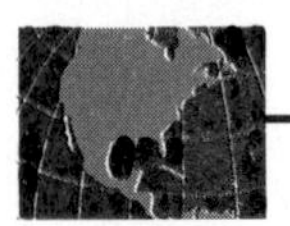

1.1 *The Information Superhighway*

Some who write about the burgeoning world of electronic media have found it useful to refer to futuristic systems of high-capacity, high-speed paths across the country and around the globe as the *information superhighway* (see Exhibit 1.a).

What It Means

One organization that explores such matters defines the information superhighway as

> . . . a system of pathways that will provide unlimited information—news, entertainment, data, personal communications—to the people who want it. That information will be available widely, easily and in mixtures of text, audio and video. Consumers will be able to send and receive information and choose when they do it. . . . [W]hether by computer, TV, phone or a hybrid of them, people will be able to "travel" the superhighway to get news, shop, socialize, work, attend school, and have fun without having to leave home. (*The Freedom Forum*, May 1994: 4)

Creative adopters of this metaphor have both expanded and subdivided it, suggesting that the superhighway has on-ramps and off-ramps, potholes and traffic jams, tolls and bridges and tunnels, speed limits or *no* speed limits, supersidewalks (for those who wish to move at a slower pace), even roadkill (to describe the unsuccessful would-be traveler).

Like any wonderfully descriptive metaphor, this one, too, has become a cliché through overuse. To avoid it, writers have fashioned substitutes and variations: *infobahn*, *I-way*, *info-highway*, *electronic highway*, *data highway*, *communicopia*—the list is endless. Detractors refer to the *info-hypeway;* visionaries speak of a *field of dreams* ("If you build it, they will come").

Although the first use of the term remains obscure, Vice President Al Gore, the Clinton administration's leading spokesperson on the subject, claims he coined "information superhighway" in 1978. The federal government laid out its expectations for the highway as a key to general economic improvement and individual well-being in a 1993 report that referred to the *National Information Infrastructure* (NII), an expression that sounds somehow more official.

But as with any good metaphor, "information superhighway" likely will be widely used until a better expression surfaces. According to a Freedom Forum study, selected newspapers, magazines, and broadcasters employed the phrase, or one of its variations, nearly 3,000 times between January 1992 and February 1994. Still, the *Miami Herald* reported three months later (while further refining the metaphor) that

> [T]he information superhighway may be one of the buzzwords of the year, but most people are still in the computerized ditch. A Louis Harris poll found that two out of three Americans had not read or heard anything about [it]. . . . Even among the 34 percent who knew the term, fewer than half said they understood it well. All the same, 27 percent thought the electronic network an excellent idea. (*Miami Herald*, 9 May 1994: 7A)

People are certainly more aware of the information superhighway today than they were half a decade ago, but many Americans know of it only superficially. They have not actually pulled onto the tarmac.

Convergence

Whatever we choose to call it, the information superhighway will result from *convergence*—the coming together of, and blurring of lines between, what until now had been essentially discrete communication forms (broadcast, cable, telephony, computers, mail). This convergence is not only of technologies, of "hardware," of means of transmission and delivery, but also of content. When complete, it will be *seamless* (without interruption) and *transparent* (with separate elements undetectable by users).

Convergence results in what most call *multimedia* (although some, in what seems a bit of a reach, refer to *telemedia* or even *telepower*)—the combination of "more than one medium, where the media can include speech, music, text, data, graphics, fax, image, video and animation. A key point is that different media are integrated—that is, they are linked, synchronized and commonly controlled. Connections are set up in an integrated way. The media can also be interactive or non-interactive" (Mayo, 1992). Multimedia, in turn, serve as foundation for the superhighway (there's that metaphor again). Exhibit 1.b offers one concept of what a future home multimedia setup may look like.

Divergence

But back to potholes and roadkill. Some observers suggest that the superhighway is not inevitable, or at least that it will not achieve wide implementation for many years. Most concerns center around basic economics and how consumer acceptance and government regulation will affect those economics.

Detractors point to *technophobia*—the fear of or disinterest in technology on the part of many Americans. Even in the late 1990s about a third of all U.S. households either have no access to or choose not to subscribe to what most young people regard as a necessity—cable. Although more than 80 percent of homes have VCRs, anecdotes abound describing the millions of units that mindlessly blink "12:00," awaiting their owner's electronic education; only about 40 percent have home computers; 34 percent have cellular phones; and fewer than 25 percent have camcorders (CEMA, 1997).

One survey found that only about 54 percent of respondents had any interest in interactive services—perhaps the most important aspect of the superhighway; 34 percent said they were not interested; 12 percent weren't sure (*Broadcasting & Cable*, 23 May 1994: 6). Of those interested, most (about 44 percent) said they would be willing to pay for *video-on-demand* (VOD), a service that would provide subscribers with a menu of movie titles and allow them to watch individual films on their home screens—for a price—whenever they wanted to. But some cable operators doubted that actual consumer use would justify the enormous investment necessary to install VOD, pointing to studies that show people are interested mostly in the latest hit movies. As for other superhighway services (on-line information services, video games, home shopping, etc.), none attracted even 25 percent of respondents. Many of the survey's predictions remain valid. Perhaps the most publicized (or hyped) lane on the information superhighway is the *World Wide Web* (WWW), a multimedia interactive feature of the "network of networks" known as the *Internet*. Yet by 1997 only about 17 percent of the population used the Web and fewer than 25 percent used the Internet at all, even for electronic mail (Nielsen, 1997).

Exhibit 1.a

The Information Superhighway

Navigators on the information superhighway may need a roadmap to find their way, as content providers—such as the examples represented on the left—deliver their services (many of them interactive) via cable systems, telephone companies, and satellites, over wires and fiber optics and through the air to individuals, businesses, and governmental and educational institutions. Interactivity provided by the World Wide Web and other components of the information superhighway blur the traditional distinction between content providers and content receivers. Individuals and institutions depicted on the right side of this exhibit increasingly create, as well as receive, content.

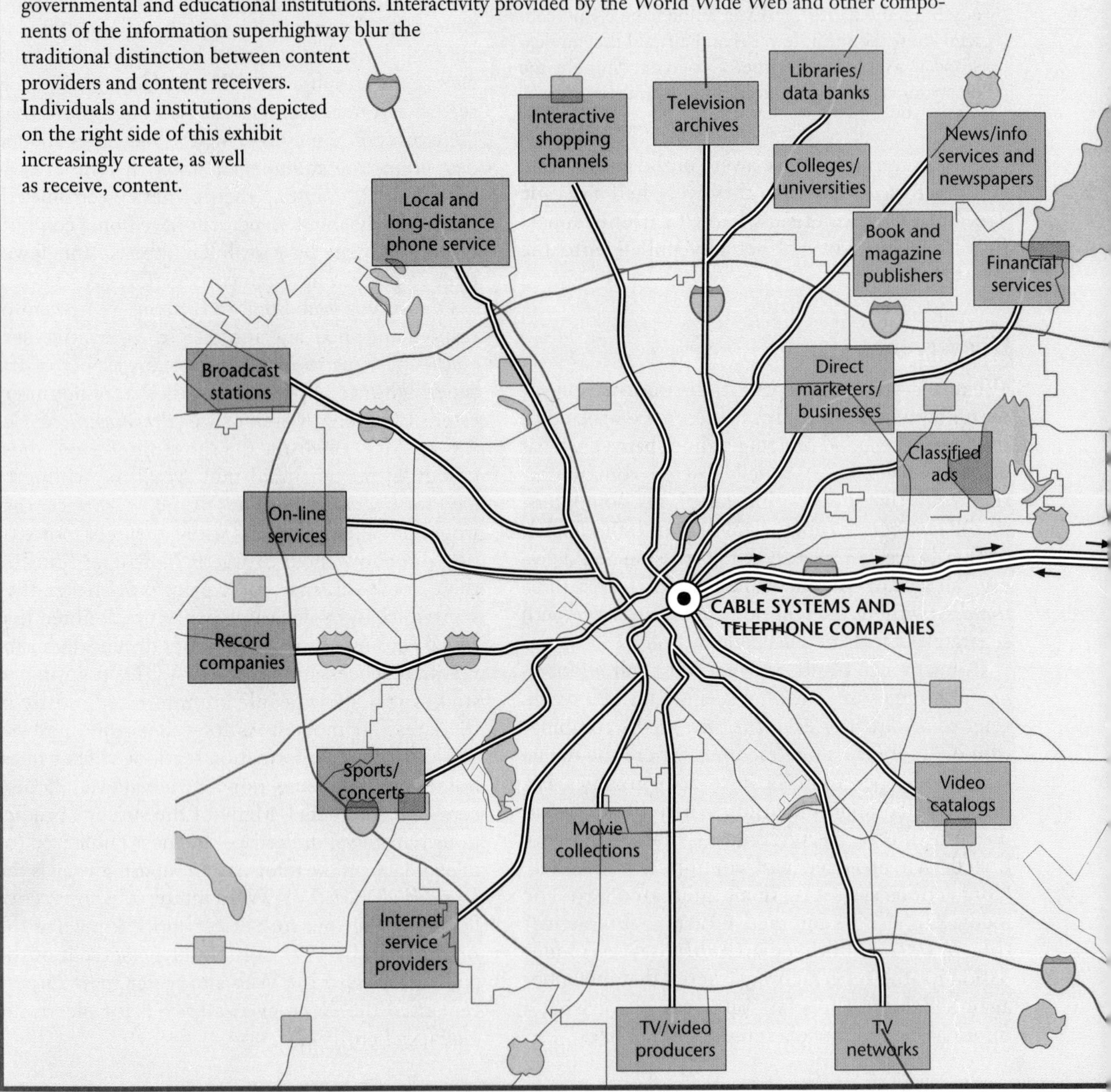

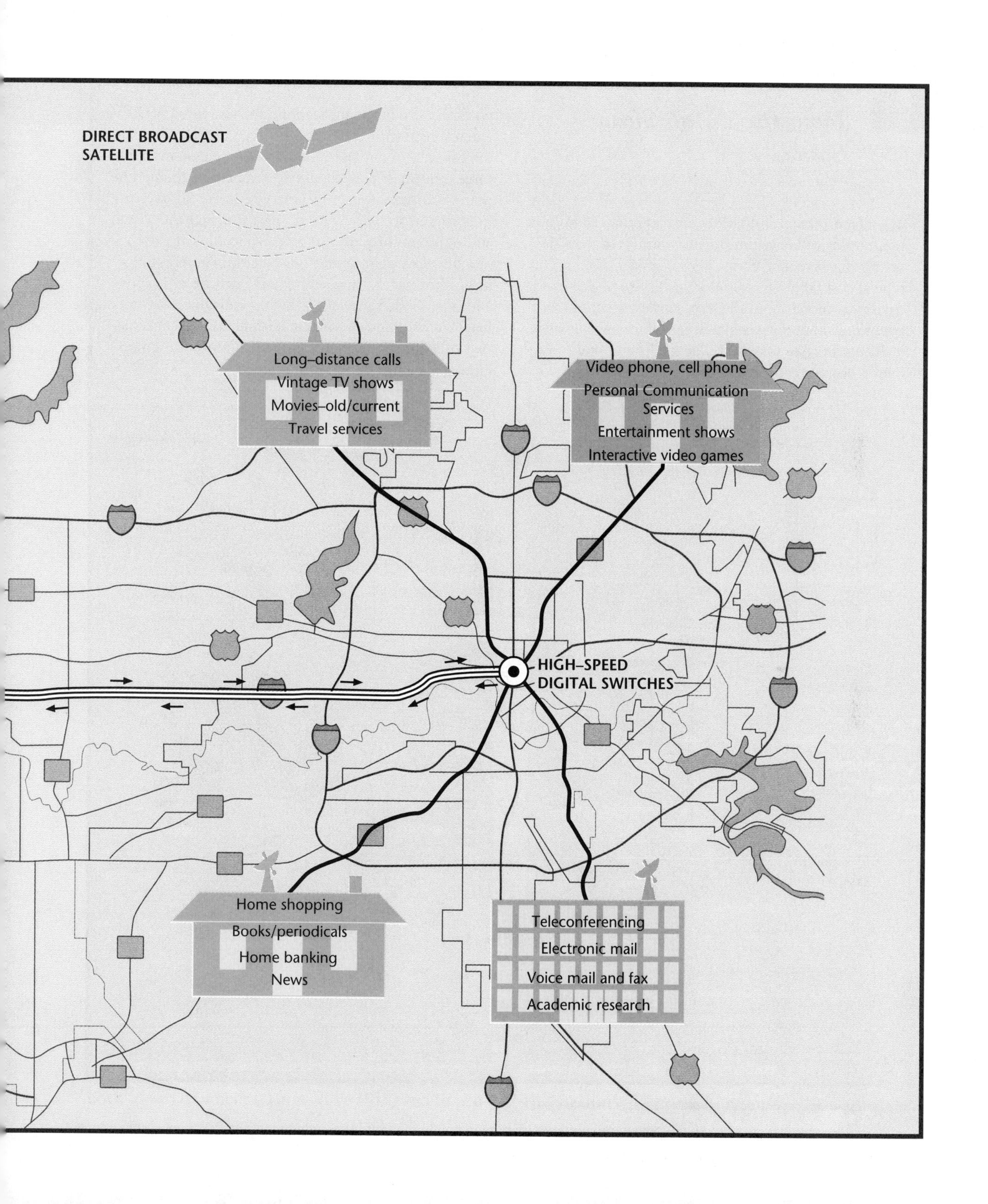

DIRECT BROADCAST SATELLITE
Long–distance calls
Vintage TV shows
Movies–old/current
Travel services
Video phone, cell phone
Personal Communication Services
Entertainment shows
Interactive video games
HIGH–SPEED DIGITAL SWITCHES
Home shopping
Books/periodicals
Home banking
News
Teleconferencing
Electronic mail
Voice mail and fax
Academic research

Exhibit 1.b

Tomorrow's Multimedia Home

Manufacturers optimistically estimate that 16 million households will have multimedia "home theaters" featuring big-screen TVs by 1998 (CEMA, 1997). The typical cost (about $3,000) for such systems is a major barrier to adoption. In the next century the number of communication devices available will increase dramatically, and prices may fall. The well-equipped multimedia home may contain many items pictured. The potential size of the projection, wall-mounted HDTV video screen (five feet across or even more), as well as the capacity of digital multi-track sound systems could force multimedia use to take over an entire room. The home multimedia center features an integrated and fully digital set of devices to offer multiple video and audio inputs through integrated decoders at the flick of a universal remote button. Not shown here, but an important part of any such room, will be interactive audio and video services allowing consumers to "talk back" to program producers. Home units for interactive video telephone and personal communication services (PCS) may be a part of such rooms.

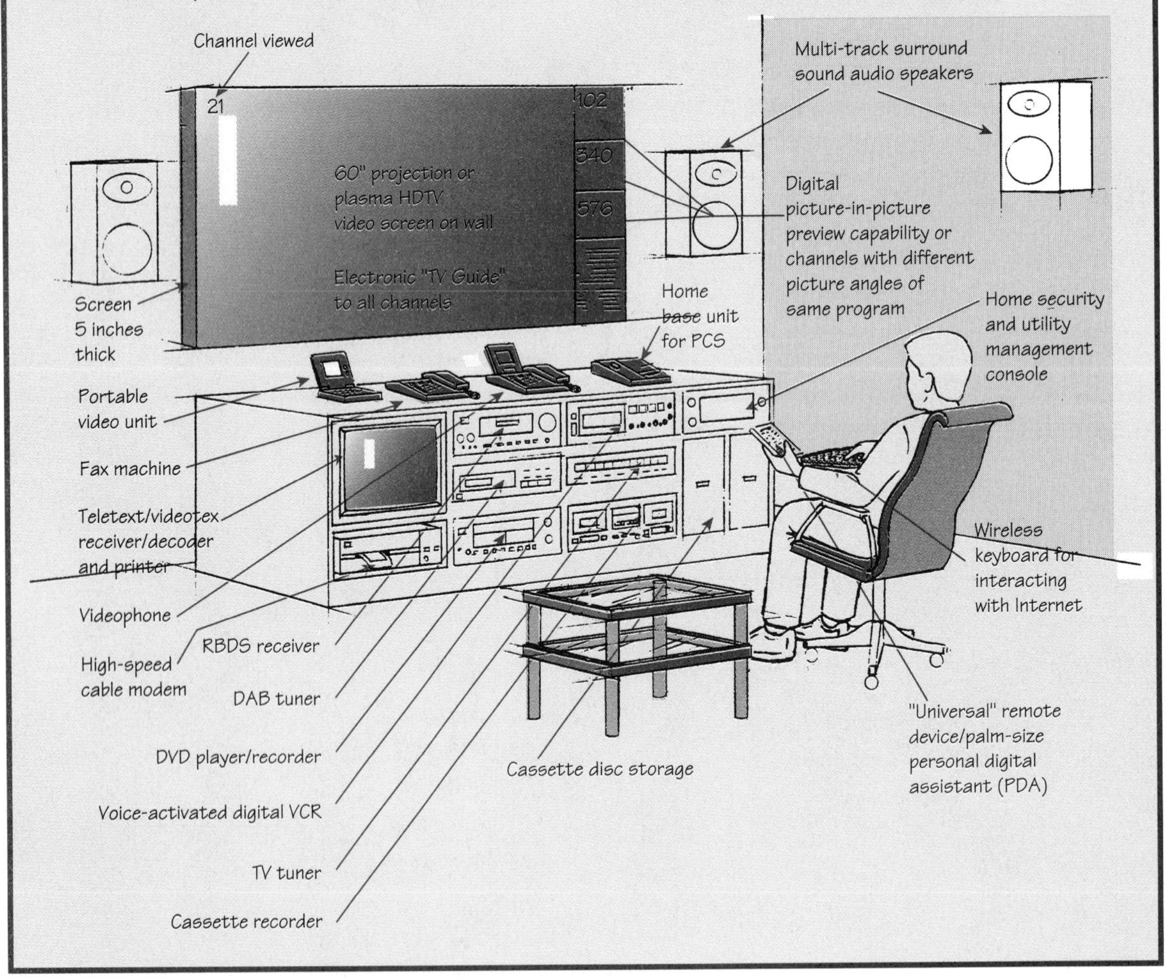

In the early 1990s the government sometimes sent conflicting messages. On the one hand, the Clinton administration in 1993 pledged $5 billion to help offset the costs of developing software and equipment necessary to build the superhighway. Additionally, Vice President Al Gore continues as the major cheerleader for bringing the benefits of the "Information Age" to all U.S. citizens. On the other hand, critics suggest that the last thing America needs is a politically backed or bureaucratically operated information infrastructure.

Some blamed government for the 1994 collapse of the historic combination of Bell Atlantic Corporation (a Regional Bell [telephone] Operating Company, or *RBOC*) and Tele-Communications Inc. (TCI, the nation's largest cable operator)—a union that offered high hopes of jump-starting the ride toward the superhighway. Valued at some $33 billion, the merger fell apart soon after the Federal Communications Commission ordered cutbacks in fees cable companies charge their subscribers. Other observers discounted the notion that government was at fault, pointing instead to ego, greed, and personality clashes of the companies' top executives as the real culprits.

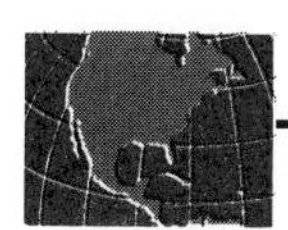

1.2 The Players

Neither Bell Atlantic nor TCI abandoned all expansion plans, however (as detailed later). Rather, they and others continued—albeit somewhat more cautiously, perhaps—to explore the best routes they might take. Although, as noted earlier, distinctions among media grow ever more cloudy, a better understanding of the convergence phenomenon requires a closer look at the major players.

Broadcast

Traditional radio and television run the greatest risk of being stalled at the on-ramp of the information superhighway. With their single, one-way-only channels, they do not fit easily into a multichannel, interactive future. But they are not dinosaurs yet. Nor need they become extinct if, as now seems the case, they move into the 21st century in their outlook and operation.

The national television broadcast networks continue to be the most-watched source of information and entertainment. Although competition from other sources has eroded the network audience, affiliates of ABC, CBS, and NBC still accounted for about 46 percent of all TV viewing in the mid-1990s. Network affiliates accounted for about 40 percent of all viewing, even in homes with cable. They likely will remain for some years as the principal providers of national news, major sporting events, and broad-based entertainment, as well as the best way for national advertisers to reach a critical mass of consumers. But the advent of interactivity may dramatically change networks from single-channel program *distributors* to multichannel *programmers.*

In fact, the concept of "channels" may disappear altogether as viewers select from menus of programs sorted by content (sitcom, drama, nature) rather than source (ABC, CBS, NBC). So may program "schedules" become obsolete as, for perhaps 50 cents, subscribers can, whenever they want, order a replay of last night's *CBS Evening News* or yesterday's episode of *General Hospital.*

In the meantime, networks have already moved into cable. Examples include NBC's CNBC and MSNBC; ABC's interest in ESPN, ESPN2, ESPNews, and Lifetime; CBS's TNN, CMT, and *Eye on People*; and Fox's fX, fXM, and Fox News Channel. More recently networks have begun such interactive applications as *on-line* information services and *CD-ROM* (compact-disc-read-only-memory) devices, already popular through home computers. In 1995 CBS became the first major broadcast television network to set up its own *home page* (a dedicated site for information) on the Internet and the other networks quickly followed on-line (see Exhibit 1.c).

The broadcast networks have considerable momentum going into the new millennium as the brand-name nationwide distributors of America's most popular TV programming. The outlook for individual broadcasting stations is less clear. The traditional broadcast model features stations operating as outlets for a single channel of programming flowing in one direction only. In a universe of videotape recorders, CDs, and both cable systems

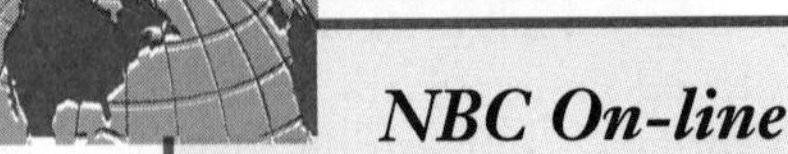

Exhibit 1.c

NBC On-line

All major networks have entered the superhighway, as this sample screen from NBC on America Online illustrates.

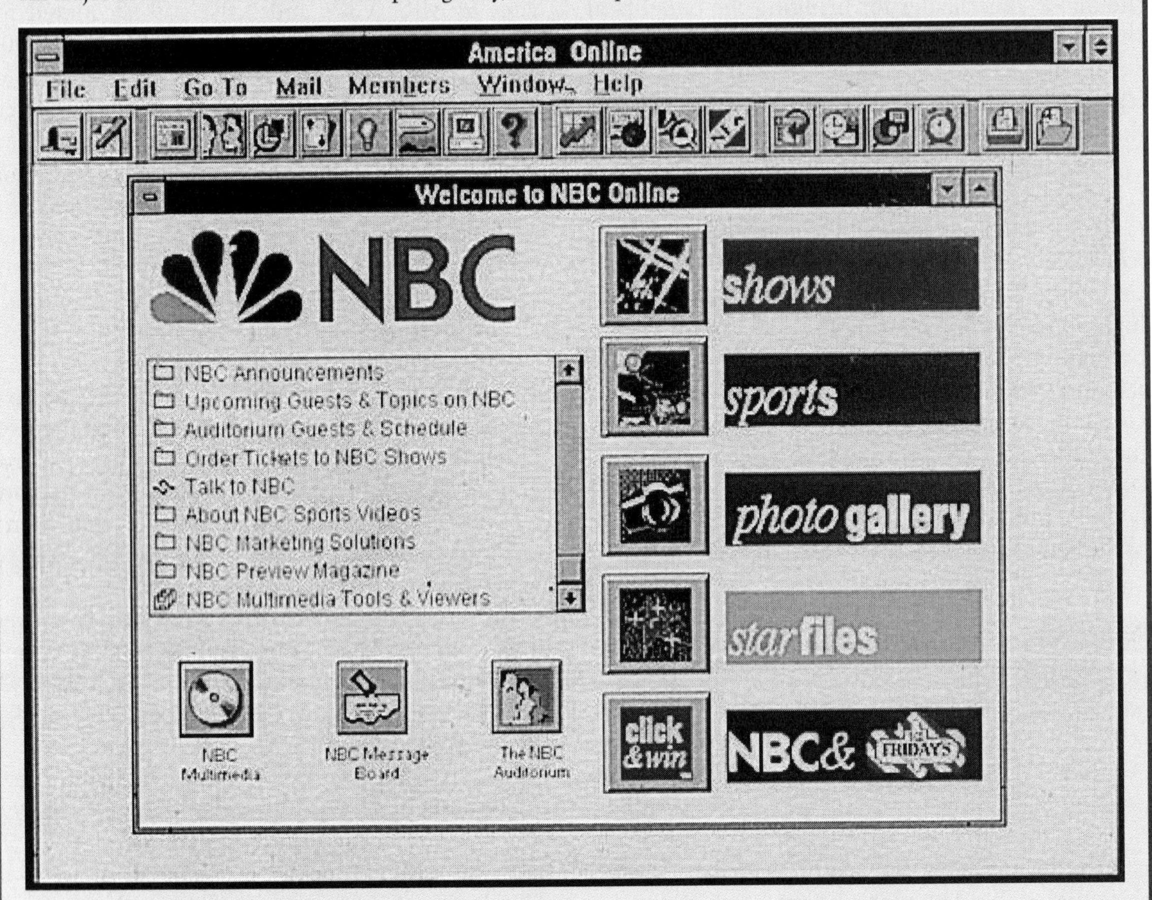

Source: Photo from the National Broadcasting Company, Inc.

and *direct-to-home* (DTH) satellites offering many dozens of program channels, broadcast stations approaching the information superhighway using this traditional model do so at a distinct technological disadvantage.

TV stations will have new tools with which to enter the fray once the transition to higher resolution pictures and digital video compression (allowing multiple channels to operate in the space formerly occupied by only one) is complete. Radio stations face a more daunting future because their signal delivery technology remains relatively unchanged.

One key to broadcasting's survival is consolidation: owning several stations in a market, each targeting a different audience, or owning many stations spread across many markets. The number of stations under consolidated ownership increased dramatically during the mid-1990s with radio and TV station deals topping $25 billion in 1996, three times as much as the previous year. Media consolidation continued in 1997 as large information and entertainment firms such as CBS, Hearst Corporation, and Tribune Broadcasting Company bought more stations and cable networks.

A second key to survival is diversification into other program-delivery technologies. Television broadcasters look forward to offering additional program and data services, opening up entirely new revenue streams, as more efficient use of their channels becomes possible with the implementation of digital technologies.

Renewed commitment to two of broadcasting's founding concepts may also help in coping with the new mass communication environment:

- *Localism.* Lifestyles in the 1990s are not what they were in earlier times, and people tend not to "put down roots" as deeply. But stations can serve the unique needs of their communities far better than a national program delivery system can. The popularity of local TV newscasts and the growth of regional cable networks indicate there is a demand for community-centered programming.
- *Universal Access.* Americans are accustomed to getting entertainment and information over the air at no additional direct expense once they purchase receivers. Broadcasting's appeal as a "free" service is strong, and, for many Americans, any delivery system charging a subscription fee will be out of reach.

Cable

Today more people receive their television programs by cable than by over-the-air broadcast. But cable also offers much more than TV.

Many cable operators are experimenting with, or are actually offering, interactive systems that permit subscribers to order movies or TV program reruns whenever they want (video-on-demand) or almost when they want (near-video-on-demand), to play video games, and to order merchandise without calling a toll-free number. As technologies advance, as systems replace coaxial cable with more efficient optical fiber, applications will expand even further. Indeed, one cable expert argued: "Forget about the over-hyped two-way interactive service. The most economic use of the cable system is to compete against the local telephone company" (*Broadcasting & Cable*, 11 April 1994: 6). But in the late 1990s, cable companies are focusing more on defending themselves against the threat from direct-to-home satellite delivery and preparing for digital broadcasting than on expanding aggressively into the telephone market.

Using video compression, some cable program services (HBO, for example) are already *multiplexing* their signals, sending out multiple channels. Nearly every day someone announces plans for a new service, specializing in everything from animals and antiques, through cowboys and golf, to parents and women's sports. Again, basic questions of economics—not technology—will determine which of these channels will survive: How many subscribers will pay to see them; how many advertisers will pay to have commercials carried by them? Because cable already has a direct, physical connection to the homes it serves, however, it is a frontrunner on the information highway.

Telephone

But a newcomer has entered the race. For years telephone companies *(telcos)* satisfied themselves by delivering *plain old telephone service* (POTS), with occasional bursts of modern technology such as call waiting. All that has changed. Now telephony is moving ahead—as either competitor to or partner with cable and others—to build the superhighway.

In 1992 the Federal Communications Commission (FCC) began relaxing restrictions on telcos, allowing them to provide video services over their networks in competition with cable companies. This led to some interesting ventures. Deals came from everywhere, strange alliances were forged, and bizarre competitions began: Nynex (another RBOC) backed Viacom (a major cable operator and program supplier) in the latter's acquisition of entertainment giant Paramount; Time Warner (operator of the local cable system) and Rochester (NY) Telephone (an independent telephone company) entered into an agreement whereby each would compete head-to-head with the other, in the same city, in both cable and telephone services; and Southwestern Bell (yet another RBOC) bought two Washington (DC) cable companies and later sought to compete in the telephone business in that same area, which was already served by Bell Atlantic. The head begins to swim.

In passing the Telecommunications Act of 1996, Congress eliminated all restrictions on telcos providing cable services to their local service areas. But telcos, legally freed to compete with cable and other multichannel video providers, began having second thoughts, at least in the short term. The economic risks of entering a new market area, unexpected technical difficulties with transitional video delivery systems, and renewed interest in expanding telephony services combined to slow expansion into video delivery. In 1995 three RBOCs (Bell Atlantic, Pacific Telesis, and Nynex) formed Tele-TV to spearhead their entry into broadband wireline networks and wireless cable. The next year the telcos dramatically scaled back support of Tele-TV and canceled plans to launch new wireless cable systems. The announcement was symptomatic of the telcos' on-and-off approach to the video and interactive media market. Although renewing their commitment to offer a full range of video and interactive services, telcos acknowledged that these services will evolve more slowly as they upgrade their telephone networks.

But the telcos have many advantages in the battle for dominance on the information superhighway. Like cable, the telephone industry already has a physical connection to the home. Indeed, telcos serve far more homes than do cable operators. In addition to their huge financial base, what gives telephone companies—at least for a time—an advantage over cable in building the superhighway is their experience with sophisticated *switches* that direct signals to and from designated locations and that are essential to any full-scale interactive service (see Exhibit 5.p, p. 147). Offsetting this advantage to some degree, however, is cable's experience in programming.

Computers

Regardless of whether program delivery is by cable systems or telephone companies, or both, computers will play a central role in building and operating the superhighway. Indeed, electronic media already employ them in a variety of ways. For example, the now-familiar set-top *converter box* that facilitates multichannel cable is a computer. Television stations and cable systems use computers to insert commercials and program materials at prescribed times. High-volume *video servers* that store and release programs for video-on-demand are computers. Personal computers have handled audio, video, and data since the early 1990s. Now the PC is merging with the TV (and the radio receiver as well), to combine broadcast, cable, and DBS reception with computing functions in a single appliance (see Exhibit 1.d).

The computer industry, however, sees itself as a far more important player in the new information age. Software companies such as Microsoft want to build operating systems that will serve as the data highway's traffic cops, controlling the flow of information to and from each viewer's screen. Meanwhile, *TV Guide* and others are perfecting electronic navigators that will tell viewers what's on television and where to find it.

But there's more. PC owners now receive audio and video programming directly on-line through the World Wide Web. The broadcast networks and many individual stations have Web sites providing interactive "streaming" (real-time delivery) of either their on-air signals or prerecorded supplementary multimedia materials. The technical quality, especially the video portion, does not yet equal that of broadcast and cable, but it is improving rapidly.

The form that future home communication equipment may take remains unclear. Some say all communication and information functions—telephone, fax, radio, television, multimedia disc-based players/recorders, even home security and utility management—will merge into a single system. The "loaded" PC/TV is a step in that direction, and Exhibit 1.d suggests how such a highly integrated stand-alone device might be configured. Personal computer manufacturers say it is relatively easy and inexpensive to add digital TV (DTV) broadcast reception capability to such PCs.

Others, however, believe people will prefer having a greater number of less expensive, more specialized, and easier-to-use communication appliances rather than a "loaded" PC/TV. For these consumers

Exhibit 1.d

Multimedia PC

Today personal computers are at the heart of media convergence. The modern PC/TV combines the features of a conventional TV or radio receiver with the interactive power of a multimedia computer. Additional features not shown in this illustration include, among others, a video camera and mike for teleconferencing, a video capture card (which allows frame-grabbing from a TV or VCR for use on the Internet or in video editing), a color printer (ink-jet or laser) for providing hard copy of on-screen materials, and a removable high-capacity storage disk drive.

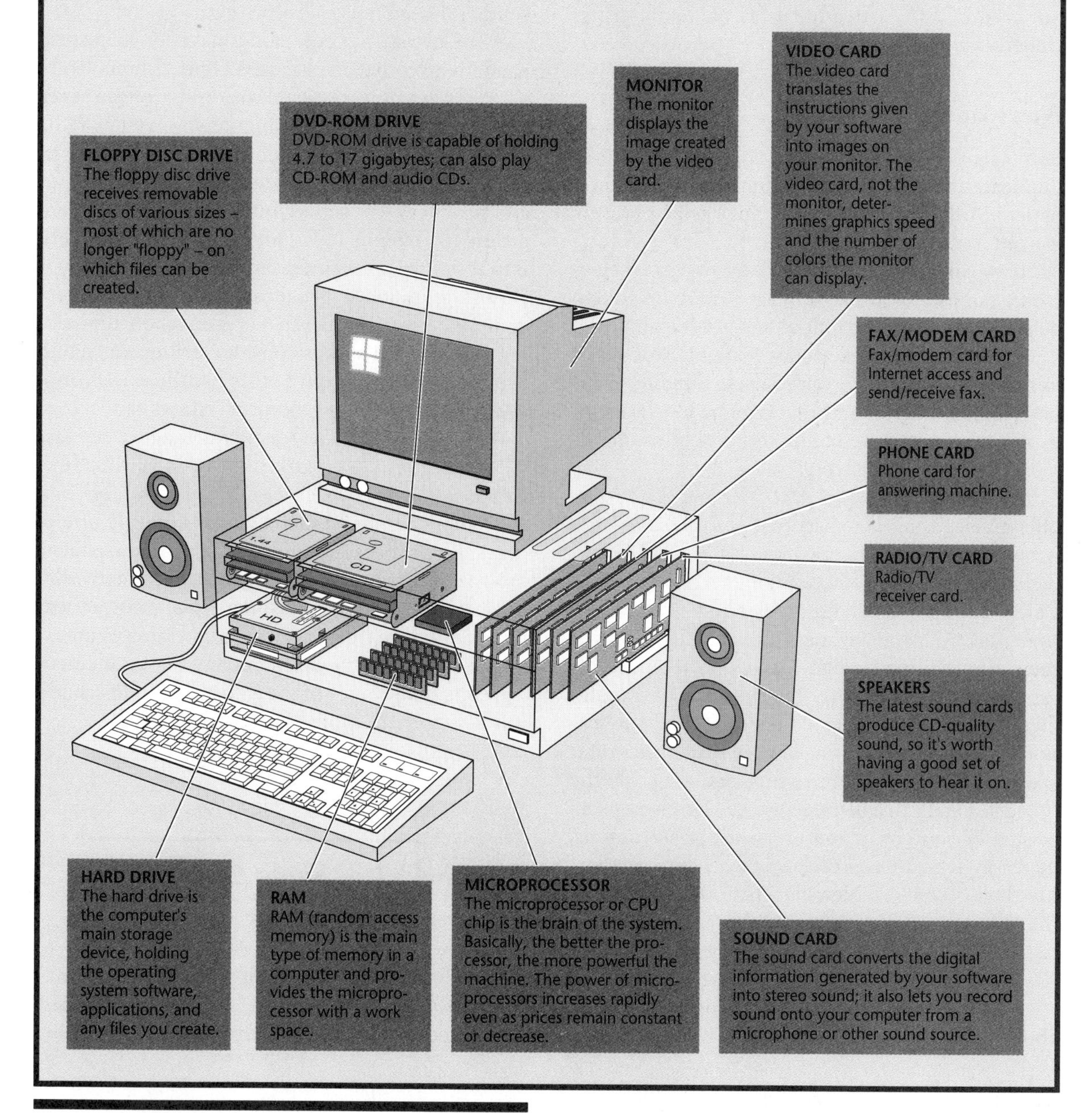

FLOPPY DISC DRIVE
The floppy disc drive receives removable discs of various sizes – most of which are no longer "floppy" – on which files can be created.

DVD-ROM DRIVE
DVD-ROM drive is capable of holding 4.7 to 17 gigabytes; can also play CD-ROM and audio CDs.

MONITOR
The monitor displays the image created by the video card.

VIDEO CARD
The video card translates the instructions given by your software into images on your monitor. The video card, not the monitor, determines graphics speed and the number of colors the monitor can display.

FAX/MODEM CARD
Fax/modem card for Internet access and send/receive fax.

PHONE CARD
Phone card for answering machine.

RADIO/TV CARD
Radio/TV receiver card.

SPEAKERS
The latest sound cards produce CD-quality sound, so it's worth having a good set of speakers to hear it on.

HARD DRIVE
The hard drive is the computer's main storage device, holding the operating system software, applications, and any files you create.

RAM
RAM (random access memory) is the main type of memory in a computer and provides the microprocessor with a work space.

MICROPROCESSOR
The microprocessor or CPU chip is the brain of the system. Basically, the better the processor, the more powerful the machine. The power of microprocessors increases rapidly even as prices remain constant or decrease.

SOUND CARD
The sound card converts the digital information generated by your software into stereo sound; it also lets you record sound onto your computer from a microphone or other sound source.

the PC industry promises a basic PC/TV appliance that looks like a conventional TV and functions with a simple remote control. PC manufacturers estimate it will sell for around $500, thousands less than either fully configured PCs or the DTV sets envisioned by the consumer electronics industry.

It is possible, of course, that the consumer will want it all: a big-screen DTV set in the living room as part of a home multimedia center (see Exhibit 1.b), a basic PC/TV in the bedroom, and a high-powered PC/TV with a full array of computer peripherals in the home office.

Newspapers

Some people say that no electronic device can ever duplicate the convenience and utility of the newspaper. Still, many in that industry have wisely elected not to take a chance.

Newspaper giant Knight-Ridder first tried electronic publishing in 1978. But its Viewtron system proved too costly and slow and folded in 1986. Undaunted by this experience, Knight-Ridder has joined with other major publishers including Gannett, Times Mirror, and the Tribune Company in exploring ways to use computers and telecommunications for news delivery.

The *Chicago Tribune* in 1992 became the first to offer an electronic newspaper—*Chicago Online*—as a national service. Within two years at least 60 dailies, including most of the nation's biggest newspapers, were offering or working on electronic services of some sort. Many newspapers provide the content of their print edition on-line. Others find it more advantageous to use their on-line presence to complement rather than duplicate their hardcopy publication. In addition, many on-line newspaper sites offer audio and video clips of newsmakers, just as do the broadcast and cable news networks. This is not surprising because print and broadcast organizations, increasingly under consolidated ownership, are sharing their resources. Newspapers also capitalize on the WWW's unique interactive power to search the classified ads, stock quotes, previously published stories, and other "data bases" on-line (see Exhibit 1.e).

In 1995 some nine major publishers owning about 225 newspapers formed the New Century Network (NCN), linking the WWW pages of "affiliated" publications at a central site. Launched in 1996, by early 1997 NCN had more than 80 participating newspapers, and nearly all the nation's major newspapers are expected to be accessible on-line through NCN by 1998.

Many newspapers also provide *audiotext* services, whereby the public calls a published number and selects from a menu of items that include news and sports headlines, weather, even samples from recently released musical recordings.

Some newspaper companies have long since expanded into other media areas. The Tribune Company, for example, owns television stations in seven of the nation's 11 largest markets as well as radio stations in New York, Chicago, and Denver. The company originates *ChicagoLand Television News* and produces TV shows, including *Geraldo*, in addition to owning newspapers in major markets. Based on gross revenue, the Tribune Company is almost as much a broadcast and entertainment company as it is a newspaper publisher. Interestingly, many Americans consider Tribune a "newspaper company" in spite of its broadcasting properties and think of Rupert Murdoch's News Corporation as a "broadcasting company" because it owns the Fox Network. However, as the News Corporation Web page (see Exhibit 1.f) reveals, Murdoch's company encompasses not only broadcast stations, broadcast networks, and newspapers, but also direct broadcast satellites, cable networks, book publishing, and on-line news, information, and gaming services. News Corporation is an example of a multimedia company simultaneously traveling many lanes of the information superhighway on an international scale.

1.3 Some Essential Terms

Given all this, one might well argue that this book's title, *Broadcasting in America*, is misleading. Indeed, its content is not limited to America but rather

Exhibit 1.e

Media Convergence On-line

This World Wide Web page combines three Atlanta mass media properties owned by Cox, an example of media convergence in the late 1990s. Note that the page provides a way to search the archives on-line and encourages user interactivity via electronic mail.

Source: © Cox Interactive Media.

examines other national systems as well. Nor is it limited, in precise terms, to broadcasting. The trouble is, electronic sources of information and entertainment have grown so varied that, quite frankly, no single word describes them all. The Federal Communications Commission (FCC), as do some others, uses the phrase "broadcasting and electronic media" to distinguish radio, television, and related services such as cable television from common carriers. These terms need exploration here to show how classifications result in legal and financial consequences.

Common Carriers vs. Electronic Mass Media

One way to determine whether a communication service is an electronic mass medium is to ask who has responsibility for its content. Telephone companies are the best example of common carriers. A telephone company may neither edit conversations nor restrict who will make use of its facilities. A broadcast station does both.

Electronic mass media select programs, performers, and speakers and edit what they say and do. In fact, many have a legal *responsibility* to do so. In exchange for assuming this responsibility, electronic mass media come under the protection of the First Amendment to the Constitution, which guarantees freedom from government interference with "speech" and "the press." Operators of common carriers, however, supply communication facilities such as telephones and satellites without assuming any responsibility for what is normally transmitted on those facilities. Having no "speaking role," they have no need to invoke constitutionally protected freedom to speak at will by telephone or satellite.

These fundamental differences between electronic mass media and common carriers have several other practical results, usually involving tradeoffs—each advantage tends to be offset by a disadvantage. The public can demand, as a right, equal access on equal terms to a common carrier, such as a telephone system. No one can assert such a right to use a broadcasting station or cable channel (except for political candidates, who are entitled to *equal opportunities* to use broadcast and cable television facilities). It is easier to obtain a common carrier license than one for a broadcasting station or cable system. The government, however, usually

Exhibit 1.f

Riding in Style

Few media companies embrace the information superhighway with greater zeal than Rupert Murdoch's News Corporation. News Corporation has global interests in traditional broadcast and print as well as newer media technologies such as the Internet and DBS.

Source: The News Corporation Limited.

controls common carrier profit levels and the fees they charge their customers. Electronic mass media (except cable, as we shall see) may charge whatever the traffic will bear and may make as much profit as they can, so long as competition prevails.

Broadcasting Defined

Many assume that *any* program or message transmitted over the air, or even over cable, constitutes "broadcasting." But to qualify as broadcasting, programs or messages must pass specific tests. As defined legally, to *broadcast* means

> to send out sound and pictures by means of radio waves through space for reception by the general public.

The phrases "by means of radio waves through space" and "reception by the general public" are crucial: Cable television systems do not broadcast through space, but rather they send programs through cable.

Communication "by means of radio waves" includes many nonbroadcast services, such as CB radio, police calls, taxi dispatching, beepers, and mobile telephones. However, these services operate *point-to-point* (from a source to one intended recipient) or *point-to-multipoint* (from a source to more than one specific recipient). Anyone with a suitable receiver can intercept such transmissions, but they are not intended "for reception by the general public" and therefore do not count as broadcasting.

The verb *to broadcast* was adopted to distinguish the new radio communication method from the previous point-to-point orientation of radiotelegraphy and radiotelephony. "Broadcasting" expresses the idea of scattered dissemination to anonymous, undefined destinations (listeners/viewers). The term comes from the farmer's act of hand-sowing grain by *casting it broadly*. The sower lets seeds fall where they may. (Students should be careful to remember, by the way, that the past tense of the verb *broadcast* is also *broadcast*—not *broadcasted.*)

Hybrid Services

Such broad distinctions between common carriers and electronic mass media become less clear with new services that have characteristics of both (referred to here as *hybrids*). Cable television, for example, at first occupied a particularly ambiguous position. Most cable systems operate much like common carriers. They enjoy local monopolies and exercise no editorial control over most of their programs, which come from broadcast stations and cable-program networks. Should these factors put cable operators in the same legal basket as common carriers?

Or, since cable systems carry broadcast programs and sometimes originate their own programs in the manner of broadcasters, should they be subject to the same laws as broadcast stations? The question was debated for three decades until Congress finally resolved it in 1984 by recognizing cable as a service that is neither a common carrier nor a broadcaster but a mixture of the two.

Cable operators continue to claim the same First Amendment rights as broadcasters. For example, they maintain that for a city to impose content limitations on a cable system it franchises would violate the cable owner's right of free speech. This claim's legality has been challenged, but so far the courts have upheld it. At the same time, however, cable is subject to governmental rate regulation (see Section 12.8).

Direct-broadcast satellite (DBS) services offer another hybrid example. Satellites usually function as common carriers, but direct-to-home satellite transmission seems equivalent to regular broadcasting, with the station located in space instead of on the ground. However, DBS program providers scramble their signals to prevent anyone other than paying subscribers from receiving them. This barrier to reception by the public puts DBS services in the point-to-multipoint rather than broadcast category. The FCC settled the question in 1987, when it defined DBS transmissions intended only for paying subscribers as *nonbroadcast* services.

Even the telephone, the classic example of a common carrier, has become increasingly hybridized. It first took on a mass entertainment information role when it offered "dial-it" services, sometimes called *mass announcement services.* For a price, listeners can hear, and in some cases participate in, audio "programs" by dialing numbers with special prefixes (usually 1-800 or 1-900). Public concern over pornographic dial-it services brings into question the common carrier's traditional hands-off policy with regard to message content.

As telephone companies move more and more into multichannel audio and video delivery of entertainment programs and information services—both their own and those of others—they become less and less distinguishable from cable systems. Indeed, telcos have the characteristics of both common carriers (when delivering switched voice/data) and electronic mass media (when delivering cable-like video services).

1.4 Back to Basics

The information superhighway metaphor proves most useful in describing those converging technologies that serve as conduits or delivery systems. It is less helpful in discussions of content providers: program producers and information suppliers.

Cable operators, telephone companies, even computer hardware and software manufacturers want to build the superhighway and control what travels over it. But, in the final analysis, corporate successes in the information age will be determined not by *how* but rather by *what* they deliver. Most people simply have no interest whatsoever in whether they receive information and entertainment through a wire (or possibly *two* wires—one for a cable system and one for the telephone company), or over the air (by traditional broadcast or by direct-broadcast satellite), or whether the picture appears on a traditional television receiver or on a computer screen. What they *do* have interest in is

- the variety and quality of the programs and services they receive,
- how easy it is for them to receive them, and
- how much they have to pay for the package.

The student of electronic communication, however, must understand how it has developed, how it works, how it is controlled, how it affects and is affected by its several audiences, and—in the next chapter—how it all began.

RECORDS
RADIOS
Music
TELEVISION
APPLIANCES

Chapter 2

From Radio to Television

CHAPTER CONTENTS

As guideposts in broadcasting's development, six dates stand out:

- 1896, the original Marconi wireless patent
- 1920, inception of radio broadcasting
- 1927, passage of the Radio Act (most of which carried over to the 1934 Communications Act)
- 1948, television network broadcasting begins
- 1975, first home VCR is sold and HBO announces plans to use a satellite for program distribution, setting the stage for cable's expansion
- 1996, FCC adopts technical standards making the transition from analog to digital television broadcasting possible.

In addition, both world wars helped define important technical phases in radio's development:

- From 1914 to 1918, World War I stimulated rapid evolution of wireless technology and set the stage for radio's debut
- From 1939 to 1945, World War II temporarily delayed television's emergence, though wartime research developed ultrahigh frequencies that TV's postwar expansion would require. Today's electronic media, however, are based on even earlier cultural and technological precedents.

2.1 *Cultural Precedents*

Public appetite for mass entertainment developed decades before radio. Popular newspapers, home phonographs, and motion-picture theaters all encouraged a mass media habit, making it easier for broadcasting to achieve success in a very short time.

Urbanization

These older media grew out of fundamental social changes encouraged by the Industrial Revolution (roughly 1740–1850). For centuries, most people worked on farms or in related agricultural roles. But industry, based initially on steam power, increasingly drew people away from the land to live and work in cities.

These urban populations became the target of what we now call *mass media*—technology-based means of communication that reach large numbers of people, delivering news and entertainment that most people find interesting, and at a price they can afford.

Penny Press

Urban concentration and increasing literacy and leisure time all helped to transform newspapers. Originally aimed primarily at a wealthy and educated elite, they became a product for the masses. The "penny press" signaled this transformation when, in 1833, the *New York Sun* began a trend toward mass-interest, mass-produced papers. Copies sold for a penny, first in the thousands and eventually in the hundreds of thousands. The penny press exploited news of everyday events, sensational crimes, gossip, human-interest stories, sports, and entertaining features—all presented in a breezy style that contrasted with the flowery essay approach of the past. Popular newspapers appealed across lines of class, gender, age, political party, and cherished beliefs. By the 1890s some mass-oriented newspapers had circulations of more than a million.

Vaudeville

Vaudeville was a popular form of entertainment featuring "variety" acts in which performers sang, danced, played musical instruments, and did comedy routines. The first U.S. vaudeville house opened in Boston in 1883, and vaudeville remained immensely popular well into the 1920s, at its peak selling more tickets than all other forms of entertainment combined. New York City alone once had 37 vaudeville houses. The development of motion pictures, and especially the "talkies" in the late 1920s, spelled the end of American vaudeville.

But vaudeville developed an audience for forms of entertainment that reappeared later in network radio programming. Vaudeville also served as a training ground for early radio performers such as Jack Benny, Gracie Allen, George Burns, and Milton Berle, all of whom also migrated successfully from radio to TV. Berle's *Texaco Star Theater* on NBC TV was phenomenally popular during the late 1940s and early 1950s, earning Berle the nickname "Mr. Television."

The Phonograph

Around the turn of the century, owning a phonograph (often housed in a handsome wooden cabinet) accustomed people to investing in a piece of furniture that brought entertainment into the home. By the end of World War I, on the eve of radio's introduction, some 200 phonograph manufacturers turned out more than two million players each year.

But records used acoustic recording methods and were little different from those that Thomas Edison invented in 1878. Radio, in contrast, continued improving its audio quality. By 1930 radio provided better sound than phonographs, as well as "live" music, and it was free, an important consideration during the Great Depression. Broadcasting companies later bought failing phonograph firms, shared radio's advanced audio technology, and brought them back to life. When TV challenged radio in the 1950s, radio broadcasting and the recording industry formed a close interdependent relationship.

Like vaudeville, the phonograph helped create a popular demand for music that radio then supplied, contributing to radio's rapid growth after 1920.

Motion Pictures

Like the phonograph industry, movies were well established by the time broadcasting began in the 1920s. Moviegoers returned weekly for serial dramas, which eventually had their counterparts in broadcasting. Movies also benefited from radio's electronic technology.

The lack of synchronized sound—the precise matching of sound and action—long stymied progress toward acceptable "talkies." They finally began in earnest in 1928, with several rival sound systems competing for acceptance. One of these sound-on-film systems had been developed by RCA, the owner of the first national radio network—an example of the many links between broadcasting and motion pictures in the 1920s, long before television brought the two visual media into an even closer relationship.

2.2 *Technological Precedents*

The "penny press," the phonograph, and the motion picture industry evolved from technological advances in mass communication, but broadcasting's technological precedents focused on point-to-point communication, which may have delayed broadcasting's development. It is also worth noting that broadcasting's technological precedents were electronic, whereas earlier mass media—newspapers, phonographs, and motion pictures—employed mostly mechanical technologies.

The Telegraph

The British first developed electrical telegraphy in the 1820s as an aid to early railroad operations. An American artist/inventor, Samuel F. B. Morse, conducted extensive telegraph experiments in the 1830s—at a time when westward expansion put a high premium on rapid communications over long distances. Morse made significant telegraph improvements, including a receiver that recorded messages on strips of paper. He and a partner devised the Morse code to translate numbers and letters into the dots and dashes of telegraphic language. That same code served for early prebroadcast wireless communications.

In partnership with the federal government, Morse installed the first operational U.S. telegraph line from Washington to Baltimore in 1844. Three years later the government sold its interest to private investors, retaining only the right to regulate telegraphic services. By the end of the Civil War, Western Union had emerged as the dominant telegraph company.

The decision to sell—or privatize—the telegraph started the United States on a telecommunications path different from that of most other countries. Typically governments own and operate national telecommunications systems under Post and Telegraph (PTT) administrations, which later often also had some control over broadcasting. However, in recent years PTT monopolies have come under increasing attack as countries adopt the more efficient American model of privately owned, competitively operated telecommunication services.

Communications over transatlantic telegraph cables began in 1866 (after a short-lived attempt in 1858), enabling the exchange of information between Europe and North America in minutes instead of weeks. Newspapers immediately seized on both submarine and land telegraph as a means of sharing information in ways not previously possible. The telegraph made news agencies such as Havas (France, 1835), Reuters (Great Britain, 1851), and the Associated Press (United States, 1856) more important players in the newspaper and magazine business.

The telegraph had a revolutionary influence because it allowed communication with distant locations at a pace literally millions of times faster than anything before it. The shrinking of the Earth into an electronically-interconnected "global village" began with the telegraph.

The Telephone

Inventors next turned their attention to wire transmission of speech itself, seeking a way to eliminate tedious encoding and decoding of telegraph messages. Investigators worked simultaneously in many nations on developing what became the telephone. Several appeared on the verge of a solution when Alexander Graham Bell delivered his designs to the U.S. patent office in 1876, beating fellow American Elisha Gray by only a few hours. Gray contested Bell's patent but Bell won in the courts. Bell organized the original Bell Telephone Company a year later, when he secured a second essential patent. Though control over Bell's patents soon passed to others who went on to develop the company known today as AT&T (until 1994, American Telephone & Telegraph Company), Bell's name still means "telephone" to many people.

AT&T developed a "long lines" network connecting local exchanges with one another, soon owning the more important local companies. In 1913 AT&T ensured its supremacy in long distance by acquiring a license to a crucial electronic invention, the Audion (see Section 2.3), which made coast-to-coast telephone service possible. AT&T's control of long-distance voice communication by wire had an important bearing on broadcasting, which after 1927 depended on such wire links for network operations.

Three giant manufacturing firms would also play key roles in radio broadcasting: Western Electric (AT&T's manufacturing arm), General Electric (GE), and Westinghouse. Together they dominated the electrical industry and helped shape its participation in early wireless communication.

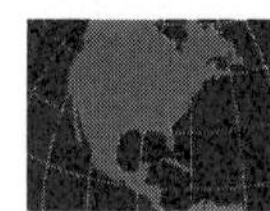

2.3 Wireless Communication

Simultaneous with the development of the telephone, other inventors were working on wireless systems. American dentist Mahlon Loomis transmitted electrical signals almost twenty miles in 1866 and six years later received a patent for his wireless telegraph system. Other Americans tinkering with wireless transmission included William Henry Ward, Nathan B. Stubblefield, and even the "Wizard of Menlo Park," Thomas Edison. France, Germany, Great Britain, and other nations have similar lists of investigators experimenting with wireless communication in the late 1800s.

Hertzian Waves

Many early experimenters succeeded by "accident" but lacked the understanding of scientific principles required for systematic development of their discoveries. Others, such as Edison, became distracted by projects they considered more rewarding. Their failure to develop a viable commercial wireless communication system was not due to a lack of published information about electromagnetic energy.

In 1873 James Clerk Maxwell theorized that an invisible form of radiant energy must exist—electromagnetic energy. He described it mathematically, foreseeing that it resembled light radiating out from a source in the form of waves.

In 1888 German physicist Heinrich Hertz reported his brilliant laboratory experiments that conclusively proved the validity of Maxwell's theory. Hertz generated electromagnetic waves, transmitted them down the length of his laboratory, detected them, measured their length, and showed that they had other wavelike properties similar to those of light.

In effect, Hertz demonstrated radio. But he sought only to verify a scientific theory, not to develop a new way of communicating. Indeed, when asked whether his discovery could be used for communication, Hertz provided theoretical explanations of why it could not! It remained for other experimenters to communicate with what they called Hertzian waves.

"The Right Releasing Touch"

A young Italian experimenter, Guglielmo Marconi, is credited with developing radio as a viable means of communication. He supplied "the right releasing touch," as a Supreme Court justice put it in upholding Marconi's primacy in a later patent suit (US, 1942: 65). Stimulated by Hertz's paper, Marconi experimented with Hertzian wave signals in the early 1890s, first sending them across the space of an attic, then for greater distances on the grounds of his father's estate.

It was characteristic of Marconi's aggressive personality that he decided his work was "ready for prime time" when he succeeded in transmitting Hertzian waves a mere two miles. He patriotically offered it to the Italian government, which couldn't see the potential. Still only 22, Marconi next went to England, where, in 1896, he registered his patent. He soon launched his own company to manufacture wireless-telegraphy equipment and to offer wireless services to ships and shore stations.

To a remarkable degree Marconi combined an inventor's genius with that of a business innovator. As inventor he persisted tirelessly, never discouraged even by hundreds of failed attempts at solving a problem. In 1909 Marconi shared the Nobel Prize in physics (with Germany's Ferdinand Braun) for achievements in wireless telegraphy. As a business manager, Marconi demonstrated a rare entrepreneurial talent plus a flair for effective public relations. In the early 1900s, to prove his system's value, he repeatedly staged dramatic demonstrations of wireless to skeptical officials, scientists, investors, and equipment buyers. Exhibit 2.a shows the Newfoundland station where Marconi received the first transatlantic wireless signals. Marconi's famous "four sevens" (No. 7777) patent for *syntony*, or "tuning," contributed significantly to making this dramatic event possible. Tuning allowed Marconi to receive one desired signal while rejecting dozens of others transmitted simultaneously.

Among his business ventures, Marconi's American branch had a decisive influence on the development of broadcasting in America. Founded in 1899, American Marconi developed a virtual monopoly on U.S. wireless communication, owning 17 land stations and 400 shipboard stations by 1914. All these facilities used a wireless extension of the

Exhibit 2.a

Guglielmo Marconi (1874–1937)

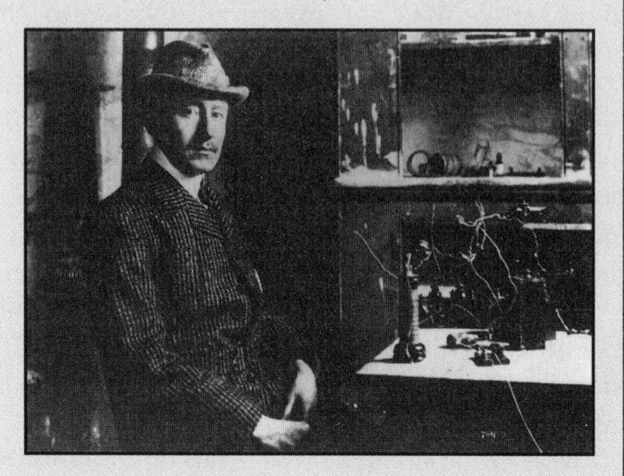

Marconi in the Newfoundland station where he received the first transatlantic radio signal in 1901. Marconi was unique among wireless pioneers for his promotional talent. Realizing the valuable publicity that bridging the Atlantic with radio signals would create for his company, Marconi established a station on Signal Hill, 600 feet above the harbor of St. John's, Newfoundland. On December 12, 1901, Marconi heard three short dots—Morse code for the letter "S"—on the crude equipment shown in this photo of his Newfoundland receiving station.

Source: Corbis-Bettman.

telegraph principle—point-to-point communication between ships and shore stations, between ships at sea, and to a lesser extent between countries.

This limited application of radiotelegraphy dominated the first two decades of radio service because the relatively crude Marconi equipment could only transmit Morse code dots and dashes. Telephones could have been used as microphones to transmit wireless speech, but methods had not yet been found to impress such complex information onto radio waves.

The Vacuum Tube

The solution to this and related radio problems came with the development of an improved *vacuum tube*, capable of electronically amplifying the changing volume and pitch of speech. American inventor Lee de Forest followed up on research leads suggested by Edison's 1883 electric lamp and the original two-element vacuum tube (*diode*) that Ambrose Fleming patented in 1904.

De Forest's crucial improvement consisted of adding a third element to a vacuum tube, turning it into a *triode*. He positioned his new element, a tiny grid or screen, between the diode's existing filament and plate. A small voltage applied to the grid could control with great precision the flow of electrons from filament to plate. Thus a weak signal could be enormously amplified yet precisely modulated. De Forest first experimented with his triode, or *Audion*, in 1906.

It took more than a dozen years to develop the Audion and the new circuits to go with it—and much of the necessary work was done by others. De Forest spent much of his time defending his Audion patents in court. Ambrose Fleming claimed de Forest had infringed his 1904 diode patent. Other British inventors, including H. J. Round and C. S. Franklin, claimed credit for developing electronic amplification devices. In America General Electric's Irving Langmuir and AT&T's Harold D. Arnold also made claims against de Forest.

The most debated issue involved the *regenerative* or *feedback circuit*, a revolutionary design allowing the Audion to generate radio signals in transmitters and greatly improve receiver sensitivity and selectivity. Both de Forest and Edwin Armstrong filed patents for this critically important application of the amplifying vacuum tube in 1914. The engineering community thought Armstrong had a better understanding of the principles involved, but the Supreme Court ruled in favor of de Forest.

In spite of these legal challenges, de Forest sold seven of his Audion patents to AT&T in 1913. AT&T used these patents to develop vacuum-tube repeaters (amplifiers in telephone lines), achieving the first coast-to-coast telephone service two years later.

The Audion marked a great leap forward—out of the age of mechanics and into the age of electronics. Electrons (particles of energy smaller than atoms) could be controlled. This transforming ability to manipulate electrons would revolutionize not only communications but also virtually all science and industry.

Commercial Uses of Wireless

Into the 1920s wireless firms made money by supplying telegraphic communication among ships, between ships and shore stations, and to link continents. Overland wireless-telegraphy services had less appeal because telephone and telegraph lines satisfied existing needs.

Wireless offered unique advantages to maritime commerce. In fact, Marconi chose international yacht races for some early public wireless demonstrations. Wireless also gained invaluable publicity from its life-saving role in maritime disasters, one as early as 1898. Each year the number of rescues increased. Exhibit 2.b describes the most famous of them all.

Radio had commercial possibilities as a long-distance alternative to submarine telegraph cables, which were enormously costly to install. Because of technical limitations, however, transoceanic-radio service did not become strongly competitive until vacuum tubes came into general use in the 1920s.

On the eve of U.S. entry into World War I, General Electric developed its *alternator*, a new type of radio-frequency generator, to send transatlantic messages reliably—a major improvement in long-distance radio communication. During the 1920s vacuum-tube transmitters displaced alternators. Tubes enabled use of the short-wave (high-frequency) portion of the spectrum—more efficient than low frequencies previously used for long-distance communication. A sharp rise in transatlantic radio traffic followed.

Military Wireless

Naturally navies took an interest in military applications of wireless from the outset. The British Admiralty used Marconi equipment as early as 1899, during the Boer War, but with little success. Technology advanced rapidly during the next decade, however, and wireless telegraphy became an important military weapon.

In April 1917 as America entered World War I, the U.S. Navy saw wireless as a threat to national security. Enemy spy agents could (some did) use radio to send information about ship movements, for example. Therefore the Navy took over all wireless stations in the country, commercial and amateur alike. It dismantled most, operating a few for its own training and operational needs.

The Army Signal Corps also used radio, as did the fledgling Air Service. Some 10,000 soldiers and sailors received training in wireless. After the war they formed a cadre of amateur enthusiasts, laboratory technicians, and electronics manufacturing employees. They helped popularize radio, creating a ready-made audience for the first regular broadcasting services that began two years after the war ended.

To mobilize America's wireless resources for war more rapidly, the navy decreed a moratorium on patent lawsuits over radio inventions. This allowed manufacturers to pool their patents and use any patent needed. Patent pooling and the pressure of war encouraged companies to move forward with innovations they might not have risked during peacetime.

By the end of the war, big business had developed a stake in wireless and was ready to branch out into new applications. AT&T had added wireless rights to its original purchase of telephone rights to de Forest's Audion. GE owned patents for its powerful alternator and had the ability to mass-produce vacuum tubes. Westinghouse, also a

Exhibit 2.b

The Titanic *Disaster, April 1912*

A luxury liner advertised as unsinkable, the *Titanic* struck an iceberg and sank in the Atlantic on her maiden voyage from Britain to the United States in April 1912. One heroic Marconi radio operator stayed at his post and went down with the ship; the second operator survived. Some 1,500 people died—among them some of the most famous names in the worlds of art, science, finance, and diplomacy—in part because each nearby vessel, unlike the *Titanic*, had but one radio operator (all that was then required), who had already turned in for the night. Only by chance did the operator on a ship some 50 miles distant hear distress calls from the *Titanic*. It steamed full speed to the disaster site, rescuing about 700 survivors.

Radiotelegraphy's role as the world's only thread of contact with the survivors aboard the rescue liner *Carpathia* as it steamed toward New York brought the new medium of wireless to public attention as nothing else had done. Subsequent British and American inquiries revealed that a more sensible use of wireless (such as a 24-hour radio watch) could have decreased the loss of life. Because of such findings, the *Titanic* disaster influenced the worldwide adoption of stringent laws governing shipboard wireless stations. The *Titanic* tragedy also set a precedent for regarding the radio business as having a special public responsibility. This concept carried over into broadcasting legislation a quarter of a century later.

Source: Illustration from UPI/Bettman Archive.

vacuum tube producer, joined in seeking new ways to capitalize on wireless. But before any of these firms could move forward, they had to come to terms with the Navy, which also had claims on wireless.

2.4 *Emergence of Broadcasting*

Throughout the early 1900s eager experimenters constantly sought an effective means of *radiotelephony*, the essential precursor of broadcasting.

Early Radiotelephone Experiments

In 1906 Reginald Fessenden made the first known radiotelephone transmission resembling what we would now call a broadcast. Using an ordinary telephone as a microphone and an alternator to generate radio energy, Fessenden made his historic transmission on Christmas Eve from Brant Rock, on the Massachusetts coast south of Boston. He played a violin, sang, read from the Bible, and played a phonograph recording. Ships' operators who picked up the transmission far out at sea could hardly believe they were hearing human voices and musical sounds from earphones that previously brought them only the monotonous drone of Morse code.

In 1907, hard on the heels of Fessenden, de Forest made experimental radiotelephone transmissions from downtown New York City. Some of his equipment appears in Exhibit 2.c.

De Forest, of course, used continuous waves produced by his Audion. A year later in Great Britain, H. J. Round used yet another device, the *arc transmitter*, to produce the continuous waves needed for voice transmission. Round sent voice messages a distance of 50 miles, but clearly de Forest's tube technology was superior to either Fessenden's alternator or Round's arc transmitter for radiotelephony. In 1915 AT&T engineers connected 500 tubes together and transmitted voice successfully, even if only experimentally, from Arlington, Virginia, to the Eiffel Tower in Paris.

By 1916 de Forest had set up in his Bronx home an experimental transmitter over which he played phonograph recordings—and even aired election returns in November of that year, anticipating by four years the opening broadcast of KDKA. With some justification, de Forest called himself *Father of Radio*, the title of his 1950 autobiography.

Government Monopoly: The Road Not Taken

Though World War I ended in November 1918, the Navy did not relinquish control of radio facilities for another 18 months. The critical decisions made at this time profoundly affected the future of broadcasting in America.

Was radio too vital to entrust to private hands? The Navy thought so, and it supported a bill in Congress late in 1918 proposing, in effect, to make radio a permanent government monopoly. Despite strong arguments from Navy brass at the hearings, the bill failed. Thus radio took the road of private enterprise in the United States, though in many other countries governments remained in charge as radio expanded in the postwar years. Only recently has the global trend been toward privatization of at least some aspects of broadcasting.

Origin of RCA

Restoration of private station ownership in 1920 would have meant returning most commercial wireless-communication facilities in the United States to foreign control. Disturbed at the prospect of British Marconi (parent of American Marconi) consolidating its U.S. monopoly by purchasing exclusive rights to GE's alternator, the Navy strongly opposed the deal.

Exhibit 2.c

Lee de Forest (1873–1961)

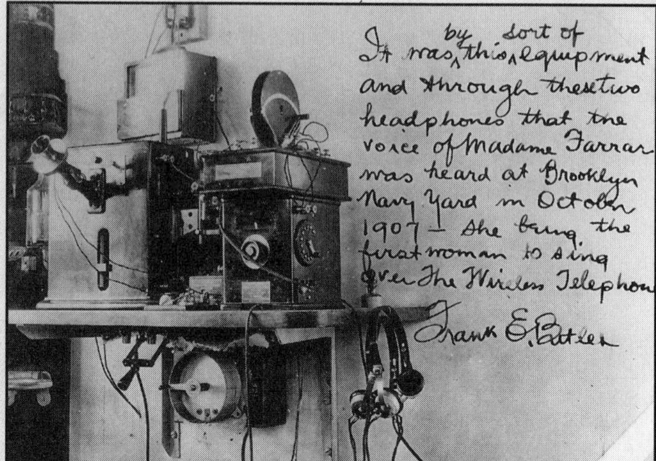

by sort of
It was this equipment
and through these two
headphones that the
voice of Madame Farrar
was heard at Brooklyn
Navy Yard in October
1907 — she being the
first woman to sing
over The Wireless Telephone
Frank E. Butler

The invention by de Forest of the amplifying vacuum tube, the Audion, was critical to the development of modern electronic communication, including broadcasting. The shipboard receiver also pictured is typical of the type used to receive de Forest's experimental radiotelephone transmissions in 1907.

Source: De Forest photo from UPI/Corbis-Bettmann; transmitter photo from Smithsonian Institution, Washington, D.C.

GE's alternator was the most advanced technology for intercontinental wireless, and the U.S. government had learned during the recent war how ridiculously easy it was for enemy nations to slash undersea cables. The government feared losing a vital national defense technology to a foreign-controlled corporation. American wireless manufacturers simply feared Marconi would renew aggressive legal actions to protect patents pooled during the war, limiting their ability to maximize profits in a booming postwar economy. Caught in a squeeze play between the U.S. government and the giants of American industry, the British-based company could neither expand nor effectively operate. With tacit government approval Owen Young, head of GE's legal department, quietly negotiated the purchase of all American Marconi stock in a "semi-friendly" takeover.

In October of 1919, GE created a subsidiary to take over and operate American Marconi—the Radio Corporation of America (RCA), shortly to become the premier force in American broadcasting. Under RCA's charter, all its officers had to be Americans, and 80 percent of its stock had to be in American hands.

Westinghouse and AT&T joined GE as investors in the new corporation. AT&T sold its interest in 1923, but RCA remained under GE and Westinghouse control until 1932, when an antitrust suit forced them to make RCA an independent corporation. Thus it remained for some seven decades until, in an historical irony, GE bought RCA back again (see Section 3.6).

If RCA became broadcasting's leading corporation, RCA executive David Sarnoff became broadcasting's leading corporate figure. From 1930, when he became president, to 1969, when he retired as board chair, he played a major role in the creation of network radio, the evolution of television, and the conversion of television to color. Today a top-flight research laboratory bears his name. Exhibit 2.d details his career—and that of his chief rival.

Cross-Licensing Agreements

RCA and its parent companies each held important radio patents, yet each found itself blocked by patents the others held. From 1919 to 1923, AT&T, GE, Westinghouse, RCA, and other minor players worked out a series of cross-licensing agreements, modeled after the Navy-run wartime patent pool. Under these agreements, each company had its own slice of the electronics manufacturing and services pie. Each parent company and RCA could build equipment for its own use, but AT&T got exclusive rights to build transmitters for sale to "outsiders." Both GE and Westinghouse received exclusive rights to build receivers sold through RCA. These arrangements were mutually advantageous given the state of wireless communication at the time.

Within a few years, however, these carefully laid plans fell into disarray because of the astonishingly rapid growth of a new use for radiotelephony—broadcasting.

The First Broadcasting Station

Stations in many nations claim to be world's oldest broadcast station. Certainly there were stations transmitting forms of information and entertainment we now consider broadcasting following World War I. But these stations typically operated sporadically and often the "program" was just something to make a noise, the real purpose of the transmission being to test the equipment.

Many believe the first broadcasting station began as 8XK, an amateur radiotelephone station operated in a Pittsburgh garage by Westinghouse engineer Frank Conrad. Dr. Conrad's equipment (Exhibit 2.e) was crude, with various components connected by a jumble of exposed wires.

In the fall of 1919, Conrad fell into the habit of transmitting recorded music, sports results, and bits of talk in response to requests from other amateurs. These informal transmissions (hardly "programs" in the formal sense) built up so much interest that newspapers began to comment on them.

Managers at the Joseph Horne Company, a Pittsburgh department store, noticed a growing public interest in wireless and decided people might buy receiving sets to pick up Conrad's "broadcasts." They installed a demonstration receiver and ran a newspaper display ad in September 1920: "Air Concert 'Picked Up' by Radio Here . . . Amateur Wireless Sets made by the maker of the Set which is in operation in our store, are on sale here $10.00 up."

Horne's ad caught the eye of Westinghouse vice president H. P. Davis, who immediately saw a novel merchandising tie-in: The company could increase demand for its receivers and promote Westinghouse's name by regularly transmitting programs. Westinghouse adapted a transmitter, installed it in a makeshift shack on the company's tallest building in East Pittsburgh, and at 6 P.M. on November 2, 1920, began broadcasting as KDKA.

By voluntary international agreement, all U.S. broadcasting stations have an initial *K* or *W* in their call signs. Although there are exceptions, such as KDKA, most stations east of the Mississippi River have W calls, and those to the west have calls starting with K. Some pioneer stations, such as WWJ (Detroit) and WHA (Madison, Wisconsin), received three-letter calls and retained them over the years. Most stations have four-letter calls. Because KDKA's opening coincided with election day 1920, the maiden broadcast reflected public interest in

Exhibit 2.d

Sarnoff and Paley: Giants of Early Broadcasting

Both network broadcasting pioneers came from immigrant Russian families, but there the similarity ceases. Sarnoff (left) rose from direst poverty, a self-educated and self-made man. In contrast, Paley (right) had every advantage of money and social position. After earning a degree from the Wharton School of Business at the University of Pennsylvania in 1922, he joined his father's prosperous cigar company.

The differences between Sarnoff and Paley extended to their personalities and special skills. Sarnoff was a "technician turned businessman, ill at ease with the hucksterism that he had wrought, and he did not condescend to sell, but Bill Paley loved to sell. CBS was Paley and he sold it as he sold himself" (Halberstam, 1979: 27).

Source: Photos: Sarnoff, courtesy RCA; Paley, courtesy CBS.

Sarnoff had been introduced to radio by way of hard work at the telegraph key for American Marconi, Paley by way of leisurely DX (long-distance) listening: "As a radio fan in Philadelphia, I often sat up all night, glued to my set, listening and marveling at the voices and music which came into my ears from distant places," he recalled (Paley, 1979: 32).

Paley's introduction to the business of radio came through sponsoring of programs. After becoming advertising manager of his father's cigar company in 1925, he experimented with ads on WCAU in Philadelphia. Impressed with the results, he explored getting into the radio business and late in 1928 took over the struggling CBS network.

Both men, shown here in about 1930, were highly competitive and pitted their companies against each other for 40 years before Sarnoff's retirement in 1969.

Exhibit 2.e

Conrad's 8XK and Its Successor, KDKA

Dr. Frank Conrad (left) and his amateur radio setup typified the improvised wireless stations operated by inventors and experimenters. It contrasts with the first KDKA transmitter facilities (right), with which the Harding-Cox election returns were broadcast on November 2, 1920.

Source: Conrad photo from UPI/Corbis-Bettmann; KDKA photo courtesy Westinghouse Broadcasting Company.

presidential vote results. KDKA offered brief news reports of election returns, fed to the station by telephone from a newspaper office, alternating with phonograph and live banjo music. The announcer, Leo Rosenberg from Westinghouse's public relations department, repeatedly asked anyone listening to contact the company. After the election KDKA began a regular daily hourlong schedule of music and talk.

KDKA met five criteria that qualify it as the oldest U.S. station still in operation, despite many other claims based on earlier experiments. KDKA (1) used radio waves (2) to send out uncoded signals (3) in a continuous, scheduled program service (4) intended for the general public and (5) was licensed by the government to provide such a service (Baudino & Kittross, 1977). However, no broadcasting licenses as such existed at the time; KDKA had the same kind of license as ship-to-shore radiotelegraphic stations.

Unhampered by competing signals, KDKA's nighttime signal could be picked up at great distances. Newspapers in the United States and Canada printed the station's program schedule. As other stations came on the air, some observed a "silent night" once a week to enable listeners to receive weak "DX" (long-distance) signals from far-off stations whose transmissions would be otherwise drowned out by local stations.

Westinghouse created KDKA to spur demand for Westinghouse receivers, but many early listeners tuned in on homemade sets. As early as 1906 experimenters discovered that certain minerals (crystals) had an amazing ability to detect Hertzian waves when touched in "hot spots" with a thin metallic wire, popularly called a "cat whisker." Almost anyone could assemble a complete "crystal set" for about ten dollars. Best of all, crystal sets required no electrical power for their operation.

In its first year of operation, KDKA pioneered many types of programs that later became standard radio fare: orchestra music, church services, public service announcements, political addresses, sports events, dramas, and market reports. But KDKA lacked one now-familiar type of broadcast material—commercials. Westinghouse paid KDKA's expenses to promote sales of its own products. The company assumed that each company wanting to promote its wares over the air would operate its own station.

2.5 *Broadcasting Becomes an Industry*

Westinghouse did not have the field to itself for long. Department stores, newspapers, educational institutions, churches, and electrical equipment dealers all soon operated their own stations. In early 1922 the new industry gathered momentum; by May more than 200 stations had been licensed, and the number passed 500 early in 1923.

But the number of broadcasting stations did not necessarily increase from year to year during the 1920s. Exact numbers are hard to determine, but some sources show fewer stations on the air at the start of 1924 than had been operating at the start of 1923, for example. The declining numbers in some years probably reflect the fact that some pioneer station owners discovered building and operating a broadcasting station was more expensive than they had anticipated. Receiver ownership, however, increased dramatically and consistently. In 1922 about 60,000 households had a radio; only two years later more than a million had radios.

Conflicting Approaches

RCA entered the field in 1921, purchasing WJZ, a second station that Westinghouse built to reach the New York market. WJZ assumed responsibility for producing its own programs, as had KDKA. But when AT&T put its WEAF on the air in 1922, it took a different approach. The telephone company explained that it would furnish no programs whatsoever. AT&T perceived broadcasting as another *common carrier*—merely a variation of telephony. In a 1922 press release about WEAF, AT&T explained its approach:

> Just as the [telephone] company leases its long distance wire facilities for the use of newspapers, banks, and other concerns, so it will lease its radio telephone facilities and will not provide the matter which is sent out from this station. (Quoted in Banning, 1946: 68)

Strange as it seems today, AT&T could not see the difference between a telephone booth and a broadcasting station. It soon became clear, however, that filling radio schedules entirely with leased time would not work. As with print media, advertisers would have to foot the bill. Advertisers, however, had no idea how to prepare programs capable of attracting listeners. To fill its schedule, the telephone company found itself getting into show business after all—a decidedly uncomfortable role for a regulated monopoly bent on maintaining a serious and dignified public image.

Soon the industry broke into two conflicting groups with different ideas about the way broadcasting would work. The "Radio Group" consisted of Westinghouse, GE, and RCA; the "Telephone Group" was made up of AT&T and its Western Electric subsidiary. In the end, it turned out that each group was partly right.

The Telephone Group correctly foresaw that economic and technical limitations made it impractical for every company wanting to communicate with the public to operate its own radio station. Instead, each station would need to make its services available to many different advertisers. AT&T miscalculated, however, in emphasizing primarily those interested in sending messages instead of those receiving them. Public goodwill had to be earned with listenable programs to pave the way for acceptance of advertising. Here the Radio Group's strategy of providing a program service prevailed. It took about four years for these conflicting approaches to sort themselves out.

Radio Advertising

AT&T's WEAF called the sale of commercial time *toll broadcasting*, an analogy to telephone long-distance toll calls. It first leased facilities for a toll broadcast in August 1922, when a Long Island real estate firm paid for ten minutes of time to extol the advantages of living in an apartment complex in New York. AT&T would not allow mentioning anything so crass as price.

Advertising on something as "personal" as radio was considered a bit scandalous. Many radio pioneers, including Lee de Forest and David Sarnoff, initially abhorred the practice. Potential advertisers must have had similar thoughts, because net income from advertising after four months was only about $5,000. Broadcasting was not making a large contribution to AT&T's profits.

In 1923 the first weekly advertiser appeared on WEAF, featuring a musical group the client called "The Browning King Orchestra"—which ensured frequent mention of the client's name, although the script carefully avoided mentioning that Browning King sold clothing. By 1928, under pressure of rising station operating costs and advertiser demand, more blatant advertising had become acceptable.

Few broadcasters had developed programming and production skills, leaving a vacuum that advertising agencies filled. They took over commercial program production, introducing the concept of sponsorship. Sponsors did more than simply advertise—they also owned the programs that served as vehicles for their advertising messages. Later on, during the height of network radio's popularity, advertisers and their agencies controlled most major entertainment shows, a surrender of control that broadcasters lived to regret.

Networks

AT&T found a long-lasting broadcast role in its interpretation of the RCA cross-licensing agreements, claiming exclusive rights to interconnect radio stations with its telephone lines. Thus, at first only WEAF could set up a network. It began in 1923 with the first permanent station interconnection—a telephone line between WEAF in New York City and WMAF in South Dartmouth, Massachusetts. By October 1924 AT&T had a regular chain or network of six stations and assembled a temporary coast-to-coast chain of 22 outlets to carry a speech by President Calvin Coolidge.

Denied the use of AT&T telephone interconnection for its rival network, the Radio Group's stations turned to Western Union telegraph lines. Intended only for the simple on-off pulses of Morse code, the Western Union wires could not carry music and voice with the fidelity of telephone lines, which were specifically designed for audio. Nevertheless, in 1923 WJZ formed a wire link to a station in Washington, DC, and by 1925 had succeeded in organizing a network of 14 stations.

Cross-Licensing Revisited

The growing demand for broadcasting equipment, especially receivers, upset the careful balance that the cross-licensing agreements had devised. The agreements had not included broadcasting, which was seen to be of only minor importance. Further, a federal suit alleged that the patent pool violated antitrust laws by aiming to control the manufacture and sale of all radio equipment. The suit added urgency to the need for change. At the same time AT&T concluded that its original concept of broadcasting as a branch of telephony was mistaken.

Accordingly, in 1926, the partners in the cross-licensing agreements redefined the parties' rights to use their commonly owned patents and to engage in the various aspects of the radio business. Briefly, AT&T refocused on telephony, selling WEAF and its other broadcasting assets to the Radio Group for $1 million, agreeing not to manufacture radio receivers. RCA won the right to manufacture receivers while agreeing to lease all network relays from AT&T. It may appear that this concession was small compensation for AT&T, but such was not the case. Radio and, later, television networking became a highly profitable business, a business monopolized by AT&T for many decades.

This 1926 agreement had a defining influence on the future of broadcasting in America. As long as two groups of communications companies fought over basic concepts, broadcasting's economic future remained cloudy at best. The 1926 agreements removed that uncertainty.

Origin of NBC

A few months after the 1926 settlement, the Radio Group created a new RCA subsidiary, the National Broadcasting Company (NBC)—the first American company organized specifically to operate a broadcasting network. NBC's 4½-hour coast-to-coast inaugural broadcast in late 1926 reached an estimated five million listeners and was reported to have cost $50,000 to produce. It took another two years, however, for coast-to-coast network operations to begin on a regular basis.

In 1927 RCA expanded NBC into two semi-autonomous networks, the Blue and the Red. WJZ (later to become WABC) and the old Radio Group network formed the nucleus of the Blue; WEAF (later to become WNBC) and the old Telephone Group network formed the nucleus of the Red.

History suggests NBC's networks took their names from the red and blue lines drawn on a map showing interconnecting links among affiliates. Of the two networks, NBC-Red was predominant, with more powerful stations, more popular programming, and a larger income from advertising. NBC-Blue affiliates were weaker stations, or those in less populous cities, and the Blue network offered more public service programming. Until the 1940s NBC had both a Red and Blue network affiliate in most major cities, allowing NBC to program two stations in the largest markets during network radio's golden age.

Origin of CBS

In 1927, soon after NBC began, an independent talent-booking agency, seeking an alternative to NBC as an outlet for its performers, started a rival network. It went through rapid changes in ownership, picking up the name Columbia Phonograph Broadcasting System as a result of a record company's investment. In September 1928 William S. Paley purchased the "patchwork, money-losing little company," as he later described it. When he took over, CBS had only 22 affiliates. Paley quickly turned the failing network around with a new affiliation contract. In his autobiography a half century later he recalled:

> I proposed the concept of free sustaining service. . . . I would guarantee not ten but twenty hours of programming per week, pay the stations $50 an hour for the commercial hours used, but with a new proviso. The network would not pay the stations for the first five hours of commercial programming time. . . . [T]o allow for the possibility of more business to come, the network was to receive an option on additional time. And for the first time, we were to have exclusive rights for network broadcasting through the affiliate. That meant the local station could not use its facilities for any other broadcasting network. I added one more innovation which helped our cause: local stations would have to identify our programs with the CBS name. (Paley, 1979: 42)

Paley's canny innovations became standard practice in network contracts. From that point on CBS never faltered, and Paley eventually rivaled Sarnoff as the nation's leading broadcasting executive (see Exhibit 2.d).

2.6 *Government Regulation*

One final piece remained to complete the structure of broadcasting in America—national legislation to impose order on the new medium. With the inception of telegraph systems in the 1840s, most governments recognized the need for both national and international regulation to ensure their fair and efficient operation.

The first international conference devoted to wireless communication issues took place in Berlin in 1903, only six years after Marconi's first patent. It dealt mainly with the Marconi company's refusal to exchange messages with rival maritime wireless systems. The conference emphasized the importance of subordinating commercial interests during emergencies. Marconi's aloof style of operation, aggressive commercial practices, and obvious desire to achieve a wireless monopoly had not made his company popular in international circles. Three years later another Berlin Convention reaffirmed the 1903 provisions, encouraged maritime nations to require wireless equipment on more ships, and established *SOS* as the international distress call.

1912 Radio Act

Congress confirmed American adherence to the Berlin rules with a 1910 wireless act requiring radio apparatus and operators on most ships at sea. In the wake of the *Titanic* disaster, the wireless act was modified to require at least two operators on most ships. A few weeks later, the Radio Act of 1912, the first comprehensive American legislation to govern land-based stations, required federal licensing of all radio transmitters. Specifically, the act required the Secretary of Commerce to grant licenses to U.S. citizens on request, leaving no basis to reject applications. Congress had no reason to anticipate rejections; it presumed that all who needed to operate radio stations could do so.

The 1912 act remained in force until 1927—throughout broadcasting's formative years. The law worked well enough for the point-to-point services it was designed to regulate. Point-to-multipoint broadcasting, however, introduced demands for spectrum never imagined in 1912. Unregulated growth of broadcast stations in the mid-1920s soon created intolerable interference. Secretary of Commerce Herbert Hoover called on licensees to restrict themselves to certain frequencies, limit their transmitter power, and operate only at certain times.

An ardent believer in free enterprise, Hoover hoped that the new broadcasting business would discipline itself without the need for government regulation. To that end, he called a series of four annual national radio conferences in Washington, DC. At the first in 1922, about 20 broadcast engineers attended; by the fourth in 1925, participants had increased to 400 and included station managers and attorneys.

Each year interference grew worse and, to Hoover's chagrin, broadcasters increasingly demanded he impose regulations the industry was unable to impose on itself. Recognizing that broadcasting could not survive without greater discipline imposed by someone, Hoover agreed, although he said he found it remarkable that any industry would request greater government regulation.

Finally, in 1926 things came to a head. A federal court confirmed that Hoover had no right (under the 1912 act) to establish regulations. He had tried to sue a Chicago station for operating on unauthorized frequencies at unauthorized times. The court found in favor of the station, remarking that, lacking specific congressional authorization in the law, Hoover could not "make conduct criminal which such laws leave untouched" (F, 1926: 618). In less than a year, 200 new stations took advantage of the government's inability to enforce licensing rules. President Coolidge urged Congress to pass a new law to regulate broadcasting. As he put it, "the whole service of this most important public function has drifted into such chaos as seems likely, if not remedied, to destroy its great value" (Coolidge, 1926). Early in 1927 Congress finally complied.

1927 Radio Act

The Radio Act of 1927 adopted most of the recommendations of Hoover's Fourth Radio Conference. Not a government move to limit free enterprise, the 1927 act provided what most broadcasters wanted. It established a temporary Federal Radio Commission (FRC) to bring order into broadcasting. The technical and licensing problems continued, however, and Congress eventually made the Commission permanent.

The FRC's first duty was to eliminate chaotic interference among broadcasting stations. The FRC quickly closed stations operating on frequencies reserved for other nations under international agreements and eliminated portable broadcasting transmitters. It specified transmitter power, frequency, and operating hours for the remaining stations, and increased the number of frequencies available for AM broadcasting. The FRC also established "clear channel" frequency assignments, greatly reducing interference while ensuring that people without a local station could enjoy radio programs at night, when signals on these frequencies travel long distances.

These actions would have been ineffective without the muscle to back them up. The FRC stepped up its inspections of stations and began checking broadcast signals closely for technical violations. An elaborate "constant frequency monitoring station" built during 1929–30 in central Nebraska became the focal point of the FRC's monitoring activities.

The courts helped by interpreting the Radio Act language requiring stations to operate in the "public interest, convenience, or necessity" in ways that allowed the fledgling FRC to act decisively. These factors combined to reassure investors and advertisers that broadcasting would develop in an orderly fashion.

2.7 *Depression Years, 1929-1937*

Two years after passage of the 1927 act, the Depression settled on the nation. A third of all American workers lost their jobs. National productivity fell by half. None of today's welfare programs existed then to cushion the intense suffering that unemployment, poverty, and hunger caused. Given the depressed state of the economy and the FRC's house cleaning, the number of radio stations on the air actually decreased (see Exhibit 2.f).

Role of Broadcasting

In these difficult years, radio entertainment came into its own as the only free (after one owned a receiver), widely available distraction from the grim daily struggle to survive. Listener loyalty became almost irrational, according to broadcast historian Erik Barnouw:

> Destitute families, forced to give up an icebox or furniture or bedding, clung to the radio as to a last link of humanity. In consequence, radio, though briefly jolted by the Depression, was soon prospering from it. Motion picture business was suffering, the theater was collapsing, vaudeville was dying, but many of their major talents flocked to radio—along with audiences and sponsors. (Barnouw, 1978: 27)

Roosevelt and Radio

President Franklin D. Roosevelt, coming into office in 1933, proved to be a master broadcaster, the first national politician to exploit the medium to its full potential in presidential politics. His distinctive

Exhibit 2.f

Growth in Number of Radio Stations, 1920–1997

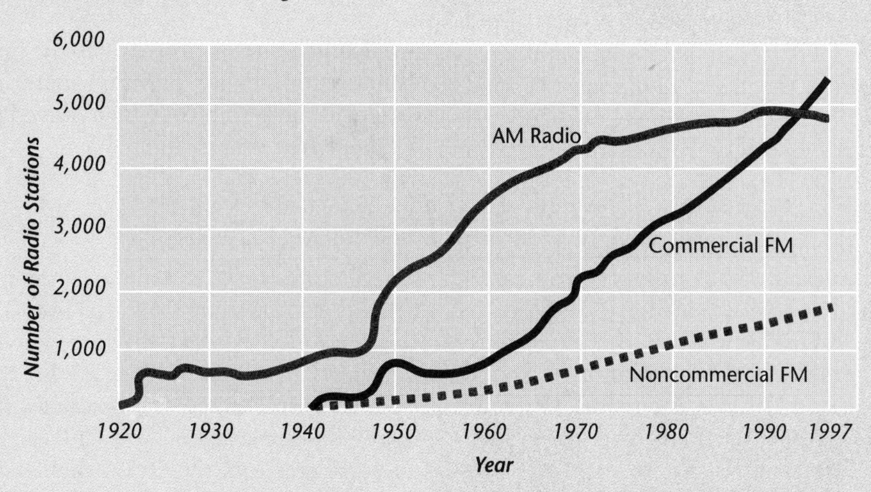

Note that significant downtrends in the growth curves occurred in 1930s AM, when the FRC imposed order on the pre–Radio Act chaos, and in 1950s FM, when its initial promise seemed not to be paying off. The sharp upward trend in the AM growth curve in the late 1940s occurred after the removal of World War II's restraints on consumer goods. In the mid-1990s, the number of commercial FM stations first matched, then surpassed, the number of AM stations.

Source: Adapted from *Stay Tuned: A Concise History of American Broadcasting* by Sterling, C., and Kittross, J. © 1990 by Wadsworth Publishing Company, Inc. Reprinted by permission of Wadsworth, Inc. 1997 data from FCC.

delivery soon became familiar to every listener who tuned to his "fireside chats," the term used to suggest the informality, warmth, and directness of these presidential radio reports to the people.

Roosevelt's political opponents also used radio effectively. Louisiana Senator Huey Long and Catholic priest Charles E. Coughlin roundly criticized FDR's policies and drew large radio audiences during the early 1930s. Roosevelt, however, emerged as undisputed master of the new medium. By the 1940s many Americans said they considered radio a more important source of political information than newspapers.

Roosevelt made a significant impact on communication regulation as well. In 1934 the president asked Congress to establish a single communication authority to pull together separate federal responsibilities for regulating both wire and wireless. After making its own study of the situation, listening to industry lobbyists, and largely ignoring a small but vocal minority seeking greater access to radio stations, Congress passed the Communications Act of 1934. It created a new regulatory body, the Federal Communications Commission (FCC), one of many agencies set up in the flurry of New Deal government activity.

Network Rivalries

During the Depression CBS tried to chip away at NBC's privileged position as the first network. Big advertisers and star performers usually selected the bigger NBC over CBS when given a choice. It did little good for CBS to build up successful programs: "We were at the mercy of the sponsors and the ad agencies," wrote Paley. "They could always take a successful show away from us and put it on NBC" (Paley, 1979: 174).

The Mutual Broadcasting System (MBS) began in 1934 with a premise different from that of CBS or NBC. In the early 1930s only two big radio stations lacked network affiliation—WGN-Chicago and WOR-New York. In 1934 they formed a cooperative network organization with WXYZ-Detroit and WLW-Cincinnati. The four stations pooled some of their own programs, most notably *The Lone Ranger*, to form the nucleus of a network schedule. But MBS attempts to expand into a national network were frustrated because most stations had already committed themselves to NBC or CBS. MBS complained to the FCC about this virtual monopoly of the older networks.

The FCC began an in-depth investigation of network practices. After three years of study, the Commission issued its *Chain Broadcasting Regulations* in May 1941, so sweeping in scope that NBC and CBS argued the new rules would devastate the industry. The FCC sought to give network affiliates more control over their own schedules and the networks less clout. Though the networks fought the rules all the way to the Supreme Court, in 1943 the Court upheld the FCC—a watershed ruling affirming the constitutionality of the Commission's rule-making powers (US, 1943).

The decision forced NBC to wrap up its dual network operation, and it sold the weaker Blue Network, which became the American Broadcasting Company (ABC) in 1945. Thanks to the new regulations, MBS and the other networks expanded rapidly after the war. Radio's four-network pattern endured into the late 1960s.

2.8 Early Radio Programs

As detailed in Section 2.1, early radio drew on familiar sources for program material, especially vaudeville acts, some of which proved readily transferable to the new medium.

Comedy

The first network radio entertainment program to achieve widespread popularity was a prime-time, five-days-a-week situation comedy, *Amos 'n' Andy*. Charles Correll ("Andy") and Freeman Gosden ("Amos") came to radio from vaudeville as a typical song-and-patter team. At a Chicago station manager's suggestion, they tried their luck at a radio-comedy series. The two white performers developed a black-dialect show in ghetto English, featuring the ups and downs of the "Fresh Air Taxicab Company of America, Incorpulated."

Amos 'n' Andy became radio's first big network hit in the early 1930s. Traffic stopped across the country and movies halted in midreel at 7 P.M. for the nightly 15 minutes of chuckles over the antics of Amos, Andy, the Kingfish, Lightnin', Madam Queen, and a host of minor characters, most of whom the versatile Correll and Gosden played themselves.

Such an impersonation of African Americans by white actors using stereotyped dialect and situations based on ghetto poverty would not even be considered today. As early as 1931 a Pittsburgh newspaper called on the FCC to ban the series, alleging racism, but its defenders argued that many blacks seemed to enjoy the program just as much as did whites. CBS dropped a later television version (with black actors) in 1953 because of NAACP opposition, but the series continued in syndication until 1966.

Comedy-variety and situation comedies reached their greatest popularity near the end of the Great

Depression, just before the beginning of World War II. Comedy became a kind of general anesthetic to ease the pain caused by years of economic hardship and the growing specter of global conflict. Many popular programs featured real-life husband-and-wife combinations: Jack Benny and Mary Livingstone (*The Jack Benny Program*), George Burns and Gracie Allen (*Burns and Allen*), Jim and Marian Jordan (*Fibber McGee and Molly*), and Fred Allen and Portland Hoffa (*The Fred Allen Show*). Ventriloquism and radio may seem incompatible, but NBC had a successful comedy-variety program featuring popular ventriloquist Edgar Bergen and his dummies. (Modern audiences may be more familiar with Bergen's daughter Candice, star of TV's *Murphy Brown.*)

Music

Radio relied heavily on music from the beginning. In the mid-1930s more than half of all radio programming consisted of music, three quarters of it carried on a sustaining (nonsponsored) basis. Large stations often hired their own musical groups, while networks supported symphony orchestras. Besides having their own large studios in which groups could perform, networks and some stations aired performances originating live outside the studios. Often these "big band remotes" originated in the ballrooms of famous metropolitan hotels. In its early years CBS devoted a quarter of its entire schedule to music. NBC began regular broadcasts of the Metropolitan Opera in 1931 and hired Arturo Toscanini out of retirement to head the NBC Symphony Orchestra in the late 1930s.

Musicians and composers welcomed such live-performance opportunities, but many stations could afford only recordings. In a single playing, recordings might reach more listeners than would hear live performances in a year. This rapid consumption *via* radio alarmed musical creators and performers. The musicians' union tried to keep recordings of all kinds off the air, and copyright holders demanded heavy payment for performance rights.

Under copyright law, playing a record in public for profit (as in a broadcast) constitutes a "performance." As such, it obligates the user (in this case a radio station) to pay copyright holders (who may include composers of the music, lyricists, and music publishers) for performing rights. These music copyright holders rely on music-licensing organizations to act on their behalf in monitoring performances and collecting copyright fees for the use of both live and recorded music. The first such U.S. organization, the American Society of Composers, Authors, and Publishers (ASCAP), dates back to 1914.

Just eight years later ASCAP began making demands for substantial payments by broadcasters for using musical works in its catalog, whether broadcast live or from records. These demands threatened stations with an unexpectedly heavy financial burden. In 1923 they formed the National Association of Broadcasters (NAB) to negotiate ASCAP's demands on an industry-wide basis.

In 1937, when ASCAP announced yet another fee increase, broadcasters decided to boycott ASCAP music and later created their own cooperative music-licensing organization, Broadcast Music, Inc. (BMI). Eventually BMI built up a comprehensive library. Competition from BMI moderated ASCAP demands, but licensing terms have been a cause of lawsuits ever since.

Recordings

Musicians were not the only ones to oppose airing recordings. In 1922 the U.S. Secretary of Commerce prohibited large stations from using recordings, arguing that records only duplicated programming listeners could receive without radio. Five years later the FRC issued orders that all stations identify as "mechanical reproductions" any programming on phonograph records to avoid having the public think the program was live. Networks banned recordings because they considered their ability to distribute live broadcasts a unique asset but made exceptions for especially newsworthy events, such as Herb Morrison's emotional description of the fiery *Hindenberg* crash in 1937. Concerns about audio quality also fueled the networks' opposition to recordings, especially for entertainment programming.

Early discs ran at 78 revolutions per minute (rpm), allowing time for only three or four minutes on a side. In 1929 broadcasters began using better-quality 16-inch ETs (electrical transcriptions). ETs revolved at 33⅓ rpm, permitting 15 minutes to a side. They were used in radio-program syndication and for specially recorded subscription music libraries on which many stations relied.

Network opposition to recorded news programming was swept away by World War II. This opened the door for recorded programming generally. ABC dropped its recording ban in 1946 to lure Bing Crosby away from powerful NBC. The laid-back crooner hated the tension and risks of real-time broadcasting, which were compounded by the need to repeat each live program in New York a second time for the West Coast to compensate for time-zone differences. Crosby invested in a then little-known company, Ampex, which developed magnetic-tape recording based on German wartime technology. As soon as broadcast-quality audiotape recorders became available, ABC agreed to let Crosby break the network ban by recording his weekly prime-time program. The other networks soon followed ABC's lead, appreciating the flexibility and control such technology offered.

Daytime Drama

Drama became a staple of daytime network programming in the 1930s. Many of these dramas were called "soap operas" for their typical sponsors. Soaps targeted female listeners, and scripts relentlessly reflected social values and gender roles considered appropriate for women during the period. "Adventurous" women working outside the home were often depicted as glamourous but sadly unfulfilled. According to the scripts, true happiness came only to dutiful wives, mothers, and homemakers. Program titles such as *When a Girl Marries* and *Young Widder Brown* suggest the general themes. The archetypical heroine, and the name of a popular soap, was *Ma Perkins*. Racial and ethnic stereotyping occurred mercilessly, even down to the naming of characters. Heroes usually had sturdy Anglo-Saxon names, but those given to villains had a "foreign" ring.

News

News depends both on outside sources of supply and on technology. Most news came from press associations or *wire services* (so called because they distributed news by telegraph for years). Newspapers calculated that they could limit radio's competition by denying broadcasters access to major news agencies. NBC had inaugurated regular 15-minute nightly newscasts by Lowell Thomas in 1930 on its Blue network—a sign that radio might soon seriously compete with newspapers.

Newspaper publishers, hoping to limit radio news, forced the networks (which were ill-prepared to gather their own news) to accept use of a special Press-Radio Bureau designed to funnel just enough news to radio to tempt listeners to purchase newspapers for more details.

The Bureau never worked effectively. For one thing, the press-radio agreement did not cover news commentaries, so many newscasters became instant commentators. In 1935 United Press broke the embargo on unrestricted release of news to radio, joined soon by International News Service (these two merged in 1958 to form United Press International [UPI]). The Press-Radio Bureau finally expired, unmourned, in 1940 when the Associated Press began to accept radio stations as members. The press associations later acquired more broadcasters than publishers as customers and began offering services especially tailored for broadcast stations, including audio feeds ready to go directly on the air.

Excesses

Reducing interference, a technical issue, initially spurred government's entry into broadcast station regulation. However, the congressional mandate to ensure that broadcasters serve "the public interest" lured the FRC, and later the FCC, into content areas.

In the early 1930s the FRC kicked Kansas broadcaster "Dr." John Brinkley off the air for prescribing sex-rejuvenation surgery and drugs of his own fabrication. Brinkley's "medical practice" was questionable, but especially so because he was not a licensed physician. About the same time the FRC also jerked the Los Angeles license of The Reverend Dr. Robert Shuler, a fire-and-brimstone crusader who often broadcast personal attacks against his critics.

Orson Welles created a panic in October 1938 when he produced a live drama nationwide on CBS in which killer Martians invaded the United States. What made the broadcast so devastating was that the program reported the Martians' "progress" in news bulletin format. People tuning in late—and many did—thought they were hearing legitimate live news reports. Thousands panicked, thinking they faced vaporization by the Martian "heat ray." That fear turned to anger against CBS and Welles when listeners realized the broadcast was a Halloween prank. The FCC investigated but declined to take action, deciding Welles and CBS intended no overt deception. Even so, the FCC's "raised eyebrows" put broadcasters on notice they should be careful.

Government reaction to what many consider excesses continues in the 1990s, as evidenced by huge fines levied against stations airing "shock jock" Howard Stern's controversial program.

2.9 *Radio in World War II, 1939–1945*

During World War II, which the United States entered in 1941, radio escaped direct military censorship by complying voluntarily with codes. For example, broadcasters avoided live man-on-the-street interviews and weather reports; the former could risk transmission of a secret code and the latter would be useful to potential enemy bombers. Radio had a role to play internationally in psychological warfare, and in 1942 President Roosevelt appointed well-known newscaster Elmer Davis to head the newly created Office of War Information (OWI). OWI mobilized an external broadcasting service that eventually became known as the Voice of America, which still broadcasts to foreign countries today.

Wartime restrictions on civilian manufacturing, imposed from 1942 to 1945, all but eliminated station construction and receiver production. Manufacturers of consumer goods devoted their capacity to military needs but continued to advertise their peacetime products to keep their names before the public. The government allowed them to write off these advertising costs as business expenses even though they had no products to sell.

Radio Drama

Released from the normal competitive pressures to sell products by support of mass-appeal material, some advertisers invested in their public image by supporting high-quality dramatic programs of a type rarely heard on American radio, though such programs were common in Europe.

Radio developed playwrights such as Norman Corwin and Arch Oboler, who won their chief literary fame in broadcasting. CBS commissioned Corwin to celebrate Allied victory in Europe with an hourlong radio play, *On a Note of Triumph,* in 1945. This emotional program climaxed an extraordinary flowering of radio art—original writing of high merit, produced with consummate skill, and always live, for the networks still banned recordings. With the end of the war years and the artificial wartime support for culture, competitive selling resumed, and this brief, luminous period of radio drama and comedy creativity came to an end.

Exhibit 2.g

Edward R. Murrow (1908–1965)

Source: Photo from UPI/Corbis-Bettmann.

CBS news reporter Murrow, shown here during World War II, often broadcast from the BBC's Broadcasting House in downtown London. He and other American reporters used a tiny studio located in a sub-basement. Once, when the building took a hit during a German bombing raid, Murrow continued his live report as stretcher bearers carried dead and injured victims of the raid past the studio to the first-aid station.

First employed by CBS in 1935 as director of talks in Europe, he came to the notice of a wider public through his memorable live reports from bomb-ravaged London in 1939–1941, and later from even more dangerous war-front vantage points. Unlike other reporters, he had a college degree in speech rather than newspaper or wire-service experience. The British appreciated his realistic and often moving word-and-sound pictures of their wartime experiences, and American listeners liked the way he radiated "truth and concern," as William Paley put it (1979: 151).

Widely admired by the time the war ended, Murrow became the core of the postwar CBS news organization. He served briefly as vice president for news but soon resigned the administrative post to resume daily newscasting. As an on-the-air personality, he survived the transition to television better than others, going on to appear in *See It Now* and in often highly controversial documentaries. He resigned from CBS in the early 1960s to direct the U.S. Information Agency under President Kennedy.

Wartime News

As the threat of European war loomed in the mid-1930s, NBC began developing European news operations to cover rapidly developing events. To counter NBC, Paley decided on a bold CBS stroke—a half-hour of news devoted to the 1938 Nazi incursion into Austria, originating live from such key cities as London, Paris, Rome, Berlin, and Vienna.

The networks' ban on the use of recorded material created tremendous problems of coordination and timing for the complex production. But in that historic half-hour, featuring Robert Trout, William Shirer, Edward R. Murrow (shown in Exhibit 2.g), and others, radio arrived as a full-fledged news medium. Thereafter, on-the-spot radio reporting from Europe became a daily news feature.

Later in 1938 came the Munich crisis where England and France abandoned Czechoslovakia to

Germany's Adolf Hitler, climaxing 18 days of feverish diplomatic negotiations among the great powers. During these tense days and nights, pioneer commentator H. V. Kaltenborn achieved fame by extemporizing a remarkable string of 85 live broadcasts from New York, reporting and analyzing news of each diplomatic move in Europe as reports came in by wire and wireless.

Thanks to CBS's early start, Paley's enthusiastic support, and a superlative staff of news people, CBS set a high standard for broadcast journalism during the war years, establishing a tradition of excellence that lasted into the 1980s. Then corporate takeovers, competition, and cutbacks (discussed in Chapter 3) began to erode news standards initiated in those pioneer days.

At this point in radio's story, television began a spectacular rise that was soon to affect the older medium profoundly. Radio's story picks up again in Section 2.13.

2.10 Pre-1948 TV Development

Television experimentation took place for decades before it developed into a mass medium. Early television produced crude pictures without sound, interesting only as curiosities. Exhibit 2.h shows the quality achieved in 1929, still far below an acceptable level. Public acceptance awaited pictures with sufficient resolution (detail) and stability (absence of flicker) for comfortable viewing—a standard at least as good as that of the home movies then familiar to wealthier consumers.

Mechanical vs. Electronic Scanning

Inventors long sought a way to break down pictures rapidly into thousands of fragments for transmission bit by bit, and then to reassemble them for viewing. Two incompatible methods of accomplishing this developed simultaneously, one based on a mechanical approach and the other entirely electronic. The mechanical approach initially showed the most promise, but the equipment was large, required almost impossibly precise synchronization between transmitter and receiver, and lacked the picture resolution needed for public acceptance. In spite of these difficulties, John L. Baird of Great Britain doggedly pursued mechanical TV systems well into the 1930s. Baird's ultimate mechanical system, however, came out second-best in 1936 when the British government compared it head-to-head with an all-electronic system developed by Marconi-EMI with assistance from RCA. Dr. H. E. Ives, Charles Jenkins, and others investigated mechanical approaches in America, but without Baird's zeal.

No single inventor can claim a breakthrough in electronic television, for television was a corporate rather than an individual achievement. However, two inventors are remembered for solving specific parts of the puzzle—Farnsworth and Zworykin. Philo T. Farnsworth, a self-taught American genius, devised an *image dissector*—a device for taking pictures apart electronically for bit-by-bit transmission. His patents required RCA, the corporate leader in American television development, to secure licenses from him to perfect its own system.

Vladimir Zworykin had immigrated from Russia in 1919 and worked as an engineer for Westinghouse. In 1923 he invented the *iconoscope*, the first electronic camera pickup tube suitable for studio operations (shown in Exhibit 4.h, p. 101).

As David Sarnoff focused RCA's resources into a drive to perfect television as early as 1930, he hired Zworykin to head the television research group. Entrusted with the task of producing a marketable television system, Zworykin's team tackled technical concerns as well as the subjective problem of developing picture-quality standards sufficient to win full public acceptance.

TV Goes Public

RCA and others had experimented with TV for many years, but it was not until 1939 that David

Exhibit 2.h

First U.S. Television Star

During early experiments with electronic television, RCA laboratories used, as a moving object to televise, a 12-inch papier-mâché model of a popular cartoon character, Felix the Cat, posed on a revolving turntable under hot lights. The image insert at left shows how Felix looked on television in 1929 when picture definition was still only 60 lines per frame.

Source: Photos from The National Broadcasting Company, Inc.

Sarnoff was ready to display his company's product publicly. Sarnoff at the time was maneuvering to have the FCC adopt the RCA system and wanted maximum publicity. He got it by making RCA's TV debut during the New York World's Fair opening ceremonies. The demonstration was successful, and RCA began acting as though its system were already the national standard adopted for telecasting, an attitude that alienated competing manufacturers and members of the FCC.

Almost two years elapsed between the RCA World's Fair demonstration and the FCC adoption in July 1941 of technical standards for black-and-white television transmission. The 1941 standards adopted were largely those originally proposed by RCA. They remained the technical foundation for

TV broadcasting for well over half a century, essentially unchanged until the FCC adopted digital TV standards in December 1996.

Within a year of the 1939 RCA demonstration, wartime needs halted production of most consumer electronics. Further development of civilian television was shelved for the duration. During World War II six experimental stations remained on the air—two in New York City and one each in Schenectady, Philadelphia, Chicago, and Los Angeles. They devoted their few hours of airtime a week primarily to civilian defense programs. Perhaps 10,000 receivers were in use, half of them in New York City.

Postwar Pause

The end of the war in 1945 did not, as some expected, bring an immediate upsurge in television activity, despite a backlog of 158 pending station applications. Many experts believed that all-out development should await adoption of a color system. And potential investors wondered whether the public would buy receivers that cost many times the price of radios—or if advertisers would pay for expensive television time.

Meanwhile progress behind the scenes resumed. In 1945 a new *image orthicon* tube came into use, improving camera sensitivity while eliminating the uncomfortably high levels of studio light that older tubes had required. In 1946 AT&T opened its first intercity *coaxial-cable* link between New York and Washington, DC, enabling the start of live interconnection.

2.11 *Growing Pains: Channels, Color, Networks*

By mid-1948 public demand and broadcaster interest inaugurated the long-anticipated television gold rush, the transition from mere experiments to true mass medium. Sixteen stations were on the air at the start of 1948—and nearly 100 two years later. The number of cities that television served grew rapidly. The audience expanded in one year by an astonishing 4,000 percent. Regular network service to a few eastern cities began. Major advertisers started experimenting with the new medium, and regular programs were launched.

Television's growing pains, however, still had not quite ended; the list of things yet to be resolved included increasing the number of channels available, adopting a color system, and establishing truly national networks.

The Freeze and *Sixth Report and Order*

In approving commercial television in 1941, the FCC had made available only 13 channels (later reduced to 12) to serve the entire United States. As more and more stations applied to go on the air, it became obvious that demand would soon exceed the supply of channels. And interference among the few stations on the air was growing because of a miscalculation of the distance needed to separate stations assigned to the same channel.

In September 1948 the FCC abruptly froze further processing of television-license applications pending solution of these problems. The freeze had no effect on applicants whose permits had already been approved. For the nearly four years of the freeze, 108 prefreeze stations enjoyed an enviable monopoly. The FCC held a lengthy series of hearings to settle the complicated engineering and policy questions that had precipitated the freeze. The much-anticipated decision, the basic charter of present-day American broadcast television, came on April 14, 1952, in the FCC's historic *Sixth Report and Order* (FCCR, 1952).

The new rules relocated some of the original 12 VHF (Very High Frequency) channels to reduce interference and added 70 new TV channels (channels 14–83) in the UHF (Ultra High Frequency) band. Experience with radar on these much higher frequencies during the war suggested that UHF stations would suffer serious coverage limitations compared to those on VHF channels. Disagreements also flared over *intermixture* (assignment of both VHF and UHF channels in the same market), the distance needed between stations on the same channel to minimize interference, and the importance of adopting color standards. After more than three years of deliberation the FCC combined four separate frequency bands to obtain sufficient spectrum space for TV. Exhibit 4.k summarizes their location in the spectrum and current TV channel numbers. The FCC eliminated UHF TV channels 70 through 83 in 1970.

Channel-Allotment Plan

There was great pressure to leave the 108 prefreeze stations alone. The new UHF channels and the VHF channels not already in use allowed, at least in theory, the FCC to license more than 2,000 TV stations. The FCC increased the number of communities allotted a TV channel from 345 to almost 1,300, but about two-thirds of these were in the technically inferior UHF band. That was certainly the case for the 10 percent (later increased to 35 percent) of the channels reserved for noncommercial educational use. Of the more than 360 educational TV stations on the air in 1997, about 240 were on UHF channels. Exhibit 2.i provides an example of how the allotment table separates stations on the same channel throughout the country, spacing them to avoid co-channel interference.

The number of on-air stations more than tripled in the first postfreeze year, as shown in Exhibit 2.j. Nevertheless, the new channel-allotment plan had serious defects. For one thing, there were still too few channels to give viewers in every market the same range of choices. Some had only one channel, some two or three, and so on. Viewers in any given market needed access to at least five channels to be able to choose among the three networks, an independent (nonnetwork) station, and an educational outlet. For another, the FCC's decision to make both VHF and UHF allotments in many localities placed UHF stations in such markets at a serious disadvantage.

UHF Problem

UHF waves have an inherent weakness—they cannot cover as great an area as VHF waves. The Commission tried to ensure equal coverage for both VHF and UHF stations by authorizing UHF to use higher power. Yet even if added power could have helped, years passed before maximum-power UHF transmitters became available. And long after UHF television service began, set manufacturers still built VHF-only receivers, forcing viewers in markets with UHF stations to buy difficult-to-use UHF converters and often poorly performing UHF antennas.

Faced with overwhelming disadvantages, UHF television declined to a low of only 75 stations by 1960. The FCC continued to seek ways to assist failing UHF stations. Its most useful step came in 1962, when it persuaded Congress to require manufacturers to equip all receivers with built-in UHF tuners. Despite FCC efforts to improve UHF viability, the *Sixth Report and Order* saddled the television industry with long-term problems. The frustration of viewers who wanted more program options than the FCC's allotments made possible encouraged eventual development of cable television and other substitutes for over-the-air reception.

By 1965 some of the FCC's UHF efforts had taken effect, and that part of the industry began a steady, though not spectacular, growth. FCC financial reports indicate that until 1974 UHF stations as a group continued to lose money. Thereafter their average profit margin increased each year. By mid-1994 UHF commercial outlets on the air (594) had surpassed VHF (559). That trend continued during the mid-1990s: There were more than 630 UHF stations on the air by 1997.

Color TV

The battle between CBS and NBC/RCA over a technical standard came to a head during the freeze. CBS proposed a system partly based on

Exhibit 2.i

TV Channel Allotment Plan

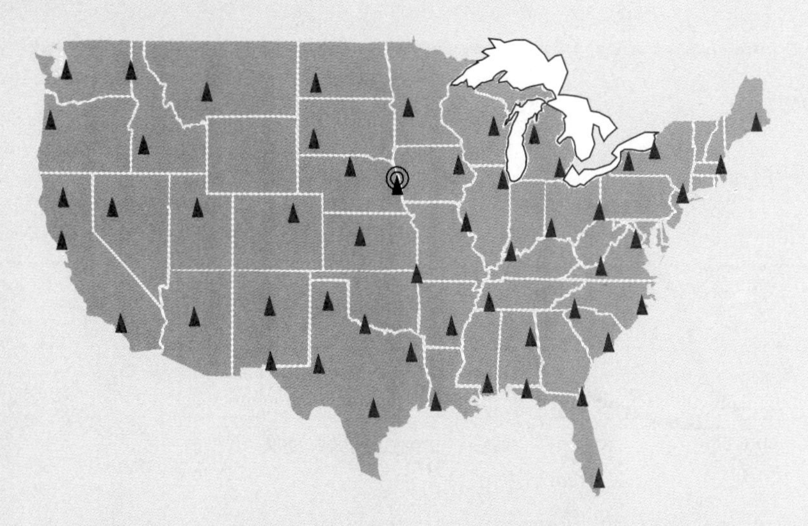

Using channel 7 as an example (a similar map could be made for any other VHF or UHF channel), this map shows the occupied channel 7 allotments (those actually on the air), except for three outside the contiguous 48 states. They are spread fairly evenly, separated from each other by a minimum of 170 miles. The list of all channels available in Omaha in 1997 is

3—occupied by KMTV, a CBS affiliate
6—occupied by WOWT, an NBC affiliate
7—occupied by KETV, an ABC affiliate
15—occupied by KXVO, a WB affiliate
26—occupied by KYNE, a noncommercial educational station licensed to the Nebraska Educational Television Commission
42—occupied by KPTM, a Fox and UPN affiliate
48—reserved for noncommercial educational use; not on air
54—not on air

This market is also served by cable, MMDS, and DBS.

mechanical scanning technology, while NBC/RCA advocated an all-electronic system. Interestingly, the FCC approved the CBS system in 1950, in the very middle of the freeze. The next year five East Coast CBS stations hooked up to broadcast the first commercial colorcast, to a very small audience. The major equipment manufacturers, led by RCA, had simply refused to make TV receivers using the

Exhibit 2.j

Growth in Number of TV Stations (1948–1997)

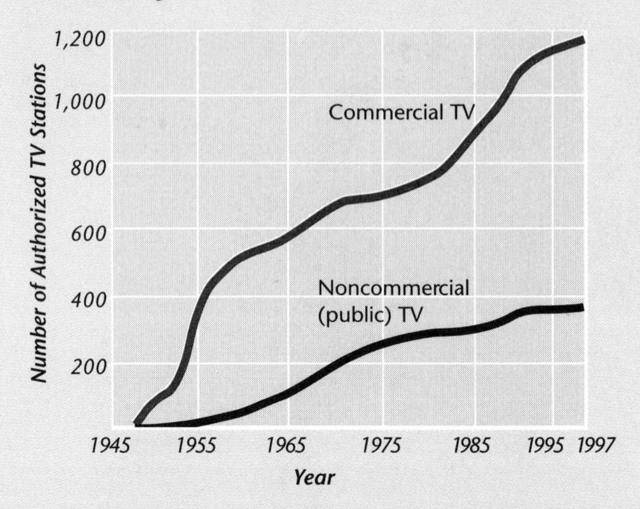

The modern TV era started in 1948, with 16 stations on the air. Only 108 stations had been authorized when the 1948–1952 freeze imposed a temporary ceiling. After that the number shot up, reaching 400 by 1955. Growth began to slow down at that point, but has never actually stopped. Noncommercial stations developed more slowly, starting with the first two in 1954. Not included here are the more than 1,900 low-power TV stations or approximately 5,000 television translators.

Source: Adapted from *Stay Tuned: A Concise History of American Broadcasting* by Sterling, C., and Kittross, J. © 1990 by Wadsworth Publishing Company, Inc. Reprinted by permission of Wadsworth, Inc. 1997 figures from FCC.

CBS system, in spite of the FCC decision. Finally, in 1953, all parties agreed to an electronic system patterned closely on RCA's. An important argument in its favor was that it had compatibility—black-and-white sets already in use could pick up the newly approved color signals and display them in monochrome.

However, color telecasts on a large scale developed slowly. Five years after the 1953 adoption of color standards, only NBC offered programs in color. Full network color production in prime time finally arrived in 1966. By 1972, nearly two decades after adoption of color standards, only half of the country's homes had color television sets. By the mid-1990s new black-and-white sets had become hard to find, except for very small "personal portable" models.

TV Relays and Recording

Full-scale national network operations had to await completion of AT&T's coaxial-cable and microwave relay links, which joined the East and West Coasts in 1951 and enabled the start of live network interconnection. Until the cable reached them, affiliates had to use kinescopes of network programs—films of television programs photographed as they were displayed on the face of a receiver tube—syndicated films, and their own live productions.

Engineers developed techniques for filming images on a kinescope—the technical name for the TV picture tube—in the late 1930s, but they became important only with the explosive growth of TV in 1948. In spite of improvements made over the years, the image quality of these "kinies" remained poor. In 1956 the Ampex Corporation demonstrated a successful videotape recorder, which saw its first practical use that fall on CBS. In a rare spirit of cooperation, competing manufacturers put aside rivalries and agreed on a compatible professional videotape standard. Videotape doomed the era of live production, as related in Section 2.12.

Ceiling on Networks

The FCC's 1952 *Sixth Report and Order* effectively limited television to three national networks because there were too few cities with four channels to allow a fourth network to compete nationally on an even footing. Even if all the independent commercial stations in every market combined to form a fourth network it would reach, at best, 85 percent of the population, whereas ABC, CBS, and NBC each could reach more than 95 percent of television homes. This situation underlines as well the FCC's greater concern for local, rather than national, television service.

Nevertheless, there have always been pressures to create a fourth national television network. MBS, the fourth radio network, lacked the money to branch into television. From 1946 to 1955, however, a fourth chain did exist—Du Mont Television Network, founded by Allen B. Du Mont, a developer of the cathode-ray tube. The Du Mont network survived only as long as lack of interconnection facilities kept networking in check. Once interconnection relays became generally available, Du Mont could not sign up sufficient affiliates to compete with the larger, older, and richer networks.

It was not until the mid-1980s that Fox Broadcasting Company emerged to become the "fourth network." In the mid-1990s the Warner Brothers Network ("The WB") and the United Paramount Network (UPN) followed Fox, each seeking a piece of the action dominated for decades by the Big Three.

2.12 Golden Age of Television, 1948–1957

As some look back with nostalgia to radio's "golden era" of the 1930s and 1940s, some feel the same way about television's first decade as a mass medium, 1948 to 1957.

Programs as Motivators

In that decade, the networks put first priority on stimulating people to buy sets. Only attractive programs could do that:

> It was the only time in the history of the medium that program priorities superseded all others. If there was an abundance of original and quality drama at the time . . . it was in large part because those shows tended to appeal to a wealthier and better-educated part of the public, the part best able to afford a television set in those years when the price of receivers was high. (Brown, 1971: 154)

Most programs, local and network, were necessarily live—a throwback to the earliest days of radio. Videotape recording was uncommon, especially at the local level, until the late 1950s. Original television plays constituted the best artistic achievements of television's live decade. "Talent seemed to gush right out of the cement," as pioneer *New York Times* critic Jack Gould observed years later.

It would be misleading, however, to suggest that all "golden age" programming achieved high artistic status. NBC's enormously popular *Texaco Star Theater*, for example, featured slapstick routines borrowed almost directly from vaudeville. Such programs appealed to less well-to-do, less elitist viewers, people for whom the purchase of a TV set was a major investment. Without such popular programs the cost of receivers would have remained high, and the growth for TV as a mass medium could have been slowed.

TV vs. Hollywood

Television programs could, of course, have been recorded from the very beginning by filming them (and many were—witness the continuing reruns of the early 1950s' *I Love Lucy*). Economic, technical, and industrial barriers delayed widespread use of this solution. Hollywood's traditional single-camera method of production is slow and cumbersome, costing far too much for general television use; and it took time to adapt film production to the physical limitations of television, with its lower resolution, smaller projected-picture area, and more restricted contrast range.

In addition, many critics opposed use of film, expecting television to escape Hollywood's straitjacket in favor of a new, live, and less trite mass-entertainment genre. The two points of view were as far apart as their two production centers—television in New York and film in Los Angeles. But the economics of both industries drove them closer together. Inevitably, as technical barriers to producing filmed television programs fell, and as Hollywood accepted use of videotape, the production base for entertainment programs shifted to the West Coast by the late 1950s.

Hollywood's old-line feature-film producers saw television as a threat, much as publishers had feared that radio would undermine newspapers. Producers and distributors refused to release their most recent feature films to television for a dozen years. At first only pre-1948 films could be shown on television. Not until the early 1960s did Hollywood conclude that the television bane could also be a boon. Movie producers realized that audiences would not stop buying theater tickets just because they could see movies on television. And the networks were ready to pay big money to show "post-48" films—movies that in most cases had little or no residual value for theatrical release. Eventually feature-film producers found they could profit even more from delayed release to cable television and the VCR market, as well as broadcasting.

Network Rivalries

In terms of program formats, many important ideas in TV's first decade sprang from the fertile imagination of Sylvester "Pat" Weaver, who became NBC's vice president for television in 1949, leaving NBC six years later as board chair. Disregarding conventional wisdom about established viewing habits, Weaver occasionally replaced regular series with one-time *spectaculars*, 60 or 90 minutes long. Weaver also foresaw that the single-sponsor show (a hallmark of network radio) simply could not last in prime-time television as costs soon exceeded what single advertisers could afford. Instead, NBC introduced participating sponsorship, which enabled several different advertisers to share costs of a program. NBC's long-running *Today* and *Tonight* shows are further testimony to Weaver's lasting impact.

Despite Weaver's NBC innovations, CBS steadily gained in the audience ratings race. Much of the CBS's success is attributable to William Paley. In the late 1940s Paley capitalized on a loophole in the tax laws that allowed him to entice talent from NBC and ABC by offering large-income performers deals they could hardly refuse. The other networks retaliated, but it was Paley who often seized the initiative.

In 1955 CBS achieved the number-one place in the ratings, a rank it would hold undisputed for 21 years. Meanwhile, the third network, ABC, found itself in somewhat the same position that CBS had occupied in the early days of network radio. Top advertisers and performers automatically chose CBS or NBC, turning to ABC only as a last resort.

A corporate breakup helped rescue ABC in 1953. When a Justice Department antitrust suit against the big Hollywood film studios forced them to sell off their extensive theater chains, one of the spun-off companies, Paramount Theaters, merged with ABC in 1953. Paramount injected much-needed funds and established a link with Hollywood that eventually paid off handsomely, helping ABC to achieve status equal to that of CBS and NBC.

In 1954 Walt Disney, the first of the major studio leaders to make a deal with television, agreed to produce a series of programs for ABC called *Disneyland* (1954–1957), later continued on NBC. The deal built ABC audiences and gave Disney free television promotion for his developing California theme park as well as for Disney feature films. Disney went on in later years to become a primary program source for broadcast and cable television. In keeping with the mergers and acquisitions characteristics of media industries in the 1990s, Disney's long association with ABC took a major turn when it purchased ABC in 1996 for around $19 billion.

In search of a different image, ABC in the 1970s began to pay more attention to shows appealing to young, urban, adult viewers. Less concerned than older networks with traditional program balance, ABC leaned heavily on action, sex, and violence in much of its prime-time programming.

2.13 Radio's Response

Television's "takeoff" year of 1948 marked both a high-water mark and the doom of full-service network radio. In that year radio networks grossed more revenue than ever before or since, excluding profits from their owned-and-operated stations. For more than 15 years, radio networks had dominated broadcasting, but television was about to end their rule.

Radio Network Decline

For the balanced news-and-entertainment network programming that prevailed before television, four radio networks seemed to offer as much as stations and audiences would use. The four supplied their affiliates a full schedule of varied programs, much as the major television networks do today.

Sponsors of major radio-network programs and stars hastened the decline in radio-network fortunes after 1948. As television captured its first audiences, it lured away major radio advertisers—and with them went star performers and their programs. By the early 1950s the radio networks were reeling. CBS's William Paley recalled that

> Although [CBS's radio] daytime schedule was more than 90 percent sponsored, our prime-time evening shows were more than 80 percent sustaining. Even our greatest stars could not stop the rush to television. (Paley, 1979: 227)

The ultimate blow came when radio stations began to decline network affiliation—a startling change, considering that previously such an affiliation was a precious asset. But program commitments to the networks interfered with the freedom to put new, tailor-made, post-television radio formulas into effect. By the early 1960s only a third of all radio stations still held network affiliations. Networks scaled down service to brief hourly news bulletins, short information features, a few public-affairs programs, and occasional on-the-spot sports events.

Today's dozens of satellite-delivered audio networks are possible because there are more stations, cheaper satellite transmission, and because each modern service or network specializes on a specific kind of programming. But this kind of radio networking differs greatly from that of the 1930s—and perhaps foreshadows the future of television.

Rock to the Rescue

With the loss of network dramas, variety shows, and quiz games, radio programming shrank essentially to music and news/talk, with music occupying by far the majority of the time on most stations—just as it had in the 1920s. Only this time the music was off records rather than live, broadcasters' aversion to recorded programming having passed into history years before. Providentially for radio, this programming transition coincided with the rise of a new musical culture, one that found radio an ideally hospitable medium.

Early in the 1950s Cleveland disc jockey Alan Freed gained national recognition by

> playing a strange new sound. A sound that combined elements of gospel, harmony, rhythm, blues, and country. He called it "rock and roll." And people everywhere began to listen. . . . It transcended borders and races. . . . Rock and roll sang to the teenager; it charted his habits, his hobbies, his hang-ups. (Drake-Chenault Enterprises, 1978: 1)

Radio proved to be the perfect outlet for this new form of expression as stations in many markets converted to a "Top-40" format in the late 1950s. The format name referred to the practice of rigidly limiting DJs to a prescribed playlist of current best-selling popular recordings. Gordon McLendon, a colorful Dallas sportscaster and station owner, helped pioneer the format. Omaha-based Todd Storz played a key role in popularizing Top-40 when he adopted it for his string of group-owned stations. Such Top-40 specialists frequently reprogrammed bottom-ranked stations and lifted them to top ratings in a matter of months.

The announcer took on a much more important role in a Top-40 format. Before Top-40, announcers typically were wallpaper figures providing continuity between programs. After Top-40, listeners could hear the same songs on many stations, so an important distinguishing attraction became the personality of the DJ. Top DJs commanded high salaries, and even those in small markets sometimes achieved local cult hero status.

Paradoxically, Top-40's spectacular success came as much from its ability to repel listeners as from its skill in attracting them. Formula programmers relied on consistency above all else; they programmed relentlessly for their target audience, no matter what other audience segments might think. Initially this often meant junking old programs attractive to older audiences. By the 1960s rock lyrics spread slogans of the disenchanted in a coded language, defying conventional, conservative standards that broadcasting had previously sought to maintain.

A second ingredient in Top-40 success was an equally single-minded dedication to constant promotion and advertising. Call letters and dial position were indelibly imprinted on the listener's mind, usually by endless repetition and promotional gimmicks, in an effort to ensure proper responses to rating surveys.

FM's Rise to Dominance

Radio's renewed focus on music brought FM to the fore. For its first quarter-century, broadcasting simply meant AM radio. In 1933 Edwin Armstrong announced and patented a much-improved alternative method of audio transmission using frequency modulation (FM). Armstrong tirelessly promoted the superiority of FM, thinking he had the support of his friend, David Sarnoff. But Sarnoff's RCA was making good money with AM, expected to make even more with TV, and devoted little attention to the development of FM.

About 50 stations made it on the air before the United States entered World War II. At the close of the war, however, the FCC shifted FM from its prewar channels to its present location at 88 to 108 MHz in the VHF band. This 1945 move made obsolete the half-million FM receivers that had been built to operate on lower frequencies.

Postwar interest in FM stations, mostly as minor partners in AM/FM combinations, peaked in 1948, with more than a thousand stations authorized. But as television's rapid climb to power began, FM declined into the background. Total station authorizations declined during most of the 1950s.

Faced with bleak prospects for FM, exhausted by continuing court struggles over his patents, and probably feeling betrayed by his one-time friend David Sarnoff, Edwin Armstrong became deeply depressed. In 1954 Armstrong put on his overcoat and stepped out the window of his 13th-floor New York apartment, plunging to his death.

In 1958 FM began to recover, not only because of greater audience interest in improved sound quality but also from the lack of spectrum space for additional AM stations.

A new pro-FM FCC policy also helped. For example, in 1961 the FCC approved technical standards for FM stereophonic sound. In 1965–1967 the FCC adopted a nonduplication rule that required AM/FM owners, first in big cities and then in smaller markets, to program their FM stations independently of AM sister stations. These measures proved effective in moving FM out of AM's shadow and created FM as an independent service with its own formats. Increased listener interest galvanized manufacturers into making FM receivers more widely available, and at lower cost. By 1974 the majority of sets included FM, and two years later so did most car radios. By the 1980s a reverse trend had set in: some radios were marketed with FM-only tuning, and 75 percent of all radio listening was to FM outlets. And ironically, when the FCC dropped the program nonduplication rules as a part of its *deregulation* in the 1980s, some FM stations allowed struggling AMs to duplicate their programming.

During the 1970s and beyond, television broadcasting, too, would find itself backed into a corner. The next chapter describes an era of new and increasing competition that threatened to undermine the entire broadcasting industry in its traditional form.

NEWS
Rollin' Ranch

Chapter 3

Cable and Newer Media

CHAPTER CONTENTS

Traditional broadcasting prospered into the 1980s, little changed from a decade or two earlier. But new multichannel delivery systems developed in the late 1970s, expanding so rapidly in the 1980s and 1990s that more American homes now receive their television programs by cable than from over-the-air stations.

Audience demand for still more viewing options, along with technical progress and important changes in regulatory thinking that favors marketplace solutions, have combined to create a fast-changing electronic media scene. Technologies introduced here are described in more detail in Chapters 4 (cable television) and 5 (on relays, recording, and the digital revolution).

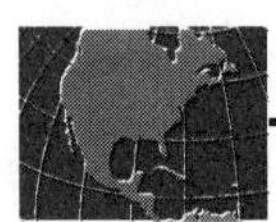

3.1 *Emergence of Cable*

Broadcasting's greatest challenge emerged from attempts to improve television-station coverage. The FCC's 1952 television channel-allotment plan (the *Sixth Report and Order* described in Section 2.11) left many "white areas"—places where people could not receive television service. Even in the biggest cities viewers might receive only five to eight channels. Most people received three or four, and virtually everyone wanted more.

Extending TV Coverage

At first low-power *repeater* transmitters met some of the demand by extending station coverage into fringe reception areas and into pockets that hills or mountains cut off from normal reception. The most common type, the *translator*, transmits on a different channel from the originating station so as not to interfere with the originating station's signal. Some stations operated several translators. Nearly 5,000 were in operation by 1997.

CATV

Another signal-extender, *community antenna television* (CATV), originally worked something like a translator, repeating the signal of one or more nearby stations. Instead of transmitting signals over the air, however, a CATV system delivers them by wire to individual subscribers. A CATV system using *wide-band* cable can deliver signals from several stations, outdoing translators by giving eager subscribers a range of program choices.

CATV systems began appearing in rural areas of both Oregon and Pennsylvania in 1948, just as regular broadcast television began. And, like radio years earlier, CATV's original economic motivation was to sell receivers. Appliance dealers who built the first systems could market television sets to consumers who otherwise would not buy them because they could not pick up over-the-air signals. Community antennas on hilltops or on high towers, unaffordable by the average viewer, obtained unobstructed line-of-sight signal paths to originating stations with viewers sharing in their cost.

CATV systems remained a local concern in the 1950s. Regulation, if any, came from municipal governments that granted cable operators franchises, permitting them to string cables on utility poles or bury them along public rights of way. Early cable systems carried only five or six channels, all of them nearby television stations. Most systems served from a few dozen to a few hundred subscribers. Fewer than 1 percent of all television households had access to cable by 1960. A decade later (1970) the percentage of homes with cable was still under 10 percent (Exhibit 3.a).

Program Augmentation

CATV operators soon sought ways to enhance—or *augment*—their service by offering subscribers something new. This took several forms as illustrated in Exhibit 3.b. Using microwave relays, some systems began delivering programs from more distant television stations. Others originated their own local programs. And some systems began to offer

Exhibit 3.a

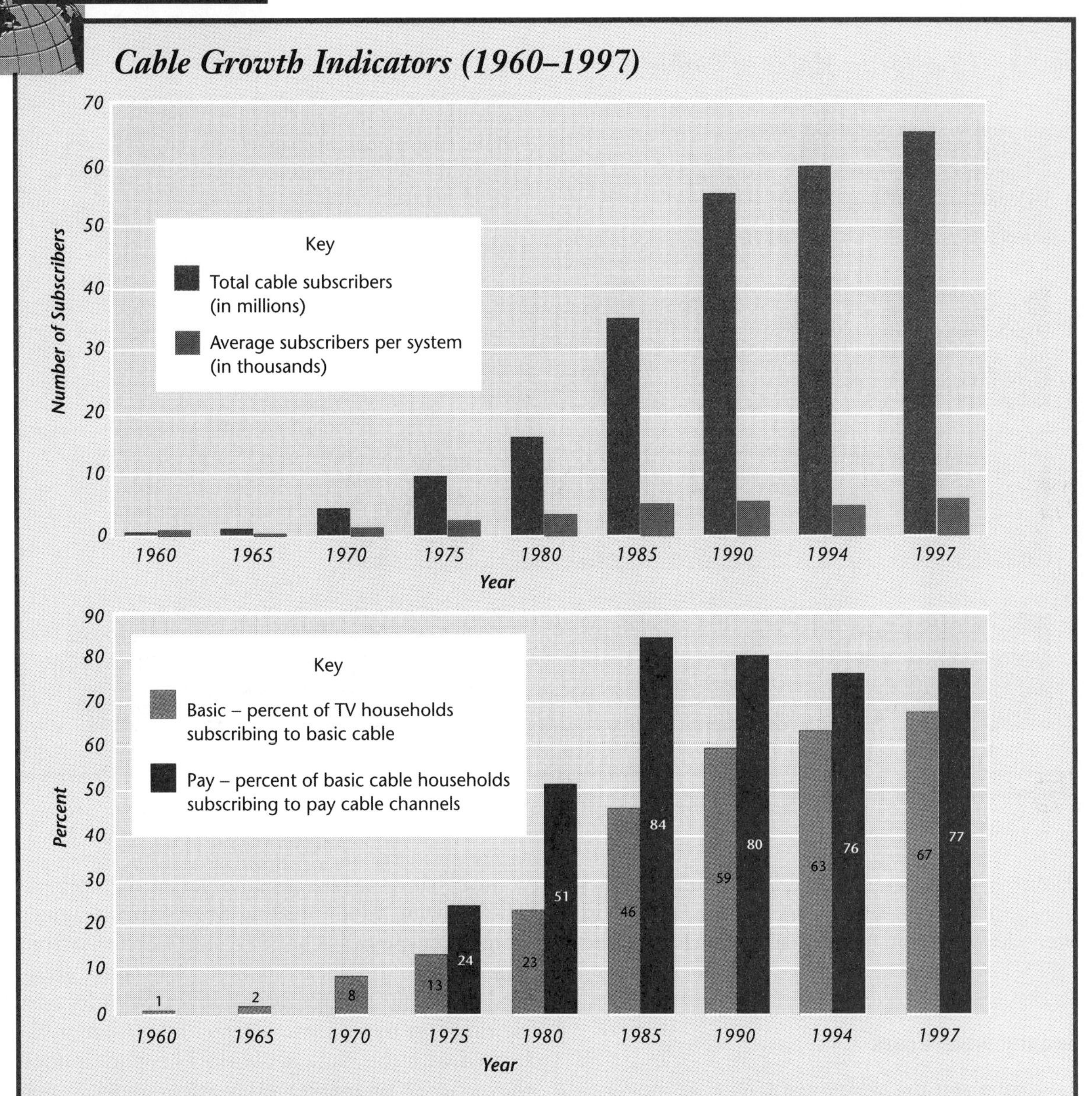

As cable grew, system size and channel capacity both increased. After 1975, pay cable spurred basic cable growth. The number of cable households continued to increase through the 1990s, but pay cable subscriptions as a percentage of cable households began declining in the mid-1980s. Recent figures show a steady to a slightly larger percentage of basic cable households subscribing to pay channels, perhaps signaling the end of a decade-long trend.

Sources: Sterling and Kittross (1990) pp. 660–661 for data through 1985; NTCA for data through 1997, citing various sources, including *Television Factbook*, A. C. Nielsen, and Paul Kagan Associates, Inc.

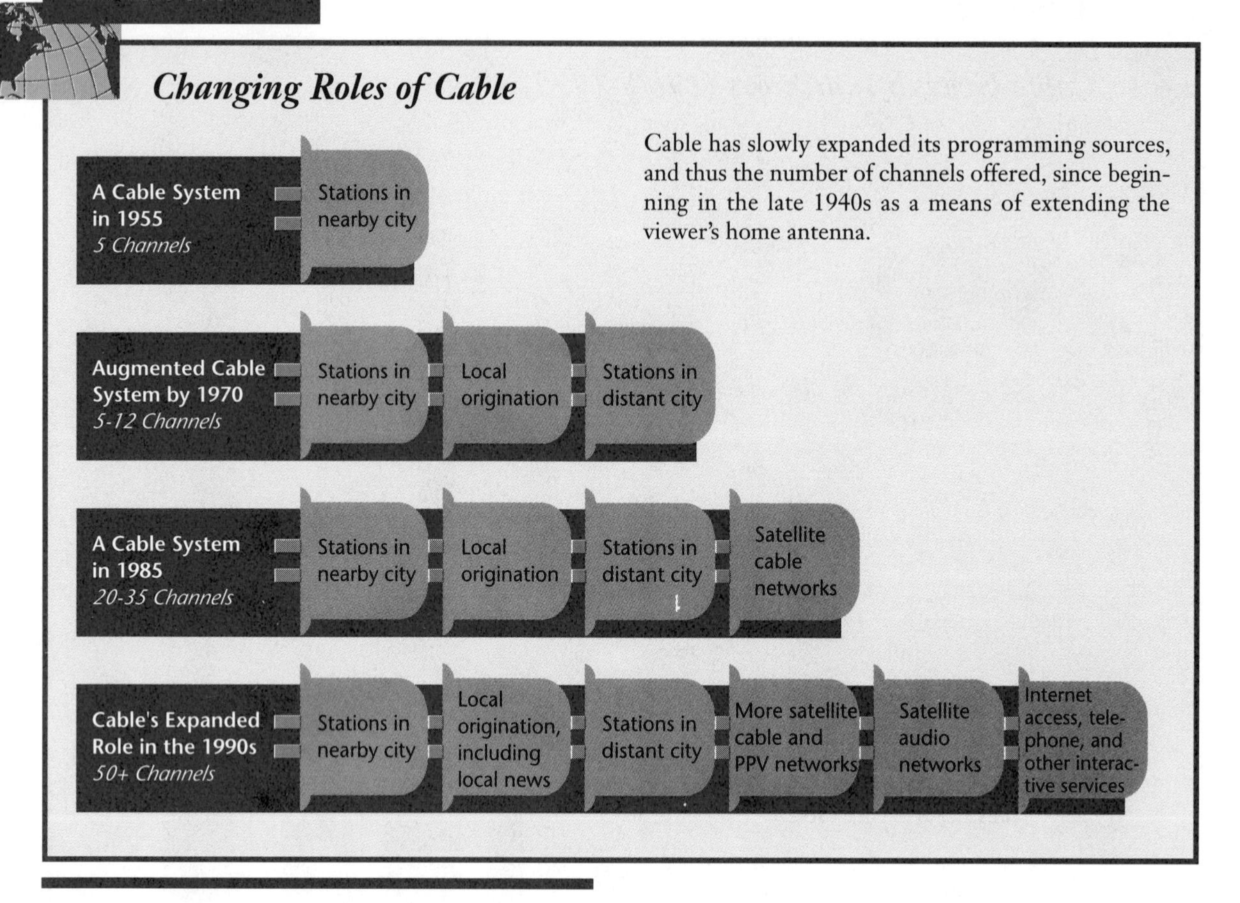

Exhibit 3.b

Changing Roles of Cable

Cable has slowly expanded its programming sources, and thus the number of channels offered, since beginning in the late 1940s as a means of extending the viewer's home antenna.

nonbroadcast programming—feature films, sports, and special events. Such augmentation, along with later additions, marked a fundamental change from CATV's original role.

Broadcaster Fears

Television stations welcomed CATV as long as it acted merely as a *redelivery service*, extending signals to unserved areas, beefing up fringe reception, and overcoming local interference. Some stations eventually found their signals carried by dozens of cable systems that substantially enlarged their audiences.

By the mid-1960s, however, broadcasters began to see augmented CATV more as a potential competitor than as a neutral extender of their signal. The growing cable practice of importing signals from distant cities tended to fuzz normally fixed boundaries of broadcast markets. A network affiliate might find its programs duplicated in its own viewing area by a cable-delivered distant station affiliated with the same network. This undermined the concept of market exclusivity, under which broadcast programs had usually been sold. Program duplication added to CATV's tendency to fragment audiences, leaving broadcast stations with lower ratings and thus less appeal for advertisers. In addition, CATV began to expand into big city suburbs, no longer merely serving those unable to receive off-air service.

Regulation Limits Cable

The FCC initially adopted a hands-off policy regarding cable regulation, noting that cable operators did not transmit signals over the air. But as cable's economic impact grew, so did broadcaster complaints. In addition to complaining to the FCC, broadcasters appealed directly to Congress, pressuring politicians to encourage the FCC to take action. The FCC reversed its earlier position and began adopting protectionist rules. In 1962 it began to use microwave relay licenses as a justification for regulating cable systems that imported distant station signals. Four years later the Commission extended regulation to all cable systems, beginning a brief period of pervasive control of cable.

The FCC soon required cable to carry all television stations within each system's areas of coverage (the *must-carry rule*) and to refrain from duplicating network programs on the same day a network offered them. No new signals could be imported into any of the top 100 markets without a hearing on probable economic impact on existing television stations. In 1968 the Supreme Court upheld this FCC role (US, 1968a).

As for cable-originated programs, broadcasters feared (long in advance of any real danger) that networks of cable systems might outbid them for transmission rights to popular programs such as major sports events. They spread scare stories about program *siphoning*—the draining away to cable of hitherto "free" broadcast programs. Broadcasters warned that fans might have to start paying to receive major league baseball and football games. They also claimed that network news and local public-service programs would suffer if cable cut into broadcaster advertising revenues.

Such tactics resulted in the FCC's so-called definitive cable regulations of 1972 (FCCR, 1972b). These rules severely restricted cable to being merely *ancillary* or secondary to over-the-air television. The new rules sharply limited the type and number of signals that cable could bring into the largest cities in competition with local stations. And cable operators had to provide, on request, access channels for local governments, educational institutions, and the general public.

Later Deregulation

Only five years after the 1972 "definitive" rules appeared, however, an appeals court held that the FCC had "in no way justified its position that cable television must be a supplement to, rather than an equal of, broadcast television" (F, 1977: 36). This rebuff, plus a change in administration at the White House (and consequently in the FCC chair), led the FCC to reconsider. It began to remove itself step by step from cable regulation.

This *laissez-faire* spirit prevailed until the early 1990s. Prompted by complaints about high subscription fees, poor service, and the arrogance of cable system operators, Congress imposed new regulations with passage of the Cable Television Consumer Protection and Competition Act of 1992. Four years later, Congress again reversed course and passed the Telecommunications Act of 1996. See Chapter 12 for additional discussion of cable regulation.

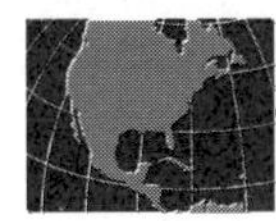

3.2 Cable Becomes a Major Player

As program augmentation expanded and cable shed its regulatory constraints, CATV began to emerge as a full-fledged competitor for television viewers. As it matured, the term CATV faded out, replaced by *cable television*. The transformation of CATV into cable combined deregulation, use of new satellite relays, and increased original programming.

Domsats' Key Role

In the 1960s communication satellites first came into use for international broadcast relays, bringing overseas events to home screens instantaneously. Demand for *domestic satellites* (*domsats*), however, was delayed because of well-developed existing national systems of microwave and coaxial-cable relays.

A 1972 FCC decision stimulated domsat demand. The commission adopted an "open-skies" policy, allowing any adequately financed and technically qualified business firm to launch satellites for domestic use (FCCR, 1972a).

Western Union's Westar I became the first American domsat in 1974. Such satellites operate as common carriers—point-to-point, rate-regulated communication services. They lease or sell satellite relay services, either to program distributors or to brokers who resell short-term access to such users as television news departments.

TVRO Antennas

A second satellite-related FCC decision, issued in 1979, deregulated *television receive-only* (TVRO) antennas, used to pick up programs relayed by satellite. This eliminated a cumbersome licensing process, opening the way to widespread satellite use for relaying program services to cable systems—and soon to broadcast stations as well. In 1977 fewer than 200 cable systems owned their own TVROs, but within a decade some 8,000 systems used them.

Turner and Superstations

Domsats and TVROs gave cable systems a more efficient and less expensive way of obtaining new national program services with which to attract subscribers. However, the programs themselves still remained in short supply. Ted Turner, an innovative entrepreneur with an almost mystical faith in cable's future, invented the superstation concept as one way to expand cable's program menu. His colorful career is outlined in Exhibit 3.c.

In 1970 Turner bought low-rated Atlanta station UHF channel 17 (now called WTBS for Turner Broadcasting System). Its schedule soon heavy with movies and sports, channel 17 was uplinked for satellite distribution to cable systems throughout the country (see Exhibit 3.d). Turner initially charged nothing for his programs to entice cable-system operators to invest in TVROs and downlink WTBS for their subscribers. Cable systems paid 10 cents per subscriber per month for the relay service. Turner's eventual profit came from higher advertiser rates on WTBS, justified by the nationwide cable audience that it acquired as a superstation. By 1995 nearly all cable systems carried TBS (as the superstation was renamed to publicize Turner Broadcasting System), representing a potential audience of some 60 million homes. Chicago's WGN and several other TV outlets also became superstations with schedules of films, syndicated fare, and sports. By the mid-1990s the superstation concept was losing its appeal. Economic factors placed superstations at a competitive disadvantage with other cable program providers. Turner converted TBS to a cable network in 1998.

HBO and Pay Cable

Domsats played a key role in expanding another major programming innovation—*pay (or premium) cable*. Home Box Office (HBO) led this development in 1972. It introduced an advertising-free channel of (primarily) feature films for which subscribers paid an additional monthly fee over and above the charge for advertising-supported channels (later called the *basic tier*).

Lack of cost-effective distribution held back pay-cable progress until 1975, when HBO leased a satellite transponder—the receive-transmit unit on a satellite. HBO offered its programs to any system in the country able to buy or lease a TVRO antenna. Satellite delivery reduced distribution costs and enabled simultaneous reception throughout the country—an essential condition for national promotion of the service.

HBO's move from ground to space relays helped create cable networks. Although only a quarter of all cable systems carried a pay-cable channel in 1977, by the mid-1980s virtually all did (see Exhibit 3.a). By 1997 HBO enjoyed a subscriber audience of almost 21 million households. In the meantime, HBO had been joined by a dozen other pay-cable networks, though none had larger audiences.

Scrambled Signals

Pay television (plus other services) created a need to prevent nonsubscribers from receiving service. Several kinds of encryption—signal-*scrambling* techniques—became available. *Encryption* distorts the signal so that only subscribers who pay for decoders can receive pay programs. The device most widely used is called VideoCipher (see Exhibit 3.e). Suppliers of encryption devices regularly update their codes to defeat pirates who peddle black-market decoders. But bootleggers can apparently solve even sophisticated encryption codes. Companies affected by pirates try to reduce losses by prosecuting unlicensed dealers—and publicizing their convictions.

Rise of Cable Networks

Until about 1980 cable remained basically parasitic, feeding off existing broadcast programs and motion pictures. In the 1980s, however, cable began to flex its programming muscle. HBO led the way with cable-specific productions—programs made especially for cable. It taped special on-stage performances, obtained rights to sports events, and commissioned original programs. In 1996 HBO ran 11 original films featuring big-name talent and won 14 Emmy awards, second only to NBC. The next year HBO set a new cable network record with 90 nominations and 19 Emmy awards.

Ted Turner made a major contribution in 1980 by introducing Cable News Network (CNN), a 24-hour schedule of news and news-related features that many broadcast critics dismissed as the "chicken noodle network" for its low budgets compared with network news departments. Turner spun off CNN Headline News 18 months later, another 24-hour service. But Turner was not alone. The cable industry in 1979 had collectively begun a full-time, public-affairs cable service, Cable-Satellite Public Affairs Network (C-SPAN), aimed at cultivating good public relations for the cable industry. Based in Washington, C-SPAN covers House of Representatives and (on a second channel begun in the 1980s) Senate sessions, congressional hearings, and other public-affairs events.

Other cable-specific program networks followed. MTV (Music Television) began in 1981 to provide 24 hours of hit recordings with matching video images, hosted by video disc jockeys. The Weather Channel, started a year later, offers 24-hour weather and environmental programming. Networks have appeared devoted to religion, country music, children's programs, home shopping, sports, ethnic programs, fine arts, and education. The growth of new cable networks continued in the 1990s, some of them competing with existing formats and others narrowly focused on subsets of older formats. A network restricted to sports seemed far-fetched when ESPN launched in 1979, but its success was followed by ESPN2 in 1994. In 1996 CNN/SI and Fox Sports appeared, along with many regional sports networks, further increasing the options for sports programming on cable. The Golf Channel proved a network devoted to a single sport could gain access, if not high viewership. The Golf Channel launched in 1995 with a potential audience of only 140,000 subscribers, but a year later it could be seen in more than 4 million households. Cable-system operators have a bewildering number of satellite-distributed program services from which to select, with no end in sight. Data on some of these are found in Exhibit 6.e.

Pay-Per-View

A more flexible cable service required subscribers to pay extra for individual programs. First appearing in 1984, *pay-per-view* (PPV) allows viewers to order individual programs from the cable company for a one-time fee. Movies, many of them "adult," account for a large number of PPV requests.

Exhibit 3.c

Ted Turner—Cable Empire Builder

Source: AP Wide World Photos/Mark J. Terrill.

Ted Turner, former board chair and president of Turner Broadcasting Systems (TBS), is a creative man. He created the first superstation (WTBS, Atlanta), Cable News Network (CNN), CNN Headline News, CNN International, CNN Airport Network, Turner Network Television (TNT), Turner Classic Movies, The Cartoon Channel, Turner Home Video, and Turner Program Services. He also likes acquiring things—things such as studio production and distribution companies (New Line Cinema, Castle Rock, Hanna-Barbera) and professional sports teams (Atlanta Braves and Atlanta Hawks). Married to screen and exercise video star and one-time political activist Jane Fonda, Ted Turner has become a symbol of the rough-and-tumble industry he helped create.

Referred to variously as "Terrible Ted," "Captain Outrageous," "The Mouth of the South," and "Man of the Year" (named by *Time* magazine in 1992), Turner is known for his outspoken opinions, his willingness to challenge the establishment, his aggressive and entrepreneurial spirit (signs reading "Lead, follow, or get out of the way" can be found on his desk, and the phrase is the title of the first Turner biography), his driving ambition, and his ego. Physically, he's the prototype of a southern gentleman: tall, lanky, with silver hair and mustache, a cigar in his mouth, and a Georgia drawl. He loves the movie *Gone With the Wind* so much he began his TNT network service with it and even named one of his five children Rhett.

Turner attended Brown University and was active in debate and the yachting club. College yachting served him well: In 1977 he won the prestigious America's Cup race. But Brown threw him out of

"Events," predominantly boxing and wrestling, are important sources of PPV revenue. Boxing matches involving either heavyweights Mike Tyson or Evander Holyfield account for most of the top PPV events. A 1996 bout between the two pugilists reportedly grossed more than $100 million, the most successful PPV event on record. A return match in 1997 generated almost two mil-

college twice for poor grades and excessive pranks—and his fraternity dropped him for burning down its homecoming display.

Turner began his business career as a salesman with his father's advertising firm in Savannah, Georgia. In 1963 he became chief executive of various Turner companies, with headquarters in Atlanta. His interest in television began in the 1970s with his acquisition of Atlanta's channel 17, then a failing independent outlet. It was there, six years later, that he dreamed up the superstation concept.

On June 1, 1980, Turner launched Cable News Network (CNN), the first live, 24-hour, all-news cable operation. Although it struggled for acceptance in its early years, CNN (including CNN Headline News, begun in 1981) now reaches virtually all of America's cable subscribers along with millions more in countries around the world. (In 1982 Turner also formed CNN Radio, a 24-hour all-news cable audio service.)

Turner does not always have the Midas touch. He stumbled in 1984 when he tried to compete with MTV. His Cable Music Channel lasted just over a month until MTV bought its assets for about $1 million plus free advertising on Turner's other cable services. His biggest battle—also an eventual failure—came in 1985 when he tried to take over CBS.

Within a year of the CBS debacle, Turner purchased the MGM studio's huge film library (including his beloved *Gone With the Wind*) for use on his TBS and, later, TNT cable networks. The purchase, however, pushed Turner deeply into debt, and he was forced to accept a consortium of large cable-system operators as partners. They received positions on his board of directors in return for their investment.

TNT began in October 1988, promoted as the first cable network designed expressly to challenge the major television broadcast networks. Initially reaching some 17 million subscribers, the service offered mostly movies but began more original programming in the 1990s. It was the fastest-growing new cable network, reaching 60 million subscribers by 1994. Among other audience-building tactics, Turner backed the computerized "colorization" of classic black-and-white films from the MGM collection. And in 1991 Turner took over the Hanna-Barbera animation studio and its film vault, which became a feeder for The Cartoon Channel begun in 1992.

By the early 1990s Turner had begun to leave most day-to-day control to managers, focusing instead on strategic planning and long-time interests—such as the environment—that he shared with Fonda. Development of the Turner cable empire continued apace. TBS helped start the first independent television station in Moscow early in 1993.

Turner was mentioned as a possible bidder for CBS once again in 1995, but CBS sold to Westinghouse and, within weeks, a merger between Turner and Time Warner was announced. Turner came out of the deal as vice chairman of the merged company with stock worth an estimated $2 billion. Media observers speculated whether Turner could adjust to being a subordinate executive in a media conglomerate. Turner associates deflected concerns about his new role, noting that Ted is the kind of guy who teaches cats to land on their feet.

lion PPV requests and an estimated $95 million in revenue. Tyson received a year's suspension from boxing for biting Holyfield's ear during the fight, and some predicted an end to large PPV profits for a tainted sport. Others predicted that a "grudge match" with Holyfield following Tyson's suspension could once again draw a large PPV audience, perhaps even setting a new revenue record.

Exhibit 3.d

Turner Pioneer Satellite Relay

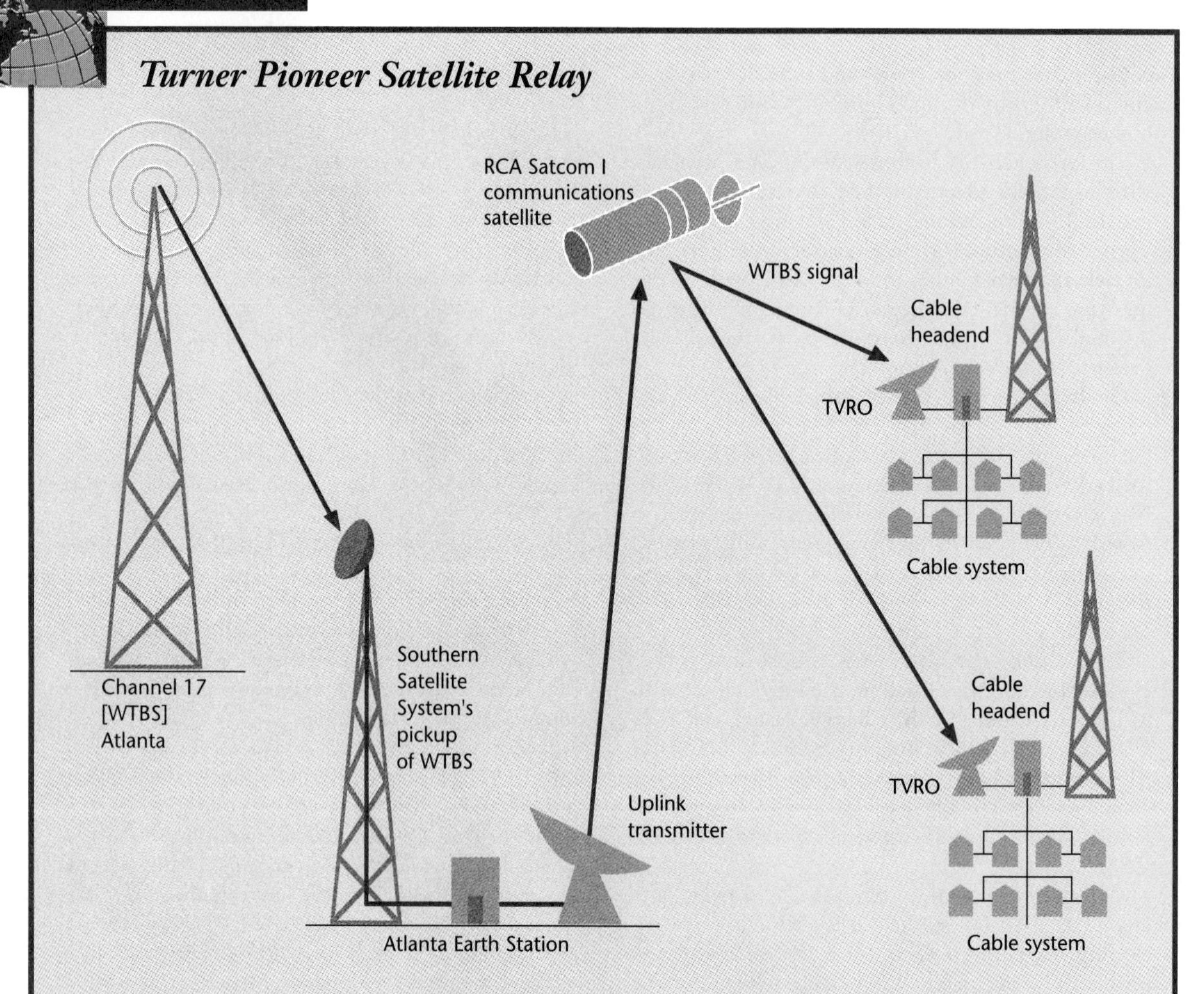

When HBO and Turner turned to domsats for distribution of their signals in the mid-1970s, they blazed a trail for more than 160 satellite-delivered cable networks. This exhibit shows how Turner initially distributed programming on his Atlanta-based UHF TV station to cable subscribers nationwide, making it the first "superstation." A changing cable market in the late 1990s made conversion from operation as a superstation to a cable network economically desirable.

Source: Turner Broadcasting Systems, Inc.

Widespread use of PPV depends on *addressability*—an efficient means for the cable system office to receive program orders and address individual PPV decoders. PPV customers use a touch-pad to communicate with a computer called a *video server* at the cable headend. The video server interacts with the customer's decoder and carries out record-keeping and billing operations. PPV developed rapidly

Exhibit 3.e

Satellite-to-Home Viewing

(A)

Orbiting satellite

Uplink

Downlink

Transmitting dish

TVRO

Descrambler

TV

(B)

(C)

(A) Satellite delivery of programs directly to viewers' homes, whether by DBS systems intended for such reception or by home pickup of relay signals designed primarily for cable system use, depends on a reliable means of scrambling the video and sound signals, to be descrambled by devices supplied only to homes that pay monthly subscription charges for the service.

(B) The descrambling device, in this case a Video-Cipher along with its hand-held remote control unit, is installed between the viewer's TVRO and the television receiver.

(C) The smaller, less expensive antennas used by DBS and the wide range of program packages available from DBS service providers make these digital satellite-to-home systems popular with millions of viewers.

Sources: TVRO courtesy Eric A. Roth; descrambler photo courtesy General Instrument Corporation; DBS antenna courtesy U.S. Satellite Broadcasting.

in the late 1980s; by 1995 nearly a third of cable households had addressable decoders. An estimated 60 percent of cable households will have addressable converters in 1998 providing access to about a half dozen PPV channels. PPV is not a service restricted to cable, however. Homes receiving direct broadcast satellite service also have access to PPV programming. DBS provides more PPV channels than are available to the typical cable subscriber.

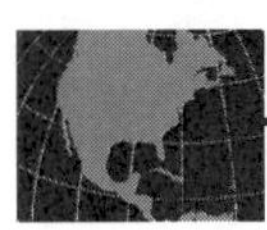

3.3 Niche Services

Over the past two decades, several alternative modes of program delivery have arisen to supplement mainstream broadcast and cable services. Each provides programming by subscription that is similar to or the same as that of existing cable channels (usually in areas lacking cable). A few offer new or different content.

Wireless Cable

A wireless *multichannel multipoint distribution service* (MMDS) can serve urban areas in competition with cable because its construction costs are far lower. Because to its viewers it resembles cable service, MMDS is ironically called "wireless cable." MMDS's spectrum location limits its coverage range to about 15 miles, and reception points must be in line-of-sight view of the transmitter. The first MMDS service began operation in Washington, DC, in December 1985.

MMDS's head-on competition with cable made program suppliers (many owned by cable operators) reluctant to serve MMDS until FCC regulations required MMDS access to such programming in the early 1990s. This change brought new life to wireless cable. There were almost five times as many systems operating in 1995 as in 1990. At mid-decade the telephone industry briefly seized on MMDS as a quick path to establishing a presence in the digital video marketplace, but fewer than a million homes received programs by wireless cable in 1997. It is possible that advancing technology, combined with a growing demand for all manner of information services, may renew interest in MMDS.

MATV/SMATV

A second niche service meets television service requirements of a building or group of buildings on a single piece of property. *Master antenna television* (MATV) and *satellite MATV* (SMATV) are special, limited-area types of cable television. An MATV installation combines a master television antenna on the roof of an apartment building with coaxial cable feeding programs to each unit. Modern multiple-dwelling buildings usually come with suitable cabling already installed.

Satellite MATV adds a TVRO antenna to the installation, (Exhibit 3.f) enabling the system to pick up satellite-distributed services. Because SMATVs operate entirely within the boundaries of private property (usually an apartment or condominium complex), and thus do not cross public rights of way, they are exempt from most cable and broadcast regulations.

Cable-system operators, who naturally want to sign up as many households as possible, oppose SMATV systems within their coverage areas. In fact, applicants for new cable franchises often press municipal authorities to assure them equal access to SMATV-equipped buildings. There are several thousand MATV/SMATV operations (mainly in motels and large apartment buildings), though they serve only a small proportion of the total television audience.

The future of MATV/SMATV was clouded with the second launch of DBS in the mid-1990s. Greatly enhanced program choices and superior picture quality may tempt apartment dwellers to install DBS receiving equipment and abandon their building's wired system provided contractual arrangements with landlords do not prohibit their doing so. An individual DBS system for each apartment is now technically practical in some locations,

Exhibit 3.f

MMDS and SMATV

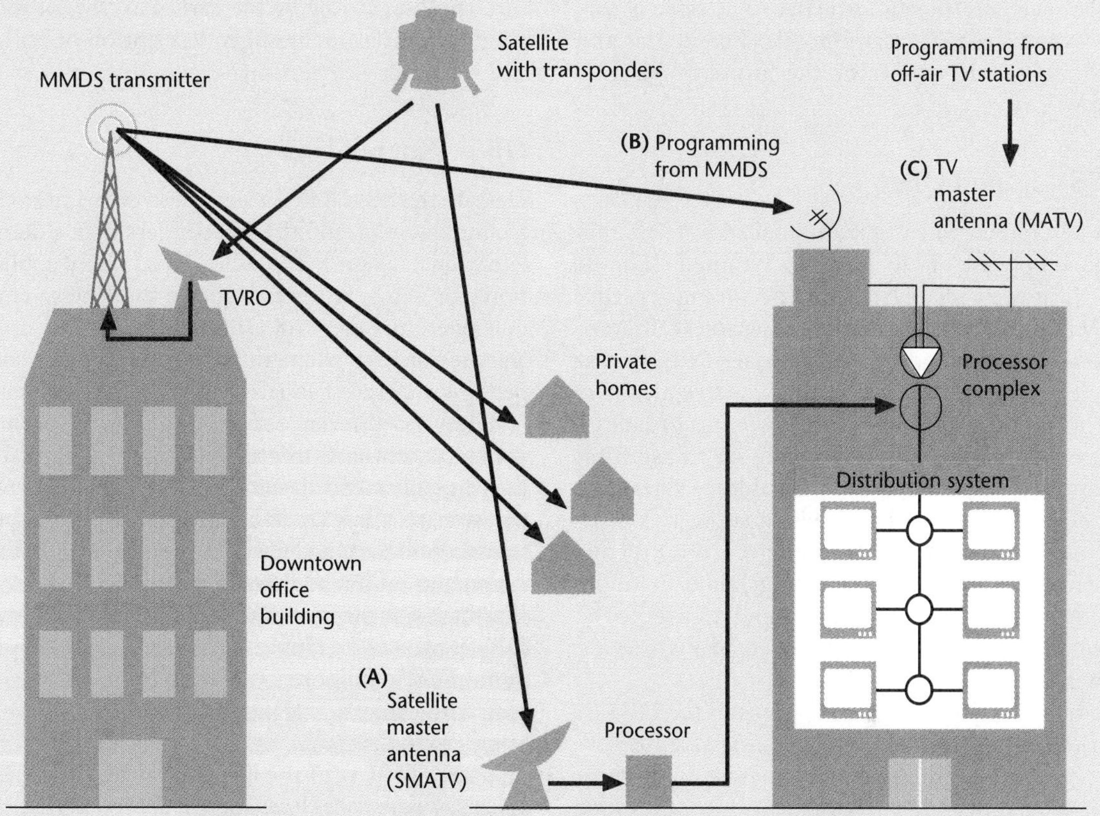

Viewers living in apartments or condominiums often receive their television signals by means of (A) a satellite master antenna television (SMATV) system, a satellite TVRO located near, as shown here, or on top of the building; (B) a multichannel multipoint distribution service (MMDS), a small terrestrial receiving antenna on the building roof; or (C) a master antenna (MATV), also on the roof, picking up off-air signals. Each of these supplies a cable distribution service within the building. Individual homes can also subscribe to MMDS services as shown here. Cable system operators pressure local franchise authorities to eliminate options (A) and (C) when they can—or at least to allow cable equal access to such buildings. DBS providers may make SMATV and MMDS less popular, but an MATV system may still be necessary to provide local broadcast signals.

thanks to the smaller dish. On the other hand, building owners may try to replace their old TVROs with DBS dishes and distribute DBS programming throughout the complex, provided the technical, legal, and economic factors involved make such conversions attractive. The same is true for owners of motels and hotels. Economics and technology will determine the future of MATV/SMATV.

DBS—Round One

Widely touted when first announced in 1980, dismissed because of high costs and limited technology just a few years later, and becoming a reality only in the 1990s, *direct-broadcast satellite* (DBS) appears at first to render any land-based delivery system obsolete by delivering programs to consumers without going through an intervening broadcast station or cable system. A broadcast transmitter, relatively cumbersome and inflexible, reaches only a limited area—and hundreds must be linked to provide national network coverage. Cable systems must hard-wire each subscribing home. A single DBS transmitter, on the other hand, can serve the entire country (with special spot beams serving Hawaii and Alaska).

The FCC first received applications for DBS licenses proposing direct-broadcasting satellite service in 1980. It authorized commercial operations in 1982, a remarkably fast regulatory go-ahead for a new service. Critics questioned whether DBS, primarily of interest to small and scattered audiences in uncabled rural areas (as elsewhere it must compete with cable and established program sources), could recoup its costs. The critics proved right—the only American DBS operation (using a Canadian satellite) was short-lived because it attracted few subscribers.

This dim record resulted from a focus on technology at the expense of programming: no DBS operator specified how its service would differ from those already available from cable, broadcasting, and home VCRs. Even in rural areas, possible DBS interest was quashed by growing use of TVROs for *C-band direct* reception from satellite channels intended as relays rather than for home reception (see p. 391). C-band satellites' requirement for larger and more expensive receiving antennas than higher-powered Ku-band DBS satellites did not discourage backyard-TVRO enthusiasts—some three million strong by the early 1990s. (For more about satellites, including a description of both C and Ku bands, see Section 5.2.)

DBS—Round Two

Despite the dismal first round, however, improving technology revived DBS interest less than a decade later. Once again the FCC received many applications for new services to begin in the 1990s. These new operators drew on DBS experience in Europe and the Far East, where the service faced fewer competing terrestrial channels than in the United States.

The very different second-round DBS projects planned eventually to utilize (1) high-power satellites and advanced-design TVROs; (2) video compression to allow multiple programs per channel (see Section 5.5); and (3) both high-definition television and digital audio (see Sections 5.7 and 5.8). But these new projects also incorporated a different DBS approach, including (1) a high priority on programming, with more specialized channels to compete with cable, and (2) projects backed by cooperating groups, *consortia*, of companies, each focusing on a specific part of the larger picture. Huge DBS expenses were thus shared across several partners.

Both U.S. Satellite Broadcasting (USSB) and Hughes DirecTV began limited operations in 1994, sharing a satellite. Only a month after DirecTV and USSB launched, a consortia of cable operators led by Tele-Communications Inc. introduced Primestar, a DBS service using a medium-power satellite requiring larger (three-foot diameter) receiving dishes. In 1996 EchoStar became a third high-power DBS service, and AlphaStar entered the market as a second medium-power service. Consumer response was good, and the new DBS providers signed up about four million households in the first two years, many of them former cable customers. DBS was attractive because it offered many more channels than cable and provided superior picture quality.

After a strong start, DBS growth began slowing. DirecTV expected to have three million subscribers by the end of 1996, but fell short by more than half a million. To attract subscribers, EchoStar tried giving away receiving equipment in Boulder, Colorado when Tele-Communications Inc. lost its local cable franchise. Months earlier the cable industry had initiated a nationwide anti-DBS advertising campaign emphasizing the high cost of connecting additional sets and the lack of local broadcast station carriage. The campaign may have had some effect because response to EchoStar's offer was less favorable than the company had hoped. DBS growth continued, however, increasing to about six million subscribers in 1997.

Recognizing the enormous appeal of a DBS provider that could offer hundreds of cable channels plus local TV stations, Rupert Murdoch's News Corporation created ASkyB and proposed just such a service. Early in 1997 Murdoch announced an agreement to merge ASkyB with EchoStar and outlined an ambitious plan of DBS development, sending shock waves through the DBS and cable industries. But things began falling apart almost immediately. Conflicts arose over operational control of the merged companies and, as critics had predicted when he announced his bold DBS plans, implementation was much more difficult than Murdoch anticipated. Only months after his initial announcement, Murdoch withdrew from EchoStar and merged ASkyB with a restructured Primestar. The arrangement was mutually beneficial because ASkyB controlled more than two dozen high-power satellite frequencies needed by Primestar, and Primestar offered Murdoch the opportunity to become quickly involved in American DBS. Murdoch assumed a subsidiary role and Primestar made no statement regarding immediate carriage of local broadcast stations, a major feature of ASkyB's original plan.

Murdoch was not alone in finding his DBS plans going awry. The parent company of AlphaStar failed to acquire rescue financing and went into receivership less than a year after beginning service.

Videotex/Teletext

Videotex and teletext services were introduced in the United States during the 1970s, and Knight-Ridder and Times Mirror test-marketed them in the 1980s. The services failed to prosper in the United States at the time but were more successful in Britain and European countries, especially France.

Teletext is a one-way, broadcast-based information service delivering print and graphic materials to television screens in homes and offices. Teletext material goes out from conventional broadcast television transmitters, occupying an unused portion of the channel called the *vertical blanking interval* (VBI). Viewers use a keypad to select specific pages or screens of data from several hundred that are continuously transmitted and updated.

Videotex, often confused with teletext, involves two-way communication over telephone or cable lines. Not restricted by the limitations of the VBI, videotex offers unlimited numbers of information pages. Videotex delivery systems are often operated by common carriers although the content is created by others.

Failure of these technologies in the United States during the 1980s sometimes is attributed to their use of an ordinary television set as the terminal device, combined with an excessive emphasis on graphics. By the early 1990s it was evident that digital computers with high resolution monitors were far better adapted than TV screens for interactive information retrieval. In the mid-1990s efforts to use the television set as a terminal device resumed. Significant advancements in digital compression, sophisticated graphic design technology, and growing demand for interactive services made use of the TV screen as a terminal device more attractive. Skeptics, however, believe most persons accustomed to two-way interactivity on the Internet will continue using PCs or hybrid TV/PCs.

Interacting with TV

Cable visionaries in the 1970s foresaw interactivity as a logical next step in evolving home communications. Customers could talk with others on the cable—teachers could interact with students, and the public could more fully communicate with stores, banks, public utilities, and safety agencies—all without leaving home.

Warner Cable gave this type of interactive service a thorough trial beginning in 1977 on its Columbus, Ohio, system. Called Qube, it offered ten channels over which viewers could respond by means of a touch-pad. Though the experiment generated a lot of media coverage, Qube failed to attract sufficient subscribers to warrant its high cost, and after three years of mounting losses Warner closed it down.

Similar viewer apathy greeted later two-way experiments, including one operated by GTE in Cerritos, California, in the early 1990s. After a brief initial fascination, viewers either lost interest or were unwilling to pay the price necessary to provide the service. However, many observers note the Internet's rapidly growing popularity and warn that cable operators failing to prepare for future interactive demands will pay a high price. Others point to the failure of Qube and suggest that interactive services may not be attractive to most cable viewers.

Whether the average cable viewer wants interactivity or not, the cable "pipeline" provides far faster data rates than phone lines, and cable operators are developing high-speed cable modems to lure dedicated Internet users. The major concern of cable operators is conversion to digital video transmission and increasing channel capacity in response to inroads made by DBS. Providing new interactive services remains a long-term objective, and conversion to digital is an important step in the process.

3.4 Electronics Revolution

The sweeping changes taking place in electronic media during the 1990s were made possible by a fundamental shift from analog to digital technology. Digital technology makes all forms of communication content—sound, images, and text—electrically equivalent. The original form of the information no longer matters to the equipment used to produce, distribute, or store it. This basic commonality of content is at the heart of the changes taking place in mass media and was made possible by the development of solid-state microelectronics.

Transistors

The "crystal and cat-whisker" detector employed in radios during broadcasting's infancy was the first solid-state electronic device, but it was not capable of amplification. Fleming's vacuum tube detector and de Forest's Audion eclipsed the crystal detector and dominated electronics for almost 50 years. But vacuum tubes were hot, bulky, fragile, power-hungry, and, worst of all, had a limited life. In 1947 a trio of Bell Laboratories engineers developed the transistor, a three-element, solid-state tube replacement that had none of these weaknesses. The engineers received the Nobel Prize in 1956 for their invention, a fitting distinction considering their transistor became the foundation on which all modern electronics is based. One of the first mass-produced transistor-based products on the consumer market, a tiny portable radio, appeared in 1954 and began the trend toward miniaturization depicted in Exhibit 3.g.

Transistors also made possible development of that prime artifact of the late 20th century, the computer. Only computers could manipulate billions of information bits with the speed and intricacy needed for digital processing. In 1945 the first large experimental electronic computer used some 18,000 vacuum tubes plus miles of wire. The tubes generated so much heat that attendants had to stand by to replace them as they blew out. Transistors solved the heat and size problem while reducing costs and enormously increasing computer speed and memory capacity.

The Chip

Tiny computers and other devices remained impractical, however, as long as hundreds of separate transistors had to be meticulously wired together

Exhibit 3.g

Changing Definitions of "Portable"

(A)

(B)

(A) The paper clip and pencil show the size of 1950s vacuum tubes in relation to 1960s transistors and current integrated-circuit chips. (B) The "father of FM radio," Edwin Armstrong, shows off an early "portable" radio that required either large batteries or household electric power. (C) Transistors began to replace vacuum tubes in the 1960s, leading to small, inexpensive portable radios. (D) In the late 1990s integrated circuits provide two-way wireless communication of voice and data from almost any location.

(C)

(D)

Sources: Photo (B) from the Bettmann Archive; photo (C) courtesy Zenith Electronics Corporation; photos (A) and (D) J&M Studios/Liaison International.

with hand-soldered connections. True miniaturization came with invention of the integrated circuit—the computer chip shown in Exhibit 3.g.

Chips are made of crystalline wafers on which microscopically small electronic components are etched. A modern chip may contain many thousands of transistors and other components. This miniaturization makes it possible to crowd all the electrical circuits needed to construct entire radios and other electrical devices into a single chip. The integrated circuit chip is responsible for such dramatic developments as inexpensive multifunction digital watches, hand-held calculators, laptop computers, and cellular telephones. It has also made possible communications satellites and *electronic news gathering* (ENG) equipment, forever changing our concept of TV news. All forms of digital video delivered via wire, fiber-optic cable, and over the air are the result of advancements in chip technology.

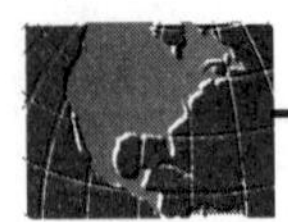

3.5 Consumer Media

Solid-state technology made possible many new consumer products that widened viewer and listener options. Some have been dramatically successful.

VCRs: Giving the Viewer Control

Except for the TV set itself, no video device is more popular than the *video cassette recorder* (VCR). As early as 1967 CBS introduced a video recorder featuring a 7-inch tape cartridge that played automatically when the cartridge was inserted. Two years later Sony Corporation of America announced its first color cassette videotape recorder. Several years passed, however, before a consumer-oriented VCR became available.

Marketing of home videorecorders began in 1975, when Sony introduced its Betamax video cassette recorder at an initial price of $1,300. Other equipment makers wanted to build Betamax machines, but Sony stubbornly refused to license Betamax and kept the price of its machines high. Sony's monopoly ended in 1977 when Matsushita brought out its technically incompatible, but longer running, VHS format. Unlike Sony, Matsushita shared VHS with other manufacturers and prices fell dramatically. A basic VHS VCR sold for around $150 in 1997. Sony's Betamax had long since become a footnote in history.

Sony promoted Betamax's ability to record TV programs for later viewing, a practice called *time-shifting*, as its major feature. TV networks and major movie studios, whose products appeared on the networks, thought off-air recording was illegal pirating of their copyrighted products, not just a convenience to viewers. But in 1984 the Supreme Court ruled that viewers would not violate copyright law by recording programs broadcast over the air for their personal noncommercial use.

Hollywood distributors eventually recognized that the VCR provided a new market for their products. Renting and selling movies on cassette became a big business. Distributors soon discovered that consumers wanted not only recent releases, but would rent or buy old movies, providing additional unexpected profits. Box office blockbusters generate the greatest profits, but home video rentals and sales help producers reduce losses on theatrical duds. By 1997 some movies rented for a few dollars and sold for $10 or less.

VCR picture and sound quality improved over the years, but operating features such as wireless remote control and easy on-screen programming are more important to many consumers. About 80 percent of American homes have at least one VCR and many people own two or more.

Video camcorders are less common, although they are in more than 20 percent of homes and prices continue to fall. Camcorders replaced 8-millimeter film cameras for shooting home movies. They produce pictures good enough that broadcast stations and networks encourage viewers to send in footage for possible use in newscasts, a practice that raises ethical questions addressed in Section 9.8.

Camcorders also contribute to entertainment programming. During much of the 1990s ABC filled an hour of prime time with *America's Funniest Home Videos* featuring recordings of people doing goofy or embarrassing things.

Audio Cassettes

Audio cassettes and the machines using them are small, making them well-suited for both use in cars and as personal portable playback devices. The audio cassette's popularity boomed following the introduction of Sony's "Walkman" in 1980. Consumers bought the Walkman and its imitators by the millions, and by 1997 some combination AM/FM/ audio cassette players sold for less than $20.

A digital audio cassette, *digital audio tape* (DAT), provides superior sound quality. Unfortunately, DAT equipment and tape are expensive. The music industry, fearing DAT-to-DAT recordings would encourage flawless duplication of their products, successfully lobbied Congress to force DAT manufacturers to include technology preventing wholesale digital copying. Although DAT is widely used professionally, cost and limits on digital duplication have combined to keep consumer interest in it low.

Philips Electronics introduced a second digital audio cassette format, the *digital compact cassette* (DCC), in 1992. Philips reportedly sank about $55 million into developing DCC. Unlike DAT machines, DCC equipment uses cassettes the same size as ordinary audio cassettes, allowing DCC machines to play analog cassettes as well as record and play digital audio. Philips considered this a significant advantage, but DCC sales were disappointing. The expense of DAT and DCC, and competition between the two formats and from other technologies, made both products commercial flops.

CDs and the MiniDisc

Introduced in 1983, the digital audio *compact disc* (CD) is one of the most successful products in consumer electronics history. The CD's audio quality, size, and durability make it ideal for use at home, in cars, and with personal portable players. The CD's success spurred development of other disc-based products such as CD-ROM. The audio cassette and CD practically killed the vinyl-based analog (LP) recording business. LPs remain popular among a devoted cadre of audiophiles, many of whom also prefer the sound produced by older tube-based audio amplifiers.

If the 5-inch CD is a compact audio disc, Sony's "MiniDisc" (MD) at approximately half its size is a subcompact. A recordable compact disc introduced in 1992, the MD gained some popularity because of its smaller size, but failed to challenge seriously the larger CD, in spite of Sony's $100-million investment. Like DAT, Sony's MiniDisc is more popular in professional circles than among consumers.

Videodisc Players

In the early 1980s both RCA and Magnavox marketed videodisc players for the home. Neither maker's machine could record video and neither machine played discs designed for the other. VCRs arrived slightly earlier and they could, of course, both record and playback. The videodisc had limited appeal for most consumers, and in the mid-1980s RCA pulled out of the videodisc business. Interest in the videodisc increased in the early 1990s. New machines appeared that could play five types of discs. Sales improved, but by the mid-1990s fewer than 1 percent of all households had videodisc players.

CD-ROM, CD-R, and CD-E

While the videodisc sputtered, consumers were buying *CD-ROM* (compact disc read-only-memory) drives for their computers. CD-ROM provided the storage capacity needed to distribute large software programs and popular interactive games.

CD-R (compact disc recordable) allows consumers to record their own CDs. CD-R discs play on any CD-ROM drive or a regular CD player.

Exhibit 3.h

TV Web Surfing

WebTV® Internet terminals and other set-top products allow the traditional TV set to provide interactivity with newer media such as the Internet. Other "information appliances" providing less challenging user interfaces may open new technologies to a wider audience.

Source: WebTV Networks, Inc.

User-friendly audio-only CD-R recorders appeared in the mid-1990 that often provide better audio. But they are more expensive and less versatile than computer CD-R drives, and they require that consumers learn to deal with mice and menus.

Another CD development is *CD-E* (compact disc erasable). Introduced in 1996, a CD-E computer drive can record, erase, and re-record blank CD-E discs. But, again, the consumer must develop computer skills to use CD-E.

Digital Video Disc

Sometimes called digital versatile disc, the *digital video disc* (DVD) provides up to 25 times as much storage space as a conventional audio CD (see Exhibit 5.n). Because of its huge capacity, DVD is an ideal medium for distributing full-length movies in wide-screen high definition with surround sound audio. Initially DVD will allow only playback, but Matsushita is developing recordable DVD-ROM and hopes to market it sometime in 1998. DVD has great potential, but much depends on the cost of players and the availability of desirable content on the high-capacity discs.

Webcasting and TV Surfing

At least 60 radio stations were "streaming" digitized audio programming in real-time over the Internet by 1996 and others quickly developed an interest in *webcasting*. Webcasting's growth is attributable in part to streaming software being available free to consumers and its low cost to broadcasters and others wanting to distribute audio on the Internet. Software for streaming video is also available, and many Internet websites feature streamed audio and/or video. Webcasting requires that consumers have a powerful multimedia computer connected to the Internet.

Another consumer product blurring the distinction between broadcasting and the PC is WebTV. Microsoft-owned WebTV uses a set-top box that lets viewers browse the Internet using a conventional TV (see Exhibit 3.h). WebTV and similar products may attract "technology-challenged" persons wanting Internet access without having to purchase a computer and learn its operation. Microsoft also joined forces with DirecTV in 1997 to offer Web site builders the opportunity to deliver audio and video to PCs using DBS signals.

Anything advancing the convergence of broadcasting with computer technology excites technophiles and futurists, but it is unclear how successful these products will be with consumers who lack their enthusiasm.

3.6 *Broadcasting—Changing Course*

Development of cable and other delivery and recording options gave people their first real electronic media choices beyond traditional broadcasting; the impact of those choices soon became evident. Dramatic changes in the fortunes of broadcast television networks provide one means of assessing such impact. In the 1980s all three changed hands—and new networks appeared.

Traditional Network Competition

In the 1970s ABC finally achieved competitive equality with its two older rivals. It gained affiliates (and thus audience), expanded its evening news to a half-hour (years after its competitors), and converted fully into color. Late in the decade ABC even managed to lead network ratings for several seasons in succession.

NBC sought the number-one slot by hiring away ABC's program chief, Fred Silverman, as president. Silverman's ABC success did not travel with him, however, and NBC remained in third place until former independent producer Grant Tinker took over the network presidency in 1980. Within five years he had piloted NBC into the top spot, where it remained for the rest of the decade.

In the 1970s CBS focused on finding William Paley's replacement. Paley restlessly hired and fired a string of potential successors. Paley finally retired in 1983, leaving Thomas Wyman as CBS chief executive officer, only the second one in the network's 55-year history. Wyman lasted but two years.

FCC Studies Networks—Again

Such boardroom and ratings battles masked more important underlying changes. In Washington the FCC, from 1978 to 1980, conducted its third (and probably last) investigation of broadcasting networks (FCC, 1980b). The Justice Department had previously signed agreements with all three networks requiring them to loosen control over independent programmers and cease serving as sales representatives for their affiliates.

The latest FCC study concluded that FCC network rules developed over the previous two decades merely restricted legitimate business interaction between networks and their affiliates without actually protecting the stations from network dominace, or improving service to cusumers. The study specifically blamed the FCC's own 1952 channel-allotment decision for the continued dominance of ABC, CBS, and NBC. So few markets have more than three VHF channel allotments that any fourth network faced a hard time in obtaining a competitive line-up of affiliates.

New Player: Fox

This barrier did not deter Rupert Murdoch, an international publishing tycoon with immense resources (see Exhibit 3.i). He launched his Fox network in 1985, the first serious attempt at a fourth national television network (since 1967 when a "United Network" failed within a month).

Initial limitations on the Fox network's size actually worked to its advantage. Fox escaped the FCC's financial-interest rule that restricted ABC, CBS, and NBC in producing their own entertainment programs (see Section 6.3). This technicality gave Murdoch freedom to use his 20th Century-Fox movie studio as a source of programs for the network.

And he was aggressive in developing Fox into a serious competitor. In mid-1994 Murdoch made a $500 million investment in a group of stations, gaining a number of major market affiliates, eight taken from CBS alone. Station values began to climb as broadcast networks were forced to increase payments for station time and a virtual bidding war for affiliates developed in larger cities.

Exhibit 3.i

Rupert Murdoch—Media Magnate

In 1985 Australian publisher Rupert Murdoch became a United States citizen almost overnight to acquire Metromedia's six independent major-market television stations (WNEW-TV, New York; KTTV-TV, Los Angeles; WFLD-TV, Chicago; WTTG-TV, Washington, DC; KNBN-TV, Dallas; and KRIV-TV, Houston) and one network affiliate (WCVB-TV, Boston) for $2 billion. He resold WCVB-TV to the Hearst Corporation for $450 million but picked up another Boston station, WFXT-TV.

Having purchased half-ownership of 20th Century-Fox film corporation in 1984 for $250 million, Murdoch acquired the remaining half in 1985 (after the Metromedia deal) for $325 million. He thus gained complete control over a company with an extensive film library (including such hits as *Cocoon* and *Aliens*) and rights to numerous television series (*L.A. Law* and *M*A*S*H*, for example). In 1985 Murdoch announced plans to form a new national television network, Fox Broadcasting Company. A core of six *owned and operated* (O&O) stations (reaching about 20 percent of all U.S. television households) served as the network.

Fox premiered in October 1986 with *The Late Show*, starring comedienne Joan Rivers. Rivers had been the primary substitute host on Johnny Carson's *Tonight Show* on NBC. The Rivers show lasted only seven months. Fox tried several other programs before giving up on the late-night time period altogether in order to concentrate on other parts of the day.

In its first season, Fox averaged between 2 percent and 6 percent of the national audience. Fox programs typically languished at the bottom of the Nielsen list, although some fared better.

The quality of the Fox affiliate line-up remained a problem. More than 120 stations carried Fox programming, but most were UHF, some only low-power outlets, and nearly all were the weakest stations in their markets. The network lost about $80 million in its first year of operation, but Murdoch marched on.

He concentrated both on programming and on improving his affiliate lineup. In 1994 Fox acquired New World Communications, which switched affiliation of its big-city stations to Fox (one from NBC, three from ABC, and eight from CBS). Before long Fox had become a major—and profitable—force in network competition, with some shows (*Married . . . With Children* and *The Simpsons*, for example) appearing on

Fox also made major investments in programming, especially sports programming. Fox successfully bid away rights to NFL football from CBS in 1993. During the last week of October, 1996, Fox's weekly ratings topped the other networks for the first time in new season, prime-time, head-to-head competition, and 30-second commercials in the Fox broadcast of the 1997 Super Bowl sold for around $1.3 million each.

Nielsen's list of top-rated programs. With the help of 22 O&O stations reaching almost 35 percent of the nation's homes, Fox programming by 1997 often tied, and sometimes beat, at least one of the three older networks on several nights of the week. In 1996 Murdoch launched the Fox News Channel (FNC), a cable network intended to compete with newcomer MSNBC as well as the firmly established CNN. FNC made immediate waves by hiring 23-year ABC News veteran Brit Hume, a recognized star. Murdoch also paid cable system operators large up-front fees to ensure FNC's access to a national audience in excess of 17 million homes. The same year Murdoch kicked off Fox Sports Net, initially available in more than 20 million cable TV homes.

Meanwhile, Murdoch in 1988 bought Triangle Publications (publisher of *TV Guide*) for $3 billion. He controlled 60 percent of metropolitan newspaper circulation in Australia, 36 percent of national circulation in Britain; he also had part-ownership in 10 book publishers, and reached more than 13 million homes in 22 European countries with Sky Channel, a satellite broadcasting service for cable-TV viewers, and still more millions with his StarTV satellite service in Asia.

Early in 1997 Murdoch announced plans to combine his U.S. direct broadcast satellite venture, ASkyB, with EchoStar. The plan included offering local TV station carriage in addition to the full range of cable networks and pay-per-view services, but the ASkyB–EchoStar merger failed to develop. Murdoch instead merged ASkyB with Primestar, assuming a less prominent DBS role and abandoning, at least temporarily, plans to offer local TV station signals via DBS. Simultaneous with the ASkyB–Primestar merger, Murdoch spent almost $2 billion to acquire shared ownership of International Family Entertainment Inc. and announced plans to program IFE's Family Channel with kids' shows 12 hours daily, competing directly with Nickelodeon, The Cartoon Network, and the Disney Channel.

Sources: Logo © 1995 Twentieth Century Fox Film Corporation. All rights reserved. Used by permission. Murdoch photo from AP/Wide World.

Networks Change Hands

The three established networks had already undergone traumatic corporate changes. After more than three decades of stable ownership, all three changed hands within a two-year period.This abrupt break with a 30-year tradition coincided with retirement of long-time leaders, declining ratings in the face of cable competition, weak financial performance, and concern about increasing future multichannel competition.

ABC went first. In 1985 Capital Cities Communications announced that it would acquire American Broadcasting Companies, parent of the ABC network, in a friendly deal valued at more than $3.5 billion (the 1953 merger of ABC with Paramount Theaters had a $25 million value—one indication of how network importance had grown). The combined firm had to shed its cable interests and several TV stations to meet FCC ownership rules (discussed in Section 13.6).

Within weeks after the Cap Cities/ABC announcement, Ted Turner revealed plans for an unfriendly takeover of CBS. The network managed to thwart Turner by repurchasing nearly a fifth of its own stock from the public. The battle left CBS deeply in debt and torn by internal dissension. Severe cutbacks followed, and many CBS employees seemed relieved when Laurence Tisch (chairperson of Loews, the entertainment and investment conglomerate), purchased nearly 25 percent of CBS's stock and became chief executive. In just two years, however, Tisch closed down CBS laboratories and sold off virtually all the network's nonbroadcast subsidiaries, including extensive publishing interests, culminating in 1988 with the sale of Columbia Records (the world's largest record company) to Sony for $2 billion. He focused all CBS efforts on traditional broadcasting. Despite its initial ratings success in the 1990s, however, Tisch seemed to tire of the business and began considering potential buyers.

NBC's turn had come in December 1985. Its parent company, RCA, weakened by years of inept management and huge losses from failed computer and videodisc ventures, welcomed a friendly buyout offer from General Electric. RCA and its NBC subsidiary sold for $6.28 billion.

And Change Again

For several years after the takeovers, news from all three networks went from bad to worse. Profits fell, audience shares continued to slip, and thousands of employees lost their jobs as networks slimmed down to cut costs and compete more effectively. Competition from cable and independent stations lessened the Big-Three networks' once dominant role.

By the 1990s, however, stabilized audience levels, growing advertising expenditures, and the approaching demise of fin/syn rules (see Section 8.2) had restored to television networking much of its former luster. Two new national networks, Warner Bros. and United Paramount, premiered in January 1995 (see Section 6.3). And later that year two of the Big Three disclosed that they would once again change ownership.

On July 31 the Walt Disney Company surprised both the broadcast industry and the financial community by announcing that it would acquire Capital Cities/ABC in a deal valued at $19 billion—the second-largest takeover in history. Brief and confidential negotiations had produced the most powerful entertainment empire in the world, combining the most profitable TV network and its ESPN cable service with Disney's Hollywood film and television studios, its theme parks, and its Disney Channel.

The next day, after somewhat longer and certainly more public bargaining, CBS said it would abandon its nearly 70-year legacy of corporate independence by accepting a $5.4 billion buyout offer from the broadcasting and industrial conglomerate Westinghouse Electric Corporation. Together they would own 15 television stations and 39 radio stations, as well as program production and distribution facilities.

Radio

Despite intense competition, the number of radio stations continued to climb after 1970, dividing listeners into ever-smaller segments. FM station numbers rose to match AM by the mid-1980s as FM's audience appeal overcame AM's former attraction. FM's growing popularity with audiences and advertisers persuaded the FCC to squeeze in more than 700 channel allotments to allow still more FM outlets.

Meanwhile, looser FCC technical oversight, combined with a growing number of stations on the air, had weakened AM's already-inferior signal quality. Under pressure from AM interests, the FCC adopted several technical and ownership rule-changes in an attempt to halt AM's decline. But when the FCC approved AM stereo in 1982, it declined to specify a specific technical standard, arguing "the marketplace" could resolve that issue. Although five competing systems soon dwindled to two, the resulting confusion among manufacturers and consumers led to few stations adopting stereo by the early 1990s. Congress finally required the FCC to pick a standard, but this seemed a decade too late to help AM reverse its downward spiral.

The FCC also added 10 new channels to the top of the AM radio band. This *expanded AM band*, located between 1605 and 1705 kHz, should reduce AM interference by moving selected AM stations to the new frequencies. But the FCC had to revise its assignment plan several times, delaying implementation.

More important to radio in the 1990s than technical improvements were changes in the rules governing ownership. In the early 1990s many owners operating only one station—especially an AM station—were destined for bankruptcy. In late 1992 the FCC loosened its rules to allow a single owner to operate two AM and two FM stations in markets with at least 15 stations. The Telecommunications Act of 1996 further lifted radio's fortunes by eliminating the national cap on radio station ownership, although restrictions on the number owned in a single market remain. Interest in radio station ownership increased dramatically as a result. By 1997 several groups owned more than 100 stations. Chancellor Media Corp./Capstar Broadcasting Partners Inc. topped the list with more than 300 stations, a number unimaginable only a few years before.

Although stronger than it was earlier in the 1990s, the radio industry still faces challenges. There will be increased competition from audio services delivered by cable, DBS, the Internet, and new satellite digital audio radio services (see Section 5.8) available to listeners nationwide at home and on the road. Radio also must eventually convert to digital transmission if it is to match the sound quality of its competitors and enhance revenues by distributing digital data. But economic efficiencies resulting from greater concentration of station ownership should improve the chances of radio meeting these challenges successfully.

Television in Transition

TV broadcasters also face challenges, the most immediate being the shift from analog to digital signal transmission. In 1996 the FCC adopted technical standards for *digital television* (DTV) after eight years of discussion. The following year Congress passed legislation instructing the FCC to allow a station to retain its analog channel until 85 percent of the station's viewers can receive its digital signal over the air, by satellite, or via cable. The legislation otherwise requires the return of analog channels to the FCC by December 31, 2006. Rather than establish a single transmission standard as it did for analog TV in the 1940s, the FCC allows broadcasters to use any DTV picture format. Most broadcasters applaud flexibility, but lack of a single government-mandated format also creates uncertainty. See Section 5.7 for further discussion of DTV.

Digital video compression (discussed in Section 5.5) may make broadcasters more competitive by allowing them to transmit multiple programs simultaneously. However, a potential conflict exists between stations wanting to offer multiple *standard definition* (SDTV) programs and networks demanding that affiliates broadcast their shows in *high definition* (HDTV). HDTV currently fills the entire digital channel, forcing affiliates to downgrade network video quality if they offer other programs while transmitting network shows. Affiliates also worry that locally produced SDTV news and commercials will look inferior to network HDTV programs. Equipping a station to pass through HDTV network programs can cost several million dollars. Gearing up for local HDTV production costs much more.

In spite of DTV's uncharted waters, new competition from DBS, and an expanding number of cable channels, TV stations are highly valued properties. Investors spent about twice as much acquiring TV stations in 1996 as they did in 1995. The promise of digital channels for existing stations and relaxed ownership limits made stations more attractive. Broadcasting's audience continues to erode, but the broadcast network affiliates remain the dominant choice of viewers, especially during prime time.

3.7 *Sorting It Out*

Development of new delivery systems and dramatic changes in older services placed enormous pressures on manufacturers, broadcasters, competing systems, and policymakers. In the 1990s even sympathetic trade publications used terms such as "merger mania" and "feeding frenzy" to describe the wild scramble among media industries trying to position themselves in an unpredictable environment where plans made one month often seemed foolish the next.

Competitive Confusion

Some observers argue that having too many competitors overtaxes program creativity while confusing consumers with constantly changing technical standards and an increasing number of media outlets. They say competition is desirable if it improves the diversity of media content, but some feel the proliferation of delivery systems has failed to achieve this objective. With obvious exceptions such as the Weather Channel, MTV, CNN, and even ESPN, many of the new channels provide little that is new in the way of program innovation. If new technology only means consumers have an extra 100 ways to view the same episode of *Gilligan's Island*, what is accomplished? Even innovative programming has its limitations. How many ESPN clones will the market support?

Competition is also desirable if it challenges creative minds to provide a higher quality of content. Again there are exceptions, but critics suggest the need to attract viewers in a more competitive environment has debased content. They charge that producers increasingly use violence, sex, and foul language to make their shows stand out in a crowded field. They point to public demands for more detailed program ratings (see Exhibit 9.d) as an indication that many viewers find TV content repulsive.

A New Player?

One factor contributing to the pressure felt by broadcast and cable leaders during much of the 1990s was continuing uncertainty regarding the role of the telephone industry.

AT&T bowed out of direct competition with broadcasters in 1926, although it retained its lucrative business networking stations. By the 1970s changing technology and competitive forces renewed telephone interest in entertainment/information media.

New telephone competitors, such as MCI and Sprint, charged that AT&T used its great size to hamper their efforts and renewed Justice Department concern over the Bell System monopoly. In 1982, after an eight-year court battle, AT&T ended the resulting antitrust suit by signing a consent decree. AT&T agreed to split off its 22 local operating companies, restructuring them into seven *Regional Bell Operating Companies* (RBOCs), or "Baby Bells." AT&T withdrew from local telephone service, focusing on increasingly competitive long-distance, manufacturing, and information services.

Meanwhile technology opened exciting new vistas for *telco* (telephone company) operations. Fiber-optic cable, for example, offers dramatic advantages over both copper cable and radio. As shown in Exhibit 5.b, it is small and easily installed, yet it has enormous broadband channel capacity and freedom from environmental interference. Fiber-optic cable is a natural for heavy-traffic telephone routes and for main distribution trunks of cable systems.

This new technology created new opportunities for telco operations in video delivery, but rules established when the Baby Bells split off prevented them from providing their own "information services." By the mid-1990s, court decisions had removed most of these restrictions and the Telecommunications Act of 1996 (discussed in Chapter 12) opened the field further.

Both broadcasting and cable saw expansion of *plain old telephone service (POTS)* into information and entertainment as a threat, although cable foresaw how its own broadband capacity could carry new services in competition with telcos. At mid-decade it appeared the telcos would move aggressively into direct competition with cable by providing interactive television. The telcos found, however, that interactive television technology was extremely expensive and, in spite of their enormous technical resources, they had problems getting it to work. They shifted away from true interactive programming and tried digital wireless cable as a way to get at least multiple video channels into the home. But they found digital signals on wireless cable frequencies did not penetrate obstacles such as trees as well as expected. Meanwhile the Internet shifted interactivity from the television to the computer, and competition for local and long-distance phone service intensified. The telcos scaled back their earlier plans significantly, but continue to expand and improve their fiber-optic networks and say they will carry video programming and high-speed data to the home sometime in the future.

As the decade comes to a close, the telcos seem more interested in developing cooperative alliances and mergers with cable than in competing with it directly.

VHS VCR
FS VID
ARPS RTN 1
ARPS RTN 2
ARPS RTN 3
ARPS RTN 4
NBC NEWS
SPECIAL REPORT
MG 2
VTR Y
MG 3
MG 4
Late Night with Conan O'Brien
NBC SDO 6A
PGM/CLOCK
M/E 2 P
M/E 3 PG
DVE CH 1 VID
DVE CH 2 VID
BARRY S
MG 1
TD PVW4
Grass Valley Group
Model 3000

Chapter 4

How Electronic Media Work

CHAPTER CONTENTS

Technology largely determines both the potential and the limitations of electronic media. Where and how far signals will travel, how much information they can carry, their susceptibility to interference, and the need for technical regulation—all of these factors depend primarily on the signal's physical nature.

This chapter surveys basic concepts involved in using radio energy for broadcasting and cable systems, the most widely used means of program delivery.

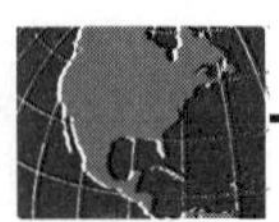

4.1 Electromagnetism

A basic natural force, electromagnetism, makes possible a host of communication services. All forms of electromagnetic energy share three fundamental characteristics:

- They radiate outward from a source without benefit of any discernible physical vehicle.
- They travel at the same high velocity.
- They have the properties of waves.

Radio Waves

Visible waves that ruffle the water when a stone is dropped into a pond radiate outward from the point of disturbance caused by the stone. Unlike water waves, radio waves have the ability to travel through empty space, going in all directions without benefit of any conductor such as wire. This "wirelessness" gives broadcasting its most significant advantage over other ways of communicating. Radio waves can leap over oceans, span continents, penetrate buildings, pass through people, go to the moon and back.

Students learn in elementary school that the "speed of light" is 186,000 miles per second. Radio shares this characteristic with light energy, as do all forms of electromagnetic radiation. This means a radio signal can circle the Earth more than seven times in one second! Radio waves also share two other characteristics with other forms of electromagnetic energy, *frequency* and *wavelength*.

Wave frequency refers to the fact that all electromagnetic energy comes from an oscillating (vibrating or alternating) source. The number of separate waves produced each second determines a particular wave's frequency. Differences in frequency determine the varied forms that electromagnetic energy assumes.

Wavelength is measured as the distance from the origin of one wave to the origin of the next. Waves that radiate out from the point of disturbance—the stone in the water or the radio transmitter—travel at a measurable velocity and have measurable frequencies and lengths.

Frequency Spectrum

A large number of frequencies visualized in numerical order constitutes a spectrum. The keyboard of a piano represents a sound spectrum. Keys at the left produce low-pitch sounds (low frequencies). Frequency (heard as pitch) rises as the keys progress to the right end of the keyboard. Exhibit 4.a diagrams the electromagnetic spectrum and some of its uses.

As frequency goes up, the practical difficulty of using electromagnetic waves for communication also increases. Frequencies suitable for radio communication occur near the low end of the spectrum, the part with the lower frequencies and longer wavelengths.

Communication satellites employ the highest frequencies currently useful for broadcasting, mostly in the range of 3 gigahertz (3 *billion* oscillations per second) to 15 gigahertz. Even these frequencies are nowhere near as high as those of light.

Only a small portion of the electromagnetic spectrum can be used with present technology. The threat of interference among stations, increasing demand for spectrum by new services, and further growth of established services make efficient spectrum management increasingly difficult.

In 1993 Congress decided to reallocate for civilian use some 200 MHz of spectrum controlled by

Exhibit 4.a

Electromagnetic Spectrum Uses

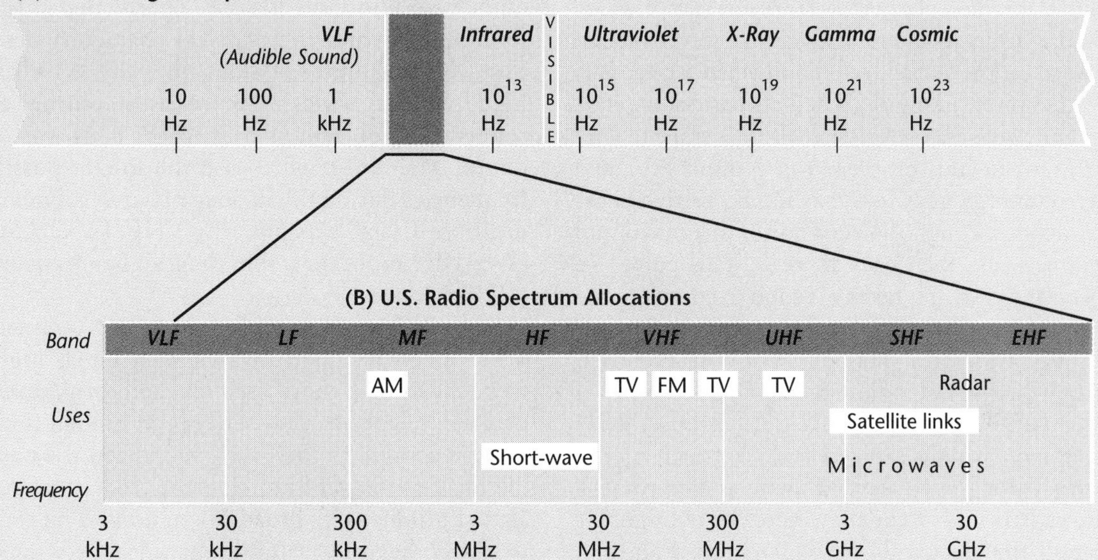

This diagram is not drawn to a uniform scale—as the frequencies rise, the scale gets smaller (otherwise it would be impossible to show this much spectrum in a single diagram). The usable radio spectrum with which we are concerned (B) is but a small part of the larger electromagnetic spectrum (A). The higher the frequencies, above the usable radio spectrum, the more potentially dangerous the output of electromagnetic radiation can become (infrared and ultraviolet rays), though even these can be useful (X-rays).

The radio spectrum is divided into bands by international agreement. Low frequency radio, a long-distance form of AM broadcasting, is used in Europe but not the United States. Though not shown here, virtually all usable spectrum is assigned to some use (aviation, land mobile, ship-to-shore, CB radio, cellular telephone, satellite uplinks and downlinks, and military applications). There is no unallocated spectrum, meaning that new services have to move older services aside (and usually upwards) in spectrum. See Exhibit 4.k for more detail on broadcast service allocations.

Changes in frequency nomenclature avoid awkwardly long numbers. Thus one kilohertz (kHz) = 1,000 Hz (hertz, or cycles per second); a megahertz (MHz) = 1,000 kHz; a gigahertz (GHz) = 1,000 MHz.

the military and other agencies. While allowing for a long period of transition (up to a decade or more), the law required that the government *auction* some of the reallocated spectrum, earning money for the U.S. Treasury. In 1994 the FCC initiated its first auctions—for personal communication and interactive video data services. Congressional pressure for more spectrum auctions remained a concern for broadcasters during the latter half of the 1990s, as discussed in Chapter 12.

4.2 *Radio and Audio Waves*

A burst of radio energy consists of a series of waves radiating from an antenna, much like waves radiate outward on the surface of a pond when a rock falls into the water. Exhibit 4.b depicts basic wave motion and wave characteristics. From a smooth surface (zero in illustration (A) in Exhibit 4.b), the wave of energy goes to a maximum positive peak, back down through the zero point, to a maximum negative peak, then back to zero. This complete process is a *cycle*. If it takes a second to complete the cycle, the wave has a *frequency* of one cycle per second, or one *Hertz* (Hz), in honor of Heinrich Hertz, who proved the existence of radio waves. If an electrical current in a wire goes through these cycles increasingly rapidly, the electrical energy eventually begins to leave the wire as electromagnetic radiation. If frequency increases enough, almost all the electrical energy is radiated and the wire becomes an antenna.

Notice that the distance the wave travels, its *wavelength*, depends on how fast the wave moves and the number of new waves created each second. Because the speed of radio energy is constant (186,000 miles a second), *frequency* alone determines wavelength. The more waves created each second, the shorter the wavelength. Either wavelength or frequency can be used to distinguish one radio wave from another. Initially the common practice was to describe a radio wave in terms of its length, and many international broadcasting stations still use wavelength to describe their signals. Modern practice in the United States is to use frequency. Bands of frequencies have names such as MF (medium frequency), HF (high frequency), and VHF (very high frequency). Exhibit 4.a shows the various frequency bands and their associated designations.

Because radio frequencies involve very large numbers of Hertz it is convenient to use abbreviations: (k)ilo instead of 1,000; (M/m)ega instead of 1,000,000; and (G)iga instead of 1,000,000,000. Thus a radio wave with a frequency of 650,000 Hz becomes 650 kHz (650 × 1,000).

Radio waves lose their energy as they radiate outward, a phenomenon known as *attenuation*. How quickly a radio wave loses its energy depends on its frequency and the physical characteristics of obstacles encountered along the way. As a rule, high frequency waves are more easily absorbed and reflected by objects in their path. This was one reason Marconi tried to use the lowest possible frequencies during the infancy of wireless communication. It also explains why VHF TV channels (2–13) are more desirable than higher frequency UHF channels (14–69).

Scientists have known for years that X-rays and other electromagnetic waves with much higher frequencies than those used for radio communication can have hazardous effects on human tissue. Only recently has the FCC recognized the possible health effects of non-ionizing radiation, such as that produced by broadcast stations. The threat to public health created by broadcast stations is considered minimal, but stations must take precautions to protect employees whose duties require they work near the transmitter and antenna system.

Radio Frequencies and Channels

Broadcasters know most people are not interested in terms like *kilohertz* and *megahertz* when they want to hear their favorite song or watch *ER*. Still, people must know where to tune their receivers and that means remembering a number. Most stations simplify the number as much as possible and associate it with a promotional phrase. An AM radio station featuring nostalgic songs of the 1940s may call itself "Mellow 66," for example. The number 66 really means 660 kHz, the station's frequency. The same is true for FM radio stations. Sometimes a radio station may call itself "Channel 82" (820 kHz) or something similar. This is correct in one sense because the government allots each AM and FM

Exhibit 4.b

Wave-Motion Concepts

(A) Starting at zero, energy moves up to a positive peak. This represents one-quarter of a full cycle and one-quarter of the final wavelength.

(B) Passing the peak, energy moves down to zero, the point at which the wave began. This represents one-half of a full cycle and one-half of the final wavelength.

(C) Having gone to a positive peak, energy falls to a maximum negative value. This represents three-quarters of a full cycle and three-quarters of the final wavelength.

(D) Starting at the negative peak, energy moves up to the zero point, the point of origin. This completes a full cycle and the distance the wave travels is one full wavelength.

These illustrations show a wave completing a full cycle in one second, a frequency of one cycle (1 Hz). If the energy were traveling at the speed of light, the wavelength would be 186,000 miles!

(E) The effect of doubling the frequency to 2 Hz is illustrated. Note that the wave now completes two cycles (2 Hz) in a period of one second. Doubling the frequency has caused the wavelength to be half as long, "only" 93,000 miles.

(F) As frequency is doubled again (4 Hz), the wavelength is again halved, making it a quarter as long as when the frequency was 1 Hz.

Because electromagnetic energy travels at a constant speed, increasing frequency reduces wavelength proportionally. Wavelength, and therefore frequency, is critical to the way a radio wave behaves once it leaves an antenna.

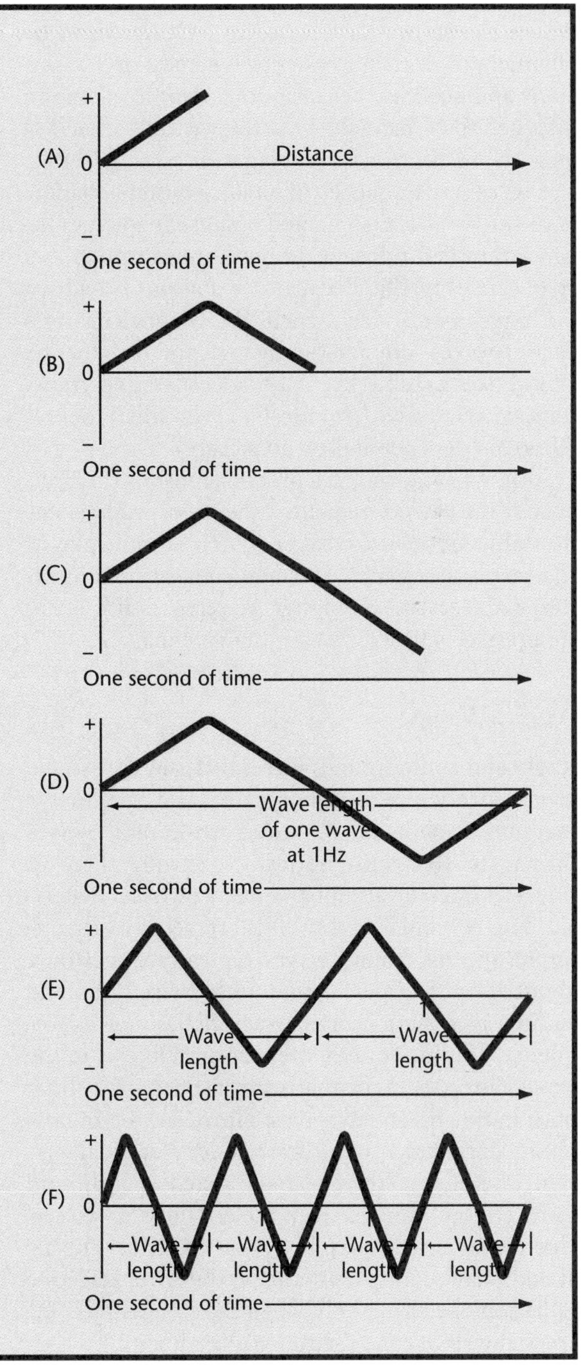

radio station a range of frequencies, a *channel*, in which to operate. The nominal frequency, such as 820 kHz, is the center of the frequency range in which the station can broadcast. Designation by channel rather than frequency is typical of TV stations and accurate because, for example, a station assigned to "Channel 6" operates within a band (a channel) of frequencies ranging from 82 to 88 MHz.

The center frequency of a radio station's channel is the *carrier wave*, so named because it conveys information (sound) superimposed on it, much as a road carries traffic. Because TV stations broadcast both audio and video, each TV channel has two carrier waves, one for the picture and one for the sound (see Exhibit 4.i). A broadcast station transmits a carrier wave (two for TV) even when there is "dead air" (silence and/or no picture).

Although transmitters produce most of their energy at the carrier frequency, they also produce undesirable *harmonics*, weaker signals at multiples of the carrier frequency. Harmonic energy can interfere with services on higher frequencies if it is not suppressed before it reaches the antenna.

Audio Waves

Radio and audio (sound) waves differ in that sound waves propagate mechanically and require a medium, such as air or water, while radio waves propagate electromagnetically, allowing them to travel through a vacuum and many physical materials. The two kinds of waves also travel at vastly different speeds. Sound waves typically travel only about 1,100 feet per second. Otherwise radio and audio waves share many characteristics such as frequency, amplitude, and attenuation. Objects in the propagation path may absorb, reflect, or refract both radio and sound waves. However, if a microphone converts a sound wave to an electrical current, the energy travels at the speed of light and some energy radiates from the conducting wire as electromagnetic waves. But most electrical energy at audio frequencies remains in the wire, effective radiation occurring only when the frequency is increased well beyond the range of audibility.

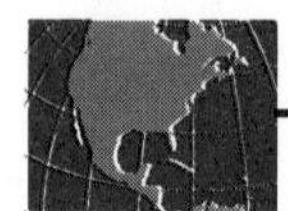

4.3 Modulation

Modulation refers to ways of imposing meaningful variations on a transmitter's carrier wave to enable it to carry programs.

Energy Patterns

Turning a flashlight on and off modulates the flashlight beam. A distant observer could decode such a modulated light beam according to any agreed-upon meaning. Modulation produces a *pattern* that can be interpreted as a meaningful signal. Morse code is, of course, one of the most universally understood patterns, involving long and short transmissions. Just as a flashlight can send Morse code, so can a radio transmitter. Marconi's wireless telegraph did nothing more than turn on the transmitter for long and short periods, the simplest kind of modulation.

When radio waves carry words, they consist of *amplitude patterns* (loudnesses) and *frequency patterns* (tones or pitches). Air molecules form pressure waves in response to the amplitude/frequency patterns of the words. A microphone responds to air-pressure patterns, translating them into corresponding electrical patterns. The microphone's output consists of a sequence of waves with amplitude and frequency variations that approximately match those of the sound-in-air patterns.

Ultimately, those AF (*audio*-frequency) electrical variations *modulate* a transmitter's RF (*radio*-frequency) carrier, causing its oscillations to mimic the AF patterns. The same kinds of variations in color and brightness in an image can modulate the video carrier wave produced by a TV transmitter. At last we have a *radio signal*—patterned variations in a carrier wave that convey meaning.

AM vs. FM

Broadcasting transmitters use either amplitude or frequency modulation. Exhibit 4.c shows how they

Exhibit 4.c

How AM and FM Differ

(A) An *unmodulated* carrier wave, emitted by a transmitter, with an unchanging frequency and amplitude pattern:

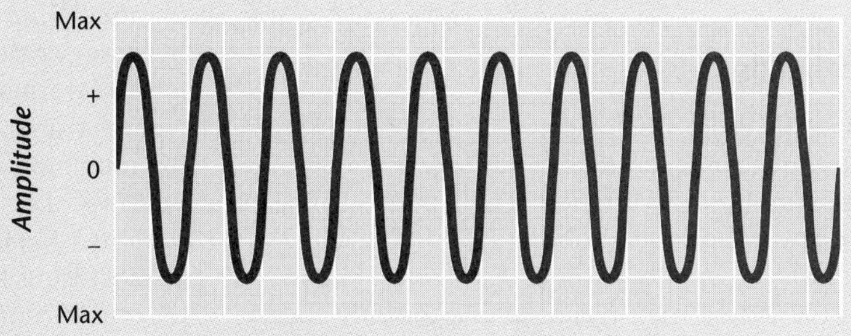

(B) An AM carrier wave, modulated by a pattern of *amplitude* changes representing a signal:

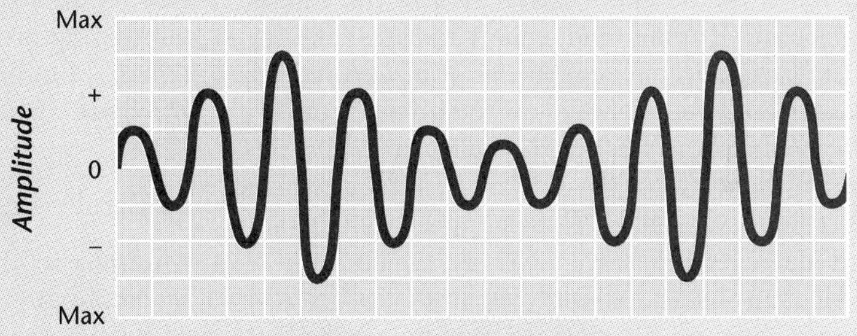

(C) An FM carrier wave, modulated by a pattern of *frequency* changes representing the same signal:

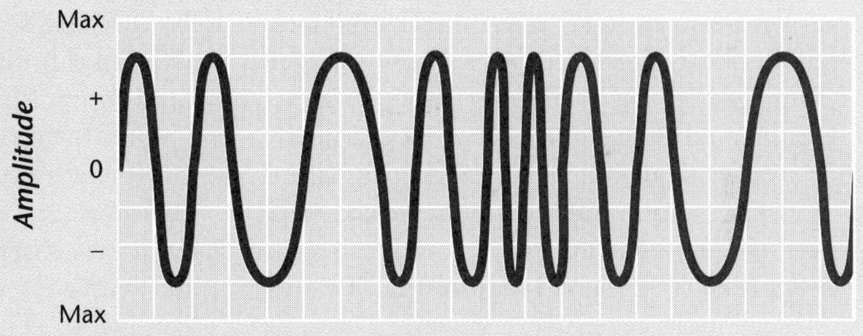

Source: From *Signals: The Telephone and Beyond* by Pierce. © 1981 by W.H. Freeman and Company. Used with permission.

work. AM (amplitude-modulated) stations are called *standard* stations because the technology of amplitude modulation was developed before that of FM (frequency modulation). Television uses both types, AM for picture signals and FM for sound.

Electrical interference caused by lightning or machinery affects AM signals because stray energy from interference distorts wave amplitude patterns. Listeners hear electrical interference as *static*. Overlapping signals from more distant stations on the same channel also easily distort AM signals.

Electrical interference has no effect on FM patterns, which rely on changes in frequency rather than changes in amplitude. FM also rejects interference from other stations more readily than does AM.

Sidebands

A specific radio frequency, not a channel or band of frequencies, identifies a station's carrier wave. For example, an AM station at 600 on the dial has a carrier wave of 600 kilohertz (600,000 cycles per second).

If the station transmits no audio, the only energy radiated is the 600 kHz carrier. When the station begins transmitting sound, energy radiates not only on the carrier frequency, but also on *sideband* frequencies both above and below the carrier. For AM stations, the audio frequencies being broadcast determine the width of the sidebands. Here is where AM radio has a problem. Popular music contains audio frequencies well above 5,000 Hz, the limit where the sidebands start spilling over into another AM station's channel. AM stations have the unpleasant option of either broadcasting the music's full audio range and risking interference, or filtering out the music's higher frequencies, making it sound muffled.

The sidebands above and below the carrier frequency are identical and, technically, no "information" is lost if a station transmits only one sideband. However, this is not an attractive solution because a *single-sideband* (SSB) signal requires that listeners buy a special, more expensive, receiver.

Some services economize on spectrum usage by suppressing one sideband. Television needs such wide channels that transmitting both sidebands would use up too much spectrum space and reduce the number of usable channels. Therefore one television signal sideband is suppressed, leaving only a vestigial lower sideband, as shown later in Exhibit 4.j.

Bandwidth

Just as the quantity of water a pipe delivers depends on the pipe's diameter, the channel's bandwidth limits the capacity of a communication system. The bandwidth needed depends on the application. A taxi dispatch service needs little bandwidth because simple intelligibility, not sound quality, is necessary. Video is an infamous "spectrum hog" requiring much more bandwidth than audio. Bandwidth becomes critical when it is necessary to transmit lots of information "right now," in real time.

Government policy usually dictates providing the minimum bandwidth required to achieve a particular goal at the time of allocation. Allocating AM stations 10 kHz channels made sense at the birth of broadcasting because the goal was to license the maximum number of stations, not broadcast high-fidelity audio. But in the 1990s broadcasters face competing program distributors with much larger "pipes." Finding ways to push more product down the same channel is a critical concern.

Multiplexing and Compression

Multiplexing is a process that conserves resources by combining related functions into a single entity. Multiplex theaters, for example, conserve money by offering several movies simultaneously within the same building. Electronic media use several kinds of multiplexing.

One form of multiplexing involves modulating two or more separate carriers in the same channel. Broadcasters turned to multiplexing when it became necessary to add color information to the signal without exceeding the existing TV channel's bandwidth. The solution was to add a color subcarrier to the signal. Exhibit 4.i shows the TV color subcarrier between the video and audio carriers. FM radio stations also use multiplexing to broadcast in stereo and provide non-broadcast auxiliary services.

In the 1990s *compression*, a technique popular in the computer industry, became extremely important to all communication industries struggling with bandwidth limitations. Rather than adding new information to the channel, compression seeks to eliminate all information not essential to the communication task being undertaken. Section 5.5 explains compression in greater detail

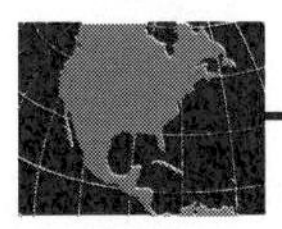

4.4 *Wave Propagation*

Modulated signals are piped from a transmitter to its antenna, the physical structure from which signals radiate into surrounding space. The traveling of signals outward from the antenna is called *propagation*.

Coverage Contours

In theory, an omnidirectional (all-directional) antenna would propagate signals over a circular coverage area. In practice, coverage patterns usually assume uneven shapes. Physical objects in the transmission path, interference from electrical machinery and other stations, and even the time of day can affect propagation distances and patterns. The sun also influences propagation. *Sunspots,* blotches on the sun's surface that increase and decrease over an 11-year cycle, dramatically influence some radio frequencies.

The higher the frequency of waves, the more the atmosphere absorbs their energy and therefore the shorter the distance they can travel. Objects wider than a wave's length tend to block propagation, causing "shadows" in coverage areas. Later in this chapter, Exhibit 4.m shows how large buildings cause "shadows" in television-station coverage. VHF, UHF, and still higher frequency waves (used for microwave relays and satellites) have such short lengths that relatively small objects can interfere with their propagation. In fact raindrops can block the shortest waves.

Because many variables affect wave propagation, broadcasters define their coverage areas in terms of reception probability. Ideally, engineers measure the signal strength at various distances from the transmitting antenna. They then plot these measurements on a map and draw lines connecting those having a minimal strength level, as defined by the FCC. The result is a map showing the station's *coverage contour.* More often engineers use FCC-approved formulas for calculating the coverage contour rather than determining it by taking measurements.

Two coverage contours usually are drawn on the same map, as illustrated in Exhibit 4.d. The flat Iowa terrain accounts for the almost perfect circularity of WHO-TV's coverage contours. The growth of cable has made coverage contour maps less important for TV, although stations continue to include them in their sales and promotional literature.

The type of modulation (amplitude or frequency) has little effect on propagation. It may be convenient to contrast the way signals of FM, AM, and TV broadcast stations behave, but it is the *frequency of the carrier wave* rather than the type of modulation superimposed on the carrier that largely determines its propagation characteristics.

Frequency-related propagation differences divide waves into three types: direct, ground, and sky waves. Each type has advantages and disadvantages that must be considered in matching frequency bands with service needs.

Direct Waves

At FM radio and television frequencies, in the VHF and UHF bands, waves follow a line-of-sight path. They are called *direct waves* because they travel directly from transmitter antenna to receiver antenna, reaching only as far as the horizon, as shown in Exhibit 4.e. Line-of-sight distance to the horizon from a 1,000-foot-high transmitting antenna propagating over a flat surface reaches about 32 miles. The signal does not cease abruptly but after reaching the horizon fades rapidly.

Engineers locate direct-wave antennas as high as possible to extend the apparent horizon. Raising the height of a receiving antenna also extends the horizon limit.

Ground Waves

AM radio broadcast stations in the United States use medium-frequency (MF) waves. Waves in the MF band travel as *ground waves,* propagated through and along the surface of the Earth. They can follow the Earth's curvature beyond the horizon, as shown in Exhibit 4.e.

Exhibit 4.d

Predicting Station Coverage

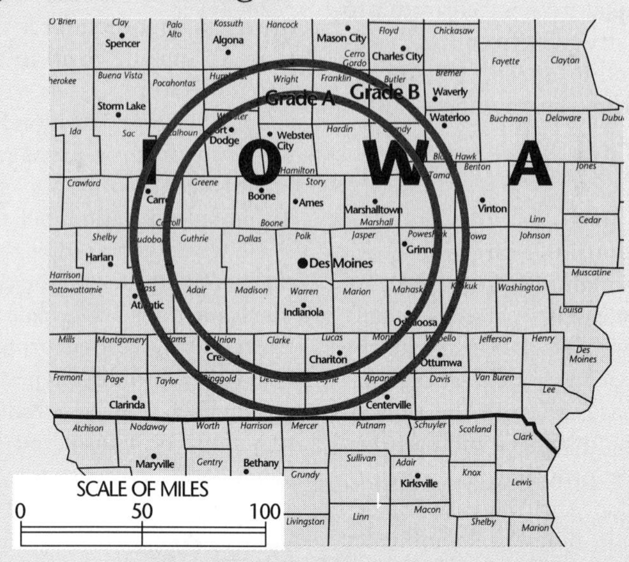

This contour map of WHO-TV, Channel 13, Des Moines, Iowa, shows Grade A and B contours. The Grade A contour includes the area within which satisfactory service is projected for 70 percent of the receiving locations 90 percent of the time; within the Grade B contour such service is projected at 50 percent of the receiving locations.

From *Television & Cable Factbook*, 1996 edition, page A-441. Used by permission of American Map Corporation and Television & Cable Factbook.

Ground waves can cover wider areas than can direct waves. In practice, however, a ground wave's useful coverage area depends on several variables, notably *soil conductivity*—the degree to which soil resists passage of radio waves—which varies according to dampness and soil composition. The area in which the ground wave provides a signal free of objectionable interference and fading defines an AM station's coverage contour.

Sky Waves

Most radio waves that radiate upward dissipate their energy in space. However, waves in the medium-frequency band (AM radio) and the high-frequency band (short-wave radio) when radiated upward tend to bend back at an angle toward the Earth when they encounter the ionosphere.

The ionosphere consists of several atmospheric layers located from about 40 to 600 miles above the Earth's surface. Bombarded by high-energy radiation from the sun, these layers take on special electrical properties, causing *refraction* (a gradual type of reflection, or bending back) of AM and short-wave signals. Refracted waves are called *sky waves*.

Sky waves can produce some interesting situations. For example, listeners in Florida may hear a Chicago radio station that listeners in St. Louis

Exhibit 4.e

Ground, Sky, and Direct Wave Propagation

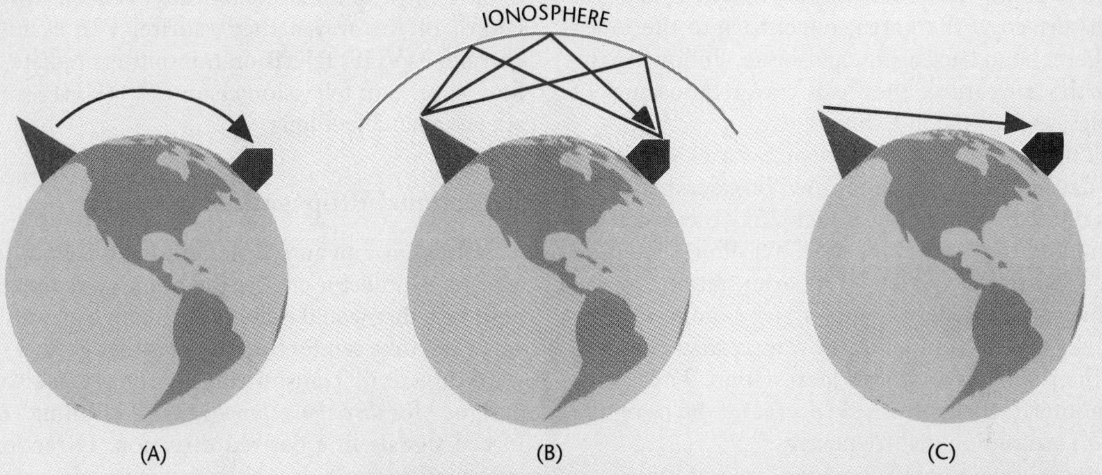

IONOSPHERE

(A) (B) (C)

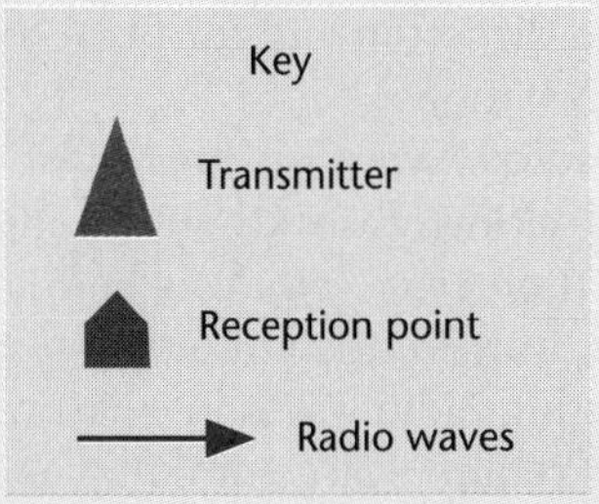

Key

Transmitter

Reception point

Radio waves

	(A)	(B)	(C)
Service	AM-day and night	AM-night only	FM and TV stations
Frequency	MW	MW	VHF/UHF
Type	Ground wave	Sky wave	Direct wave

Because of their location in the spectrum, radio and television services have different signal propagation paths.

AM radio is most complicated because its *medium-frequency* spectrum location gives it two different means of wave propagation: ground waves in the daytime (A), and both ground and sky waves at night. Ground waves travel along the surface of the Earth. Given good soil conductivity and sufficient power, ground waves outdistance direct waves, reaching well beyond the horizon. As discussed in the text, sky waves (B) bend back toward Earth when they encounter the ionospheric layer of the atmosphere at night. Sky waves may bounce off the Earth and back to the ionosphere several times, and different frequencies are affected at different times of the day and night. In short, they are unpredictable and thus cause more interference than they provide useful service. Because of potential sky-wave interference, some AM stations leave the air at night. *Short-wave* international services make use of sky waves, changing frequencies to best match predicted ionospheric effects on sky waves.

FM radio and television stations, because of their *VHF* and *UHF* spectrum location, are subject only to direct line-of-sight wave propagation at any time of day or night (C). These waves are sent out from elements on top of antenna towers and directed downward toward the reception area, blocking off radiation that would otherwise scatter upward and out into space. The "horizon" limit, of course, depends heavily on local terrain.

cannot, in spite of St. Louis being closer to the station's transmitting antenna. St. Louis is in the *skip zone*, the area between where the Chicago station's ground wave ends and its sky wave returns to Earth.

Under the right frequency, power, and ionosphere conditions, returning sky waves bounce off the surface of the Earth, travel back to the ionosphere, bend back again, and so on. Following the Earth's curvature, they can travel thousands of miles, as depicted in Exhibit 4.e.

The ionosphere's effectiveness varies with time of day and with frequency. AM broadcast stations produce sky waves only at night. Sky waves can extend a station's coverage area, but often they have an opposite effect. Sky waves from stations on the same carrier frequency may arrive simultaneously, or sky waves from a distant station may compete with ground waves from a local station. The resulting interference effectively decreases the coverage of all stations on that frequency.

Short-wave stations, however, are not limited to using a single carrier frequency as are AM broadcasting stations. They can switch frequency several times throughout the day to take continuous advantage of the ionosphere's changing refractive abilities. These stations must be careful to tell listeners where to tune when they change frequencies. Shortwave broadcast stations estimate propagation conditions weeks or months in advance and publish the frequencies they plan to use. They also announce, on-air, the new frequency to which they are moving.

Antennas

Whichever type of propagation may be involved, antennas are needed, both to transmit signals and to receive them. Antenna size and location can have a critical influence on the efficiency of transmission and reception.

Small antennas built into receivers can pick up strong AM and FM signals. The higher the frequency, however, the more elusive the signal and the more essential an efficient antenna becomes—for radio perhaps a retractable whip antenna or, for best results, an outdoor antenna. In many locations, indoor "rabbit ears" and loop antennas suffice for VHF and UHF television signals, respectively, although some UHF signals require an outdoor antenna.

Transmitting antennas (see Exhibit 4.f) vary in size because for them to work efficiently their length must be mathematically related to the length of the waves they radiate. For example, channel 2 (VHF) television transmitters radiate 20-foot waves, but television channel 48 (UHF) waves are less than 2 feet long.

Directional Propagation

A flashlight or an auto headlight focuses light into a beam. A reflector creates the beam by redirecting light rays that would otherwise radiate backward or sideways, thus reinforcing the rays that go in a forward direction. Transmitting antennas can also be designed for *directional propagation*—beaming reinforced signals in a desired direction. Directional propagation has value both for increasing signal strength in a desired direction and for preventing interference between stations.

Concentrating radio waves by means of directional propagation increases their strength. This increase, called *antenna gain*, can be many-thousandfold. Gain, however, is not always easily achieved. Concentrating radio waves in a particular direction sometimes requires large or complex antenna designs. AM radio stations need physically large antennas because of their long carrier waves. A directional system requires at least two tall towers, perhaps more, so a directional AM broadcast antenna requires space and is expensive. AM stations need directional antennas more than FM and TV stations, to reduce sky wave interference.

Receiving antennas are also directional; this is easily demonstrated by turning a portable AM radio in different directions. Signals from a given station come in most strongly when the radio's built-in antenna points in the direction of that station's transmitter. Outdoor antennas used for TV (and FM radio) reception also display directional characteristics. Reception of signals arriving from several directions often requires outdoor antennas mounted on rotatable masts.

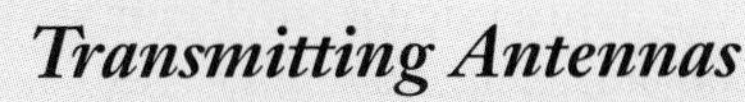

Transmitting Antennas

(A)

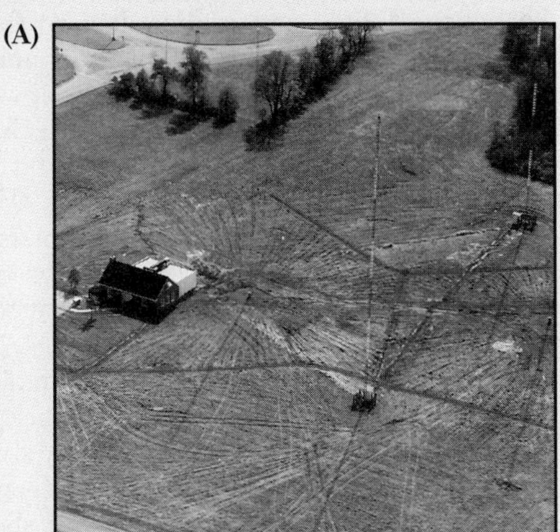

(B)

(C)

Antennas differ depending on the service for which they are built.

(A) *AM antennas* use the entire steel tower as a radiating element. For efficient propagation, the height of this tower must equal a quarter the length of the waves it radiates. Propagation of ground waves depends on soil conductivity, among other things. The photo shows heavy copper ground cables being buried in trenches radiating from the tower base to ensure good ground contact. There are several antennas here for sending out a directional signal.

(B) *Short-wave antennas* are very complex. Each transmitting site has many antennas to feed different HF frequencies, along with other suspended wires that act as reflectors to beam signals in specific directions to reach designated overseas target areas. The Voice of America uses scores of antennas like this one.

(C) FM and TV broadcasters often mount transmitting antennas atop tall buildings, providing the height needed for effective signal propagation without the expense of building supporting towers.

Sources: Photo (A) courtesy of Stainless, Inc., North Wales, PA; photo (B) courtesy of Voice of America, Washington, DC; photo (C) © Bruce Jaffe/Liaison International.

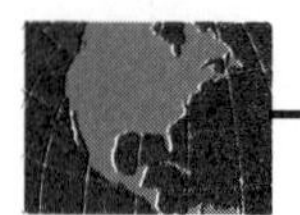

4.5 *Mutual Interference*

Mutual interference between stations limits the number of stations that can be licensed to operate in any particular community.

Co-Channel Stations

The primary troublemaker, *co-channel interference*, comes from mutual interference between stations operating on the same frequency. Co-channel stations must be located far enough apart geographically to prevent their coverage contours from overlapping. Separation rules must take into account that signals too weak for useful reception by an audience may nevertheless cause co-channel interference. A station's *interference zone* therefore extends far beyond its service area.

In keeping with U.S. policy of allowing as many local communities as possible to have their own stations, the Federal Communications Commission (FCC) makes co-channel separation rules as liberal as possible. Changing daytime versus nighttime coverage areas of AM stations, caused by sky-wave propagation at night, complicate the problem. The FCC seeks to prevent such sky-wave interference by requiring many AM stations to use lower power at night. Some stations have to reduce power and use a directional antenna system. Sometimes the authorized power level (measured in watts) is less than that of a bedside lamp. The resulting coverage area can be so small the station owner may find it more profitable simply to leave the air after dark. Relocating some stations in the expanded AM band (see Section 4.6) will help reduce these interference problems.

Adjacent-Channel Stations

Engineers design TV and radio receivers to be selective, to tune in one station and reject those operating on different frequencies or channels. But building highly selective receivers would be expensive, so set manufacturers balance *selectivity* against cost. The result is that receivers sometimes pass not only the desired signal but also those of strong stations on adjacent channels, those just above and below the desired station. To reduce *adjacent-channel interference* the FCC avoids placing stations on adjacent channels in the same market. But what about communities that have TV stations on channels 6 and 7 or 13 and 14? As seen in Exhibit 4.j, there is a big jump in frequency between these channels. This allows assignment of both pairs of channels to the same market without danger of adjacent-channel interference, in spite of their consecutive numbering.

Adjacent channel signal strength falls off rapidly with increasing distance, so the separation needed between adjacent channel TV stations is only about 60 miles. Some TV stations operating on the same channel (co-channel) need a separation of about 220 miles. Maximum-power FM stations using antennas high above the ground must be about 180 miles apart to avoid co-channel interference and 150 miles to prevent adjacent channel interference. Because of sky wave propagation, AM radio stations can suffer both co-channel and adjacent channel interference from stations many hundreds of miles away.

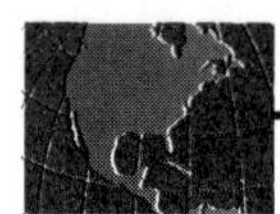

4.6 *AM Stations*

As a broadcast-station class name, AM is somewhat misleading because the term means simply *amplitude modulation*, and many services other than standard radio broadcasting use such modulation. The video component of the television signal, for example, is amplitude modulated.

Band and Channels

For years the government assigned AM broadcast stations to frequencies ranging from 535 to 1,605 kHz. This band provided 1,070 kHz of spectrum in which to assign stations. Each AM station received a channel 10 kHz wide, so there were 107 channels (1,070 divided by 10) on which AM stations could operate. In 1988 an international agreement expanded the U.S. AM broadcast band to 1,705 kHz. This expansion made available 10 new regional channels.

Rather than adding new stations in this expanded band, the FCC decided to use the space to reduce AM broadcast station interference by moving existing stations. Using a complex formula, the FCC in 1997 identified 88 AM stations for "migration" to the new frequencies. Stations allowed to move will have five years in which to do so and, during that period, may transmit on both their old and new frequencies. This will help listeners follow these stations up the band. AM stations will still operate in a restrictive 10 kHz channel, but spreading them out should reduce co-channel and adjacent channel interference.

Channel and Station Classes

The FCC has a rather complex method of categorizing AM stations according to the type of channel assigned and their class. There are three types of channels: clear, regional, and local. Clear channels provide stations assigned to them a considerable degree of freedom from interference, allowing them to cover large geographical areas. Regional channels cover smaller areas, and local channels typically serve only a single community.

There are four classes of AM stations, ranging from Class A to Class D. Class A stations have the most "privileges" in terms of maximum transmitting power and permissible hours of operation, while Class D stations have the least. There is considerable overlap, and the FCC rules devote several paragraphs to explaining the differences (see 47 CFR 73.21). Exhibit 4.g outlines the method of categorizing AM stations in greater detail.

Transmission

High AM radio transmitter power improves efficiency of both ground-wave and sky-wave propagation, getting greater coverage and overcoming interference. Power authorizations for domestic U.S. AM broadcasting range from 250 watts to a maximum of 50,000 watts, as indicated in Exhibit 4.g. Cincinnati's WLW used "superpower" (500,000 watts) during the second half of the 1930s, but the experiment ended shortly after the beginning of World War II.

AM radio stations have wavelengths between 593 and 1,823 feet long, depending on their exact carrier frequency. Because antennas at least a quarter-wave long are necessary, AM stations must use vertical steel towers several hundred feet tall to radiate their signals.

In choosing sites for AM antennas, engineers look for good soil conductivity, freedom from surrounding sources of electrical interference, and distance from aircraft flight paths. Ground waves are critical to the coverage provided by AM stations, so the station's antenna system must include an effective ground system. Much of an AM radio station's antenna consists of many buried copper wires radiating from the base of the vertical steel tower. Engineers often have trouble convincing station owners that the expensive buried parts are no less important than the tower above ground.

Low-power AM signals can be fed into building steam pipes or power lines, which serve as distribution grids. The signals radiate for a short distance into the space surrounding these conductors. Services using this propagation method are called *carrier-current* stations. Colleges and universities often use carrier-current stations, which require no licensing.

Unfortunately, carrier-current signals on power lines will not pass through standard transformers. Modified transformers will pass the signals, but "jumping" every transformer on campus can be difficult and expensive, often negating the advantages of unlicensed operation. *Travelers Information Service* (TIS) is a licensed application of carrier-current sometimes used to help motorists navigate in congested areas such as busy airports. Roadside signs tell drivers where to tune their AM radios to get helpful information.

Exhibit 4.g

AM Radio Station Classification Structure

AM Radio Stations and Channel Classes

Station Class	Channel Class	Number Channels in Class	Power Range (kW)	Percentage in Each Channel Class
A, B, D	Clear	60	.25–50	35
B, D	Regional	51	.25–50	44
C	Local	6	.25–1	21

AM Radio Stations by Station Class

Station Class	Number of Stations*	Percentage of Total Stations†
A	77	1
B	1730	35
C	1023	21
D	2072	42

*Data provided by FCC in early 1977. The numbers vary as new stations sign on and others leave the air.
†Percentages do not add to 100 percent because of rounding.

Class A stations operate unlimited schedules only on clear channels with powers ranging from 10 to 50 kW. Class A stations are protected from other stations on the same or adjacent channels.

Class B stations operate unlimited schedules on clear and regional channels with powers ranging from .25 to 50 kW. Class B stations in the extended AM band (1605–1705 kHz) operate with a power of 10 kW.

Class C stations operate only on local channels with powers ranging from .25 to 1 kW.

Class D stations operate either daytime, limited time, or unlimited time with nighttime power less than .25 kW. Daytime powers range from .25 to 50 kW. Nighttime operations of Class D stations are not protected from interference, and Class D stations must protect all Class A and Class B stations from nighttime interference.

AM Stereo

Because of its narrower channel, AM took longer to develop stereo than did FM. AM stereo uses a *matrix mode* that melds left and right channels in a way somewhat similar to the two channels of FM stereo but without the full high-fidelity benefit. The few AM stereo stations in existence in the early 1990s had mostly chosen the Motorola C-QUAM standard finally adopted by the FCC (see Section 12.9). Early excitement regarding AM stereo had largely dissipated by the middle 1990s. Few listeners were aware of its existence, and programming on many AM stations benefited little from being offered in stereo.

Short-Wave AM

Short-wave broadcasting, which also uses AM, has been allocated parts of the HF (high-frequency) band between 6 and 25 MHz. The ionosphere refracts waves in this band both day and night, enabling round-the-clock coverage in target zones thousands of miles away from the originating transmitter.

The U.S. public makes little use of short-wave broadcasting. However, some foreign countries use short waves extensively for domestic services. A few privately operated American HF stations, mostly evangelistic religious outlets, broadcast to foreign audiences. Short waves are used extensively for international diplomacy.

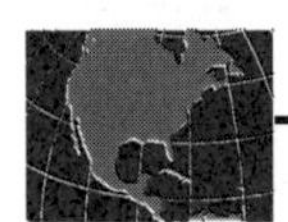

4.7 *FM Stations*

Referring to AM as *standard broadcasting* has become sorely out-of-date because FM's inherently superior quality has enabled it to forge ahead of AM in numbers of both listeners and stations.

Bands and Channels

U.S. frequency-modulation broadcasting occupies a 20-MHz block of frequencies running from 88 to 108 MHz in the VHF band. This 20-MHz block allows for 100 FM channels of 200 KHz (.2 MHz) width. Exhibit 4.k shows their spectrum location in relation to other broadcast channels.

The FCC numbers FM channels 201 to 300, but licensees prefer to identify their stations by their midchannel frequency (in megahertz) rather than by channel number (88.1 for channel 201, 88.3 for channel 202, and so on). The lowest 20 FM channels are reserved for noncommercial use. The FM channel width of 200 kHz is generous—20 times the width of the AM channel.

Transmission

The coverage area of an FM station depends on the station's power, the height of the transmitting and receiving antennas above the surrounding terrain, and the extent to which obtrusive terrain features or buildings block wave paths. In any event, FM uses direct waves, which reach only to the horizon.

Once determined by these factors, an FM station's coverage remains stable, day and night. This stability is one reason the FCC had an easier time preventing FM interference than it did AM interference.

The FCC divides the country into geographic zones and FM stations into three groups according to coverage area: Classes A, B, and C, defined by power, antenna height, and zone. Classes B and C are further subdivided, but generally the maximum coverage areas for FM stations by class are as follows: Class A about 15 miles; Class B about 30 miles; Class C about 60 miles. The maximum power/height combination permits 100,000 watts (twice the maximum AM station power) and a 2,000-foot antenna elevation.

Frequencies used for FM broadcasting are much higher than those of AM stations, so they have proportionally shorter wave lengths and much smaller antennas. FM stations use tall towers only to place the small antenna high above the ground, not to actually radiate the signal as do the towers of AM stations. FM stations take advantage of this by mounting their antennas on mountains or tall buildings. FM stations depend on direct, not ground, waves to provide primary coverage and do not need expensive buried radial systems. Finally, the small size of FM antennas allows engineers to design radiating systems that exhibit *antenna gain.* Using an FM antenna with gain increases signal strength as does boosting transmitter power but, over time, is much less expensive than a larger transmitter's monthly electric bill.

Reception

FM receivers with quality loudspeakers can reproduce the audio frequencies essential for high-fidelity sound. In addition, FM radio has greater *dynamic range* (the loudness difference between the weakest and strongest sounds) than does AM, making for more realistic reproduction of music.

FM's inherent immunity to static and interference gives FM radio a significant quality advantage over AM radio. FM's static-free reception is particularly noticeable in the southern part of the country, where subtropical storms cause much natural interference. FM also has an advantage in large cities, where concentrations of electrical machinery and appliances cause static.

FM Stereo

The left and right channel audio off stereo tapes and CDs pass through the FM radio station's equipment as entirely separate signals until they reach the station's stereo generator, a subsystem of the transmitter. Here the two audio channels combine with supersonic (above the range of human hearing) pilot and subcarrier signals. The result is a composite signal that modulates the station's carrier wave. FM receivers designed to reproduce stereo audio use the subcarrier and pilot signals to separate the two audio channels once again. Just as it is possible to watch color TV signals on a monochrome set, a monaural FM radio reproduces the audio in full fidelity, minus the separation between audio channels required for stereophonic effect.

Subsidiary Communications Service (SCS)

FM stations can also multiplex additional information to provide a non-broadcast service received using special equipment. This multiplexed signal most frequently provides background ("elevator") music to stores and offices, or a reading service for the blind. SCS can accommodate a wide range of other applications, including paging, distribution of stock and bond prices, detailed weather reports, and traffic information. SCS may also provide station promotional information (such as music format, call letters, or location) in a visual format on special receivers, as discussed in Section 5.8.

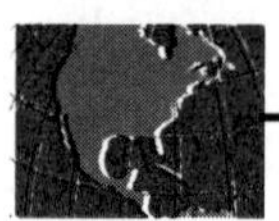

4.8 Electronic Pictures

The basic structure of a TV station remains much as it was in the late 1940s (see Exhibit 4.l). Cameras and microphones still convert sounds and pictures into electrical signals, which are fed to a transmitter, where they modulate RF carrier waves eventually radiated from an antenna. Transition to solid-state technology and microelectronics has, however, changed the way modern cameras operate. The TV receiver still uses a vacuum tube for display of the video image (see Exhibit 4.n), and most broadcasters continue to encode the analog signal much as they have for more than five decades.

Camera Technology

From the 1940s until the early 1990s tube-type cameras dominated the broadcasting industry. It was not until the 1980s that broadcasters began using cameras with a solid-state *charge-coupled device* (CCD) rather than tubes to convert light energy to an electrical signal. The video produced by these "chip" (CCD-equipped) cameras relegated them mainly to shooting news footage, where picture quality is less critical. Solid-state technology advanced dramatically in the 1990s, and cameras using CCDs are now commonplace. Many excellent studio cameras using tubes remain in service, but broadcasters are replacing them with CCD cameras when they fail. CCDs offer the usual advantages of solid-state devices compared to tubes: less weight and bulk, longer service life, greater reliability, and reduced energy consumption.

Exhibit 4.h shows a photo and simplified functional diagram of the CCD used in modern cameras. CCDs use hundreds of thousands of tiny photodiodes to convert light to electricity. The electrical charge depends on the intensity of the

Exhibit 4.h

The TV Camera

(A)

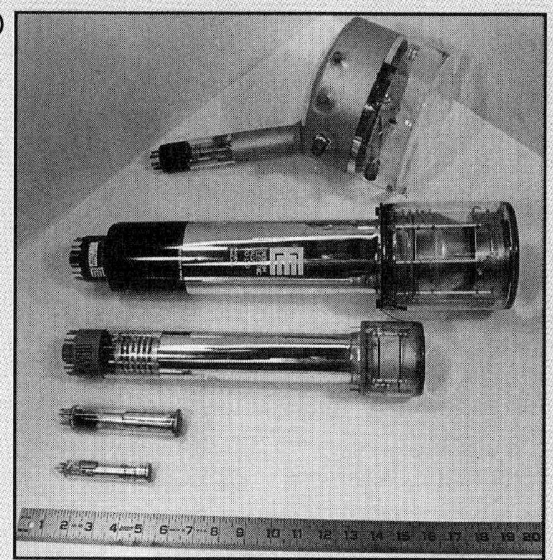

(B)

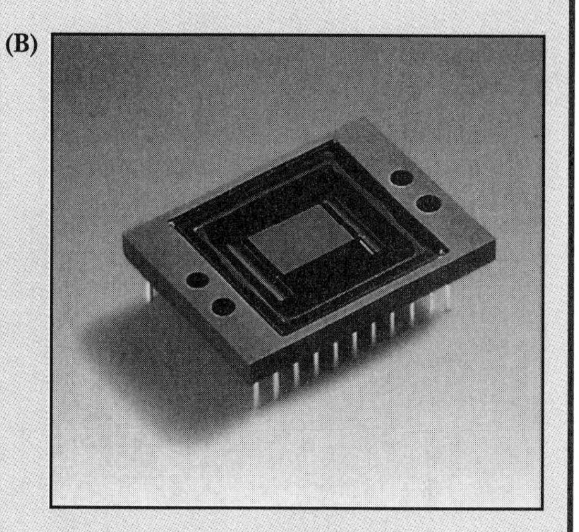

(C)

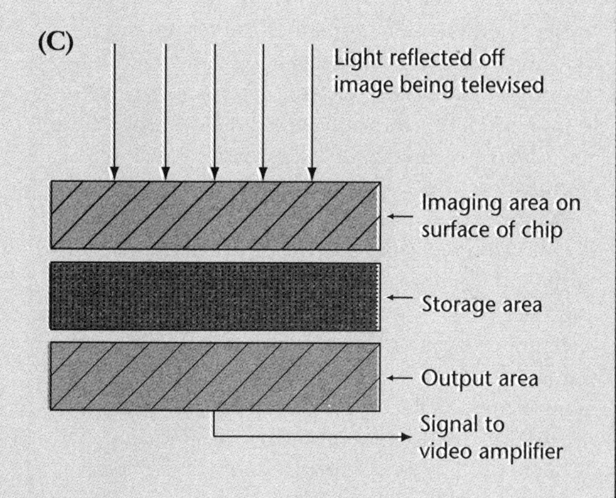

(A) As technology improved, tubes became smaller. Commercial broadcasters first used the *iconoscope*, the odd shaped tube at the top. Then *image orthicons*, with 3-inch and 4.5-inch faces, replaced iconoscopes. Smaller tubes such as the *vidicon* and *plumbicon* in turn replaced the image orthicon. (B) In the 1990s smaller, more rugged and energy-efficient chips largely replaced tubes as imaging devices for cameras. (C) Light striking the surface of the CCD is converted to electrical charges. These charges are then collected in the storage area before being passed to the output area. From the output area the signal representing the light and dark parts of the televised image are clocked out to the camera's video amplifiers.

Sources: Photo (A) courtesy Frank Saverwald, Temple University. Photo (B) courtesy of Sony Electronics Inc.

light striking the photodiode, so the CCD creates a pattern of electrical charges analogous to the bright and dark portions of the image. Each photodiode produces a *picture element (pixel)* that combines with others to form the complete image. The electrical charges produced by the photodiodes are transferred and stored in another part of the CCD and, finally, read out in a line-by-line order corresponding to the frame and line scanning frequencies required by TV receivers.

Exhibit 4.i

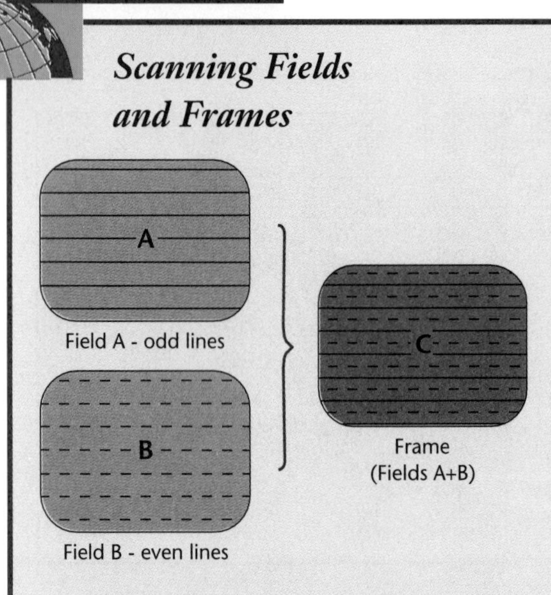

A complete electronic picture, or frame, is made up of two fields, each of which makes up half the picture definition. One field (A) scans the odd-numbered lines and the other (B) scans the even-numbered lines (shown here as broken). Together the lines of the two fields "interlace" to make up a complete frame (C).

At the end of each scanned line, and each field, the electron beam has to fly back to start a new line or field. During these retrace times, a blanking signal prevents the beam from activating any picture elements. When the beam returns to the top of the screen at the end of each field, the time it takes, equivalent to scanning 21 normal lines, is called the vertical blanking interval (VBI). It is during this VBI that nonpicture information can be transmitted, such as closed captions and teletext.

Broadcast-quality cameras use three CCDs, one for each of the primary colors of light, which combine their output to produce all the colors we see on TV. Consumer quality cameras typically have only a single CCD and deliver a less crisp color image, but one that is still amazingly good considering their price.

Scanning and Frame Frequency Standards

The technology of tube-based display devices requires that the video information be transmitted in a pattern that allows the receiver to "paint" images on screen one line at a time using a moving electron beam. The U.S. television system (standards differ from country to country) calls for a nominal *525 lines* of video information to be painted in horizontal rows from top to bottom to create the image on the TV screen. The number of lines making up the images helps determine the *resolution*—the clarity—of the image. The more lines, the greater the resolution. However, not all 525 lines of the U.S. analog standard are used to convey pictures. Some time—and thus some lines—must be devoted to the auxiliary signals used by the receiver to control the scanning.

As illustrated in Exhibit 4.i, TV uses a method called *offset* or *interlaced scanning* in which all the even-numbered lines are scanned first, followed by the odd-numbered lines. Each of these scans constitutes an odd or even *field*. Two fields combine to make a complete *frame* of video. This odd-even scanning of lines was a way to prevent the image from appearing to flicker without having to increase the channel bandwidth required for TV transmission. The U.S. analog TV system paints the screen with 30 complete frames (60 fields) every second. The human eye and brain cannot distinguish individual images flashed in rapid succession. This *persistence of vision* causes viewers to see a smoothly moving image rather than a series of individual frames of video.

Progressive scanning display devices paint the lines on the screen one after the other, in numerical sequence. Computer monitors use progressive scanning, and the digital TV transmission standards adopted in 1996 allow digital television (DTV) display devices to use either interlaced or progressive scanning methods.

Auxiliary Signals

Several types of *sync (synchronization) pulses* included with the video information are needed to ensure that each pixel in each line appears on the receiver screen in the same location it had leaving the camera. *Horizontal sync* pulses tell the receiver where to begin and end each horizontal line of video. *Vertical sync* pulses tell the receiver when an odd (or even) field is complete and the next field should begin. Additional *blanking pulses* turn off the flow of video information while the "paint brush" (actually the electron beam in the receiver's picture tube) is repositioned to begin the next scan line or the next field. In the NTSC standard the time required to position the beam to begin a new field is equivalent to that needed to paint 21 lines of video. No picture (video) information flows during this relatively long *vertical blanking interval* (VBI), allowing broadcasters to insert other information.

The VBI is often used to provide *closed captioning* for hearing-impaired viewers, market reports, or on-screen *electronic programming guides* (EPG) such as *TV Guide Plus+*. Capturing and displaying information embedded in the VBI requires special decoder circuits either built into the TV receiver itself or in a separate set-top box designed for the purpose.

TV Sound

U.S. television uses FM sound. The sound subchannel is located in the upper part of the television channel, as shown in Exhibit 4.j. Sound needs no synchronizing signals to keep it in step with the picture. Sound and picture occur simultaneously and go out in real time.

Television's sound channel is designed to respond to a sound range from 10 to 15,000 Hz. It also has room for subcarriers that enable multiplexing stereophonic sound and alternative-language soundtracks within the channel.

Color TV

Signal specifications for color television provide for *compatibility*—black-and-white (monochrome) receivers can reproduce a color signal without the aid of an adapter. Multiplexing permits adding a color component without enlarging the television channel. Exhibit 4.j shows the position of the color subcarrier in the television channel.

RCA developed this color system, and, with National Television System Committee (NTSC) approval, the FCC adopted it in 1953. The NTSC color system has served the industry well for more than 40 years, but broadcast engineers are critical of it, saying NTSC really means "Never The Same Color," a reference to the difficulties of getting colors on screen to match exactly those seen in the studio.

Color cameras pick up images in the three primary colors of light: red, green, and blue (RGB). This is sometimes confusing because red, blue and yellow are the primary colors familiar to artists who mix pigments that *reflect* light. TV pictures are created by mixing light itself. TV starts with a black screen, and the primary colors of light combine to create white; artists start with a white canvas and the pigments mix to create black.

The camera divides light energy entering the lens into the three primary colors before it reaches the tubes or CCDs that convert light energy to electrical energy. The fidelity of *hues* (colors) and *saturation* (richness of each color) in the transmitted image depends on the camera's ability to process the three primary colors properly. Additionally, the color signal has a brightness attribute called *luminance*. The luminance component in the signal provides most of the fine detail in the picture, and luminance alone suffices for monochrome receivers.

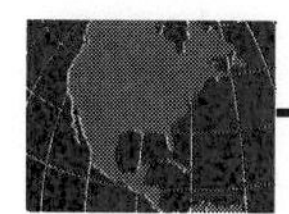

4.9 TV Channels

U.S. broadcast standards call for a 6-MHz television channel—600 times the width of an AM radio broadcast channel. Indeed, *all* AM and FM broadcast channels together occupy less spectrum space than only four television channels. Exhibit 4.j shows how the 6 MHz of the channel are utilized.

Exhibit 4.j

The TV Channel

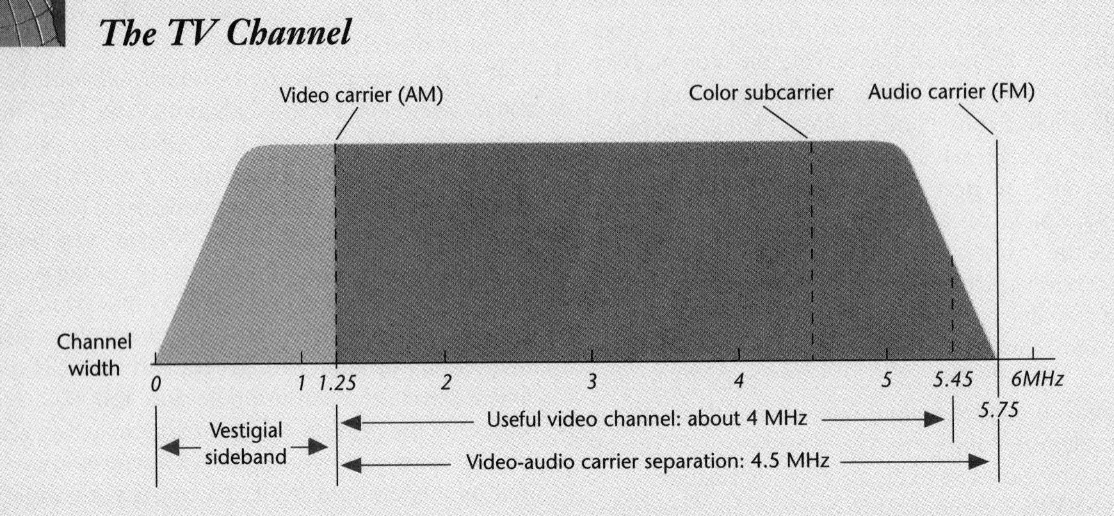

The architecture of the 6-MHz television channel includes a vestigial (sort of an "electronic leftover") lower sideband that takes up the 1.25 MHz below the video carrier frequency. Note that .25 MHz is unused above the audio subcarrier. These seemingly "wasted" spaces reduce adjacent channel interference. When modulated by the black-and-white video (luminance) information, the main video carrier's upper sideband overlaps the sideband of the color subcarrier, but *interleaving* minimizes conflict. Uneven distribution of frequencies carrying the luminance signal makes such interleaving possible. If you visualize the frequencies of the monochrome information as the teeth of a comb, the color information occupies frequencies represented by the spaces between the teeth.

Note the audio carrier near the top of the TV channel. A multichannel television sound (MTS) station can use this area to accommodate the main audio signal, a stereo audio signal, a secondary audio program (SAP) signal, and a closed-circuit "professional channel" for transmitting telemetry signals or other data requiring only 3 kHz of bandwidth. The SAP signal may be used to provide a foreign-language version of the program being broadcast.

Source: Adapted from 47 CFR 73.699.

Channel Width and Resolution

Even with such wide channels, television achieves relatively low picture resolution by the standards of theatrical motion pictures and good-quality still photography. The average home receiver displays about 150,000 pixels per frame, but a projected 35-mm film frame contains about a million.

Television resolution standards represent a compromise. Higher quality has to be sacrificed to avoid using too much spectrum space. The same spectrum-saving goal motivates suppression of the lower sideband of the television picture signal.

The resolution required varies with the application. Early PC users tried to use TV sets as computer monitors. They found that, at a viewing distance of 18 inches, the TV lacked adequate resolution for word processing. Such work required a more expensive, higher-resolution monitor. Yet most people happily watch rented movies on VHS tape—a delivery system providing less resolution than broadcast TV—using their "inadequate" TV sets, but at typical viewing distances of 10 feet. This suggests some programming may not require high resolution with its associated high bandwidth demands. Broadcasters struggle with this issue as they weigh the advantages of using their digital TV channels for a single program in high definition or for several lower-definition programs requiring the same channel bandwidth instead.

Location in Spectrum

Exhibit 4.k shows the location of television channels in the spectrum relative to the location of channels other broadcast services use. An FCC table allots specific channels by number to specific towns and cities throughout the United States. The map in Exhibit 2.i gives an example, showing all the places to which one particular channel has been allotted.

Exceptions to the allotment tables are *low-power television* (LPTV) facilities. In 1982 the FCC gave final approval for LPTV service. LPTV stations operate on the same VHF and UHF channels and have the same technical characteristics as other TV stations, but they use greatly reduced power, from 10 to 1,000 watts. Coverage under ideal conditions is only about 10 miles but in congested areas is often much less. LPTV stations are considered secondary services and, provided they do not cause interference to full-power TV stations in the area, may operate on any channel. In 1997 there were more than 1,900 licensed LPTV stations in the FCC's database, but only a fraction of them were believed to be on the air. LPTV's secondary status became a serious liability when the FCC began assigning channels for digital television (DTV). The Commission made some concessions and pledged continuing efforts to accommodate LPTV, but the service's future in a digital world remains uncertain.

4.10 TV Transmission

Synchronization

The *sync generator* keeps the camera's output in step with the receiver display tube. Exhibit 4.l shows the sync generator in relation to other control-room equipment.

In addition to providing the horizontal and vertical sync pulses, the sync generator also inserts blanking signals during retrace periods and synchronizes inputs from external picture sources such as recordings and network feeds. Synchronizing and blanking cues become part of the composite picture signal.

Combining Picture and Sound

The picture (video) and sound (audio) components travel from the studio to the transmitter as independent signals. The transmitter site may be miles away from the studio so that it can be near the antenna. At the transmitter, video and audio signals modulate separate audio and video transmitters.

A television station's power is usually stated in terms of the *effective radiated power* (ERP) of its *video* signal. The video transmitter operates on higher power than does the audio transmitter because it has a much greater information load, requiring up to 20 times as much power.

Video and audio signals meet for the first time in the *diplexer*—a device that combines video and audio signals before feeding them as a composite signal to the antenna for propagation.

Propagation

Engineers elevate television-transmission antennas as high as possible—on mountain peaks, roofs of tall buildings, or the tops of tall steel towers. The antenna towers themselves do not radiate signals as do AM towers. As is true of FM stations, they only support the radiating elements, which are relatively small, in keeping with the shortness of television's VHF and UHF waves. Television (and FM radio)

Exhibit 4.k

Summary of Broadcast Channel Specifications

This table pulls together information scattered throughout the text to allow comparison of the major channel specifications of radio and television services. Note especially the allocation of television channels into four separate frequency allotments in two different bands.

Service	AM Radio	FM Radio	VHF Television	UHF Television
Frequencies	535–1705 kHz (MF band)	88–108 MHz (VHF band)	54–72 MHz (2–4) 76–88 MHz (5–6) 174–216 MHz (7–13) (VHF band)	470–806 MHz (14–69) (UHF band)
Total channels	117	100	12	56
Bandwidth (single station)	10 kHz	200 kHz (equivalent to 20 AM channels)	6,000 kHz (equivalent to 600 AM or 30 FM channels)	6,000 kHz
Classes of stations and power limits	Class A 10–50 kW Class B .25–50 kW Class C .25–1 kW Class D <.25–50 kW	A .1–6 kW about 15 miles B >6–50 kW about 30 miles C 25–100 kW about 60 miles	Channels 2–6 about 65 miles up to 100 kW Channels 7–13 about 55 miles up to 316 kW	Channels 14–40 about 40 miles up to 5,000 kW Channels 41–69 about 30 miles up to 5,000 kW
Educational allocation	None	88–92 MHz (20 channels)	137 specific channel allotments	559 specific channel allotments
Factors affecting station coverage and signal quality	Frequency Power Soil conductivity Day/night (sky wave) Thunderstorms Directional antennas	Antenna height Power (to a degree) Unlimited time No directional antennas High fidelity Little static Line of sight range	Antenna height Frequency Power (to a degree) Rigid spacing Line of sight range Unlimited time	Antenna height Frequency Power (to a degree) Rigid spacing More limited line of sight range Unlimited time

Exhibit 4.I

TV System Components and Signals

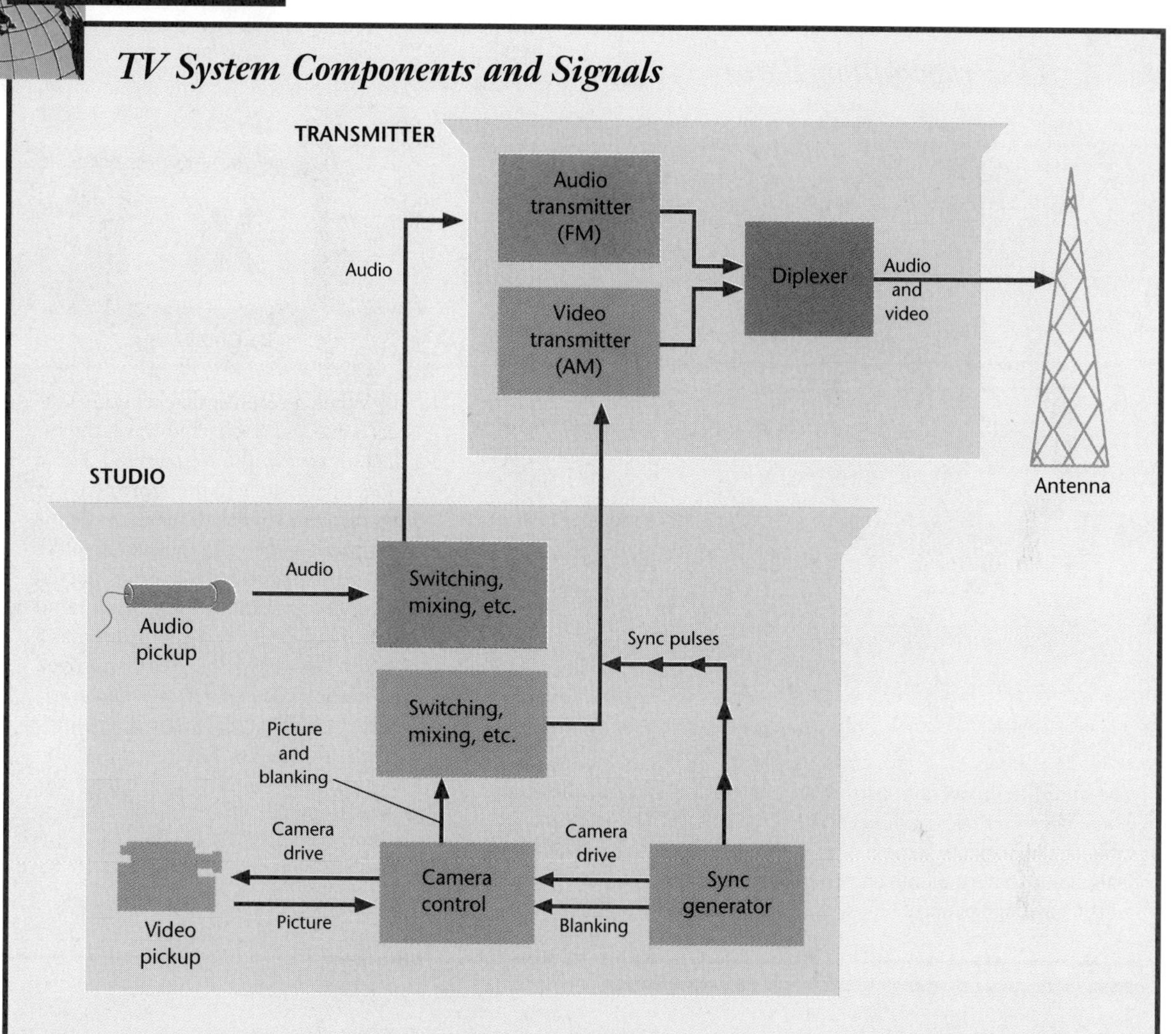

Each block stands for a function that in practice may involve many different pieces of equipment. The lower portion of the diagram represents the basic items and functions involved in studio operations; the upper portion, those involved in transmission. Separate sets of equipment handle the video and audio components all the way to the diplexer, the device in transmitters that finally marries the two signals so that they can be transmitted as a composite. (Exhibit 4.j shows how the audio and video components are multiplexed in a single channel.)

Source: Based on Harold E. Ennes, *Principles and Practices of Telecasting Operations* (Indianapolis: Howard W. Sams, 1953).

Exhibit 4.m

TV Propagation Paths

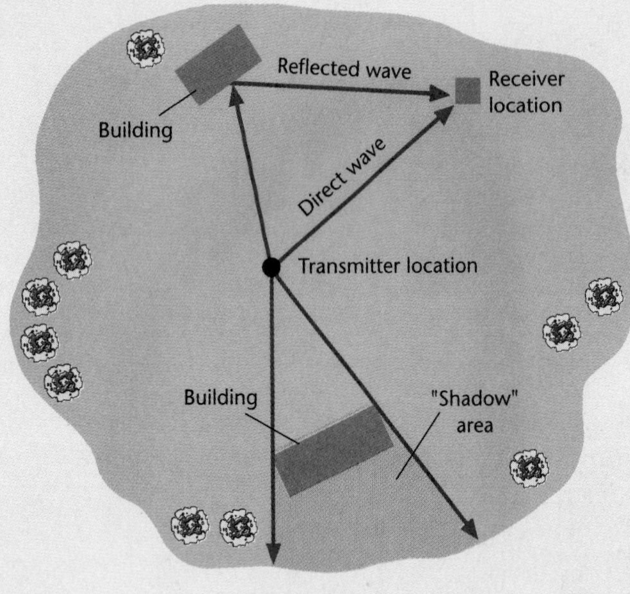

(A) Wave behavior

(A) Simplified coverage pattern of a television station, showing some characteristics of direct-wave propagation: The waves carry only to the horizon; some may encounter surfaces that reflect signals; some may encounter obstructions that cause "shadow" areas in the coverage pattern.

(B) Ghost image

(B) When a receiver detects both a direct wave and a reflected wave, the reflected wave will have traveled over a longer path and will therefore arrive at the receiver slightly later than the direct wave. When the receiver displays the signal of the delayed reflected wave, the image lags slightly behind that of the direct wave, appearing as a "ghost." In 1994 the FCC approved TV station use of *ghost-cancellation reference* (GCR) signals aimed at eliminating ghosts. GCR allows television receivers to identify and delete the second signal.

station antennas provide *beam tilt.* Beam tilt concentrates the radiated energy downward so little of it skims over the line-of-sight horizon to be lost uselessly in space. Objects in the propagation path block UHF waves more easily than they block VHF waves. UHF waves also attenuate more rapidly than VHF waves. The FCC therefore allows analog UHF stations to use very high power (up to 5 million watts) to compensate to some extent for the inherent coverage limitations of UHF waves. Digital TV stations will operate with powers ranging from 50,000 to one million watts. In setting television-power limits, the FCC uses a formula that takes antenna height into consideration.

A television station's coverage distance and the shape of its coverage area thus depend on several factors: transmitting and receiving antenna height and efficiency, obstructive terrain features, transmitter frequency, and effective radiated power. Exhibit 4.m shows how obstructions can affect propagation.

Conversion to digital TV (DTV) transmission provides new propagation challenges. The picture quality of an analog signal degrades gradually as the signal loses strength, but digital signals exhibit a *cliff effect.* A digital signal provides perfect pictures up to the point the signal becomes too weak for the system's error-checking algorithms to compensate. At that point the picture is lost entirely, as if the system had fallen off a cliff. Conversion from analog to DTV requires that many VHF stations move to UHF channels, frequencies traditionally prone to propagation problems. The migration to UHF channels combined with DTV's cliff effect has caused some to fear that stations may lose coverage with the conversion to digital broadcasting. The FCC acknowledges some stations will lose coverage but predicts that more than 93 percent of the new DTV stations will reach at least 95 percent of their old analog service area once conversion is complete. Some broadcasters say this prediction is unrealistically optimistic.

4.11 TV Reception

Antennas and Contour Grades

Simple indoor antennas often suffice if the viewer is within the station's Grade A contour. Viewers located in the station's Grade B contour probably need an outdoor antenna for acceptable reception. Antennas operate at maximum efficiency on only one frequency, so antennas used for receiving many channels require a compromise. Even so, the short wavelengths used for TV allow receiving antennas to be highly efficient and directional. Like outdoor FM radio antennas, TV antennas must point toward the station, and, if stations are at widely different points of the compass, the antenna must be rotatable.

Your TV Set

Like transmitters, receivers process the video and audio parts of the signal separately. In conventional receivers, the video information goes to a *kinescope,* a type of *cathode-ray tube* (CRT) (see Exhibit 4.n). This is the tube the viewer looks at as the receiver "screen."

On the inside face of the kinescope, *pixels* consisting of phosphorescent particles glow when bombarded with electrons. Within the neck of the kinescope resides an electron gun. It shoots a beam of electrons toward the face of the tube. Guided by external deflection coils, the electron beam "paints" an image on the screen (actually the inner face of the tube).

The beam stimulates pixels that glow with varying intensities, depending on the beam's strength. It lays down an image, line by line, field by field, and frame by frame. Synchronizing signals originated by the sync generator activate deflection coils, keeping the receiver scanning in proper sequence.

Color Reception

Phosphors that glow in the red-blue-green primaries coat the inner face of the color CRT, arranged either as narrow parallel stripes or as triads of dots. Receiver circuits decode the television signal into components representing the energy levels of the three primaries in each pixel. Three streams of electrons, one for each primary, strike the tube face, using one or more electron guns.

Only primary colors appear on the kinescope. The eye perceives varying hues in accordance with varying intensities of the primaries as delivered by the electron beams. One can confirm that only the primaries appear by looking at part of a color picture on the face of a kinescope with a magnifying glass. Only dots or stripes in red, blue, and green will be visible.

4.12 Solid-State Receivers

Most television receivers still rely on the cathode-ray tube (the CRT or kinescope tube) for picture

Exhibit 4.n

Color Kinescope Tube

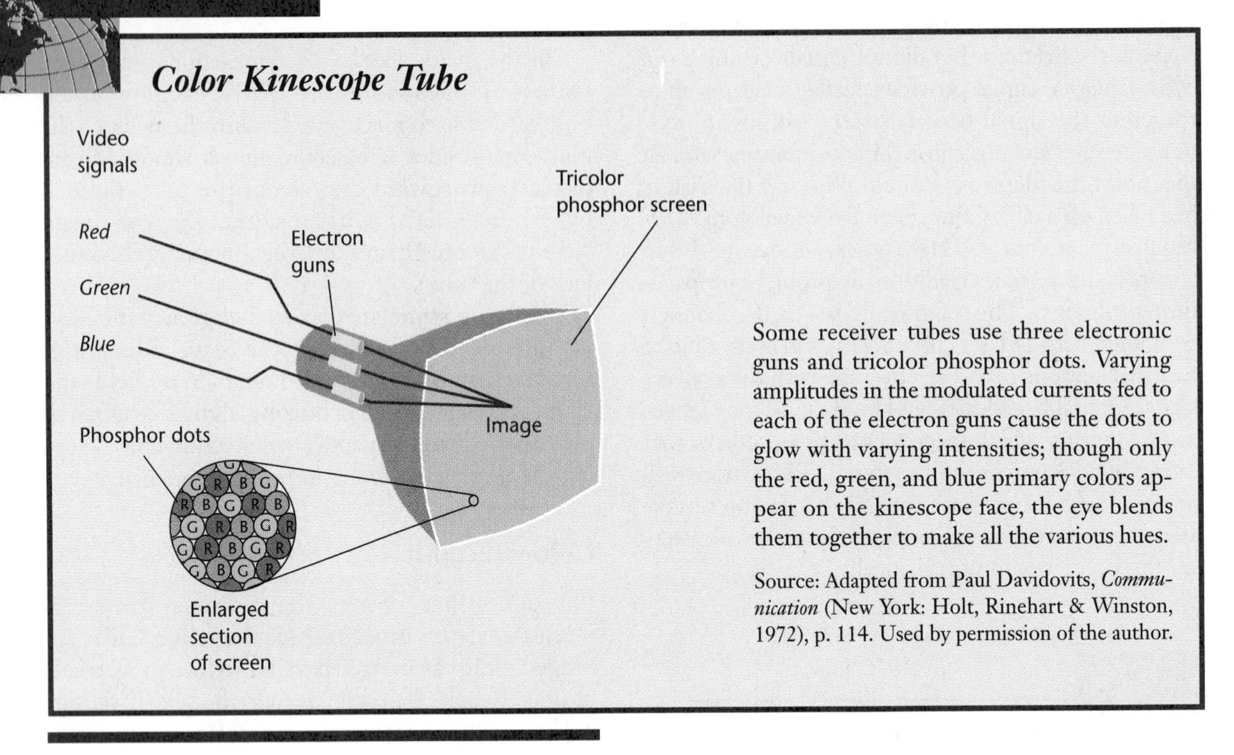

Some receiver tubes use three electronic guns and tricolor phosphor dots. Varying amplitudes in the modulated currents fed to each of the electron guns cause the dots to glow with varying intensities; though only the red, green, and blue primary colors appear on the kinescope face, the eye blends them together to make all the various hues.

Source: Adapted from Paul Davidovits, *Communication* (New York: Holt, Rinehart & Winston, 1972), p. 114. Used by permission of the author.

display. However, researchers are exploring several technologies for producing flat screens that could be hung on a wall.

Flat screens can already be found in tiny, portable television receivers and in laptop computers. One type uses *liquid crystal display* (LCD) technology. It employs a liquid crystal fluid, containing thousands of pixels, sandwiched between two plates of glass. When voltage is applied to the pixels, they twist on or off, blocking a back light that passes through a polarized filter. The quantity of passed light varies, depending on how much the pixels have twisted or untwisted.

LCDs are of two types. *Passive-matrix* LCDs continuously supply low voltage to all pixels, increasing or decreasing the voltage when an element needs to be turned on or off. This process results in flickering, ghosting, and poor contrast, limiting the use of these LCDs largely to those devices, such as digital watches and some computer screens, that display static images.

For moving video images, manufacturers use *active-matrix* LCDs, which turn each picture element on and off independently (see Exhibit 4.o). The sheets of glass between which the liquid crystal is trapped carry thin films of connected transistors etched onto their inner surfaces. Each tiny transistor is, in effect, a switch that can be turned rapidly on or off to polarize the liquid crystal at that point, and hence let light through. Three transistors are needed to create a single pixel capable of displaying a range of colors. Unlike passive-matrix LCDs, active-matrix LCDs are colorful, bright, flicker-free, and provide contrast comparable to that of a CRT.

Scientists continue to experiment in their efforts to improve flat screens. One system, for example,

Exhibit 4.o

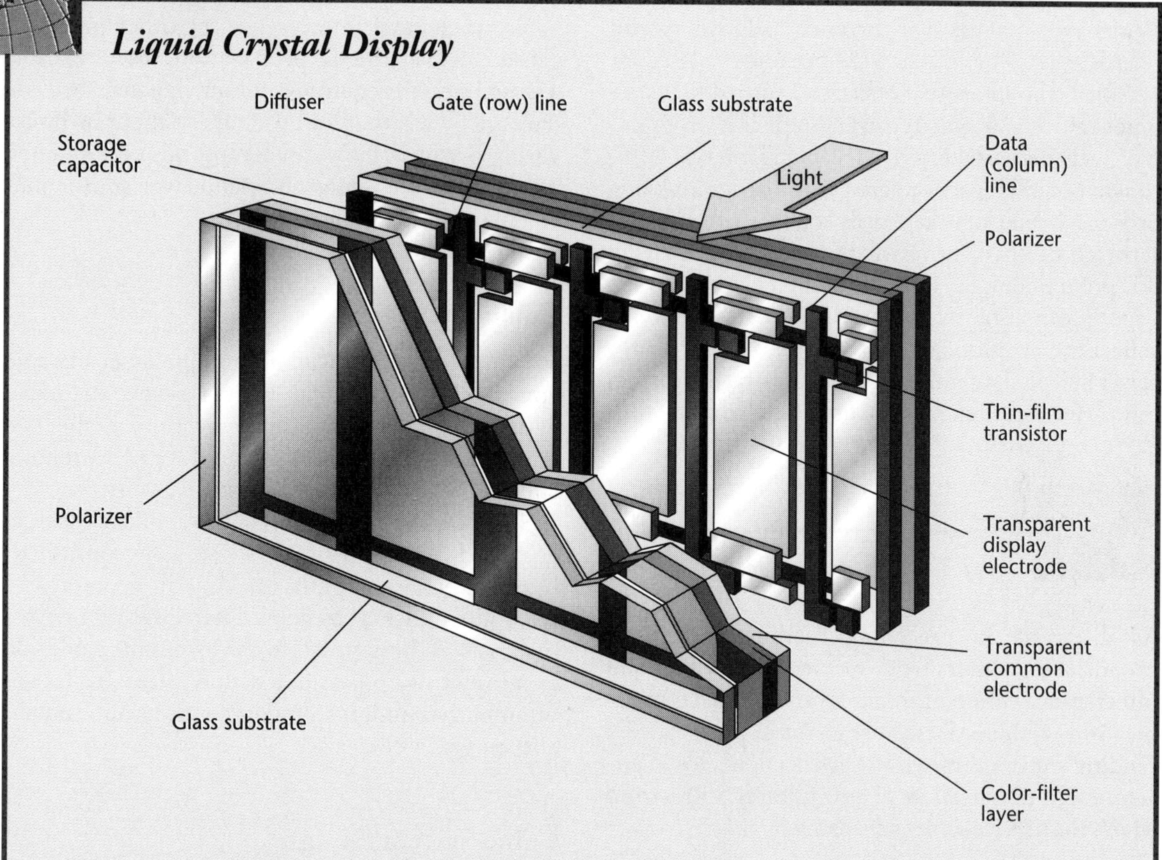

Liquid crystal display screens vary in their construction and in the materials they use, but they all sandwich some form of light-emitting elements between two plates of glass.

Source: Adapted from "Flat-Panel Displays" by Steven W. Depp and Webster E. Howard, *Scientific American*, March 1993, p. 91. Used by permission of Michael Goodman.

has replaced the LCD rows and columns of pixels with a grid of tiny channels filled with gas—similar to miniature fluorescent tubes. Another sandwiches a thin film of luminescent phosphors between sheets of glass that have rows and columns of electrodes printed on them.

Converging Technologies

Just as flat-screen displays first used in laptop computers migrated to TV sets, other computer-oriented technology is merging with TV receiver technology. Digital-analog incompatibility initially delayed progress, but $300 digital set-top boxes appeared in the mid 1990s, enabling viewers to use their TV sets to surf the Internet and view World Wide Web pages. Concurrently, add-on circuit boards selling for around $150 became available for personal computers that effectively turned the computer into a TV/radio receiver. Some TV programmers began using the vertical blanking interval (VBI) to *intercast* Web pages related to the program being broadcast. The add-on boards enable a computer to separate the two forms of information, displaying TV programming in one part of the PC screen, Web page information in another. These developments reflect the evolutionary convergence of the TV receiver and computer monitor into a single device that can perform the functions of both (see Exhibit 4.p).

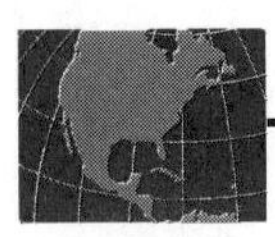

4.13 Cable Systems

As discussed in preceding sections, television broadcasting suffers from spectrum crowding and interference inherent in any over-the-air communication. Cable systems avoid these problems by sending signals through the artificial, enclosed environment of *coaxial cable* (see Exhibit 5.a)—from which the service derives its name.

Cable Advantages

Properly installed and maintained, cable is a closed system preventing cable signals from radiating beyond the cable and external signals from over-the-air services getting in. This signal isolation means cable can use many frequencies denied to broadcasting and other over-the-air services that would otherwise cause interference (see Exhibit 4.q). This extremely wide range of available frequencies allows cable systems to offer viewers many channels, compared with an over-the-air station's single channel in the same geographical area. Newer fiber-optic cable will increase channel capacity even more, as discussed in Chapter 5.

Signals conveyed by a cable system have the same characteristics as broadcast signals but do not suffer from ghosting, fading, co-channel and adjacent channel interference, or other problems caused by over-the-air propagation. Keeping cable a closed system requires constant vigilance, because damage by weather and rodents or inept in-house wiring by subscribers can compromise the system's integrity. Cable operators spend large sums annually maintaining their systems.

Cable Drawbacks

Yet cable also sacrifices the unique asset of radio communication, the ability to reach audiences without the aid of physical connections. Cable systems are also expensive to build—and to rebuild later in order to increase channel capacity.

Signals traveling through "wave guides" such as coaxial cable—other than those using optical fibers—attenuate rapidly, making it necessary to reamplify signals at frequent intervals, thus increasing costs. Where possible, firms mount cables on existing utility poles; but within cities, cables are often buried underground in conduits and tunnels at even greater expense.

Cable System Design

At the system's *headend*, cable assembles programs from various sources and delivers them via coaxial cable to subscriber homes. Besides reception facilities, a headend contains equipment for reprocessing incoming signals, equalizing and feeding them to a modulator for transmission over the system's delivery network, and assigning each program source to a specific cable channel.

Most cable systems have a *tree-and-branch pattern*. As Exhibit 4.r shows, *trunk cables* branch to lighter *feeder cables* that carry the signals to neighborhoods of homes, where still lighter *drop cables* connect the system to individual households.

A headend can feed programs over a radius of about five miles; covering wider areas requires

Exhibit 4.p

Convergence

Advancing technologies increasingly blur the traditional distinctions between the television set and the personal computer.

Source: Brad Perks.

Exhibit 4.q

Cable Spectrum Architecture

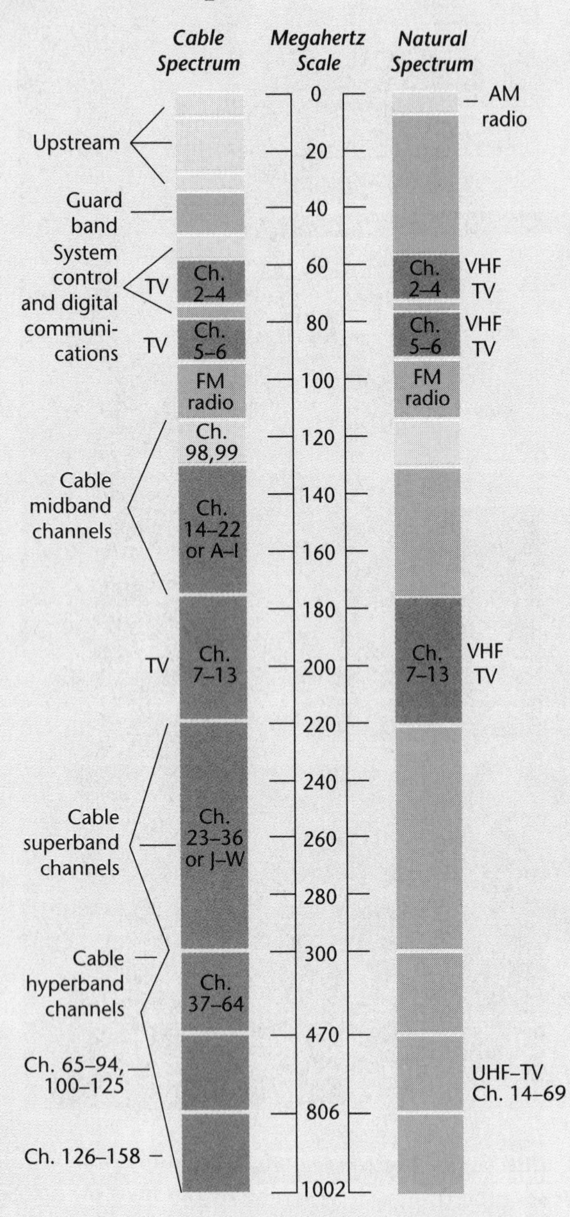

Cable television in effect isolates the spectrum from its natural surroundings by operating within the confines of coaxial and fiber-optic cables. This insulation from nature greatly reduces the possibility of interference, always a serious problem when using the natural spectrum. It also enables cable to use segments of the spectrum that have been allocated to nonbroadcast services. Note that cable retains the VHF channel frequencies and numbering system, but positions what are called *Midband* channels (numbered 14 to 22) and "Super Band" channels (numbered 23 to 36) in frequency bands allocated in the natural spectrum to nonbroadcast services, below the frequencies allocated to over-the-air UHF television. Cable can manipulate signals more easily at these lower frequencies.

Exhibit 4.r

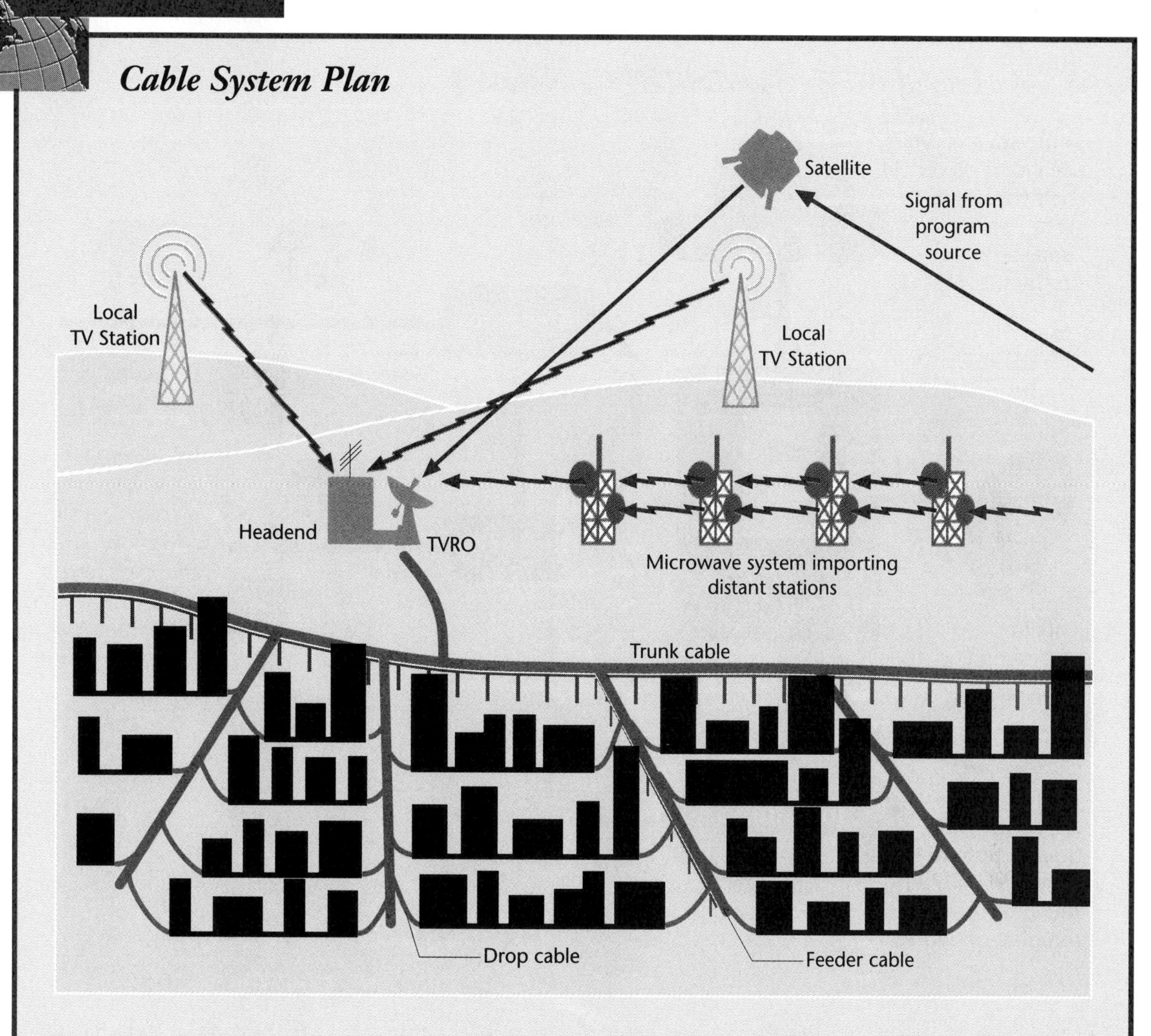

The cable system headend contains its amplifiers and sometimes studios for locally originated programming, although such facilities may be separately located. The headend receives off-the-air TV station signals picked up by special antennas, and possibly also signals from more distant stations fed by microwave relay. The most important are usually several TV receive-only (TVRO) antennas for picking up satellite signals relayed from a variety of program sources. Trunk and feeder distribution cables, shown mounted on poles in the sketch, are often run underground within urban areas.

Exhibit 4.s

A Hybrid Fiber-Coaxial (HFC) Network

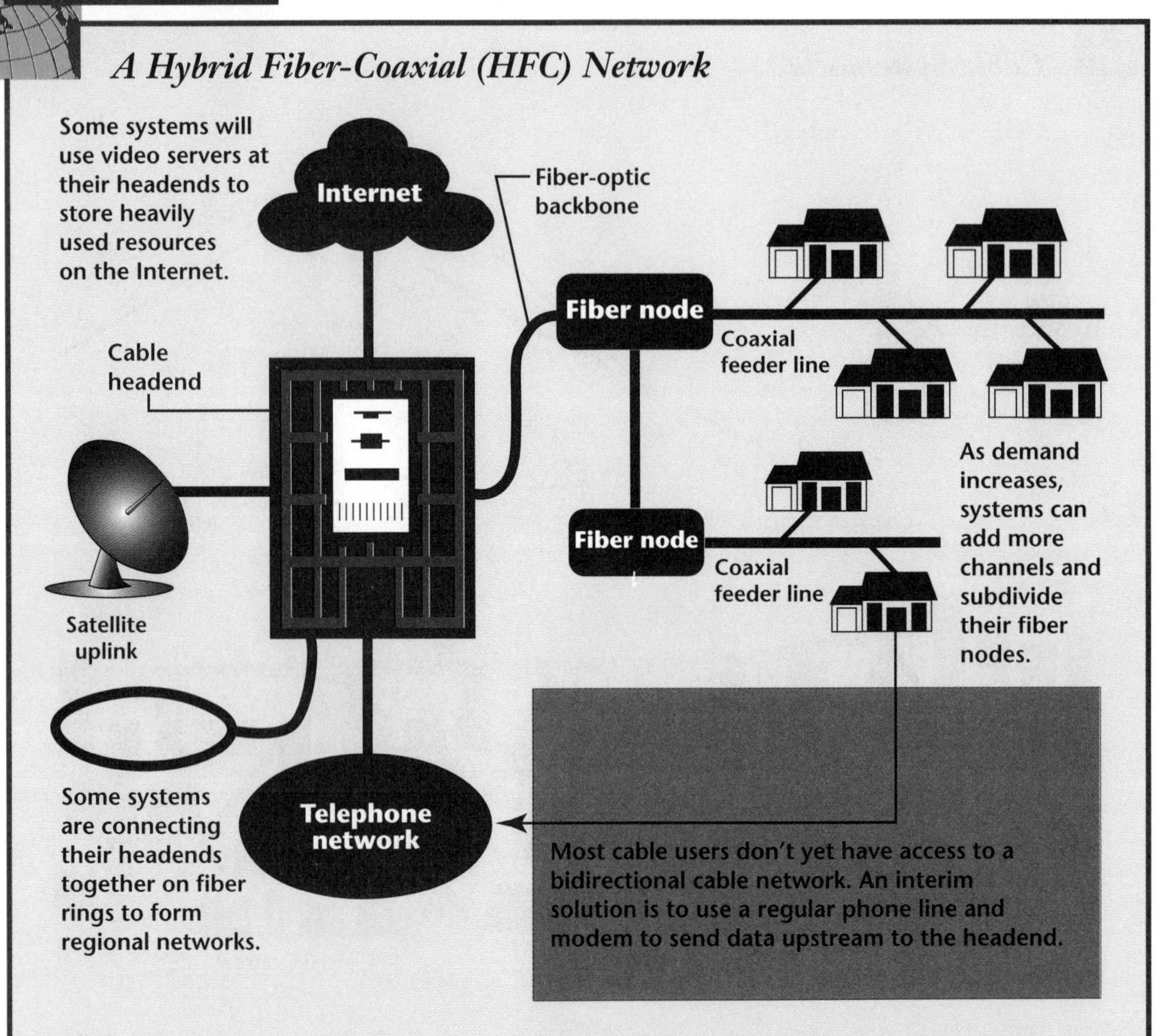
Some systems will use video servers at their headends to store heavily used resources on the Internet.
Internet
Fiber-optic backbone
Fiber node
Coaxial feeder line
Cable headend
As demand increases, systems can add more channels and subdivide their fiber nodes.
Fiber node
Coaxial feeder line
Satellite uplink
Some systems are connecting their headends together on fiber rings to form regional networks.
Telephone network
Most cable users don't yet have access to a bidirectional cable network. An interim solution is to use a regular phone line and modem to send data upstream to the headend.

Cable systems are installing fiber-optic backbones that increase bandwidth, allowing more TV channels and interactive services such as Internet access. However, many systems still lack full high-capacity bidirectional capability. Cable may provide high-speed downstream data delivery, while a less capable signal path handles upstream data moving to the headend.

From: "Break the Bandwidth Barrier," by Tom R. Halfhill, BYTE MAGAZINE, Sept. 1996 © by the McGraw-Hill Companies, Inc. Reprinted by permission.

subsidiary headends that receive the programs via special *supertrunk coaxial* (or fiber-optic) cables or via microwave relays. Newer systems adopt a star pattern with a series of local hubs. This restructuring is especially useful in installing fiber trunk lines.

As suggested in Exhibit 4.r, programs for cable systems come to the headend from five sources: over-the-air reception of nearby television stations, relays of more distant stations, locally produced or procured material, satellite-to-cable networks, and special audio and text services.

Tuning Cable Channels

Cable television uses the same radio-frequency energy as do over-the-air transmitters. Exhibit 4.q shows the broad band of frequencies fed through a coaxial cable. A few older systems still carrying only 12 channels rely on VHF tuners in subscribers' television sets. Systems with more channels supply customers with an adapter unit—or *converter box*—that has its own tuning facility. It feeds into a receiver channel (usually 2, 3, or 4, whichever is *not* an over-the-air channel locally).

However, virtually all television receivers now come "cable ready" or "cable compatible," meaning that they can receive all channels offered on a cable system, including those used by cable but not by over-the-air broadcasters without the need for a converter box.

Systems offering many channels typically have several encrypted (scrambled) channels that require set-top converters even with cable-ready receivers. An *addressable converter box* is necessary to receive pay-per-view (PPV) programming. A cable subscriber orders PPV programs, and the cable headend sends a special signal allowing that specific subscriber's converter box to pass the program.

Interactive Cable Services

Note in Exhibit 4.q that a small portion of the cable spectrum is dedicated to upstream signals flowing from the subscriber's end of the system to the headend. This provides limited two-way capability and holds promise for interactive services delivered by cable. But until upgraded hybrid fiber-coaxial (HFC) systems provide greater upstream bandwidth, or new technologies use existing bandwidth much more efficiently, cable operators may join DBS operators in providing an optional upstream data connection. These interactive systems use high capacity DBS or cable channels to send data downstream, but a separate, relatively narrow bandwidth channel, such as a telephone line, handles upstream data, as seen in Exhibit 4.s. Although not ideal, this asymmetrical arrangement works in many interactive situations where the amount of data moving downstream is greater than that to be moved upstream.

SBS SBS·4
HUGHES
NASA

Chapter 5

Relays, Recording, and the Digital Revolution

CHAPTER CONTENTS

The most dramatic recent changes in electronic media technology have come in how programs are stored and moved from place to place. A useful way to relate the many technologies involved is to examine three basic functions:

- *relay* or distribution of programs from one or more central sending points to many receiving points (stations or systems);
- *recording* or storage of programs for repeated and shared use by networks and stations or systems; and
- *delivery* of programs to the home (by broadcast station, cable system, wireless cable, satellite, or the Internet), as described in Chapter 4.

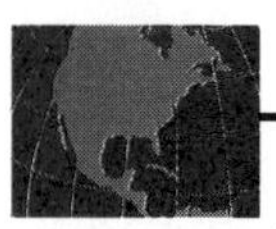

5.1 Terrestrial Relays

Electronic means of relaying media signals include those that use physical interconnections (wire and coaxial/fiber-optic cables) and those that operate over-the-air (microwave relay systems).

Wire Relays

Any point-to-point or point-to-multipoint linkage, whether by wire or radio, can function as a relay. A relay's channel bandwidth determines which types of material the relay can distribute.

Early broadcasters tried using telegraph lines in the 1920s when the telephone company refused to lease lines for networking (see Section 2.5). But telegraph lines lacked the needed bandwidth. Voice required lines designed for that purpose. In turn, lines designed for voice were inadequate for the demands of television's full-motion real-time video signal. Internet users face a similar bandwidth limitation when they use conventional modems and the "twisted pair" of the *plain old telephone system* (POTS) to send and receive video. Television requires a special form of transmission line known as *coaxial cable*, illustrated in Exhibit 5.a. The coaxial cable's outer

Exhibit 5.a

Coaxial Cable

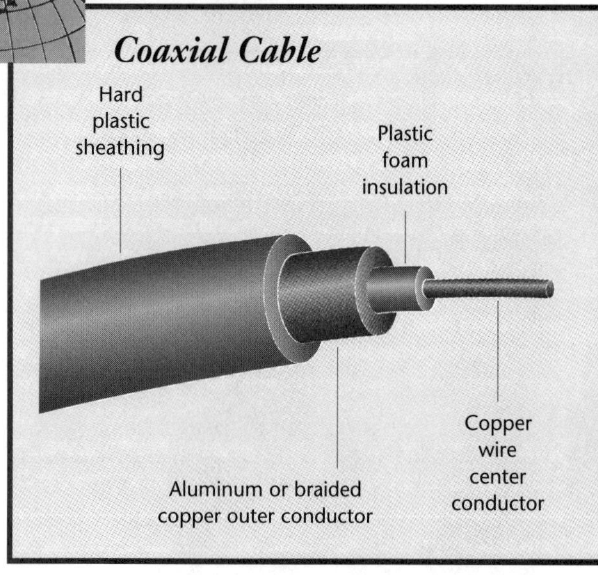

The name *coaxial cable* derives from its two conductors with a common axis: a solid central conductor, surrounded by a tubelike hollow conductor. Radio energy travels within the protected environment of the space between the two conductors. Cable television relies on this type of conductor, as do many terrestrial relay links that convey television signals, telephone calls, data, and other types of information. Fiber-optic cable is increasingly taking over trunk communications for these industries, as discussed in Exhibit 5.b.

Source: Adapted from Walter S. Baer, *Cable Television: A Handbook for Decisionmaking,* R-1133-NSF, Santa Monica, CA: RAND, 1973, p. 4. Used by permission.

conductor provides a barrier that both confines the signal within the cable and protects it against external electromagnetic interference. Attenuation still occurs, but repeater amplifiers inserted at intervals compensate for the loss. Cables' broad bandwidth is useful not only for television, but any time it is desirable to relay lots of data from point to point rapidly.

Fiber-Optic Cable

The high capacity of fiber-optic conductors makes them ideal for handling digitally processed information.

In fiber-optic cable, hair-thin strands of extremely pure glass convey modulated light beams with a bandwidth in the thousands of megahertz (see Exhibit 5.b). A single glass filament has more than 600 times the information-carrying capacity of a coaxial cable.

Ordinary light will not travel efficiently through an optical fiber. Instead, *lasers* (light amplification by stimulated emission of radiation) or *light-emitting diodes* (LEDs) must be used to generate light consisting of a very narrow band of frequencies.

Fiber-optic cables have many advantages as relay links, especially for very heavy traffic. Little loss occurs from attenuation, reducing the number of repeater amplifiers required. Cables are small and lightweight. They neither radiate energy to interfere with other circuits nor receive interference from the outside.

Fiber-optic cables have been permanently installed on heavy-traffic telephone routes and are increasingly used for the main distribution lines of cable systems. Eventually fiber-optic cables will probably replace most conventional copper wire, at least for major trunk relays. Indeed, cable and telephone crews were racing to install fiber in the 1990s, driven by their need for stronger bargaining positions in Washington policy battles about what roles the two industries would play in information delivery.

Microwave Relays

Another relay method uses radio waves in the form of microwave links. Microwaves vary in length from 1 meter to 1 millimeter. Microwave relays usually employ the UHF band (only a portion of which is allocated to broadcast television channels).

Terrestrial microwave relays use UHF waves that attenuate rapidly in the atmosphere and can only travel to the horizon. Higher transmitter powers and antennas that concentrate the energy into a narrow beam (see Exhibit 5.c) compensate to a considerable degree for losses from atmospheric attenuation. Increasing the height of the transmitting and receiving antennas extends the line-of-sight path over which UHF waves can travel, but there are practical limits on microwave tower heights.

Because of atmospheric attenuation and height limitations, relay towers are typically spaced about 30 miles apart. Repeater equipment on one tower receives transmissions from the previous tower, amplifies the signals, and retransmits them to the next tower in the series. It takes more than a hundred towers to span the continental United States from coast to coast.

5.2 Satellite Relays

Like coaxial cable, satellites first served a relay function and only later began to deliver programs directly to consumers. Most satellite traffic still consists of relays, such as network programs relayed to affiliates and on-the-spot news events relayed to studios. But improvement in satellite technology now also accommodates another hybrid—direct-broadcast satellite (DBS) services.

Satellite vs. Microwave Relay Distribution

Microwave relay networks cannot span oceans. Live transoceanic television first became possible only when international communication satellites began to function as relay stations in space. Orbiting more than 22,000 miles above the Earth, a single satellite has line-of-sight access to some 40 percent of the globe's surface, as suggested by Exhibit 5.d.

Exhibit 5.b

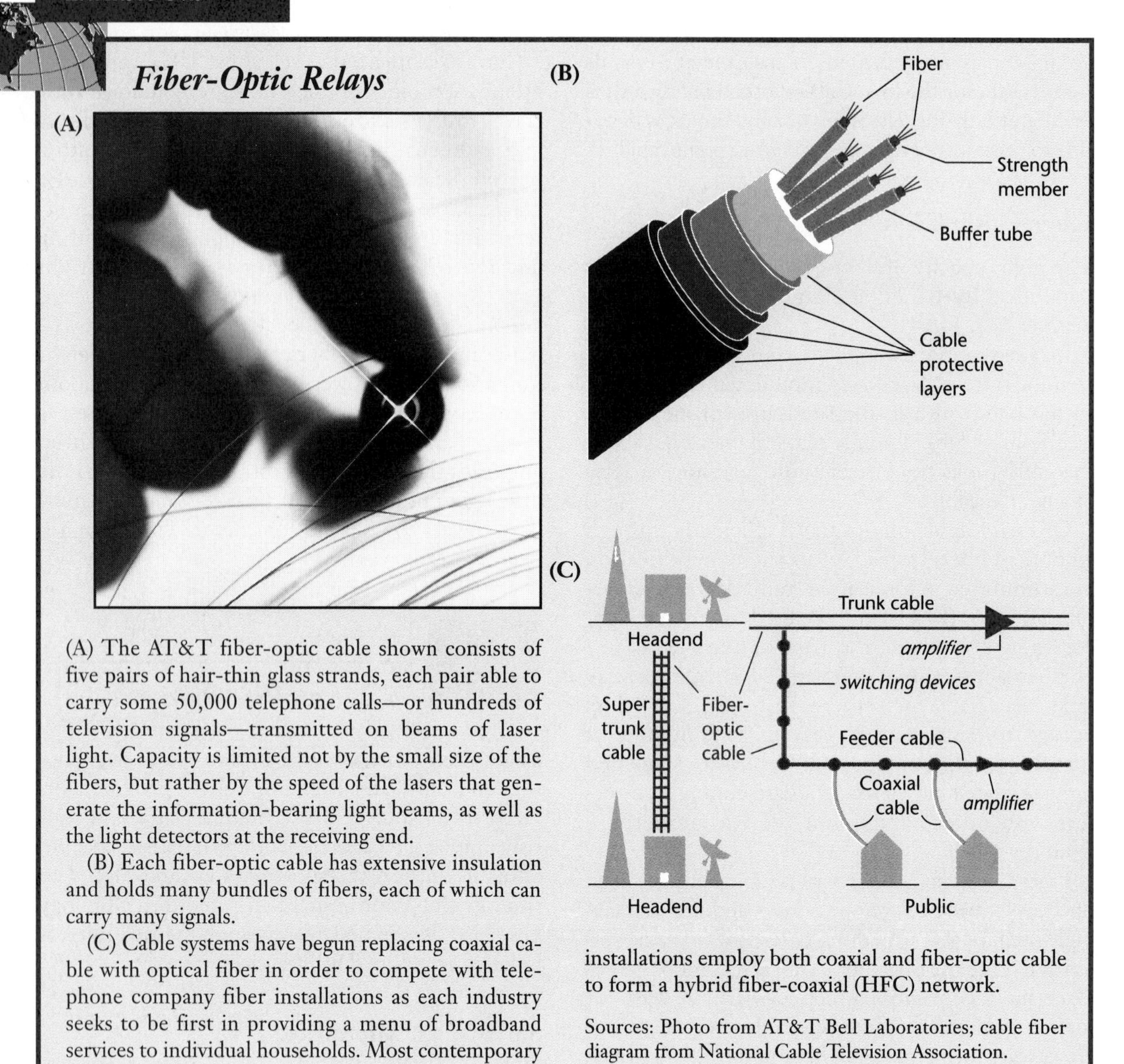

(A) The AT&T fiber-optic cable shown consists of five pairs of hair-thin glass strands, each pair able to carry some 50,000 telephone calls—or hundreds of television signals—transmitted on beams of laser light. Capacity is limited not by the small size of the fibers, but rather by the speed of the lasers that generate the information-bearing light beams, as well as the light detectors at the receiving end.

(B) Each fiber-optic cable has extensive insulation and holds many bundles of fibers, each of which can carry many signals.

(C) Cable systems have begun replacing coaxial cable with optical fiber in order to compete with telephone company fiber installations as each industry seeks to be first in providing a menu of broadband services to individual households. Most contemporary installations employ both coaxial and fiber-optic cable to form a hybrid fiber-coaxial (HFC) network.

Sources: Photo from AT&T Bell Laboratories; cable fiber diagram from National Cable Television Association.

Although often likened to microwave towers thousands of miles in height, communication satellites differ fundamentally from the older relay technology. A microwave repeater links one specific location with only two others: the next sending and receiving points in the relay network. A satellite, however, links a group of relay stations (the satellite's receive/transmit units) to an unlimited number of receiving Earth

Exhibit 5.c

Microwave Relay Antennas

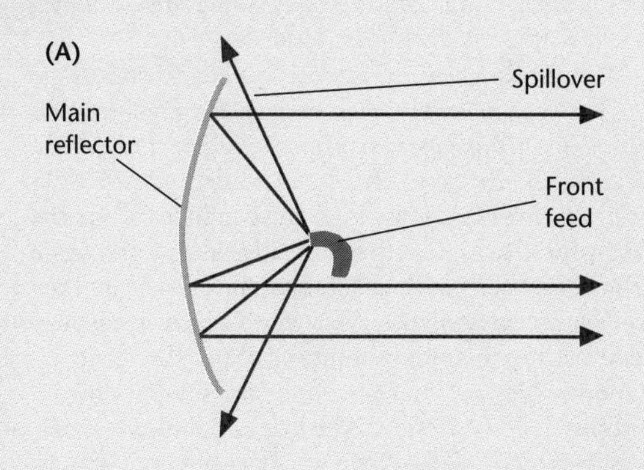

(A) An outlet positioned in front of the dish-shaped parabolic reflector delivers the radio waves, which are then reflected outward in a narrow beam resembling that of a searchlight.

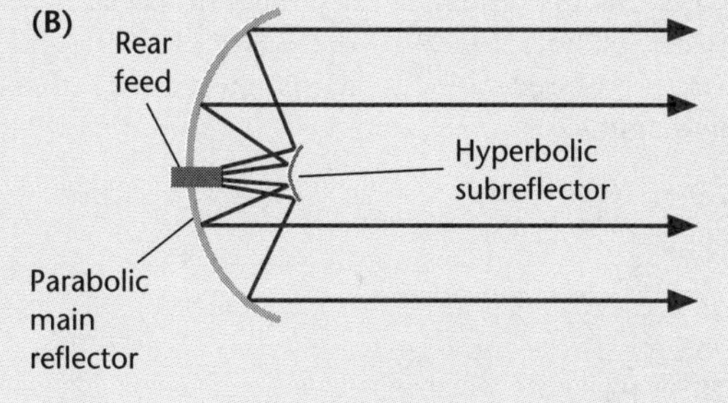

(B) A variation of the parabolic reflector directs the energy from the center of the reflector toward a small subreflector that sends the waves back to the main reflector for transmission.

(C) Dish-shaped reflectors lose efficiency because the feed tube or subreflector cuts off part of the outgoing beam. The horn reflector avoids this problem.

(D) Microwave relay towers can be seen throughout the United States, with all of these antenna types mounted at various heights, providing trunk communication to companion antennas about 30 miles distant.

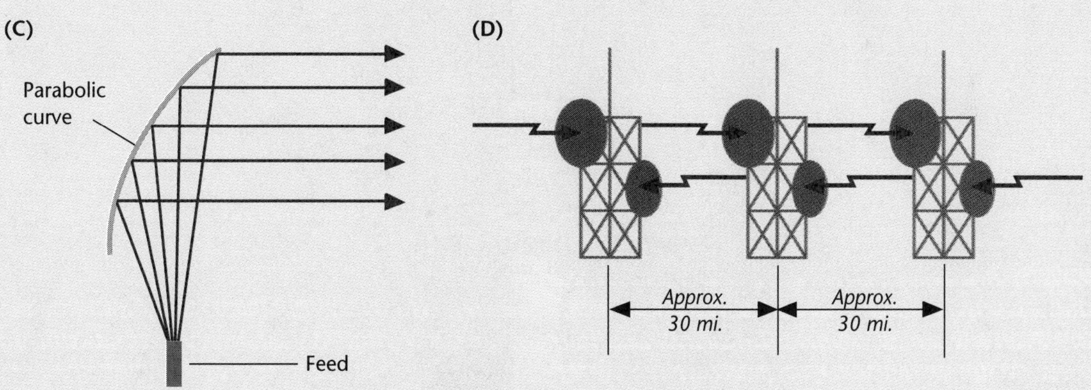

Source: From Graham Langley, *Telecommunications Primer*, 2/e. Copyright © 1986. Used by permission of Pitman Publishing.

Exhibit 5.d

Geostationary Satellite Orbit

(A) Looking down from space over the North Pole, one can see how three satellites orbiting above the equator can "cover" most of the Earth ("most" because their signals fade at polar latitudes). The satellites appear to remain stationary with reference to Earth when orbited at an altitude of 22,300 miles—the geostationary (or geosynchronous) orbit (GSO). The INTELSAT international satellite system operates satellites over the Atlantic, Pacific, and Indian oceans.

(B) American *domestic satellites* (*domsats*) orbit in the GSO above Colombia, each with an assigned position expressed in degrees west from the prime meridian at Greenwich, England. Ku-band, C-band, and DBS satellites are shown at different distances from the Earth for clarity, but they actually occupy the same orbit. Linkages between some satellites indicate they are *hybrids* carrying both Ku- and C-band transponders. Although some satellites occupy the same assigned positions, there is room for several "birds" without fear of collision. The diagram indicates satellites in orbit or planned for launch at the time of publication. Because satellites age or fail unexpectedly in mid-life, as did Telstar 401 on January 11, 1997, and must be replaced, the diagram is illustrative rather than definitive.

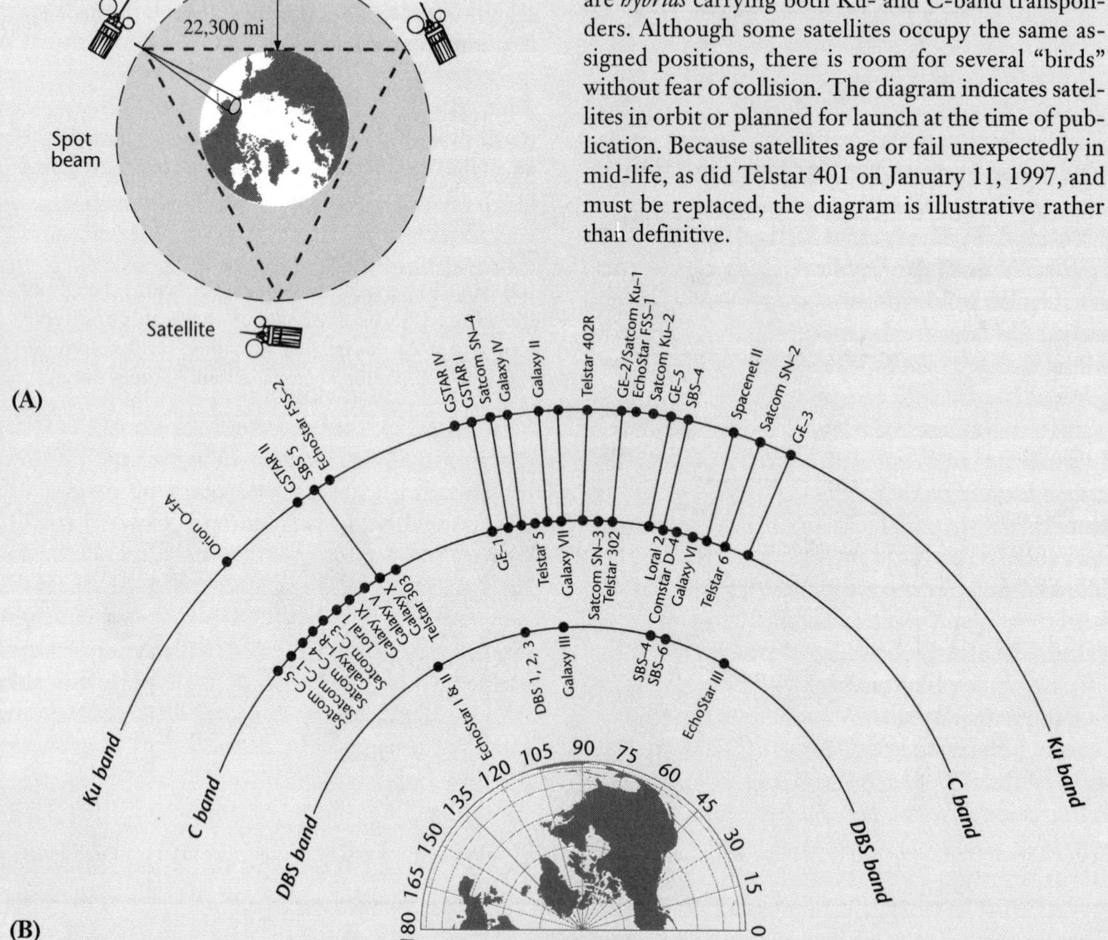

Source: Based on diagram in *Broadcasting & Cable* (13 March 1995: 36–37); 1997 data from *Broadcasting & Cable* (27 January 1997: 66); *Satellite Times* (March/April 1997: 43–53), and FCC International Bureau.

stations. Adding more Earth stations adds nothing to transmission costs, whereas linking up new destinations in microwave relay networks does.

Satellites are also distance insensitive. They can reach Earth stations at any distance within the satellite's *footprint* (coverage area). Distance adds no transmission expense as it does with microwave networks. In addition, microwave signals lose quality as they go through scores of reamplifications in being passed on from one repeater station to the next. But satellite relays amplify a signal only once before sending it down to Earth stations.

Geostationary Orbit

Early satellites moved across the sky like the sun and moon, requiring huge receiving antennas with costly tracking mechanisms to keep them pointing toward the moving signal source. But today's satellites can be positioned so that they remain stationary with respect to their target area on the Earth below. Once a receiving antenna has been adjusted to point in the right direction, it needs no further attention.

Satellites that appear to stay in one location above the Earth operate in *geostationary* (or *geosynchronous*) *orbit*—an orbital position directly above the equator at a height of about 22,300 miles. At that height, objects revolve around the Earth at the same rate that the Earth turns on its axis. Moreover, the centrifugal force tending to push a satellite outward into space cancels the gravitational force tending to pull it back to Earth, keeping it suspended in space for years.

The geostationary orbit consists of an imaginary circle in space. Satellites in that enormous orbit actually move through space at about 7,000 miles per hour. From the perspective of an observer on Earth, however, they seem to stay in one place, keeping in step with the Earth as it rotates. In practice, geosynchronous satellites tend to drift out of position, but ground controllers can activate small on-board jet thrusters to nudge a satellite back to its assigned orbital slot.

Through the International Telecommunication Union (ITU—see Section 14.3), the nations of the world have allotted each country one or more specific slots in the geosynchronous orbit for domestic satellites. Exhibit 5.d shows slots occupied by U.S. domestic satellites. The ITU identifies positions in degrees of longitude, east or west of zero longitude, called the prime meridian. Zero longitude arbitrarily runs through Greenwich, England.

Allotments in the segments of the orbit suitable for "looking down" on areas of high traffic density, such as continental North America, are in high demand. This demand has created a potential slot scarcity. However, video compression (described in Section 5.5) provides a partial solution to this dilemma.

Low Earth Orbit

Scarcity of slots for GSO satellites and enthusiasm for communication services are causing some firms to reconsider low Earth orbit (LEO) satellites. All early satellites were LEOs, "birds" in non-stationary orbit hundreds, rather than thousands, of miles above the earth. Geostationary satellites are superior because they have longer lives in orbit than LEOs, and a single GSO covers much of the Earth's surface. But in the mid-1990s Microsoft chairman Bill Gates and cellular telephone pioneer Craig McCaw formed the Teledesic Corporation to develop a global, broadband, interactive "Internet-in-the-sky" using LEOs. They proposed blanketing the Earth with more than 800 LEO satellites in 21 north-south orbits so that one would be overhead of practically every part of the Earth at all times. From a user's perspective on Earth it would seem that a single GSO satellite was providing the service, but, in reality, signals would be relayed by different satellites constantly moving into and out of the overhead position. Teledesic established 2002 as its target date for service to begin. The complexity and cost (estimates began around $9 billion) of building, launching, and coordinating hundreds of LEO satellites zipping around the Earth caused critics to dismiss the concept as impractical. The Teledesic proposal illustrates, however, another of the many plans industry leaders are devising to cash in on the worldwide demand for more communication services, including interactive audio and video.

Spectrum Allocations

Like Earth-based transmitters, satellite transmitters need internationally allocated transmission channels. Satellites used in broadcasting transmit on microwave frequencies in the 3- to 7-GHz region (C band) and the 11- to 15-GHz region (Ku band). Most communication satellites use the C band. More powerful satellites, intended primarily for direct reception by small home antennas, use the higher-frequency Ku band.

In 1996 the FCC announced its plan for licensing a variety of communication services in the immense Ka band (17–30 GHz). The plan set aside frequencies for those wanting to offer terrestrial as well as both geostationary and LEO satellite-based services. Opening this additional band for commercial development promised to reduce pressure on the increasingly cluttered C and Ku bands.

In areas of heavy terrestrial microwave usage, ground-based services often interfere with Earth stations receiving C-band signals. Ku- and Ka-band signals escape this drawback. But Ku- and Ka-band waves are so short that raindrops in heavy downpours can interfere with their propagation.

Each satellite needs two groups of frequencies, one for *uplinking* (on-board reception) and one for *downlinking* (on-board transmission). These frequency groups must be far enough apart in the spectrum to prevent interference between uplink and downlink signals. Thus satellite frequency allocations come in pairs—4/6 GHz, 12/14 GHz, and so on—with the lower frequencies used for downlinking.

The downlink frequency bands must be large enough to accommodate a number of different channels for simultaneous transmission by the satellite's *transponders*—combination receive/transmit units. Most relay satellites carry 24 transponders. Each can transmit two television channels (more with video compression), for a capacity of 48 television channels per satellite (or many more narrowband channels such as those for telephone or radio transmissions).

Ku-band satellites usually carry fewer transponders, because their higher power means more weight per transponder. Some satellites combine both C- and Ku-band transponders for maximum flexibility.

Transmission/Reception

Satellite transmitting antennas focus their output into beams to create footprints of varying size. Exhibit 5.e shows an example. The narrower the beam, the stronger the signal within the footprint, because directionality causes signal gain.

A satellite downlink beam is strongest at its center, growing progressively weaker at reception points farther out. Earth stations located near the margins of footprints therefore need larger-diameter antennas than those closer to the center. Besides being on the margin of the continental footprint, stations near the poles have an additional problem using geostationary satellites. The receiving antenna of an Earth station must be pointed toward the satellite in orbit over the equator. For stations near the poles, this requires that the antenna be pointed almost parallel to the Earth's surface. This, in turn, increases the likelihood that the Earth station will pick up interfering signals from land-based transmitters. Special spot beams are sometimes used to help boost signals in these areas.

The diameter of Earth antennas varies from more than 100 feet to less than 1 foot. The variously shaped antennas of receive-only satellite Earth stations, such as those shown in Exhibit 5.f, have become familiar sights. Round or square antennas about 18 inches in diameter have become the standard for direct-broadcast reception in homes.

The signals captured by a satellite receiving antenna are extremely weak. They therefore need beefing up by a special high-quality amplifier—a *lownoise amplifier* (LNA). LNAs magnify the incoming signal strength by as much as a million times. Even this enormous amplification is inadequate if the antenna's efficiency is decreased by snow or ice collecting in the dish. Manufacturers design dishes to minimize the effects of ice and snow, but Earth station dishes located in the nation's "snow belt"

Exhibit 5.e

Satellite Footprints

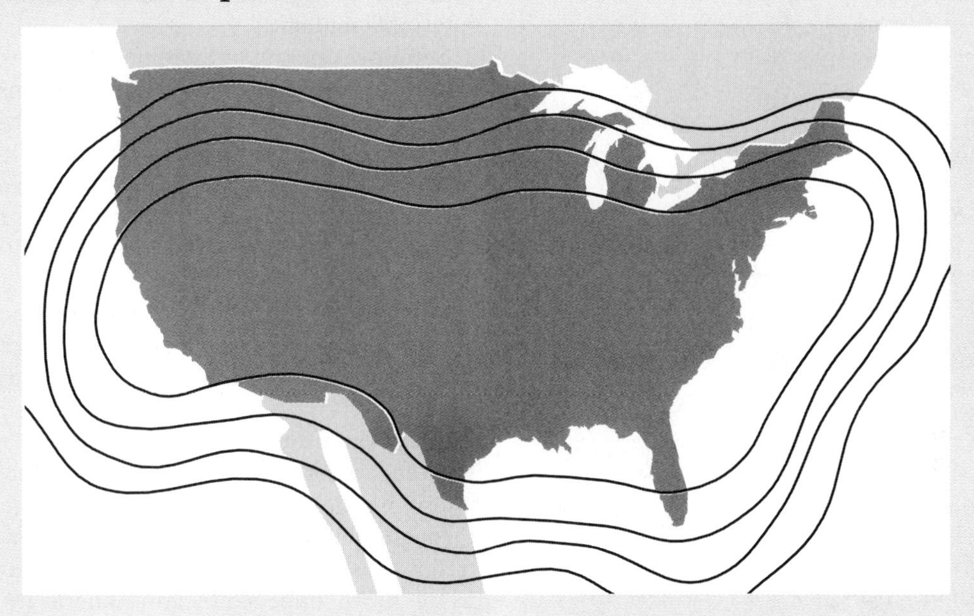

A satellite's effective coverage area, known as its *footprint*, appears on this map in terms of levels of signal strength. The inner contour defines a satellite's *boresight*, its area of maximum signal strength allowing use of smaller receiving antennas. Outlying areas beyond the borders of the country receive only *beamedge* power, beyond which satisfactory reception cannot be expected even with large receiving antennas. This map shows the footprint of the same Satcom Ku-2 satellite diagrammed in Exhibit 5.d, located at 81 degrees west longitude. The satellite actually orbits far to the south of the U.S., a position above the equator near the Galapagos Islands. Not shown are smaller regional beams that can be directed toward Hawaii and Alaska to serve antennas in those areas.

Source: Mark Long, ed., *World Satellite Almanac: The Global Guide to Satellite Transmission and Technology*, 3rd ed. © 1993 Mark Long Enterprises, Inc., Ft. Lauderdale, FL. Used by permission.

sometimes must be swept out. The LNA feeds the satellite signal to a *downconverter*, which translates the high satellite frequencies into the lower frequency range that television receivers use. Some systems use a *lownoise block converter* (LNB) that combines the functions of amplification and frequency conversion in a single device.

Satellite Construction

Communication satellites need five essential groups of hardware components (see Exhibit 5.g):

- *Transponders*, the receive/transmit units that pick up programs, amplify them, and transmit them back to Earth

Exhibit 5.f

TVRO Earth Stations

Each of the relatively expensive *television receive-only* (TVRO) Earth stations (antennas) shown here concentrates the weak satellite signals into a narrow beam directed at a small second reflector mounted on the tripod. This secondary reflector beams the signal into a horn at the center of the TVRO dish, from which it is fed, still as a very weak signal, to a lownoise amplifier (LNA) or lownoise block converter (LNB).

Source: Photo courtesy Christopher H. Sterling.

- *Antennas* for receiving uplink signals and transmitting downlink signals (both program material and telemetering information)
- *Power supplies*, consisting of arrays of solar cells and storage batteries
- *Telemetering devices* for reporting the satellite's vital signs to, and for receiving instructions from, the ground controllers
- *Small jet thrusters* for moving the satellite, orienting it, and holding it in its assigned position, activated on command from ground controllers

Orientation is vital to a satellite because its antennas must always point in the target direction and its arrays of solar collectors, located on the satellite's body or on extended wings, must be positioned to receive direct rays from the sun. These solar collectors provide electricity to operate the satellite. They also charge on-board batteries that take over during periods when the Earth's shadow interrupts sunlight.

Satellites operate at extremely low power relative to terrestrial relays. Power per transponder varies from about 5 watts to 240 watts (the higher power for Ku-band satellites designed for direct-broadcast reception). Most satellite transmitters use no more wattage than ordinary electric light bulbs, though their power is focused by directional antennas.

It seems paradoxical that, despite atmospheric absorption, satellites send signals such great distances with so little power. However, for most of their 22,300-mile journey, satellite signals travel through the near vacuum of space. When at last they encounter the Earth's relatively thin atmospheric envelope, they pass almost straight down through it, experiencing little attenuation. Terrestrial radio signals, in contrast, travel nearly parallel to the Earth, impeded by atmospheric absorption along their entire route.

Satellite Launching

Most commercial U.S. communication satellites have been launched from Cape Canaveral in Florida. Launching involves two phases: A powerful rocket vehicle overcomes the initial drag of gravity and air resistance, carrying the satellite into low orbit. After being released from the rocket, the satellite's own, less powerful, on-board rockets propel it into the high, geostationary orbit. Exhibit 5.h illustrates the sequence of events.

Direct-Broadcast Satellite Service

Cable television systems, broadcast stations, and other satellite-relay users can afford relatively large, expensive TVROs (television receive-only antennas). Large antennas compensate for the low power of C-band satellite transponders. An unexpected bonanza for TVRO antenna manufacturers

Exhibit 5.g

Satellite Components

(A) shows that the largest part of most communication satellites is an array of solar panels (which power on-board batteries)—in this case two huge wings on either side of the electronic core of the satellite.

(B) labels the major parts of a satellite as defined in the text: transponders, antennas, power supplies (batteries and the "wing" solar panels, shown here folded up for launch), and thrusters for minor orbit adjustments.

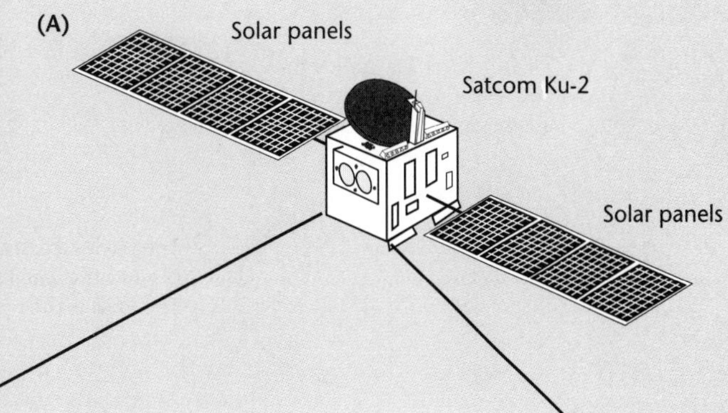

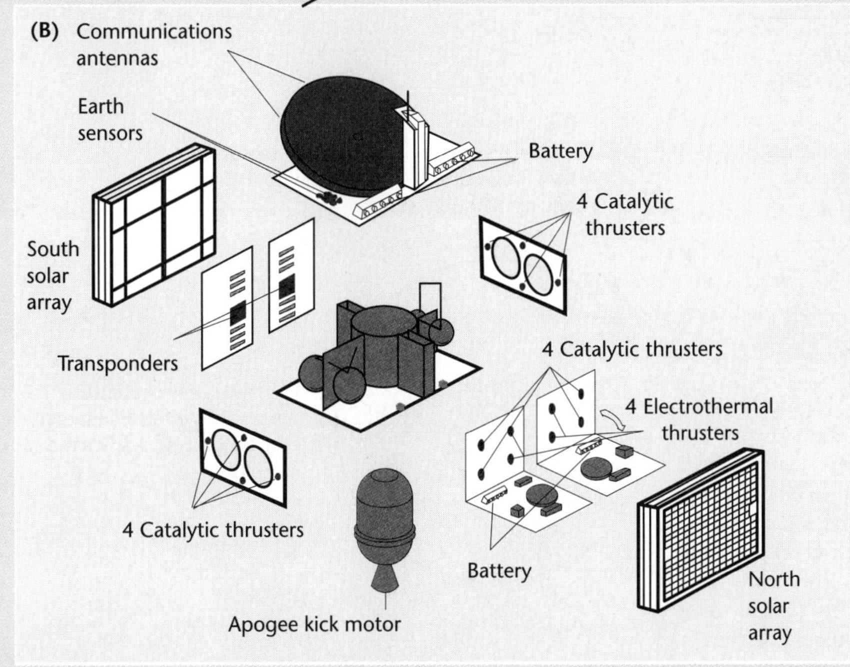

General Electric built the Satcom Ku-2 satellite and the National Aeronautics and Space Administration (NASA) launched it into orbit. With a useful life of about a decade, Satcom Ku-2 used Ku-band frequencies for—among other services—distribution of NBC television signals to the network's more than 200 affiliates, satellite news gathering (SNG) transmissions, and Conus news cooperative delivery transmissions to some 120 affiliated stations in the U.S. and abroad. A number of nonbroadcast services also used its transponders.

Source: Mark Long, ed., *World Satellite Almanac: The Global Guide to Satellite Transmission and Technology*, 3rd ed. © 1993 Mark Long Enterprises, Inc., Ft. Lauderdale, FL. Used by permission.

Exhibit 5.h

Satellite Launches

1 Launch vehicle carries the satellite into low Earth orbit, where it is released.

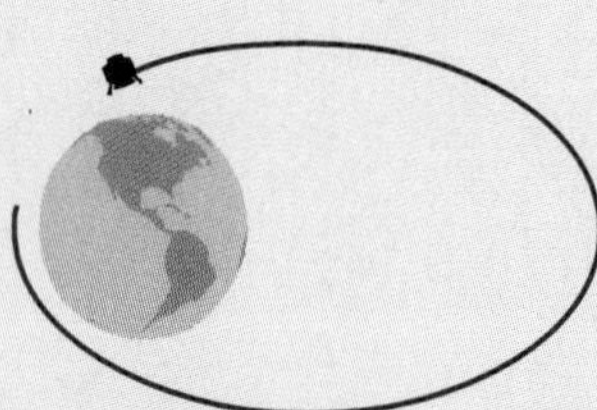

2 The payload assist module (PAM) fires, placing the satellite in a transfer orbit. The PAM is then jettisoned.

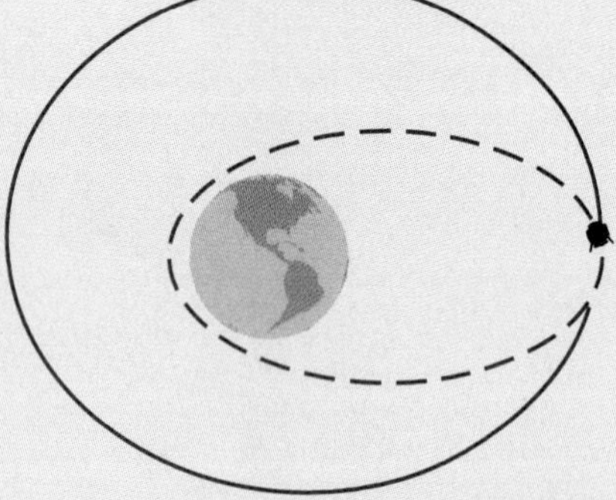

3 Several days later, the apogee kick motor (AKM) fires, putting the satellite into a nearly circular drift orbit.

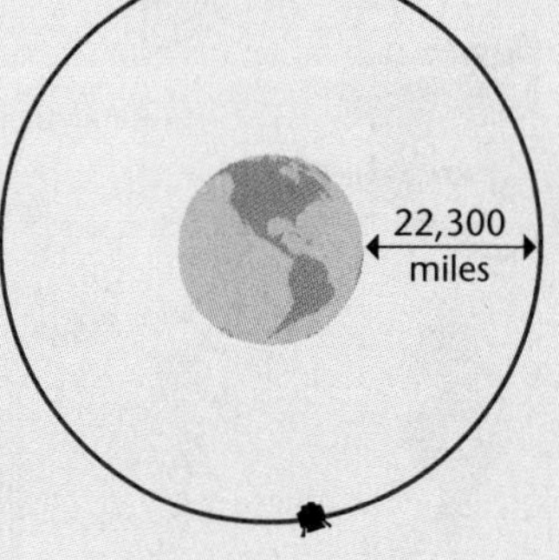

4 A final series of adjustment burns from the AKM settle the satellite into a circular geosynchronous orbit.

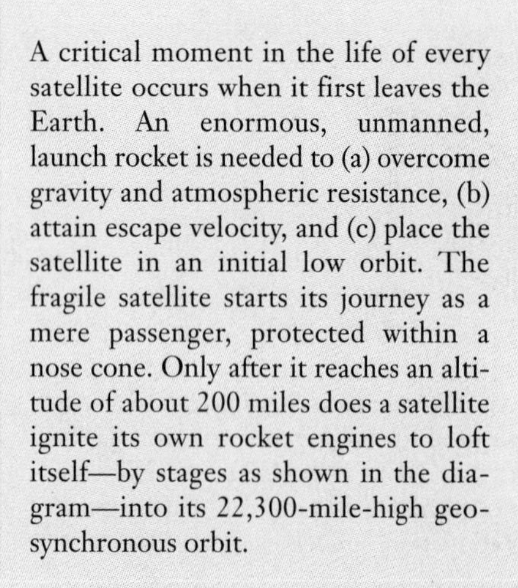

A critical moment in the life of every satellite occurs when it first leaves the Earth. An enormous, unmanned, launch rocket is needed to (a) overcome gravity and atmospheric resistance, (b) attain escape velocity, and (c) place the satellite in an initial low orbit. The fragile satellite starts its journey as a mere passenger, protected within a nose cone. Only after it reaches an altitude of about 200 miles does a satellite ignite its own rocket engines to loft itself—by stages as shown in the diagram—into its 22,300-mile-high geosynchronous orbit.

Source: Mark Long, ed., *World Satellite Almanac: The Global Guide to Satellite Transmission and Technology*, 3rd ed. © 1993 Mark Long Enterprises, Inc., Ft. Lauderdale, FL. Used by permission.

came when people hungry for video programs but beyond the reach of either television stations or cable began buying TVROs. They found they could intercept C-band satellite-relay signals with somewhat smaller, less expensive "backyard dishes."

About 4 million such dishes on the order of 6 to 10 feet in diameter had been installed in the United States by the mid-1990s. They can pick up as many as 150 different programs from domestic and foreign satellites, most of them private relays, such as news feeds, not intended for public consumption. Such home pickups became known as *C-band direct reception* because the general public received the signal directly from the satellite rather than from an intermediary cable system or broadcast station.

Ku-band *direct-broadcast satellite* (DBS) vehicles, on the other hand, are designed specifically for home reception. The uplink leg of a DBS transmission acts as a broadcast relay, sending signals from a distribution point to the satellite—the equivalent of a broadcast station. The downlink leg in turn acts as a broadcast signal, delivering programs directly to consumers without the need for terrestrial transmitters. The shortness of the Ku-band waves and the high power of DBS transponders favor the use of receiving antennas suitable for mounting on private dwellings—in some cases only 18 inches in diameter but in any case not more than 3 feet. The development of DBS systems in the United States is discussed in Section 3.3.

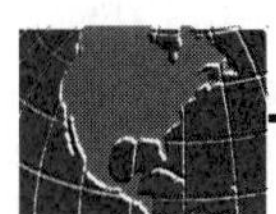

5.3 Analog Sound Recording

We turn now from relays to recording technologies, beginning with the earliest to emerge—the art of sound recording. We deal first with *analog* recording and move next to digital.

Discs

In analog disc recording, a sound source causes a *stylus* to vibrate as it cuts a concentric groove in a revolving master disc. The stylus transforms frequency and amplitude patterns of sound into corresponding minute wiggles in the grooves. Molds derived from pressings of the master disc can be used to mass-produce copies. In playback, the grooves cause vibrations in a pickup-head stylus, which converts (transduces) the movements into an equivalent electrical signal that is amplified and transduced again to cause vibrations in a loudspeaker or earphones—the sounds we hear.

Until the late 1940s no alternative to discs existed, other than experimental recording on thin strands of wire. For a time, discs began to lose favor as users turned more and more to tape, though digital recording methods today have revived use of discs.

Tape

Magnetic-tape recording, unlike analog disc technology, makes recording as well as playback readily available to consumers and combines both functions with portability. The tape itself is plastic, coated with tiny particles of metallic compound. The number and tiny size of the particles available per second of running time, as determined by the tape's width and speed, define storage capacity.

Master sound recordings on half-, one-, or two-inch-wide tape usually call for a tape speed of 15 or even 30 inches per second (ips). In broadcasting, a playback speed of 7½ ips usually suffices. Much lower speeds can be used when quality is less crucial, as in office dictation and station output monitoring. Multitrack master recording and other specialized tasks call for tape stock that is wider than the standard quarter-inch.

Audio signals from a microphone or other source cause electrical variations in a recording head over which tape passes. These variations create patterned arrangements of the metallic particles. On playback the tape passes over another electromagnetic head, where the tape's magnetic patterns create a modulated electric current for delivery to amplifiers and speakers. Running tape over a third electromagnet, an erase head, rearranges particles to neutralize (erase) any stored magnetic patterns so that the same tape can be reused repeatedly.

Exhibit 5.i

Analog Sound Recording Formats

(A)

(A) *Reel-to-reel* tape remains the standard for professional audio production.

(B)

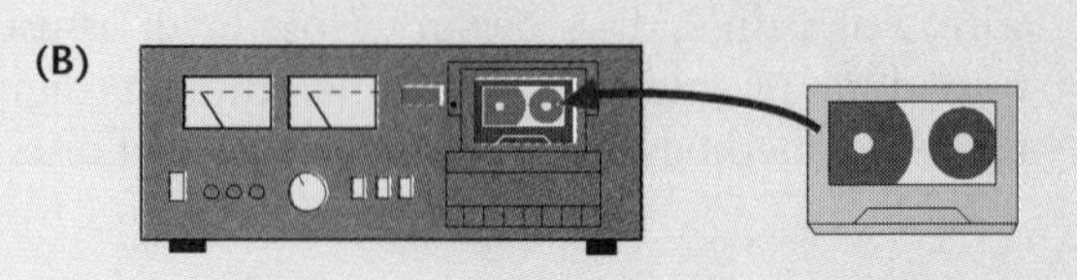

(B) A mini-version of the reel-to-reel, enclosed within a plastic casing, is the *cassette* tape, a worldwide analog recording standard.

(C)

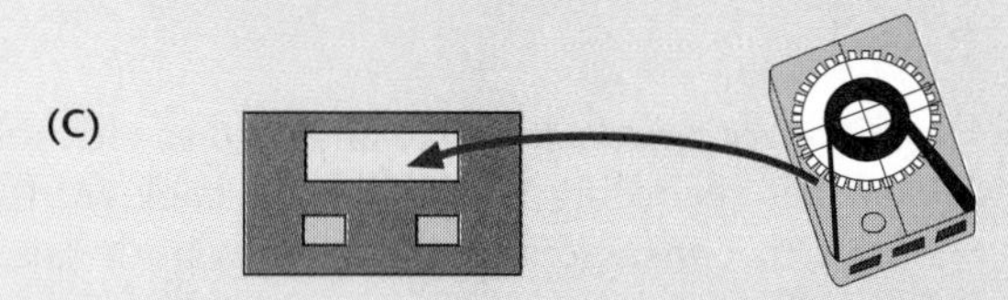

(C) Briefly popular as a consumer product in the 1960s and 1970s, the single-hub tape *cartridge* is still used professionally for short announcements and commercials, but is being replaced by disc-based recording media.

Source: Diagrams based in part on those in Alkin, *Sound Recording and Reproduction* (1992): 141.

Originally all tape recorders—both professional and consumer—had a reel-to-reel configuration, with each reel separate and accessible. Now, however, enclosed cassettes or cartridges protect the tape and are more convenient to use than open reels.

Cartridges

A cartridge, often called a *cart*, has a single hub and contains an endless recordable tape loop that repeats itself (see Exhibit 5.i). Carts recognize the next recorded program or commercial on the tape and come to a stop until the operator manually presses the play button again. This self-cuing feature freed radio disc jockeys to spend more time answering the phone and performing other chores. Carts also encouraged automated operation of radio stations. Many carts were loaded into an automated player with each cart containing a single program item such as a commercial, station ID, or promotional announcement. Musical selections often played off reel-to-reel tape machines built into the automation equipment because they provided audio quality not matched by carts. Inaudible cues on the various tapes instructed the on-board computer when to play a particular cart. Cart-based automation systems were electro-mechanincal monsters with dozens of motors, relays, and other gizmos. It was not unusual to hear two commercials running at the same time when something in the system malfunctioned. In the mid-1990s carts rapidly lost ground to smaller, cheaper, and less error-prone tapeless automation systems.

Cassettes

A cassette incorporates double hubs, one each for feed and take-up reels, in a single small housing (again, see Exhibit 5.i). After playing, the cassette must be rewound. Alternatively, in the case of half-width recordings, some equipment reverses the tape direction, or the cassette may be manually flipped over to play a second "side." The cassette lends itself equally to both analog and digital recording methods.

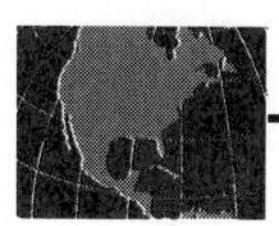

5.4 *Analog Video Recording*

Until 1956 the only way to record TV programs was to film them off the screen of a TV set as they aired using a specially adapted camera. These *kinescope recordings* lost much of television's already skimpy detail. When used for rebroadcast, the filmed programs looked flat and hazy.

Videotape

Early videotape recorders (VTRs) were costly *quadruplex* studio recorders. VTR designers needed to increase greatly the speed at which tape passes over the recording and playback heads to capture the large amount of information contained in pictures-plus-sound. They solved the tape-speed problem by mounting *four* recording heads on a revolving drum (hence the name *quadruplex*). The drum rapidly rotated *transversely* (across the width of tape) while the tape itself simultaneously moved longitudinally, as it does in sound recording (see Exhibit 5.j). The combined movements of heads and tape produced an effective head-to-tape speed of 1,500 inches per second.

Later, simpler and cheaper professional videotape recorders using one-inch and smaller tape stock came on the market. They retain the principle of combining head and tape movements but use fewer heads. Instead of laying down the track transversely, the heads cross the tape at an angle, producing a *slanted* track, as shown in Exhibit 5.j. The heads spin on a disc mounted inside a stationary drum or capstan. The tape wraps around the drum in a spiral (helical) path—hence the name *helical* for slant-track recorders.

For portable equipment, professionals opted first for three-quarter-inch tape (called *U-matic*), but as technology improved, half-inch, 8-mm (about one-third of an inch), and even quarter-inch video tape evolved.

Consumer VCRs

Like many videotape recorders used by professionals, those designed for the home use tape enclosed within cassettes—hence the term *videocassette recorder (VCR)*.

Home VCRs depend on the user's television set for playback, but they contain their own tuners to enable users to record one channel off the air or from cable while watching a different program. VCRs can also play or record the output of home videocameras (*camcorders*) as well as display rented or purchased tapes. Most VCRs come equipped with many sophisticated computer-assisted features, such as slow motion and the ability to be programmed days in advance to record a sequence of shows on different stations. Some provide freeze-frame storage "windows" to monitor as many as nine channels at once on screen, along with a "mosaic" function that changes video images into patterns of colored squares.

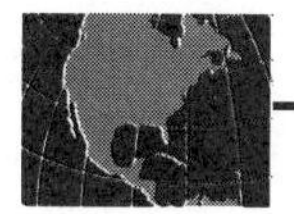

5.5 *Digital Signal Processing*

Satellites rely heavily on digital signal processing. Broadcasting, however, is based on an older processing method that creates signals analogous (similar) to original sounds and images. Analog processing imitates nature. For example, the continuously rotating hands of a clock imitate movement of

Exhibit 5.j

Analog Video Recording Formats

(A) *Transverse Quadruplex Format*

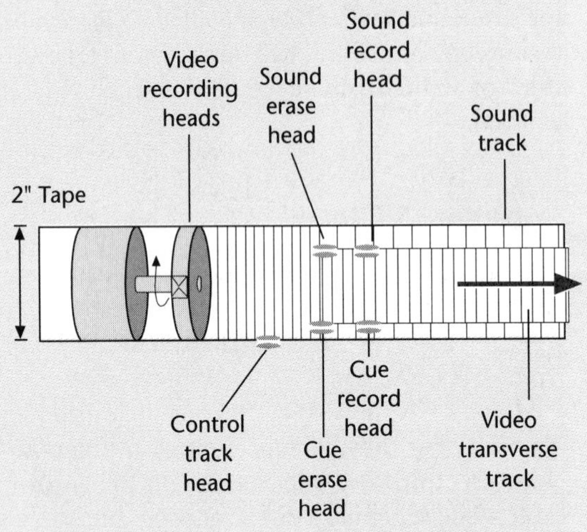

(B) *Helical Format*

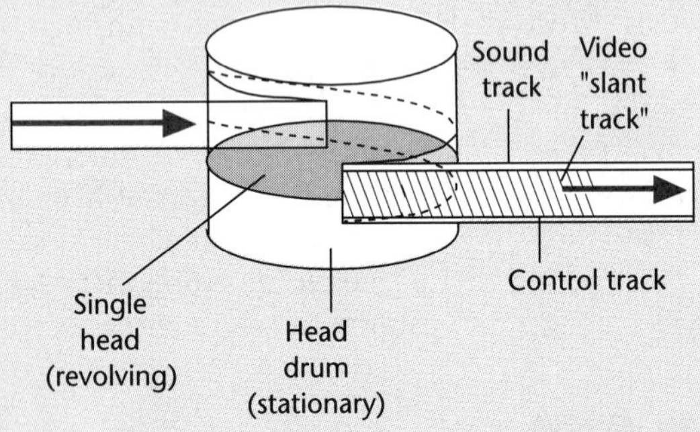

(C) *One-inch Format*

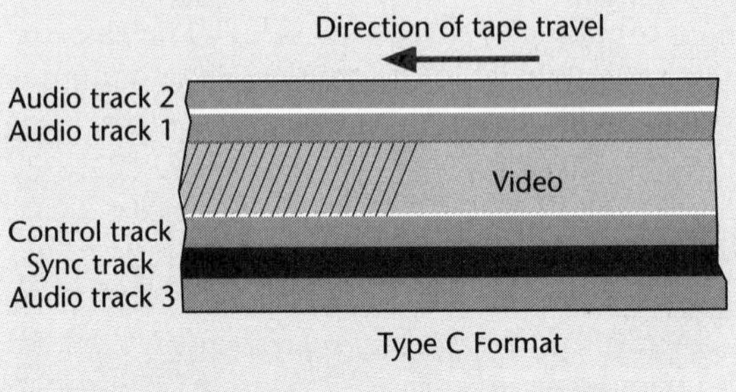

(A) *Transverse quadruplex format:* four video recording heads mounted on a rapidly spinning wheel, shown at the left, lay down transverse tracks across the width of the two-inch tape. Sound is recorded longitudinally along one edge, auxiliary information along the other edge.

(B) *Helical format:* the tape spirals around a large, stationary drum. Within the drum, the videorecording head spins on a revolving disc, making contact with the tape as it slips over the drum's smooth surface. Because of the spiral wrap, the tape moves slightly downward as well as lengthwise, so that the combined movements of tape and recording head produce a slanting track, as shown. Some helical recorders use multiple heads, some use different wrap-around configurations.

(C) *One-inch format:* still narrower VTR tapes are used— ¾-inch, half-inch, 8-mm, and even ¼-inch.

Source: Courtesy Ampex Systems Corporation. Used with permission.

shadows on a sundial. A digital timepiece, however, tells time directly in numbers, jumping from one number to the next. Digital signals bear no resemblance to natural forms because they consist entirely of digits—simple strings of numbers.

Sampling

Digital processing can be likened to cutting up a picture (the analog signal) into thousands of tiny pieces, selecting every other piece, assigning a number to each of those pieces representing its amplitude, transmitting the numbers, then using the numbers to reassemble the picture with every other piece missing. If the pieces are tiny enough, the absence of the missing pieces will not be noticed.

Cutting up an original analog signal and leaving out some pieces is done by high-speed sampling. Each sample consists of a short pulse of energy, proportional in strength to the original signal's amplitude at that point. Each energy pulse is then *quantized*—labeled with a number representing the momentary amplitude of the analog signal.

Encoding

The numbers attached to each sample are encoded by converting them from the decimal number system (0, 1, 2, 3 . . . through 9) into the binary (two-part) number system (0 and 1 only). Computers have made the binary system familiar along with the term *bit*, which stands for *binary digit*.

Numbers expressed in binary digits consist of nothing more than strings composed of only two digits. They are conventionally expressed as "zero" and "one" but can also be regarded as equivalent to "on" and "off."

As an example, the output of a microphone consists of an analog signal, a continuously varying electrical amplitude (that is, voltage). A digital processor samples this continuous amplitude pattern, breaking it down into a series of small, discrete amplitude values. An encoder quantizes each value by assigning it a number in binary form representing its momentary amplitude. The digitized output consists of a string of "power-off" signals (zeros) and "power-on" signals (ones).

Bit Speed

Capacity of a digital channel is measured by *bit speed*—the number of bits per second that one channel can handle. Digitally processed signals inherently need wider channels than do the same signals in analog form. Sampling takes place many thousand times per second. Those thousands are in turn multiplied by the number of binary digits it takes to quantize each one. Exhibit 5.k shows how expressing a three-digit decimal number digitally results in 12 binary digits. An analog telephone channel requires a bandwidth of only about 3 kHz. When converted to digital code, however, a telephone call requires a 32-kHz transmission channel to accommodate its bit speed of 64,000 bits per second.

Exhibit 5.k offers more detail. It suffices to sum up here by saying that digital processing converts a continuous signal into a series of samples that are given numerical values encoded as binary numbers.

Advantages

The extreme simplicity of digitized signals protects them from many extraneous influences that distort analog signals. A digital signal cannot be distorted or misunderstood as long as the elementary difference between "off" and "on" can be discerned. In contrast, each new manipulation of analog signals lowers their quality. Recording, relaying, and other processing of analog information inevitably introduce noise, causing quality loss. However, each new digital copy of a digitally encoded sound or picture is in effect an original.

Because digitally processed information exists as binary digits, the stored signal can be taken apart and reassembled at will in infinitely varied forms. The possibilities for manipulation are endless.

Spectrum Disadvantage

Digitally processed signals need wider channels than do the same signals in analog form because a string of binary digits is needed to identify each tiny sample. The first communication modes to use digital processing, therefore, were those with relatively simple signals that made no great demands on the spectrum—data processing and telephone calls, for example.

The need for high-capacity channels delayed the application of digital methods to broadcast transmission and reception. Dramatic advances in

Exhibit 5.k

More on Digital Signal Processing

Digital signal processing has become so pervasive in contemporary life that it's worth a little effort to learn how it works.

Actually, digital signal processing began with the first electrical communication system, the 19th-century telegraph. Telegraph operators sent messages in Morse code by means of an on/off key that controlled electricity going down the telegraph wire. The code consists simply of varying lengths of "on" and "off," presented to the ear or eye as dots, dashes, and spaces, which in turn represent letters of the alphabet, punctuation marks, and numbers.

Modern digital signal processing also employs simple on/off signals, although the transmitter's carrier wave always remains on. They represent the elements of a *binary code*, a two-digit number system that requires only two code symbols, conventionally written as 0 and 1. All communication content can be reduced to nothing more than strings of zeros and ones.

A system that communicates digitally needs to make only one elementary distinction. "On" and "off" differ so obviously that they leave little chance for ambiguity. That simplicity makes digital signals extremely "rugged"—able to withstand external interference and imperfections in transmission and copying systems.

The familiar ten-digit *decimal* system (0 through 9) is used in everyday life. In that familiar system, the values of digits depend on their *positions* relative to one another, counting from right to left. Each new position increases a digit's value by a multiple of 10.

Thus the number 11 means (counting from right to left) one 1 plus one 10 (1 + 10 = 11). The binary code also relies on position, but each digit's position (again counting from right to left) increases by a multiple of 2. Thus in binary code the decimal number 11 becomes 1011, which means (counting from right to left) one 1 plus one 2 plus no 4 plus one 8 (1 + 2 + 0 + 8 = 11). Here's an example converting the three-digit number 463 to Binary Coded Decimal (BCD) form:

Multipliers:	8 4 2 1	
Binary numbers	0 1 0 0	= 0 + 4 + 0 + 0 = 4
	0 1 1 0	= 0 + 4 + 2 + 0 = 6
	0 0 1 1	= 0 + 0 + 2 + 1 = 3

As the examples indicate, it takes more digits to express a number in the binary system than in the decimal system. Thus, although the simplicity of digital transmissions makes them less subject to error, they need larger channels than analog transmissions.

Conversion of an analog signal to a digitized signal involves *quantizing* the analog signal, that is, turning it into a number sequence. Quantizing consists of rapidly sampling an analog waveform and assigning a binary numerical value to the amplitude of each momentary item in the sample.

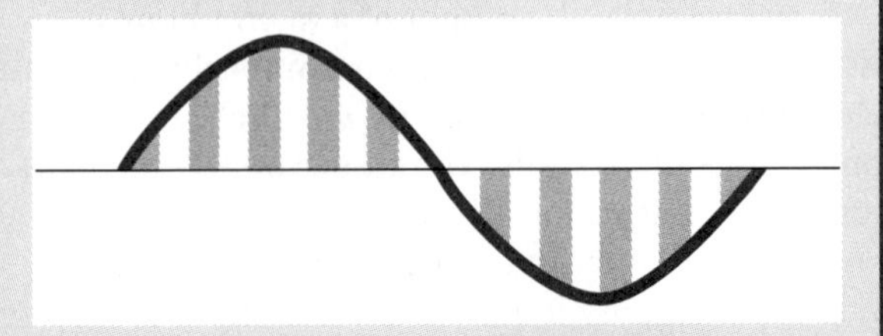

A digitized waveform looks something like the diagram above. The bars indicate the points at which the wave's amplitude is sampled. The higher the sampling rate, the greater the fidelity of the digitized signal. There is an equation for calculating the sampling rate necessary to avoid distortion; it usually calls for sampling thousands of times per second.

Exhibit 5.1

Video Compression

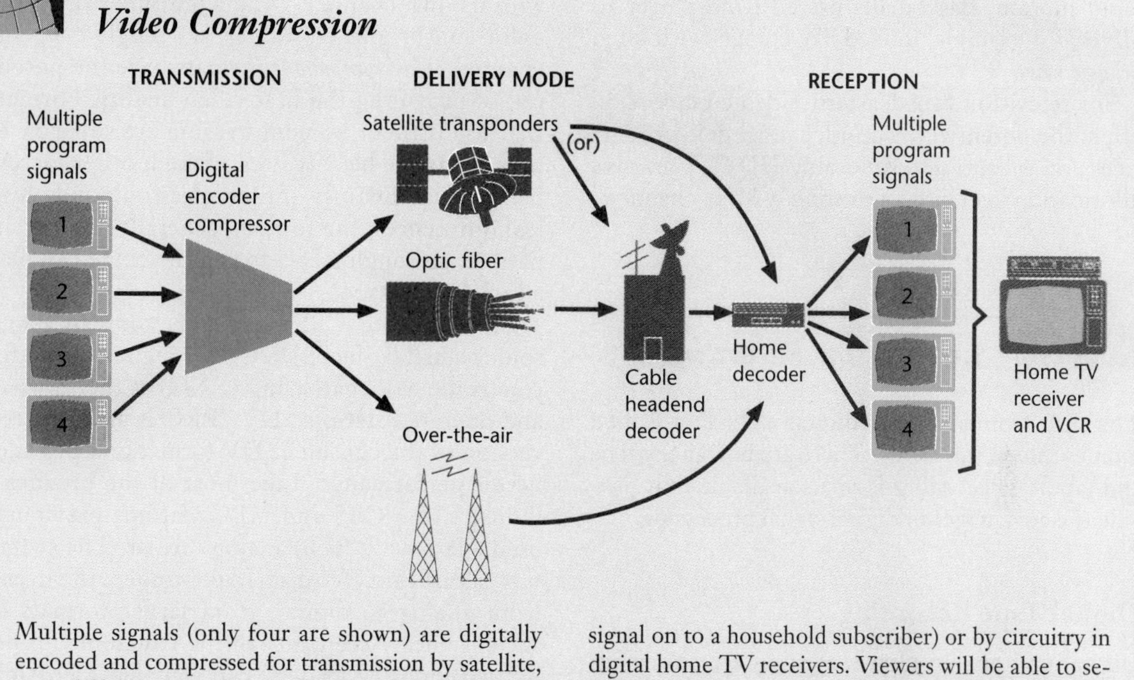

Multiple signals (only four are shown) are digitally encoded and compressed for transmission by satellite, optical fiber, or over-the-air services. The compressed signal is *de*compressed either by an intermediary (such as a cable system headend decoder before passing the signal on to a household subscriber) or by circuitry in digital home TV receivers. Viewers will be able to select from dozens and eventually hundreds more signals than they now receive and may be able to record from compressed signals as well (see Section 5.6).

compression technologies during the 1990s now allow the industry to begin conversion. Indeed, digital compression enables broadcasters to transmit multiple programs in the same size channel required for one analog signal. Broadcasters applaud this benefit resulting from the conversion to digital, but some are concerned that the DTV signals may not duplicate their analog service area, as discussed in Section 4.12.

Compression

Digital signals' hunger for frequency bandwidth drives ongoing development of *signal compression* (Exhibit 5.1). Compression economizes on frequencies by offering tradeoffs—more signals per channel at the cost of marginally reduced picture resolution or color fidelity.

In any video transmission, not all pictures change totally from frame to frame; some picture elements remain the same over a series of frames. One video compression system transmits information about only those elements that change, thus reducing the average amount of new information that has to be processed each second. Other methods achieve compression by marginally reducing picture quality (resolution) or color quality, or by slowing the pace of perceived movement (as in some video conferencing systems).

Some early experiences with compression proved less than "picture perfect," however. The process proved unable to handle some scenes that involved rapid motion. Basketballs passed from player to player, for example, looked like comets with long, orange tails.

For television broadcasters, video compression offers the potential for multichannel delivery services and is central to squeezing HDTV's massive information load into an existing 6-MHz channel.

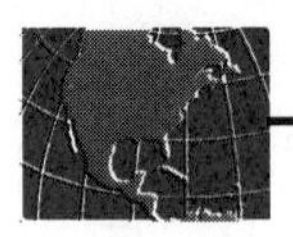

5.6 Digital Recording

The shift from analog to digital recording gained momentum in the 1990s. There are now many tape and tapeless recording formats available that provide the advantages of digital signal processing.

Digital Tape Recording

Digital and analog tape recording are similar in many ways, the primary difference being that digital recorders store audio and video as binary numbers rather than in analog wave form. Tape, because it is a linear technology, often requires shuttling through many yards of tape to play a desired selection. Like analog tape, digital tape requires that the tape make physical contact with the playback head(s) of the recorder, wearing away the coating that stores the information with each use. Eventually the tape wears out and the recorded information is lost.

As noted in Section 3.5, neither digital audio tape (DAT) nor Philips' Digital Compact Cassette (DCC) is popular with consumers. Although many professionals like DAT, most people continue to use analog cassettes or a tapeless digital format for audio.

In the mid-1990s a small, consumer-oriented digital video (DV) format using tape only ¼-inch wide reached the market, and professionals have more than half a dozen digital videotape formats from which to choose. All are cassette-based but otherwise differ significantly. Some handle the color and brightness components separately, each with its own channel. Others multiplex (mix) the color and brightness values in a single *composite* channel. The *component* formats have the potential of providing the best video quality. Formats that use little or no compression are superior to those that are heavily compressed, other factors being equal. Because of their electronic and physical differences, the formats generally are incompatible, although a D-5 format machine can play a D-3 format tape.

In 1997 both CBS and NBC adopted Panasonic's small (¼-inch) DVCPRO digital format for electronic news gathering (ENG) at their owned and operated stations. DVCPRO is an enhanced version of the consumer DV format that provides better performance. Like most of the broadcast industry, the CBS and NBC stations previously used a larger (½-inch) analog format. The switch was significant because it continued the trend from analog to digital, from larger formats to smaller ones. Acceptance of the Panasonic format by two industry leaders also signaled a possible end to the domination of Sony in the ENG equipment market.

Tapeless Digital Recording

Digital recording on disc became much more popular during the 1990s. As noted in Section 3.5, the compact disc (CD) has practically replaced analog LP discs for the distribution of audio. The audio CD and its variants provide random access to anything on the disc in microseconds. They also use a laser beam to read the information on the disc, which means the CD never wears out regardless how often it is played. The technology of the CD is illustrated in Exhibit 5.m. Laser technology may eventually displace most other types of information storage, including the magnetic floppy disks used in home computers.

A computer version of the CD used to distribute software, interactive video games, and other types

Exhibit 5.m

Compact Disc System

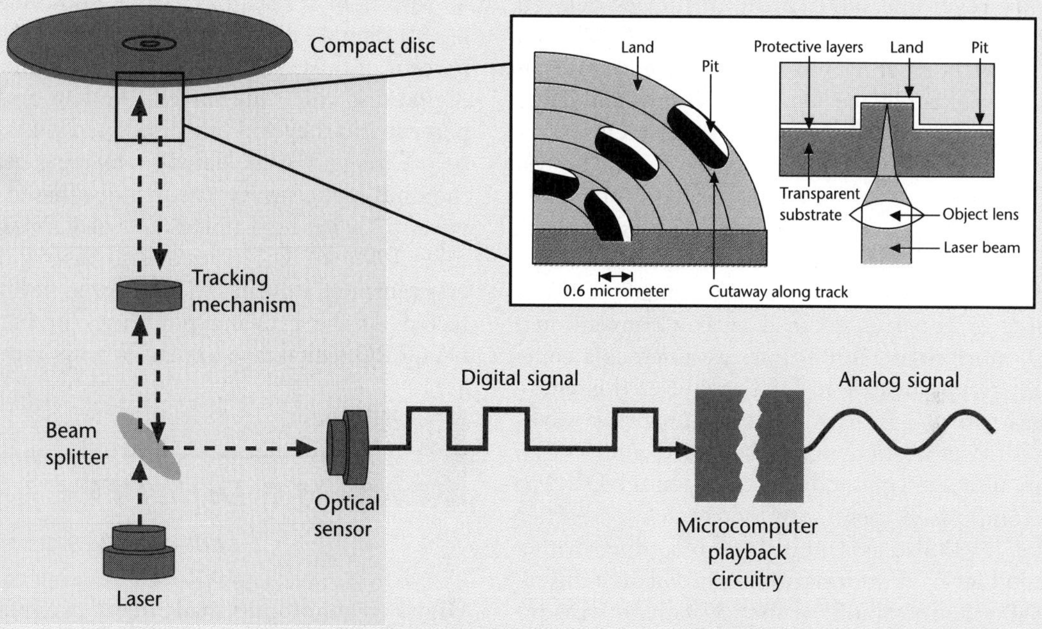

Sound is recorded on an optical disc in digital form. Microscopic pits alternate with a shiny reflective surface to create "on/off" binary digital signals encoding the original sound. A laser beam, guided by a supersensitive tracking mechanism, strikes the pits, which reflect the beam back (as a modulated digital signal) to an optical sensor. The sensor feeds its digital output to a microcomputer that converts the digital information into analog form for delivery to speakers or earphones.

Source: IEEE Spectrum.

of information, is the CD-ROM (compact disc read-only memory). The CD-ROM is similar to the audio CD and theoretically shares the audio CD's approximately 680 megabyte capacity. CD-ROM manufacturers often use only the 550 megabytes nearest the center of the disc to ensure reliability in critical applications. Once considered a luxurious computer peripheral, CD-ROM drives were in more than 40 percent of all home computers by 1996, and today practically every computer sold comes with CD-ROM.

Another CD family member is CD-R (compact disc recordable), which allows users to record on the disc only once but play back any number of times. In late 1996 CD-E (compact disc erasable) was introduced. CD-E allows repeated erasing and playback similar to magnetic tape, computer floppy disks, and hard drives, as well as Sony's MiniDisc, briefly discussed in Section 3.5.

Conflicts arising first over technical standards and, later, objections from the entertainment industry regarding the copying of movies delayed the development of the *digital video disc* (DVD). In 1995 two competing and incompatible versions of DVD were announced—one by Toshiba and Time Warner, the other by Sony and Philips. Observers saw in the conflict a replay of the 1970s VHS vs. Beta battle over videocassette formats. The competitors later compromised on DVD standards. The standards provide for disc capacities ranging from 4.7 to 17 gigabytes. A 17-gigabyte DVD stores 25 times as much data as a conventional CD, more than 11,000 times as much as a computer floppy disc. The DVD achieves this enormous capacity in part by using a shorter wavelength laser beam than that used with audio CDs. This allows the pits and data tracks on the DVD to be compressed even more tightly than on earlier discs. DVD also uses multi-layer recordings with a second layer of information recorded, at a different depth, on top of the first. Finally, DVDs are two-sided devices, with data stored on both sides of the disc. Audio CDs and CD-ROMs store information on only one side. See Exhibit 5.n for a comparison of CD and DVD technology. The least-expensive DVD players sold for about $500 in 1997, but prices are expected to drop below $300 by 1999.

Recordable (DVD-R) and erasable/recordable (DVD-RAM) versions of the super-capacity DVD are expected sometime in 1998. Capacity of the DVD-R and DVD-RAM will be around 8 gigabytes and 5 gigabytes, respectively, because only single-layer recording will be possible.

In the future there may be more digital recording devices that use neither tape nor discs. Early in 1996 NEC of Japan announced development of a prototype video player with no moving parts. The tiny Silicon View player uses a solid-state flash-memory card a little larger than a pack of paper matches. NEC says the device will play about an hour of stored programming when it reaches consumers around the year 2000.

Digital tape recorders will be around well into the next century, but the trend toward tapeless technology is clear. A mid-decade survey suggested almost 80 percent of TV stations will have non-linear (disc-based) digital editing systems by 1998. Almost 60 percent of those surveyed said they planned to buy digital disc video file servers. And an amazing 42 percent said they will buy disc-based video cameras before the end of the decade—amazing because, at the time of the survey, the first disc-based cameras were still being field-tested (*Television Broadcast*, October 1996:86). U.S. sales of disc-based video cameras remained slow in 1997, however, negatively affected by the growing popularity of Panasonic's DVCPRO digital tape format for news gathering.

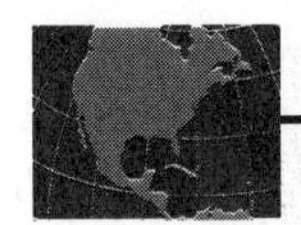

5.7 *Digital TV Transmission*

Digital technologies made great progress in the acquisition, editing, and storage of TV programming during the 1980s and early 1990s. In the distribution area, however, the calendar often seemed stuck at 1941, when American TV broadcast standards were developed. Movement toward more modern standards was slow. The major impetus for change emerged from a desire for TV pictures and sound rivaling that found in movie theaters. The initial goal was a ***high-definition television*** (HDTV) system providing larger screens, vastly improved picture resolution, and multichannel sound. What has emerged is a new digital television (DTV) broadcasting environment in which HDTV as originally conceived is an important, but not exclusive, component.

Analog Beginnings

Engineers in Japan, beginning work in the 1960s, made the first significant progress with HDTV, de-

Exhibit 5.n

Comparison of Conventional Audio CD and DVD

Figure (A) shows the surface structure of a CD with a storage capacity of about 680 megabytes. Notice that the disc is single-sided and each side has only one "layer" of pits representing digital data.

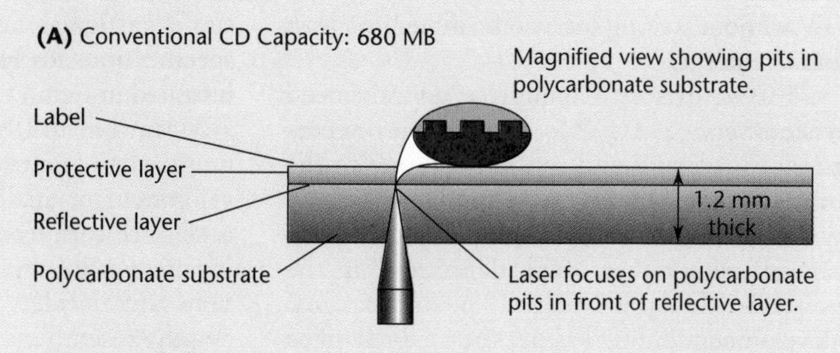

Figure (B) shows a DVD, a double-sided disc with two layers of pits on each side. This architecture allows the DVD to store 25 times as much digital information as a conventional CD.

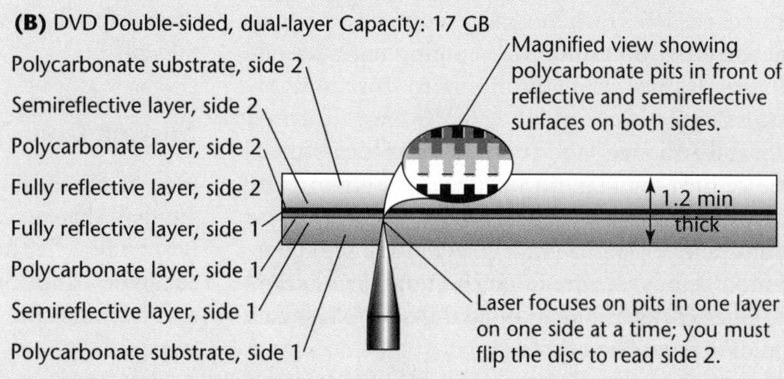

Source: Adapted from "CDs for the Gigabyte Era" by Tom R. Halfhill, BYTE MAGAZINE, Oct. 1996 © by the McGraw-Hill Companies, Inc. Reprinted by permission.

veloping a 1,125-line picture with a 16-to-9 aspect ratio. This analog system, called MUSE, required 36 MHz channels—six times wider than the U.S. 6-MHz standard. Japanese engineers later developed a broadcast MUSE version using a 9-MHz channel.

By late 1991 the Japanese were providing satellite delivery, eight hours a day, of MUSE service to selected viewing sites in Japan. However, initial cost of receivers (about $18,000) drastically limited HDTV penetration. By the mid-1990s Japanese analog HDTV sets were still twice as expensive as ordinary sets, and fewer than 350,000 households had HDTV receivers.

America's Response

The FCC began in 1988 to consider U.S. HDTV standards to ensure U.S. terrestrial broadcasting's ability to keep up with Japanese and European efforts in the race toward improved television. Also, nonbroadcast services (domestic cable, VCR, and DBS), not being subject to the NTSC 6-MHz channel constraint of broadcast television, could forge ahead with their own nonbroadcast version of HDTV without waiting for a compatible broadcast version to emerge.

The FCC Advisory Committee on Advanced Television Service (ACATS), established to compare HDTV systems and recommend one to the Commission, sparked creative thinking among many domestic and foreign companies. The term "advanced" rather than "high-definition" in the committee's title was prophetic. The original goal was development of HDTV, but the focus changed in the early 1990s when several companies began concentrating on rapidly-developing digital technologies. Rather than continuing to work as competitors in isolation, the developers formed a "Grand Alliance" and merged their ideas to accelerate development and approval of a workable digital system. In 1993 the Advisory Committee on Advanced Television Service selected a digital compression standard, agreed on the number of scanning lines (1,080), and adopted a Dolby CD-quality audio system for HDTV.

Although continuing to claim HDTV was the ultimate objective, broadcasters began urging Congress to allow them to use digital transmission technology for other services, such as interactive television, pay services, and high-speed data transmission. This shift in emphasis raised new questions and occasioned fierce debate regarding the transition from an analog to a digital broadcast service. It was not until mid-1996 that the FCC authorized three stations to begin broadcasting digital signals experimentally.

Just as a final transmission standard seemed ready for adoption, objections from leaders in the entertainment, computer, and cable television industries once more delayed the decision. They claimed some parts of the proposed standard interfered seriously with the smooth convergence of television and computer-based technologies. A compromise in late November allowed the FCC to adopt a digital television broadcast standard on Christmas Eve, 1996. Reflecting the compromise, the FCC chose not to impose a single format for all digital broadcasting. Instead the Commission adopted 18 formats that broadcasters may use to suit a particular need. One format embodies the specifications for HDTV agreed to in 1993 and illustrated in Exhibit 5.o.

Meanwhile in 1994 the Japanese government announced an end to its support for and further development of analog HDTV, but then retreated under pressure from Japanese manufacturers. The Japanese Ministry of Posts and Telecommunications later decided to promote a new digital system closely resembling that adopted by the United States, finally bowing to the inevitability of a digital world.

Moving from NTSC to Digital

One of the most difficult decisions facing the FCC amid all this fast-moving technical change was how best to handle the transition from existing NTSC receivers and broadcast and cable production and transmission equipment to the selected digital standard.

Simultaneous with the development of transmission standards for digital broadcasting, the FCC wrestled with the issue of how best to transition stations from their NTSC analog channels to their new digital channels. Knowing the transition would take years, the FCC needed to give stations new digital channels while allowing them to continue analog broadcasting on their old channels.

Early in 1992 the FCC decided on a 15-year transition period following adoption of a digital transmission standard. Broadcasters were to receive a second six-MHz channel in the UHF band for digital broadcasting while continuing conventional analog transmission on their respective existing channels. Broadcasters would have to begin digital (at that time meaning high-definition) operations six years after the FCC's selection of a digital standard. They would have to simulcast half their programming in their second year of digital operation,

Exhibit 5.o

High-Definition Systems

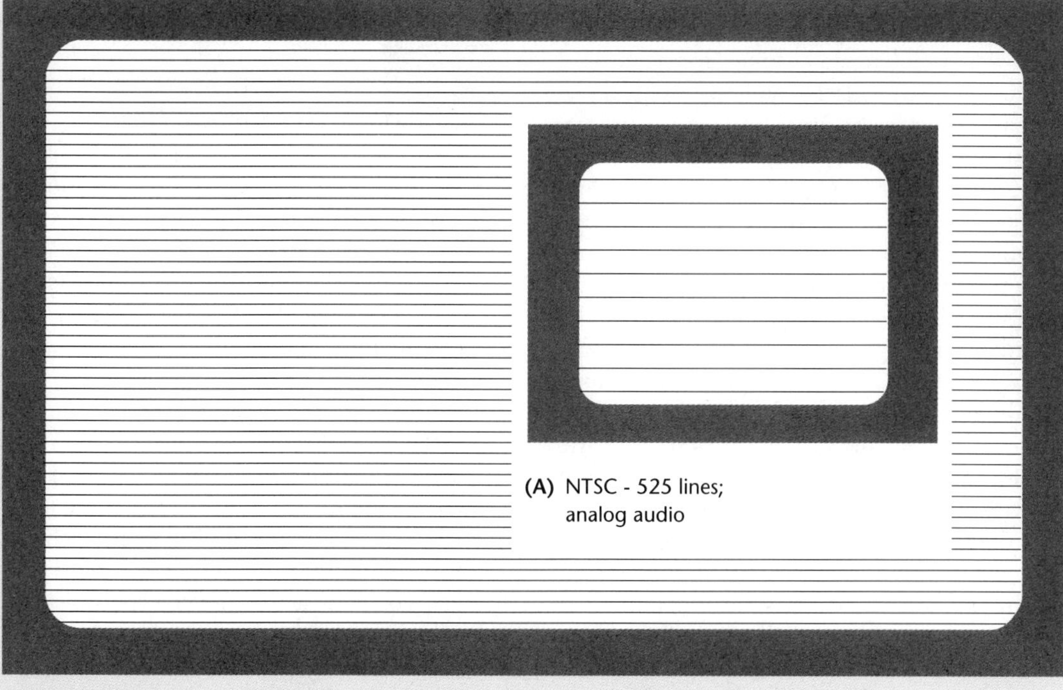

American HDTV—compared to the NTSC standard—will have more than twice as many scanning lines for improved picture definition, a much wider screen size, and both digital audio and video signal generation. The diagrams compare (A) NTSC television with its 4:3 picture aspect ratio as approved in 1941 (and modified with color in 1953) with (B) HDTV with its 16:9 aspect ratio.

Source: Diagrams based in part on Mark Long, author, *World Satellite Almanac: The Global Guide to Satellite Transmission and Technology*, 3rd ed. (Fort Lauderdale, FL: Mark Long Enterprises, Inc.). Used with permission.

and all of it by the third year. At the end of the transition, when presumably a substantial portion of television homes would have digital receivers, stations and networks would convert to digital operations exclusively. All TV broadcasting would be on UHF channels, the broadcasters having given up their old NTSC broadcast channels for reallocation to other services.

In 1996 the FCC revised its transitional plan dramatically by announcing a new channel allotment scheme. Rather than move terrestrial broadcasters to all UHF digital channels, the Commission proposed packing most of the new digital licenses into a "core spectrum" comprised of channels 7–51. When the estimated 15-year transition to digital was complete, 23 channels formerly used by broadcasters (channels 2–6 and 52–69) would be auctioned to other users providing other services. Broadcasters immediately countered that the entire range of TV channels (2–69) was necessary to avoid interference and maintain existing signal coverage patterns during the transition.

When the FCC finally assigned the new digital channels in 1997 the plan had changed again. The FCC made initial digital assignments on all available TV channels (2–69), but announced stations eventually would be assigned to a "core" DTV spectrum of either channels 2–46 or 7–51. The FCC also sharply accelerated the phase-in period for DTV. Rather than allowing stations six years to begin digital broadcasting, the FCC required at least three network-affiliated DTV stations be on the air in the top 30 markets by November 1, 1999. The mandate ensures DTV's availability to more than half the nation's TV households by the year 2000. The networks promised to beat the FCC's deadline. NBC pledged 80 percent of its owned and operated stations in the top ten markets will broadcast a digital signal by November, 1998, only 18 months after the FCC's adoption of digital transmission standards. ABC promised 60 percent, CBS 57 percent, and Fox 33 percent. The FCC shortened the transition period for all commercial stations from 15 to 10 years, until the end of 2006. Later legislation allows a station to retain its analog channel until 85 percent of the station's viewers can receive the digital signal either over the air, by satellite, or by a cable. The FCC also scuttled earlier requirements that stations simulcast digital and analog programming during the transition, except during the last few years. The plan imposed no new public interest requirements on digital broadcasters, but the FCC expects to add some during the transition. Broadcasters were pleased that the plan does not require they pay a spectrum fee for their new digital channels.

5.8 Digital Radio

Digital audio broadcasting (DAB) is developing even more slowly than the digital HDTV version of television. It ultimately promises (or threatens, depending on one's point of view) to replace analog radio broadcasting—both AM and FM—with a single new digital system of radio.

As with HDTV, the FCC would likely set up a transition period during which stations would broadcast in analog AM and FM on their existing frequencies, while simulcasting DAB on new assignments (or possibly even on the existing frequencies, thanks to new multiplexing technology). Then, when enough people had DAB receivers, stations would cease their analog transmissions.

By the early 1990s about 10 different technologies for achieving DAB existed, all of which fell into one of two categories. Some would require additional spectrum to achieve true digital sound quality, whereas others would operate in the present combined AM and FM bands, though likely with wider channels.

DAB Spectrum Needs

In 1992, the International Telecommunication Union (ITU) established frequencies (1.452 to 1.492 GHz) in the L-band for world-wide DAB despite objections from the United States. The United States used those frequencies for military and other purposes and did not want to reassign those services to accommodate DAB. Accordingly, later the same year the FCC announced that American DAB would use 50 MHz of "S-band" frequencies between 2.31 and 2.36 GHz. Both satellite-delivered and terrestrial DAB would use these frequencies if it were determined that digital and analog signals on the existing AM and FM bands could not coexist. By 1996 the FCC had reduced the range of frequencies from 50 MHz to 40 MHz because of potential interference with Canadian signals in the lower portion of the S-band. Meanwhile, Congress passed appropriation legislation mandating that many of the remaining frequencies be auctioned to a variety of wireless services. As a result, by 1997 only half (25 MHz) of the space originally reserved for DAB remained available. An industry group testing various DAB systems questioned whether 25 MHz could provide the quality and diversity of DAB services consumers expected.

DAB Progress

The National Association of Broadcasters wants to offer DAB "in-band," meaning by existing terrestrial stations using their currently assigned frequencies. NAB argues this approach will provide a smoother transition to DAB than reassigning stations to entirely different frequencies in a new band. The NAB also opposed satellite-delivered direct-to-home or direct-to-car DAB, which the FCC calls *digital audio radio service* (DARS).

In 1992 the FCC granted experimental authority for testing in-band DAB. In the mid-1990s the NASA Lewis Research Center lab tested seven systems, five of them in-band systems. Four systems emerged for field testing, but their proponents gradually withdrew their participation, claiming the testing procedure favored the European-based Eureka-147 DAB system. With the FCC's original S-band frequency allocation cut in half and proponents of in-band DAB unable to agree on test procedures that would lead to a standard, the status of terrestrial DAB in 1997 was little advanced from where it had been at the start of the decade. The FCC, however, went forward and auctioned two 12.5 MHz segments of S-band spectrum for satellite-delivered digital audio radio service. The winning bidders, Satellite CD Radio, Inc. and American Mobile Radio Corporation, each paid more than $83 million for DARS spectrum.

DAB is advancing in many European nations and Canada. Digital audio service also is available to American households on a subscription basis by both cable and DBS. But establishing digital service with access comparable to that provided by traditional AM and FM broadcasting remains an elusive goal.

Data by Radio

After long debate, both broadcasters and consumer electronics manufacturers—working through a National Radio System Committee—agreed on technical standards for a *radio broadcast data system* (RBDS), transmitted on an FM station subcarrier for receivers equipped with a special microchip.

RBDS transmits a digital signal—an ad slogan, identification of the station's format, or other information—that can be displayed in a special receiver panel. The system would also make it possible for an RBDS-equipped car to tune stations by format (any one of more than 20 preset codes), with the receiver jumping to a different but more powerful signal when the tuned signal became weak. And in an emergency situation—an approaching tornado, for example—RBDS could be used to turn on a receiver to transmit a warning message.

5.9 Networking and Switching

Whereas traditional broadcasting stations typically transmit all of their signals to all persons who have receivers, newer generations of multichannel media intend only some persons to receive only some of their signals.

Networking

Radio and broadcast television use *networking* (see Sections 5.1 and 5.2), as relays to connect themselves to program services and to each other, not for delivery to the public. Early cable systems used networking in much the same way but also for delivery of all of their channels to all of their customers.

Most newer media, however, intend to deliver only some of their program services to only some consumers. A modern cable system, for example, intends to deliver its package of basic program services only to those who pay a monthly subscription fee and to deliver other programming (from HBO, Showtime, and others) only to those who pay an extra charge for those premimum services.

Cable and DBS systems achieve selective audience delivery through *scrambling* (or *encrypting*) some signals and unscrambling them for paying customers, or through the use of electronic *traps* in drop cables, which prevent delivery until they are removed.

Interactive media complicate the process even more. Only some signals from some sources are to reach only some receivers who, in turn, are to send signals back *upstream* to the appropriate originating source. Two principal elements are required to accomplish this complicated task. First, sources must be linked with receivers, by terrestrial connection or otherwise, so that together they form a *network*. Then signals on that network must be routed, or *switched*, from source to intended receiver and, where interactivity exists, from receiver back to intended source.

The best examples of switched networks are those operated by the telephone companies. Few people appreciate the enormity and complexity of the *public switched telephone network* (PSTN). We simply push a few telephone buttons and we are—almost miraculously—connected to precisely the person we want to reach.

The public telephone network resembles a cable system (see the cable layout in Exhibit 4.o). *Drop wires*, which connect individual homes to utility poles, join others to become a *distribution cable*, which in turn joins others to form a *feeder cable* that completes the connection from each home to a *central office*. Here the similarity ends, because, unlike cable system headends, telephone central offices connect to each other by *trunk lines* and, via satellite and undersea cables, to their counterparts around the world.

Switching

This vast network interconnection, must be combined with sophisticated *switching* capability, to provide the one-to-one telephonic communication we have come to take for granted.

In its infancy, telephony accomplished its switches manually. A telephone operator asked what number you wished to reach and literally plugged a wire (called a *patch cord*) into a console so that your line would be linked to another (see Exhibit 5.p). This was an agonizingly slow process. The caller gave the long distance operator the number being called, then hung up and waited, perhaps an hour or more, for the various patches to be completed. Today switching is handled electronically by digital switches. The time needed to connect with a phone on the other side of the planet is little more than that needed to reach a friend across town.

Asynchronous Transfer Mode (ATM) is the name given to one type of digital switch. Rather than transmit information continuously (synchronously), as happens when television stations send out their signals, ATM gathers data together in groups called *cells* and transmits them as discrete packages. This asynchronous method permits more efficient use of circuits and can handle huge volumes of voice, data, and video.

As media converge, and as demand increases for voice, data, and video transmission, new systems and standards must be put in place. And the terms used sometimes can be misleading. For example, the *Integrated Services Digital Network* (ISDN), although not truly a network, is an evolving set of standards for multiple services, by

Exhibit 5.p

From Switchboard to Switch

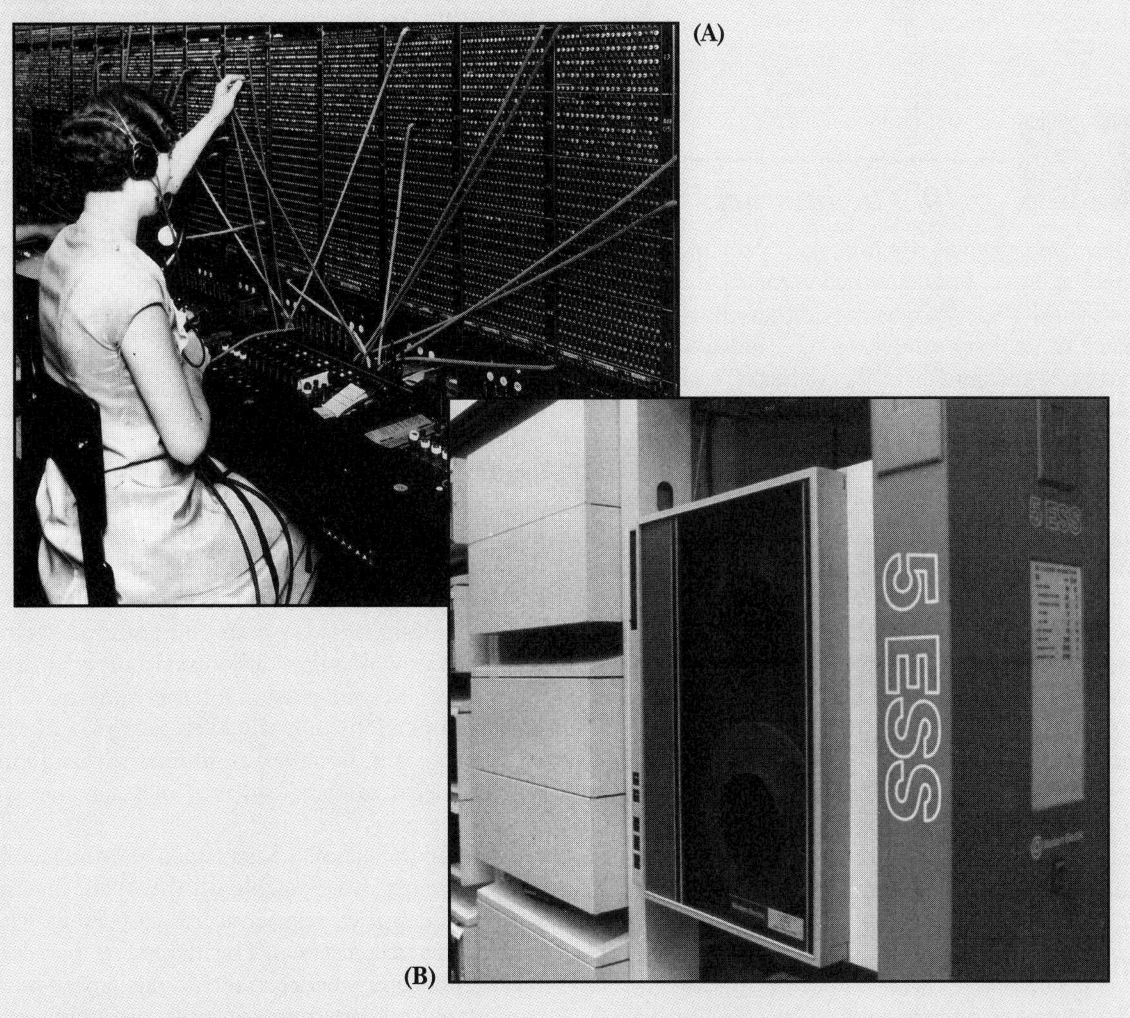

(A) Some people yearn for a return to the "good old days" when a telephone operator—a live human being—would ask each caller, "Number, please?" and then, using a manual switchboard with its many patchcords, make a physical connection of the caller's line to that of the intended receiver.

(B) But the volume and complexity of today's telephone traffic require sophisticated, though visually unarresting, digital switches, such as AT&T's 5 ESS system.

Source: Photos from AT&T Archives.

telephone and other companies, across both narrow and broad bandwidths. Similarly, *Synchronous Optical Network* (SONET) describes not a network but rather another set of standards, in this case for signal transmission over optical fibers.

5.10 The Internet

Some people believe the Internet *is* the "Information Superhighway." For them the Internet, and especially the World Wide Web, is the quintessential convergence technology combining text, audio, and video into a new medium that will, eventually, replace most traditional electronic media. Although many people think that is unlikely, the Internet is an important part of the "digital revolution" of the 1990s.

The Internet began in the late 1960s as a government-sponsored computing network called ARPANET. This network provided computer-to-computer communications among researchers working on defense-related projects at universities and research centers. Other research-oriented computer networks also developed. In the 1980s these networks became interconnected through the NSFnet (National Science Foundation), a "network of networks." Interest in sharing computer data increased dramatically with the proliferation of computers in the hands of persons outside the traditional research environment.

World Wide Web

The arcane protocols required initially limited use of the Internet to computer experts. There were attempts to make the Internet easier to use, such as providing a menu screen from which users simply selected the function they wanted performed rather than typing out complex instructions. In the late 1980s a Swiss nuclear research center (CERN) developed a new approach to linking computers, *hypertext transfer protocol* (HTTP). HTTP and *hypertext markup language* have made possible the creation of the World Wide Web (WWW) by allowing computer users to access files residing on other computers of any type in any location.

Use of the Web increased dramatically following the 1993 introduction of *Mosaic*, a revolutionary software program providing network navigation aids and featuring the icons and point-and-click features familiar to users of Macintosh computers or PCs running Microsoft Windows. The Mosaic graphical user interface opened the door to computer-based communication for millions of people. Similar browser programs offering additional features and greater power appeared later, two of the most popular being *Netscape Navigator* and *Microsoft Internet Explorer.*

Less sophisticated browser programs are available to help people using computers with limited memory or processing power. Because much of the information available on the Web is text, a text-only program allows these users to enjoy many benefits of Web access. Users of the more powerful browsers sometimes turn off the images download feature to reduce the time needed to access a file. Web site designers also sometimes offer users the option of downloading only text or a less graphically sophisticated version of the content. The graphical portions of many Web files are increasingly important, however, as designers learn to integrate visual information into message content more effectively.

As noted in Section 3.5, streamed real-time audio and video are also available on the Web, but the hardware demands restrict access to persons with well-equipped computers. The narrow bandwidth of the telephone line between the average home user's computer and the Internet host computer also makes audio, and especially video, quality marginal.

Internet Access

Most people access the Internet from their home using a modem-equipped computer hooked to a conventional voice telephone line. The modem (*mo*dulator-*dem*odulator) is necessary because computers require digital data while residential phone lines and their associated circuits are designed for analog audio. Internet access requires paying a fee to either an *Internet Service Provider* (ISP) or a commercial on-line service, such as America Online. Businesses, schools, public libraries, and some individuals often use a high-capacity digital connection rather than modems and conventional phone lines, but the cost is prohibitive for most people wanting access from their home.

Subscribers connect to the Internet by calling their ISP or on-line service where a computer answers. The user's modem and the host computer test the connection and then allow the user to log on. Once the user is identified as a subscriber, the host computer grants access to the services for which the user has paid.

Barriers to Growth

The challenge of setting up a computer, making the necessary phone line connections, loading the browser software, and learning how to use it seems an enormous bother to many people. Manufacturers are responding by bundling hardware so everything needed is already installed. Software for the Internet also is coming already installed on the computer's hard drive or on an easy-loading CD-ROM disc. The graphical user interfaces of browsers already discussed make using the systems easier. Still there are those who find the idea of connecting to the Internet daunting.

Paradoxically, others are concerned that the Internet is too easy to use. Some parents would like having home access but hesitate for fear their children will be exposed to pornographic content. This is possible because almost everything is available on the Web, from a tour of the White House to directions for making poison gas. However, software is available that allows parents to block access to objectionable content.

The lack of bandwidth provided by residential phone lines is also a barrier to the Internet's development. The bit speed of modems increased dramatically during the 1990s. Compression technology also effectively increased the rate of delivery. But many dedicated Internet users find the rate inadequate for even motionless images, especially during peak use hours when millions log on and begin accessing files simultaneously.

The Future

Estimates of Web use vary wildly. One report suggests that only about 17 percent of Americans use the Web (CommerceNet/Nielsen Media Demographic and Electronic Commerce Study, 1997). Skeptics point to such estimates as proof that the importance of the Web is exaggerated. Others note the estimate is twice what it was only 18 months earlier and that the survey did not include Web use by children and teens younger than 16.

Many believe the transition from analog to digital TV will accelerate the convergence of the TV and PC already underway. Some observers estimate that computer manufacturers can build and sell 20 to 50 million DTV-compatible computers by the time television manufacturers build a million DTV television sets. If that happens, the move to DTV also may increase Internet use because more people will have the tools available in their new DTV appliance.

Doritos

Chapter 6

Commercial Operations

CHAPTER CONTENTS

Commercial electronic mass media are businesses. Their primary motivation is to make money. And as in any business, the way to make money is to offer a salable product and to have income exceed expense. Advertising's dominant role in financing most broadcast media has had a profound impact on the types, number, and variety of program services offered. Cable and similar technologies rely mostly on subscriptions for their support, although advertising plays an increasing role in those media as well. Advertising on the Internet has grown explosively in the last few years and will probably continue to do so into the next century. This chapter focuses on commercial operations, leaving the special financial problems of public broadcasting to the next chapter.

6.1 The Basics

Exhibit 6.a summarizes the dimensions of the commercial and noncommercial broadcasting and cable industries in relation to total audience potential. The individual outlet—more than 15,000 stations and nearly 12,000 cable systems—forms the basic economic unit of these industries.

These 27,000 units function as local retailers, delivering programs directly to consumers. People sometimes use the phrases *local station* and *local cable system*, but in fact *all* stations and systems are local in the sense that each is licensed or franchised to serve a specific local community. An exception to the local outlet model, *direct broadcast satellite* (DBS) systems, market their services directly to consumers on a nationwide basis. Similarly, a growing number of "virtual" radio stations are using the Internet to reach national and international audiences.

Despite the localism of licenses and franchises, economic efficiency favors centralization of station/system ownership and program production. Still more efficiency comes from vertical integration—common ownership of production, distribution, and delivery facilities (see Section 6.4). The large organizations that result increasingly dominate the media economy.

6.2 Broadcast Stations

Definition

In the United States, the traditional commercial broadcast station can be defined as an entity (individual, partnership, corporation, or nonfederal governmental authority) that

- holds a license from the federal government to organize and schedule programs for a *specific community* in accordance with an approved plan;
- transmits those programs *over the air*, using designated radio frequencies in accordance with specified technical standards; and
- carries commercial messages that promote the products or services of profit-making organizations, for which the station receives compensation.

Within limits, an individual owner may legally control more than one station, but each outlet must be licensed separately to serve a specific community. Moreover, each license encompasses both transmission and programming functions. A station therefore normally combines three groups of facilities: business offices, studio facilities, and transmitter (including an antenna and its tower). Usually all facilities come under common ownership, although in a few cases stations lease some or all of them.

Station Functions

Exhibit 6.b offers examples of *tables of organization* (sometimes called *organization charts*) for both a radio and a television station. These vary widely from facility to facility, but all outline the station's personnel structure and indicate who reports to whom.

All commercial stations need to perform four basic functions: general and administrative, technical, programming, and sales.

Exhibit 6.a

Electronic Media Dimensions

Total U.S. Population	270 million
Total U.S. Households	98 million
Total U.S. Television Households (TVHH)	97 million

Commercial Radio Stations	
AM (47%)	4,811
FM (53%)	+5,477
	10,288
Noncommercial FM Stations	+1,889
Total Radio Stations (40% AM; 60% FM)	**12,177**
Commercial TV Stations	
VHF (47%)	558
UHF (53%)	+637
	1,195
Noncommercial TV Stations	
VHF (34%)	124
UHF (66%)	+241
	365
Low-power TV Stations	
VHF (28%)	555
UHF (72%)	+1,446
	2,001
Total TV Stations	**3,561**
Total Broadcast Stations (77% radio; 23% TV)	**15,738**
Cable Systems	11,600

These data change almost daily but are representative for 1997.

Source: *Broadcasting & Cable*, December 20, 1996, p. 62. Copyright © 1996. Reprinted by permission of Cahners Publishing Company.

General/administrative functions include the services that any business needs to create an appropriate working environment—services such as payroll, accounting, housekeeping, and purchasing. Services of a specialized nature peculiar to broadcasting usually come from external organizations, such as engineering-consulting firms, audience research companies, and program syndicators. For a network affiliate, the main such external contract is with its network.

Technical functions, usually supervised by the station's chief engineer, center on transmitter operations, which must follow strict FCC rules, and the maintenance and operation of studio equipment.

Program functions involve planning and implementation. Major program planning decisions usually evolve from interplay among the programming, sales, and management heads. Because most stations produce few non-news programs locally, the program department's main job is to select and schedule prerecorded programs. A program manager typically serves as the department head in this area at a TV station (an operations manager in radio). By the 1990s, however, many stations had dispensed with that title, leaving program decisions to the general manager or combining the function with that of the promotion department.

Promotion (increasingly called *Creative Services*) includes making its potential audience aware of a station's programs through advertising, on-air announcements, newspaper listings, even T-shirts and bumper stickers.

News, although a form of programming, usually constitutes a separate department, headed by a news director who reports directly to top management. This separation of news from entertainment makes sense because of the timely nature of news and the unique responsibilities news broadcasting imposes on management. At some TV stations news operations have taken on great importance by becoming major profit centers. On the other hand, except for all-news stations, local news appears to be declining in importance for most radio operations.

Sales functions divide into local and national aspects. Station sales departments have their own staff members to sell time to local advertisers. To

Exhibit 6.b

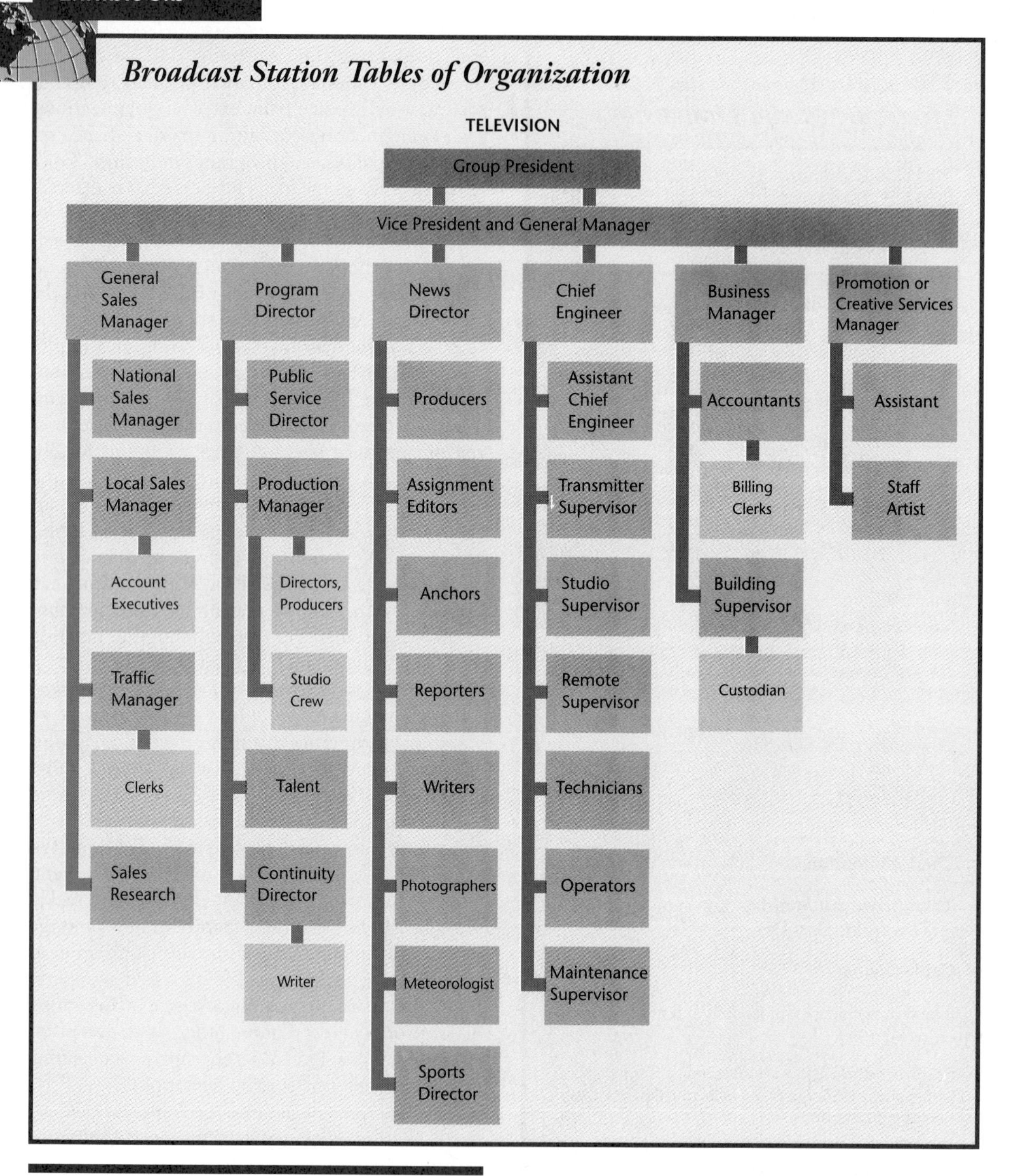

Broadcast Station Tables of Organization
TELEVISION
Group President
Vice President and General Manager
General Sales Manager
National Sales Manager
Local Sales Manager
Account Executives
Traffic Manager
Clerks
Sales Research
Program Director
Public Service Director
Production Manager
Directors, Producers
Studio Crew
Talent
Continuity Director
Writer
News Director
Producers
Assignment Editors
Anchors
Reporters
Writers
Photographers
Meteorologist
Sports Director
Chief Engineer
Assistant Chief Engineer
Transmitter Supervisor
Studio Supervisor
Remote Supervisor
Technicians
Operators
Maintenance Supervisor
Business Manager
Accountants
Billing Clerks
Building Supervisor
Custodian
Promotion or Creative Services Manager
Assistant
Staff Artist

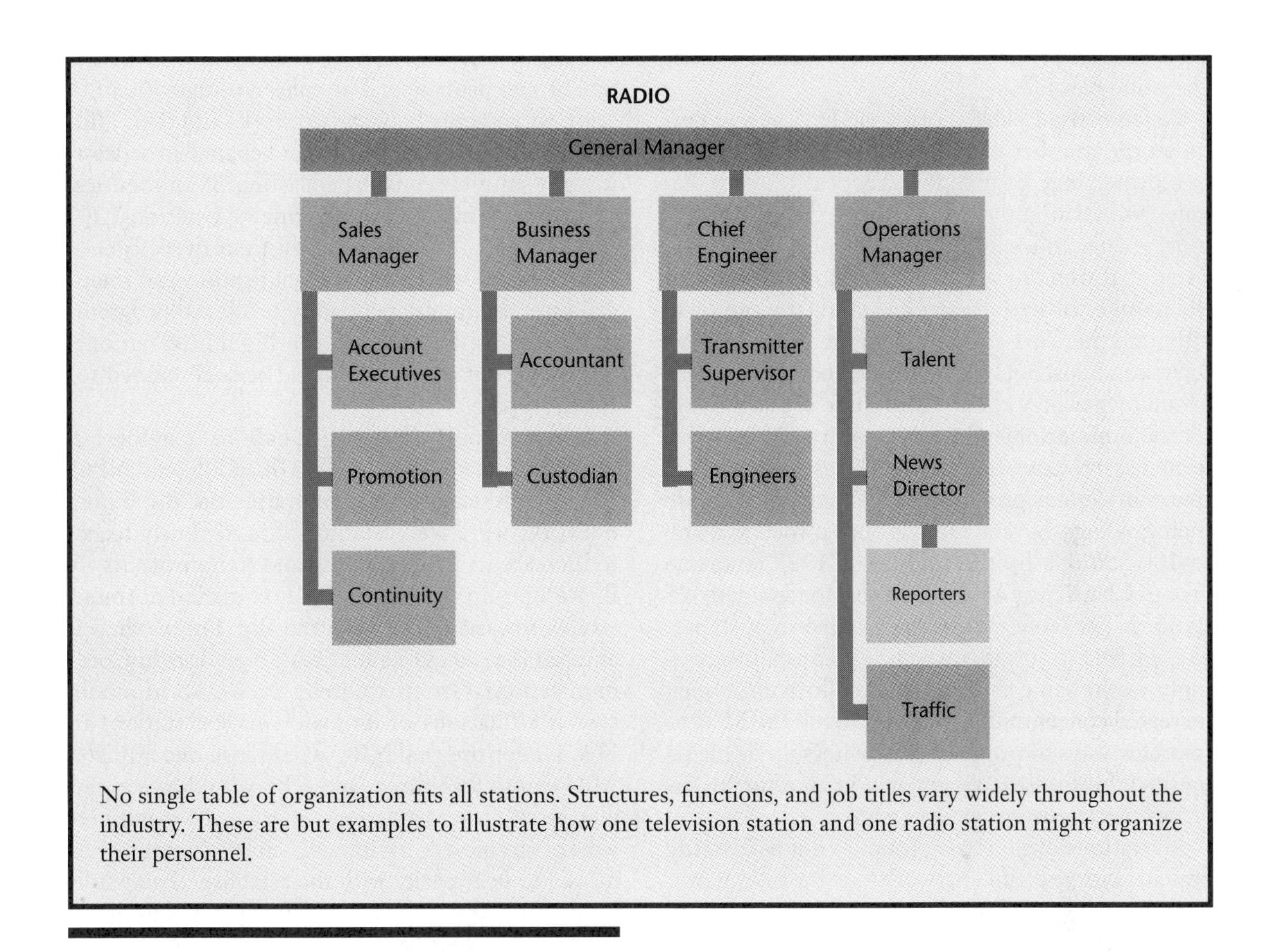

No single table of organization fits all stations. Structures, functions, and job titles vary widely throughout the industry. These are but examples to illustrate how one television station and one radio station might organize their personnel.

reach regional and national advertisers, however, a station usually contracts with a national sales representative firm that acts for the station in out-of-state business centers. A network affiliate benefits from a third sales force, that of its network. Exhibit 6.c suggests some of the tasks routinely performed by a television sales manager and other executives.

One job straddles the program and sales function—that of the traffic department. It coordinates sales with program operations, preparing the daily program *pre-log* (usually called simply the *log*), which schedules facilities, personnel, programs, and announcements. Traffic personnel ensure fulfillment of advertising contracts and arrange for *make-goods*—the rescheduling of missed or technically inadequate commercials or bonus spots when a TV program fails to deliver promised ratings. Traffic maintains a list of *availabilities* (or *avails*), keeping sales personnel up-to-date on commercial openings in the schedule. Traffic usually fills unsold openings with public-service or promotional announcements. At many stations, computers handle much of the complex work of the traffic department.

Station Groups

Like other business enterprises, broadcasting can benefit from economies of scale. A company owning several stations can buy programs, supplies, and equipment in bulk and spread the cost of consultants and attorneys across several stations, sharing experiences and new ideas.

For reasons of public policy, the FCC places limits on the number of radio and television stations one owner may control (see Section 13.6 for details). Still, within those limitations, the trend is toward greater concentration of control. Congress assisted that trend by removing all national limits on the number of stations any one company can own, although there are still limits on the percentage of television households (35 percent) that any one TV group owner may cover. Also, there are still limits on the number of stations one entity can own per local market. Consequently, there was a flurry of activity in 1996 as group owners sought to increase their holdings. Several radio groups owned as many as 100 stations by the middle of 1997, and one group, Chancellor Media Corporation, owned 325 stations. The largest television group owner, Rupert Murdoch's Fox group, owned 23 stations. Interestingly, at the same time Congress allowed for such increased concentration, it also ordered the FCC to find new ways to promote the ownership of media by small businesses, presumably in an effort to diversify electronic media ownership.

All stations owned by a group do not necessarily affiliate with the same network—or for that matter with any network. Stations owned by the networks themselves, known as network *owned-and-operated* (O&O) stations, are the exceptions. Each national television network's O&O group reaches between 25 and 35 percent of the nation's television households, ensuring ABC, CBS, NBC, and Fox their own prestigious outlets in major markets.

6.3 Broadcast TV Networks

Definition

Although the term *television network* means one of the major national commercial networks to most people, in fact there are many television networks in operation. Some are national, some regional, and some only exist for the purpose of distributing one or two programs. The rules dealing with affiliation agreements between networks and their affiliate stations define network, or "chain," broadcasting as "simultaneous broadcasting of an identical program by two or more connected stations" (47 U.S.C. 153(p)). Until recently there were special rules dealing with program syndication and financial interests in independently produced programming that applied only to the Big Three national networks, but those rules have been rescinded for the most part.

For years only three major national commercial television networks existed—ABC, CBS, and NBC. Fox premiered in 1986, long after the Big Three networks were well established, and slowly began to increase its programming and to improve its affiliate line-up. In May 1994 Fox made enormous strides toward parity with the Big Three when it entered into an agreement with New World Communications Group whereby New World would switch affiliations of its major-market stations to Fox. Under the deal NBC would lose one affiliate, ABC would lose three, and CBS would lose eight. The move created a domino effect as the major networks scrambled to line up new affiliates. Fox moved to near parity with the established networks by gaining the network rights to NFL Football (outbidding long-time rights holder CBS) and by developing a network news service, Fox NewsEdge. By the end of 1997, each of the Big Three plus Fox had about 200 affiliates, through which they could reach virtually all U.S. households.

Meanwhile a new Warner Brothers network (calling itself "The WB") premiered on January 11, 1995, followed five days later by United Paramount Network (UPN). Like Fox before them, both began with limited hours of programming, and both had aspirations of eventually becoming the fifth and sixth, fully competitive national broadcast television networks. In addition, more than a hundred regional and part-time networks are in operation.

Like stations, networks vary in their organizational structure, yet each must fulfill the same four basic functions of administration, programming,

Exhibit 6.c

A Day in the Lives of TV Executives

	General Manager	Program Director	General Sales Manager	News Director
8:30	Open mail; dictate letters and memos.	Check discrepancy reports for program and equipment problems; take appropriate action.	Check discrepancy reports for missed commercials; plan make-goods.	Meet with assignment editor and producer; plan the day.
9:30	Discuss financial statements with business manager.	Call *TV Guide* with program updates.	Local sales meeting; discuss accounts and quotas.	Meet with union shop steward; discuss termination of reporter.
10:00	Call group headquarters regarding financial status.	Prepare weekly program schedule.	Accompany local account executive on sales calls.	Read mail; screen tape of last night's newscast.
10:30	Meet with civic group angry about upcoming network program.	Select film titles for Saturday and Sunday late movies.	More sales calls.	Discuss noon news rundown with show producer.
11:00	Call network; ask for preview of questionable show.	Meet with promotion manager regarding *TV Guide* ad for local shows.	Call collection agency; discuss delinquent sales accounts.	Meet with Chief Engineer regarding SNG failure.
12:00	Lunch with major advertiser.	Lunch with syndicated program saleswoman.	Lunch with major advertiser.	Monitor noon news; lunch at desk.
2:00	Department heads meeting.	Department heads meeting.	Department heads meeting.	Department heads meeting.
4:00	Meet with Chief Engineer regarding new computer system in master control.	Meet with producer/ director to plan local holiday special.	Prepare speech for next week's Rotary Club meeting.	Meet with producer and director; plan rundown of 6:00 PM newscast.
6:00	Dinner with Promotion Director job candidate	Attend National Academy of Television Arts & Sciences annual local banquet.	To airport; catch flight to New York for meeting with national sales rep.	Monitor 6:00 PM news.

Not all television station executives work 10- or 12-hour days, but many do, especially as competition from cable and the other new media increases.

engineering, and sales. Networks, however, enjoy the luxury of a much higher degree of specialization than do stations. NBC, for example, has separate units to handle operations, entertainment, news, sports, the television network, and NBC owned-and-operated stations.

Affiliation

More than half of all full-power commercial television stations affiliate with one of the major networks. Most function as a *primary affiliate*—the only affiliate of a given network in a given market. *Secondary affiliates*, typically those in markets with only two stations, share affiliation with more than one network. A few markets have only a single station. In Presque Isle, Maine, for example, WAGM-TV has the unusual privilege of picking and choosing programs from all networks.

Affiliation does *not* mean that a network owns or operates the affiliated stations. ABC, CBS, NBC, and Fox do, of course, own some television stations. But they contract with hundreds of other affiliates, agreeing with each to offer it the network's programs before offering them to any other station in the same market.

The station, in turn, agrees to *clear* time for all or portions of the network schedule. However, it has the right to decline to carry any specific program, or it may offer to carry the program at a time other than that of network origination (a *delayed broadcast*, or DB); the network may or may not agree to this last option.

In effect, traditional television networks buy affiliates' time to deliver station audiences to network advertisers. Affiliates sell their time to networks at a rate much lower than they charge other customers. In exchange, a broadcasting network offers its affiliates

- *monetary compensation* to the stations based on audience size and composition;
- a structured schedule of *network programs* at no direct cost;
- *simultaneous program distribution* so that affiliates can receive the service at the same time;
- an *advertising environment* that appeals to the affiliates' clients; and
- a *sales organization* that finds national clients to purchase advertisements that occupy the network's portion of the affiliates' commercial time.

Network-Affiliate Contract

Networks and their affiliates formalize the economic link between them through an *affiliation contract.* A clause at the heart of such a contract defines the terms on which the network will make payments—called both *station compensation* and *network comp*—to the station for the right to use the station's time. (In contrast, the new WB network has a "reverse compensation" plan under which affiliates, rather than receiving payments, pay WB 25 percent of all incremental profits they derive from their association with the network. UPN, on the other hand, relies solely on advertising for its revenue.)

Each television network uses a different plan for calculating compensation. The amounts paid to affiliates vary from station to station, reflecting differences in market size, station popularity, and other factors. Rates in top markets such as New York and Los Angeles run into thousands of dollars per hour; they go down as low as $50, and some affiliates in the smallest markets receive no compensation at all.

Network compensation represents a surprisingly small percentage of the gross revenues of network-affiliated stations—on the average, less than 5 percent. But stations measure the value of affiliation less in terms of compensation than in terms of the audiences that network programs attract.

Affiliates profit from the sale of spots in the 90 seconds or so that the network leaves open for affiliate station breaks in each prime-time hour of network programming and the seven or eight minutes made available at other times of the day. Moreover, the stations' own programs (whether locally produced or purchased from syndicators) benefit from association with popular and widely promoted network programs.

Network Regulation

Unable to control networks directly (because it does not license them), the FCC regulates them indirectly through rules governing the stations they own and the contracts affiliates make with them.

The FCC's network regulations were originally intended to ensure station autonomy and to prevent the networks from becoming the overwhelmingly dominant force in broadcasting. For example, regulations prevent network contracts from forcing stations to clear time for network programs. However, as network power waned with the increasing strength of competing media, the FCC became less concerned with network dominance.

Nevertheless, the FCC still prohibits national broadcast television networks from preventing an affiliate from accepting programs from other networks, nor may an affiliate prevent its network from offering rejected programs to other stations in its market. A network may not influence an affiliate's nonnetwork advertising rates, nor may it function as national spot sales representative for any of its affiliates other than its O&O stations, though in early 1997 the FCC was considering dropping these two rules.

Clearance

With or without FCC regulation, networks and their affiliates experience a somewhat uneasy sharing of power, complicated by political and economic factors too subtle for contracts to define. In one sense the networks have the upper hand. Affiliation plays a vitally important role in an affiliate station's success. However, without the voluntary compliance of affiliates, a television network amounts to nothing but a group owner of a few stations rather than the main source of programming for some 200 stations.

The complex relationships between networks and their affiliates hinge on the act of *clearance*—an affiliate's contractual agreement to keep clear in its program schedule the times the network needs to run its programs. An affiliate might fail to clear time or might preempt already cleared time for several reasons. Network public-affairs and other nonentertainment offerings usually get low ratings and therefore most often fail to get clearance. Sometimes stations skip low-rated network programs in favor of syndicated shows simply to keep audiences from flowing to the competition. Often a station wants to increase the amount of commercial time available. It can run more commercials than the network allows by substituting a syndicated program or a movie.

Networks rely on affiliates not only to carry their programs but also to carry them *as scheduled.* Delayed network broadcasts erode national ratings. Networks also need simultaneous coverage to get the maximum benefit from promotion and advertising.

In practice, affiliates accept about 90 percent of all programs offered by their networks as scheduled, most of them on faith. Stations can request advance screening of questionable programs but usually feel no need to do so, even though, as licensees, stations rather than networks have the ultimate legal responsibility. Because most television programs come in series, affiliates know their general tone, so the acceptability of future episodes can usually be taken for granted.

Thus affiliates have little or no direct influence over the day-to-day programming decisions of their networks. In the long run, however, they have powerful leverage. Network programming strategists take serious note of the feedback that comes from their affiliates.

Changing Network-Affiliate Relations

Starting in the late 1980s, when rising costs and increasing competition combined to weaken ABC, CBS, and NBC, the traditional network-affiliate relationship began to crack under the strain. Affiliates felt that network compensation failed to reflect the true value of their time to their networks. Networks increased the amount of precious advertising time available to themselves in their schedules, a move affiliates resented. Pre-emptions reached all-time highs, costing the networks millions of dollars in lost revenues.

In the mid-1990s, after a flurry of affiliation switches, the networks experienced a renewed appreciation for their station "partners." In many cases they raised compensation rates. And whereas historically they had limited affiliation contracts typically to two years' duration, they began writing agreements with their most desired affiliates that ran as long as ten years. With the longer contracts, however, came network expectations—and increased pressure—for more complete program clearances by their affiliates.

Technology has freed many affiliates from their former, almost total dependence on networks for nonlocal news. Using satellites and minicams, stations can now cover not only local but also national and even international news. In fact, roles have partially reversed, making networks dependent on affiliates for coverage of local stories of national significance. An especially dramatic example of this dependence occurred in June 1994 when networks relied on local station helicopters for pictures of O.J. Simpson as his car travelled Los Angeles freeways on the way to his home, where he was arrested for the murders of his former wife and her friend. Another example of this role reversal occurred in April 1995 when the networks received much of their information and video of the Oklahoma City bombing incident from their local affiliates.

Competition from Independents

Approximately 400 stations, most of them UHF, are known as *independents.* They have no full-service network affiliation, although they may be affiliated with Fox, UPN, the WB, or one of the regional networks. As Fox continued to enlarge its presence, however, calling a Fox affiliate an independent became at best a legal technicality.

The FCC's *prime-time access rule* (PTAR) long played a major role in turning some independents into profit makers. Starting in 1971, PTAR gave independents their first chance to counterprogram effectively against network affiliates in the 7:00 to 8:00 P.M. (Eastern and Pacific time) period. They gained the advantage of rerunning network series during that hour, a program option that PTAR denied to affiliates in the major markets. Citing a changing media environment, the FCC voted to rescind PTAR effective in 1996.

Other factors favoring independent stations included their coverage of live sports events, their success in outbidding affiliates for popular syndicated program series, their aggressive promotional campaigns, and their establishment of the Association of Independent Television Stations (INTV) in 1972. Recognizing that most of its members had some form of network affiliation, the Association changed its name in 1996 to the Association of Local Television Stations (ALTV). Cable also helped by making signal quality of UHF independent stations equal (to cable viewers) to that of VHF affiliates.

The fact that so many formerly independent stations have now affiliated with one of the emerging networks demonstrates that affiliation with a national network remains one of the most valuable assets a television station can have.

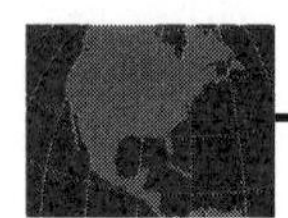

6.4 Cable

Cable/Broadcast Comparison

The economic organization of cable television systems and networks differs substantially from that of broadcast stations and networks.

Cable systems depend primarily on subscriber fees, not advertising, for their revenue. A cable system therefore owes allegiance only to the television households in its franchise area that choose to subscribe (in 1997 as high as 89 percent of the households in Palm Springs, for example, but averaging about 67 percent of all television households in the country).

A commercial broadcaster, in contrast, depends almost entirely on advertisers and has a legal obligation to serve the total audience in its market area, which is usually much larger than that of a cable system. Large cities often divide their municipal areas into several different cable franchises. Cable systems outnumber commercial television stations about 10 to 1. Whereas most viewers can tune in a number of local radio and television stations, almost without exception they can subscribe to only a single cable system.

System Organization

By the mid-1990s more than 11,600 *cable systems* (the basic units of the industry) served about 65 million subscribers in the United States. More than 95 percent of cable subscribers had access to 30 or more channels; about 47 percent could see 54 or more; some could view more than 100. Exhibit 6.d shows how a typical system structures its organization. Whether small, with a few hundred subscribers, or large, with thousands, each cable system performs the same four basic functions as do broadcast stations, as described in Section 6.2.

However, a cable system's *technical* functions differ from those of a broadcast station. Broadcast technicians' jobs end when the signal leaves the transmitter—listeners/viewers are on their own when it comes to arranging for reception. Cable technicians must be concerned with the integrity of both the sending and the receiving aspects of the system. Often this division of responsibility results in a distinction between *inside-* and *outside-*plant personnel.

The inside group operates and maintains studio facilities, as well as the complex array of equipment at the *headend,* which receives programs by various means from program suppliers and processes the signals for delivery via coaxial and fiber-optic cables to subscribers. The outside group installs and services subscriber cable connections and equipment.

As for the *programming* function, cable systems start with a more even playing field than broadcasters because the distinction between network affiliates and independents does not exist in cable. Nor does broadcasting's elaborate symbiotic relationship between an affiliate and its sole network exist for cable.

Typically cable systems fill their multiple channels with programs from both broadcasting stations and cable networks. The general manager of smaller systems makes program decisions, usually in consultation with the marketing director. Some cable systems produce local programs, often on channels programmed by community or educational organizations. Although a few cable companies have made a commitment to locally produced news, most systems defer to broadcast radio and television in this expensive and personnel-intensive area.

Cable system *sales* operations might better be called *sales and marketing.* Although some systems sell commercial time on some of their channels, marketing the cable service to subscribers ranks as the all-important function that brings in most of the revenue in the form of monthly subscriber fees.

The cable system marketing department tries to convince nonsubscribers to subscribe and current subscribers not to disconnect. These personnel also encourage existing subscribers to purchase additional service tiers and premium channels. The marketing department's customer service representatives interface with the public, answering telephones for eight or more hours a day, responding to complaints from subscribers and questions from potential customers. A system's ability to handle these contacts promptly and skillfully can have a profound effect on its financial success and, in extreme cases, on whether it keeps or loses its franchise.

System Interconnection

Often several cable systems in a large market *interconnect* so that commercials (or programs) can be seen on all of them, rather than only one, providing combined advertising coverage equivalent in reach to that of the area's television stations.

System interconnection may involve physically linking the systems by cable, microwave, or satellite (*hard* or *true* interconnections), or it may depend simply on the exchange of videotapes (*soft* interconnection).

Multiple-System Operators

Firms large enough to make the capital investments necessary to build, buy, or improve cable systems are not likely to be attracted by the limited potential of a single franchise. Thus emerged a trend toward *multiple-system operators* (MSOs)—firms that gather scores and even hundreds of systems under single ownership. Such firms have the resources to bid for high-cost, politically intricate cable franchises in metropolitan areas. Not until the 1992 Cable Act were any efforts made to establish cable ownership limits (see Section 13.6).

In the 1990s the trend toward mergers and consolidations increased. In fact, the five largest MSOs

Exhibit 6.d

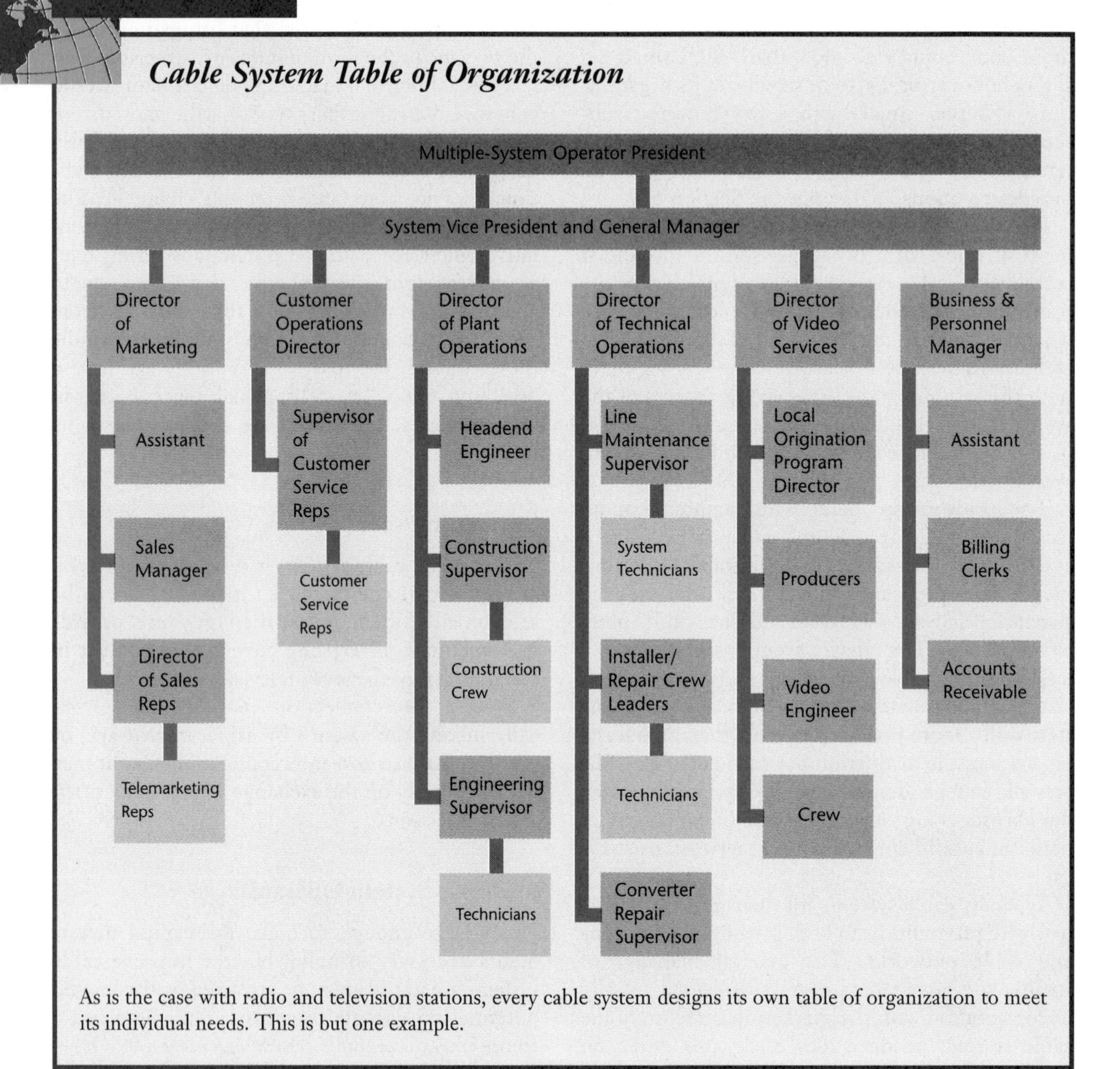

As is the case with radio and television stations, every cable system designs its own table of organization to meet its individual needs. This is but one example.

now serve almost two-thirds of all cable subscribers nationwide. Nonetheless, hundreds of smaller MSOs operate in the United States. The largest MSO, Tele-Communications, Inc. (TCI), has about 14 million subscribers, followed by Time Warner with approximately 12 million. Impressive though those numbers are, each company serves fewer than 15 percent of all television households. By contrast, the largest television station group can reach nearly 35 percent.

Vertical Integration

Cable television often involves ownership links with related businesses such as program production, distribution, and delivery. This type of linkage is called *vertical integration* and the cable-TV industry resorts to it increasingly. Tele-Communications Inc. (TCI) provides perhaps the best example. TCI operates the nation's largest MSO. It also owns parts of News Corp. (owner of Fox TV), several cable networks (Encore, Court TV, Black Entertainment Television, Home Shopping Network, QVC, American Movie Classics), and a satellite program distribution service (National Digital Television Center). TCI also owns part of an already vertically integrated program producer/distributor/station owner—the Turner Broadcasting System. With the proliferation of new networks in the late 1990s, the percentage of networks owned by MSOs is actually declining.

Enter the Telcos

For much of the history of the cable television industry, telephone companies were prohibited from providing cable television services in the same area they provided local telephone services. The Bell telephone companies were prohibited from providing cable services altogether. This changed dramatically with the passage of the 1996 Telecommunications Act, when Congress, attempting to stimulate competition in the cable television industry, allowed telephone companies to begin delivering cable services—both within and outside their local telephone service areas. According to legislation, telephone companies could offer video programming services in a number of different ways: over the air as do wireless cable operators, as common carriers, by obtaining franchises and operating as a traditional cable system, or through a hybrid operation known as "open video systems" which combine aspects of common carriage and the cable TV model of operation. (See section 12.9.)

By the beginning of 1997, several telephone companies were taking advantage of their newfound freedom. Some companies began investing in wireless cable systems, and others chose to operate as traditional cable systems. Bell Atlantic was operating a system in Dover Township, Delaware; Ameritech was supplying service to subscribers in Columbus, Ohio, and suburban Detroit; Bellsouth initiated service in Chamblee, Georgia; GTE built a system in Clearwater, Florida; and US West was active in Omaha, Nebraska. Although some local telephone companies were more interested in entering the newly opened long-distance telephone market, clearly some telcos were ready to compete head-to-head with the cable television industry. In response to the new competition, cable operators are moving premium channels to basic service packages, adding new channels, and dropping subscription fees for some service packages.

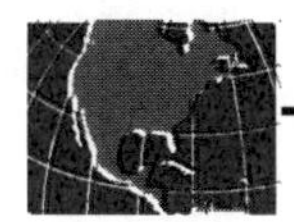

6.5 Cable Program Services

Cable systems typically carry most broadcast television stations whose signals cover their franchise areas. In addition, some systems produce limited amounts of their own programming and offer *access channels* for programs produced locally by others. Beyond that, they draw on three main types of centralized program providers: basic-cable networks, superstations, and pay-cable networks. Despite ever-increasing channel capacity, most cable systems are unable to carry all program services available to them. As the major cable networks began programming multiple channels rather than only one, competition for carriage intensified. Exhibit 6.e gives data for a representative selection of cable program providers.

Basic-Cable Networks

The 1992 Cable Act requires cable operators to offer subscribers a *basic service tier* (or level) of program sources, including, at a minimum, local television stations and all public, educational, and government access (PEG) channels. (See Section 12.8 for TV stations' *must-carry* and *retransmission*

Exhibit 6.e

Examples of Satellite-Distributed Cable Program Services

Network (launch date/owner)	Homes Reached (millions)	Content
Advertiser- and Industry-Supported Cable Program Services		
ESPN (1979/Capital Cities–ABC; The Hearst Group)	70.9	Sports events, news, and information
CNN (1980/Turner Broadcasting)	71.0	News, sports, weather, business
USA (1980/Paramount; MCA)	70.7	Series, movies, specials, sports
C-SPAN (1979/Cable industry)	67.1	House of Representatives, public affairs programs
TBS (1976/Turner Broadcasting)	70.9	Movies, sports, specials, series
MTV (1981/Viacom)	66.7	Music videos, series, pop culture
Premium (Pay) Services		
Network (launch date/owner)	**Homes Reached (millions)**	**Content**
HBO (1972; 1975 via satellite/ Time Warner)	32.0	Movies, sports, specials
Showtime (1976; 1978 via satellite/Viacom)	13.3	Movies, sports, specials

consent options.) A cable operator may add other channels of programming to this minimum, or *lifeline*, service. Subscribers pay a monthly fee to receive the basic tier.

The 1992 Cable Act defines *cable programming service* as all video programming provided over a cable system except that provided on the basic service tier. This includes programming from such *cable networks* as USA, ESPN, and MTV, nearly all of which are *advertiser-supported.* Some cable systems include such networks in their basic service tier, others on what they call an *expanded basic tier.*

Home Shopping Networks		
Network (launch date/owner)	**Homes Reached (millions)**	**Content**
QVC (1986/Comcast-TCI)	55.7	Electronics, jewelry, cosmetics, fashions
Home Shopping Network (1985/Publicly traded)	50.7	Electronics, jewelry, cosmetics, fashions

Pay-Per-View Services		
Service (launch date/owner)	**Homes Reached (millions)**	**Content**
Request Television (1985/Reiss Media)	26.0	Movies, special events
Viewer's Choice (1985/Cable consortium)	44.0	Movies, sports, specials

Sources: *Television & Cable Factbook*, Warren Publishing (Washington, DC, 1996); *Broadcasting & Cable*, 9 December 1996: 100; *Broadcasting & Cable*, 16 December 1996: 120; *Cablevision*, 9 December 1996: 126.

Some systems have several tiers, each for a separate fee.

For virtually all basic-cable network programs (home shopping networks and Video Jukebox—now named simply "The Box"—are exceptions), a cable system pays a fee directly to each network that it carries. Fees range from just a few cents up to 50 cents and more per subscriber per month. As competition for channel space increased in the late 1990s, some new cable networks, such as Animal Planet and Home & Garden, offered cable operators up-front cash payments of up to $8 per subscriber to secure carriage on their systems. This practice may well continue as the number of new cable networks increases. A basic-cable network sells commercial spots to its own advertising clients and usually leaves about two minutes of advertising time each hour for local sale by cable systems.

Advertising-supported cable networks have far smaller staffs than ABC, CBS, or NBC, and they reach smaller audiences. They maintain commercial sales departments but also must devote major attention to selling *themselves* to cable systems. Unlike major broadcast television networks, some have difficulty finding affiliates because so many cable networks compete for outlets.

Superstations

The *superstation* is a paradoxical hybrid of broadcasting and cable television—paradoxical because, although the FCC licenses each broadcast station to serve only one specific local market, superstations also reach hundreds of other markets throughout the country by means of satellite distribution to cable systems (see Exhibit 3.d).

Cable operators often include superstations in one of their basic packages of channels, paying a few cents per subscriber per month for the service. The station gets most of its revenue through higher advertising rates, justified by the cumulative size of the audiences it reaches through cable systems.

By the 1990s cable systems serving approximately 60 million subscribers—about two-thirds of all U.S. television homes—carried the nation's first superstation, Ted Turner's WTBS. In 1996, the role of the superstation began to change. WWOR, a popular superstation from New York, stopped its nationwide distribution, and Ted Turner announced that WTBS would cease operations as a superstation. Turner promised to replace WTBS with a similar cable network. These events leave WGN in Chicago as the most prominent and popular superstation still in operation.

Some cable systems also carry *radio superstations*, notably Chicago's Beethoven Satellite Network (WFMT-FM), a classical music station. Listeners pay an extra fee to the cable system operator to receive such radio stations, as well as other audio services, which go through a special cable connection to their home high-fidelity stereo systems. Subscribership to premium audio services has never exceeded 3 percent of cable households.

Before superstations altered the rules, copyright holders selling syndicated programs to individual stations had based their licensing charges on the assumption that each station reached a limited, fixed market. For example, syndicators formerly licensed programs to WTBS to reach only the Atlanta audience; but once it achieved superstation status, its programs reached audiences in hundreds of other markets, in many of which television stations may have paid copyright fees for the right to broadcast the very same programs.

Responding to complaints about the inequities created by duplicate television-program distribution on superstations, the FCC reimposed its *syndicated exclusivity* (syndex) rule in 1988. Syndex requires each cable system, at a local television station's request, to delete from its schedule any superstation programs that duplicate programs for which the station holds exclusive rights.

The rule spawned a new industry: companies that supply programs that cable systems can insert on a superstation channel to cover deleted programs. TBS, meanwhile, worked out a *blackout-proof* schedule, free of syndicated programs subject to exclusivity clauses.

Pay-Cable Networks

Subscribers pay their cable systems an additional monthly fee to receive *pay-cable*, or *premium*, networks. In exchange, the subscriber gets programs (recent movies, sporting events, music concerts, etc.) without commercial interruption. Pay-cable subscription fees average about $10 per month for each of the more popular premium channels. The cable operator negotiates the fee with the program supplier, usually splitting proceeds 50/50.

Included under pay cable are *pay-per-view* (PPV) networks. Instead of paying a monthly fee, PPV subscribers pay a separate charge for each program they watch, much as they would at a movie theater.

Although their programs resemble those of regular pay-cable networks, PPV services offer recent theatrical releases before they are available on other channels, as well as special events that regular pay-cable networks decline to carry because of the high price tag. Professional wrestling and boxing matches typically draw the largest PPV audiences. Like other program services, PPV networks take advantage of multiplexing, offering their fare on several channels. This allows them to offer more programs and to schedule them more times each day.

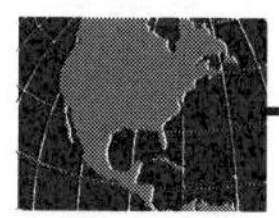

6.6 *Advertising Basics*

Unlike cable television, broadcasting in America operates primarily as an advertising medium. Dependence on advertising colors every aspect of broadcast operations. As Exhibit 6.f indicates, broadcast television ranks as the largest *national* advertising medium, though newspapers surpass television in total advertising volume. Radio comes fifth in *total* advertising dollars, after direct mail and the Yellow Pages but ahead of magazines. Advertising on cable, although rapidly increasing, remains small relative to advertising on other electronic media.

Local, National-Spot, and Network Ads

Broadcast advertising falls into three categories, defined by coverage area and image: local, network, and national spot.

Local advertising comes mostly from fast-food restaurants, auto dealers, department and furniture stores, banks, food stores, and movie theaters. When such local firms act as retail outlets for nationally distributed products, the cost of local advertising may be shared by the local retailer (an appliance dealer, for example) and the national manufacturer (a maker of refrigerators). This type of cost sharing, known as *cooperative advertising*, or just *co-op*, supplies radio with a major source of its revenue.

When a station connects to a national network, it instantly becomes a medium of national advertising. For advertisers of nationally distributed products, *network advertising* has significant advantages:

- The advertiser can place messages on more than 200 stations of known quality, strategically located to cover the entire country.
- Advertising can be placed on all those stations with a single transaction.
- The advertiser has centralized control over commercial messages and assurance that they will be delivered in the chosen times and in a preferred program environment.
- The advertiser benefits from sophisticated network market research.
- The advertiser gains prestige from the very fact of being on a national network.

Despite those advantages, some national advertisers find networks too costly or too inflexible. They have the option of using *national-spot advertising*. Advertisers using national-spot advertising work through their advertising agencies and stations' national sales representatives to assemble ad hoc collections of nonconnected stations. The commercial announcements go out to the chosen stations by mail or by satellite. As Exhibit 6.g shows, television depends heavily on national-spot sales.

National-spot advertisers can choose from several program vehicles: station-break spots between network programs, participating spots in local or syndicated programs, and sponsorship of either local or syndicated programs. National-spot advertising thus enables advertisers to capitalize on audience interest in local programs, something the network advertiser cannot do. Exhibit 6.h shows that the largest national advertisers use spot and network in combination to achieve better coverage than either could yield on its own.

Pros and Cons

As an advertising medium, broadcasting has unrivaled access to all family members under the changing circumstances of daily living. In addition, car

Exhibit 6.f

Advertising Volume of Major Media: Local versus National

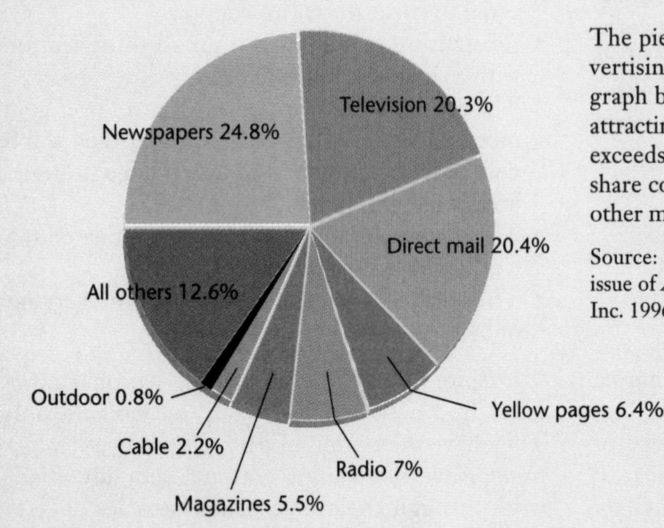

The pie chart at left shows the percentage of all advertising dollars allocated by media type. As the bar graph below indicates, newspapers retain the lead in attracting local advertising dollars, while television far exceeds newspapers in national. Cable, although its share continues to rise rapidly, still lags far behind all other major media except outdoor.

Source: Reprinted with permission from the May 20, 1996 issue of *Advertising Age.* Copyright, Crain Communications Inc. 1996.

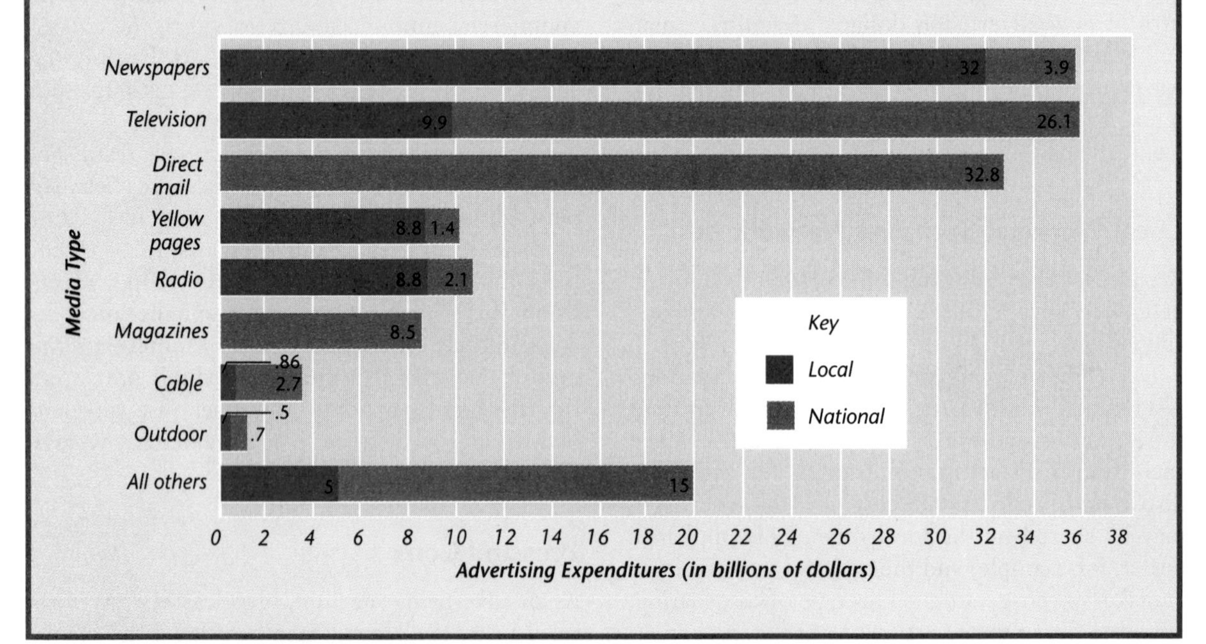

Exhibit 6.g

Broadcast Station and Cable System Revenue Sources

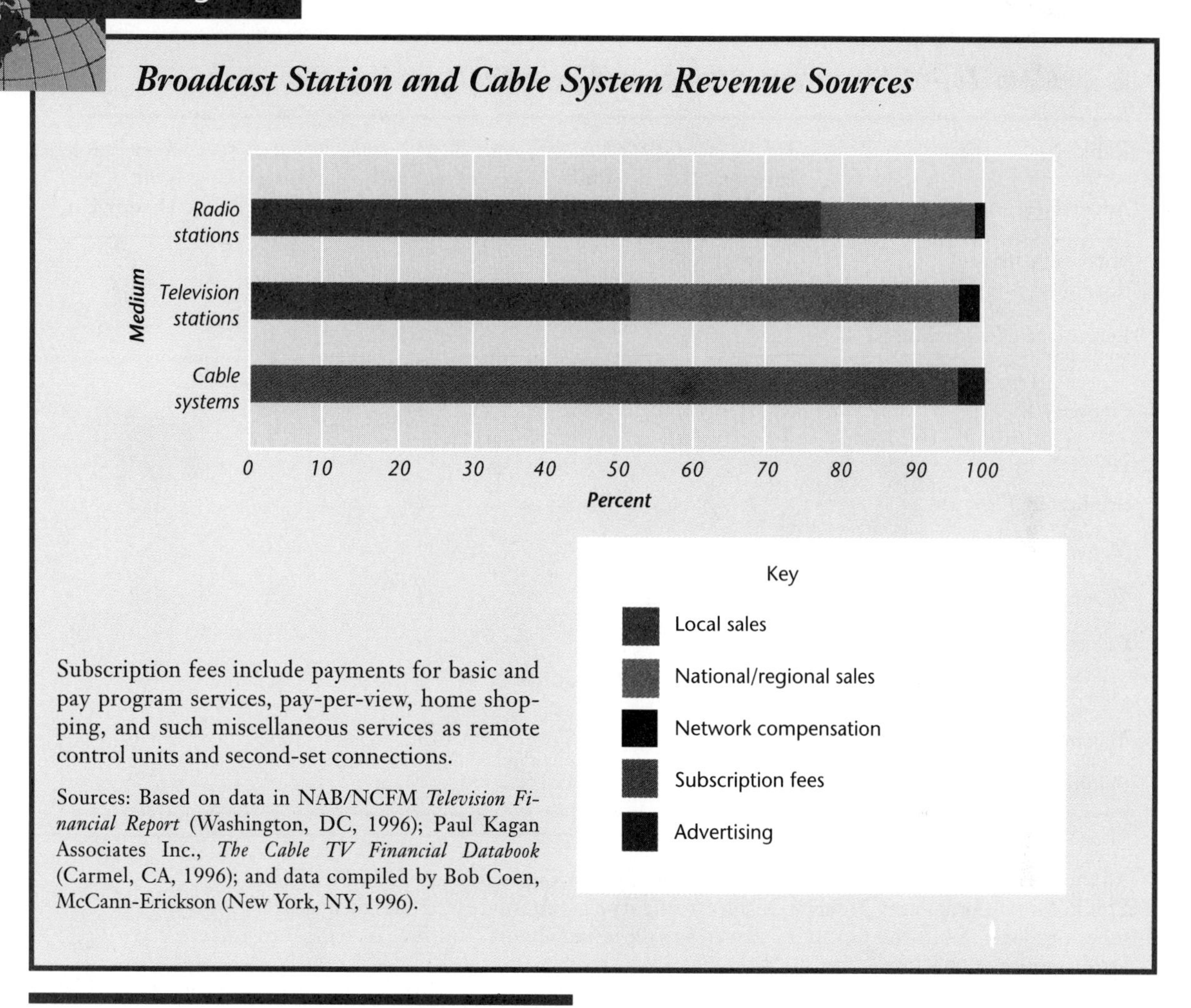

Subscription fees include payments for basic and pay program services, pay-per-view, home shopping, and such miscellaneous services as remote control units and second-set connections.

Sources: Based on data in NAB/NCFM *Television Financial Report* (Washington, DC, 1996); Paul Kagan Associates Inc., *The Cable TV Financial Databook* (Carmel, CA, 1996); and data compiled by Bob Coen, McCann-Erickson (New York, NY, 1996).

and portable radios allow broadcasting to compete with magazines and newspapers as an out-of-home medium. Television is the only medium that can vividly demonstrate how a product works. Above all, the constant availability of broadcasting as a companionable source of entertainment and information gives it a great psychological advantage.

However, commercials must make their point quickly. Normally most radio commercials last 60 seconds, and most television commercials 30 seconds, though there is a growing trend toward 10- and 15-second spots. Even the longest spot cannot duplicate the impact of a large printed department-store ad or a supermarket ad with clip-out coupons. Listeners or viewers cannot clip out broadcast ads to consult later. Nor can broadcasting compete effectively with the classified sections of newspapers, despite attempts with want-ad and home-shopping programs.

A broadcasting station has a limited commercial *inventory* of openings for commercials in its program schedule. Its single channel allows it only 24

Exhibit 6.h

How Top Advertisers Allocated Their Budgets in 1995

Rank Advertiser	1 Procter & Gamble	2 Phillip Morris	3 General Motors	4 Time Warner	5 Walt Disney Co.
Total estimated ad expenditures (in billions)	$2.8	$2.6	$2.05	$1.3	$1.3
Percentage of expenditures allocated to—					
Network TV	23	18	24	9	16
Spot TV	7	8	14	7	14
Syndicated TV	7	5	2	4	2
Network radio	*	*	1	*	*
Spot radio	*	*	1	1	2
Cable networks	6	3	4	3	2
Newspaper	*	*	4	18	17
Magazines	9	15	21	10	6
Outdoor	*	3	*	*	*

* Less than 1 percent

All these premier advertisers chose broadcast television for their major expenditures. Even Phillip Morris, prohibited by law from advertising its cigarettes on electronic media, sells other products in sufficient quantity to warrant allocating almost a third of its budget to broadcasting and cable.

Source: Reprinted with permission from the September 30, 1996 issue of *Advertising Age*. Copyright, Crain Communications Inc. 1996.

hours of "space" each day. Moreover, only so many commercials can be crammed into the schedule without alienating the audience. Print media, in contrast, can expand advertising space at will simply by adding pages.

Broadcasting also suffers the drawback of being unavailable to advertise some products. Congress forbade cigarette advertising in 1971, and most broadcasters voluntarily ban such products as hard liquor and "adult" movies.

Integrating Ads into Broadcasts

Most broadcast advertising takes the form of *announcements*, or *spots*. Advertisers usually prefer *scatter buying*—spreading their spots over several programs. That avoids risking too much on any one program and gains exposure to varied audiences.

Some programs have natural breaks where spots can be inserted without interrupting the flow—between rounds of a boxing match or between musical selections on a radio show, for example. In other cases the break must be artificially contrived. The art of writing half-hour situation comedies includes building the plot to successive break-off points for insertion of commercials. Viewers often complain about arbitrary interruptions of theatrical feature films, whose scripts do not, of course, provide seemingly natural climaxes every ten minutes to accommodate commercial breaks. Some viewers also object to television "official" time-outs during football games, taken for the convenience not of players but of advertisers.

As a matter of both law and custom, stations make *station breaks*—interruptions in programming for the insertion of *identification announcements* (IDs). Commercials inserted during these interruptions are called *station-break announcements.*

Networks interrupt their program feeds periodically to allow affiliates time to insert mandatory IDs. They also leave time for affiliates to insert commercial announcements sold by the stations to local or national-spot advertisers. Because of their close association with popular network programs, commercials inserted in these *network adjacencies* are especially valuable.

Two quasi-commercial types of announcements also occur where commercials normally appear. *Promotional announcements* (promos) call attention to future programs of networks and stations. Most broadcasters consider on-air promotion their most effective and cost-efficient audience-building tool. *Public-service announcements* (PSAs), though they resemble commercials, are broadcast without charge because they promote noncommercial organizations and causes. PSAs give stations a way of fulfilling public-service obligations and, along with promos, serve as fillers for unsold commercial openings.

Time devoted to clusters of commercial and other announcements at station breaks, once limited to a minute or two, has constantly edged upward. Advertisers as well as audiences often complain of this proliferation, which is referred to as *commercial clutter.*

Cable Advertising

Cable-system operators face more complexities than do broadcasters when inserting commercials in their programs. They have to deal with many channels, each with different requirements for advertising insertions. Modern systems use computer programs triggered by electronic signals originated by cable networks to insert locally sold commercials automatically into cable-network programming. Advertiser-supported cable networks and superstations include commercials within programs, in the same way as do broadcast networks.

Perhaps more than anything else, lack of adequate statistical information about cable audiences has impeded the growth of cable advertising. Cable's many program choices so fractionalize its audience that few channels attract sufficient viewers to yield statistically valid measurements. Although in the past rating services supplied audience information for only the most popular national cable networks, plans are under way for a variety of local cable measurement services (see Section 10.7).

6.7 *Advertising Rates*

Two sets of variables affect the price of broadcast advertising time, one set relatively stable, one relatively dynamic. Station managers have little or no control over *market size*, *station facilities* (frequency, power, antenna location), and *network affiliation*, if any. Station managers *do* control *programming*, *promotion*, and *sales*. Good management can lure demographically desirable audiences away from competitors by offering popular programs supported by effective promotion. An efficient sales department can lure advertisers away from competitors with persuasive arguments, good prices, and careful follow-up.

Most large advertisers make their electronic media buys based on ratings, the statistical measurement of

Exhibit 6.i

TV Station Rate Card

Rate Card No. C41b, Effective January 12

F = Fixed
P = Pre-emptible on two weeks' notice
Q = Pre-emptible on one week's notice
I = Immediately pre-emptible

All spots rotate horizontally and vertically

60-second = double the 30-second rate
20-second = 30-second rate
10-second = 50% of 30-second rate

Announcements between time blocks take the rate of the higher block.

			30-Second					
			F	**P**	**Q**	**I–1**	**I–2**	**I–3**
Daytime & Fringe								
Mon–Fri	7–9	Wake Up Call	$ 60	$ 50	$ 40	$ 30	$ 20	$ 10
	9–10	*Regis & Kathy*	70	60	50	40	30	20
	10–12	Morning Block	70	60	50	40	30	20
	12:30–4	Afternoon Block	80	70	60	50	40	30
	4–5	*Oprah Winfrey*	100	90	80	70	60	50
	5–6	News	140	130	120	100	80	60
	6–6:30	News	200	180	170	160	150	140
	7–7:30	*Wheel of Fortune*	240	220	210	200	190	180
	7:30–8	*Jeopardy!*	240	220	210	200	190	180
	11–11:30	News	140	130	120	100	80	60
	11:30–1:30	Movie	70	60	50	40	30	20
Weekend								
Saturday	7–12:30	Kids' Rotation	80	70	60	50	40	30
	1–2	Wrestling	90	80	70	60	50	40
	2–5:30	Various	80	70	60	50	40	30

This represents only a portion of a hypothetical small- to medium-market station's sales rate card. Many stations publish no card at all; others use them only for internal operations and do not make them publicly available. As advertiser demand increases or decreases, sales managers direct their account executives to charge customers higher or lower rates listed for each program or time period.

audiences exposed to a product's commercials (see Chapter 10). In particular they order spot schedules designed to achieve a predetermined number of *gross rating points* (GRPs)—that is, to reach an overall number of viewers—or *target rating points* (TRPs), to reach audience subgroups such as teens or women 18 to 49 years of age.

Cost per Thousand

No standard formula for setting appropriate broadcast rates exists. Market forces, however, eventually tend to bring prices into line. The industry uses *cost per thousand* (CPM) to compare relative advertising costs. CPM is the cost of reaching 1,000 (represented by the Roman numeral "M") households or other defined targets.

CPM is calculated by dividing the cost of a commercial by the number of homes (in thousands) that it reached. Advertisers typically pay an average CPM of about $10 for a prime-time spot on one of the four major television networks, and about $6 on cable networks.

CPM calculations enable advertisers to compare the cost of advertising on one medium with the cost on another, one station with another, and one program with another. These and similar measurements also enable a station or network to guarantee a specific audience in advance. Advertisers whose commercials fail to reach the promised viewer level usually receive additional commercials—*make-goods*—at no cost (almost never does a broadcaster give refunds), a practice called selling *on the come.* When the 1994 CBS miniseries *Scarlett,* for example, averaged an 18.5 rating, the network had to deliver a number of make-goods to advertisers who had paid as much as $325,000 for a 30-second commercial in the program, based on CBS guarantees of a 24 rating.

Price Variables

Broadcast advertising depends for its effectiveness on cumulative effect. A buyer therefore normally contracts for spots in groups (a *spot schedule* or a specially priced *spot package*). Prices vary according to the number of spots purchased (*quantity discounts*) and other variables, such as the following:

- *Time classes.* Typically stations divide their time into specific dayparts, and even subclasses of dayparts, with different prices for each
- *Spot position.* For a guaranteed place in the schedule, advertisers are willing to pay the premium rate charged for *fixed-position* spots. Less expensive *run-of-schedule* (ROS) spots may be scheduled by a station anywhere within the time period designated in the sales contract. Stations often *rotate* spots, both *horizontally* (over different days) and *vertically* (through different time periods) to give advertisers the benefit of varying exposures for their commercials.
- *Pre-emptibility.* Stations charge less also for pre-emptible spots, which advertisers buy on the understanding that their contract will be canceled, or their ads played at another time, if a higher-paying customer turns up.
- *Package plans.* Stations offer at discount a variety of packages, which may include several spots scheduled at various times and on various days.

A television station may offer more than a hundred different prices for spots. Most publish *rate cards,* often in *grid* form (see Exhibit 6.i), although some limit their distribution to station personnel. Rate levels can be defined quite arbitrarily, enabling the station to quote different prices for the same spot. Such a grid gives sales personnel great flexibility in negotiating deals without having to resort to under-the-table rate cutting.

The rate for the same spot position in the same program may change over the course of a season if the audience for the program rises or falls significantly. In the late 1990s the cost of a 30-second network spot in a regular prime-time television program averaged about $125,000. Highly rated programs and special events may cost much more: A 30-second spot on *Seinfeld* cost $550,000 in 1996, and Fox sold its 1997 Super Bowl spots for $1.2–$1.3 million per half-minute.

Radio-network ad rates range widely, influenced by daypart and audience reach. In the 1990s the average spot on a major radio network ran about

$1,000. Ad rates cover time charges only, not the cost of producing the commercials.

Cable Rates

Cable advertising rates vary according to three levels of audience potential: a single cable system, several interconnected systems, and national program services.

Cable systems offer many advertising opportunities in both locally originated programs and national program services. Ad rates vary accordingly, from as little as $2 for a listing in a system's classified ad channel to $400 and sometimes higher for a 30-second local commercial inserted into one of the more popular cable networks. Local cable operators are fond of marketing their advertising opportunities with the line "all the advantages of TV advertising but at radio rates."

System interconnects differ from cable networks. A network may deliver its programs to thousands of systems throughout the nation, but interconnects involve mostly the placement of commercials on a number of systems within a geographic region. The area covered by such a group of systems may represent anywhere from as few as 13,000 subscribers, as in the case of Rock Springs/Green River, Wyoming, to more than 3.5 million in the metropolitan New York City interconnect called WNYI. This wide variation in interconnect configuration results in rates that may run as low as $20 to as high as several thousand dollars for a single 30-second spot.

Cable networks and superstations, though national in scope, do not necessarily command the highest advertising prices. Again, audience size and composition determine price. In the 1990s, on average, fewer than 3 percent of U.S. television households tuned in to any advertising-supported cable program service at any given moment. Accordingly, daytime rates varied from the low hundreds to the thousands of dollars per commercial. In prime time, A&E, for example, charged as little as $3,800 per 30-second spot, while CNN and TBS charged as much as $11,000.

Alternative Ad Buys

Some types of ad-buying deals, used by both broadcasters and cable operators, fall outside normal rate practices.

- In a *trade deal* (also called a *tradeout*), a station or cable system exchanges commercial time for an advertiser's goods or services.
- In *barter deals*, stations exchange advertising time for programs. These are discussed in detail in Section 8.2.
- *Time brokerage* refers to the practice of selling time in blocks to brokers, who then resell it at a markup.
- *Per-inquiry* (PI) buys commit the advertiser to pay not for advertising time but for the number of inquiries or the number of items sold in direct response to PI commercials.

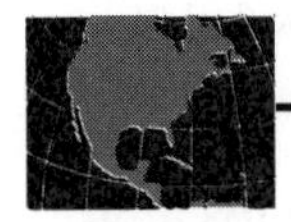

6.8 Sale of Advertising

As noted in Section 6.7, the effectiveness of its sales operation largely determines the success of any commercial enterprise.

Network Sales

The four broadcast television networks maintain their own sales departments, typically headed by a vice president, usually with offices in New York and Los Angeles and sometimes in such other cities as Chicago and Dallas. They tend to organize into specialties such as prime time, late night, news, sports, and so on, and they deal almost exclusively with advertising agencies.

Ad-supported cable networks and superstations also have sales departments. Advertisers who wish to place orders with several networks often do so through time-buying organizations, which negotiate buys on their behalf for spots on a combination of cable networks.

Local Sales

Most television station sales departments employ a general sales manager, a local (and sometimes also a national) sales manager, account executives, and support staff. The number of account executives (a fancy name for salespersons) varies from station to station. About six usually suffice for a medium-market outlet. At some stations a sales assistant or even a secretary does all the support work; some include research specialists and commercial writers on the sales staff.

Sales managers hire, fire, and train salespersons, assigning them a list of specific advertisers and ad agencies as contacts. Beginning salespeople sometimes start without benefit of such an account list (other than, by industry tradition, the Yellow Pages). They must develop their own accounts, making *cold calls* on potential new advertisers.

Salespeople usually work on a commission basis, keeping a percentage of all advertising dollars they bring to the station. This arrangement gives them incentive to make more sales to raise their income.

Spot Sales

Stations and cable systems gain access to national advertising business through *national sales representative* firms (*reps* for short). Some stations also have regional reps for nonnational sales outside the station's service area. A rep contracts with a string of stations, acting as an extension of the stations' own sales staff in national and regional markets. Television reps have only one client station in any one market, whereas radio reps often have more than one.

Reps perform many services other than sales. Their national perspective provides client stations or systems with a broader view than that of local markets. Reps often advise clients on programming, conduct research for them, and act as all-around consultants. In return for their services, rep companies collect a commission of 8 to 15 percent on spot sales they make for their clients.

So-called *unwired networks* offer an alternative way of selling national-spot advertising. Under this concept, companies buy commercial time, usually in bulk and at a discount, from television stations throughout the United States. Then they sell it, at a markup, to national advertisers. One such organization, for example, offers a package of spots in prime-time movies on independent stations around the country, selling time on many stations in a one-invoice transaction. Traditional sales reps strongly oppose the unwired-network concept, viewing it as a threat to their exclusive national representation of client stations.

Cable Sales

Advertising-supported cable networks deliver relatively small audiences compared with the four major broadcast networks. However, some advertisers target the specialized audiences attracted by cable's dedicated channels (sports fans who watch ESPN, for example).

Because of cable networks' limited audience reach, advertisers are likely to place commercials on several different networks, which can be a bothersome task. Time-buying organizations, such as Cable One, solve this problem by negotiating deals on behalf of advertisers for spots on any combination of cable networks.

Advertising on the Internet

One of the hottest areas of advertising growth in the mid-to-late 1990s involved the Internet, specifically the World Wide Web. Advertisers increasingly place ads on the Web, principally in association with browser home pages (such as Netscape), search engines (such as Yahoo and Magellan), and at Web sites that attract a desired demographic. Midway through 1996, nearly half of the leading 100 national advertisers in the United States had begun advertising on the Web. In 1995, ad expenditures on the Web totaled only $35 million, but that figure ballooned to over $300 million in 1996. Some estimates place ad spending on the Internet at over $2 billion by the year 2000.

Advertising Agencies

All regional and national advertisers and most large local advertisers deal with media through advertising agencies. Agencies conduct research; design advertising campaigns; create commercials; buy time from cable systems, broadcast stations, and cable and broadcast networks; supervise implementation of campaigns and evaluate their effectiveness; and, finally, pay media on behalf of advertisers they represent.

Agencies become intimately familiar with each client's business problems, sometimes even assisting in the development of new products or the redesigning and repackaging of old ones. Ad agencies that handle large national accounts determine the best *media mix* for their clients, allocating budgets among the major media, as indicated in Exhibit 6.h.

For decades, ad agencies traditionally received a 15 percent commission on *billings*—the amount advertising media charged. That is, an agency would bill its client the full amount of advertising time charges, pay the medium 85 percent, and keep 15 percent as payment for its services. By the 1990s, however, fewer than half of all ad-agency arrangements used this scheme. Some agencies now accept less than 15 percent or charge fees in addition to commission; some work on a straight-fee basis; and some work on a "cost-plus" basis.

In any event, that media allow a discount on business brought to them by agencies creates an odd relationship: The agency works for its client, the advertiser, but gets paid by the medium in the form of a discount on time charges. The travel business operates similarly. A travel agency works, at least theoretically, for the traveler but gets paid by the hotel or airline in discounts on charges.

Proof of Performance

After commercials have aired, advertisers and their agencies need evidence to show that contracts have been carried out. Broadcast stations log the time, length, and source of each commercial when broadcast. These logs provide documentary proof of contract fulfillment. Broadcast business offices rely on logs when preparing proof-of-performance affidavits to accompany billing statements. At many stations, computers do the logging automatically. Some stations also make slow-speed audio recordings of everything they air as backup evidence in the event of a dispute, retaining the tapes for three to six months. Cable systems rely for confirmation mostly on equipment that not only inserts commercials into various cable programs but also provides verification of proper performance.

Advertisers and agencies can get independent confirmation of contract fulfillment by subscribing to the services of such companies as Competitive Media Reporting, a joint venture of Arbitron and VNU, that conducts systematic studies of radio and television commercial performance. Its MediaWatch service (formerly Broadcast Advertisers Reports, or BAR) uses a computerized electronic pattern-recognition system to monitor television announcements on ABC, CBS, NBC, and Fox; on 12 cable networks; and on stations in 75 markets. Its Network Radio Service records and analyzes commercials broadcast on 14 radio networks.

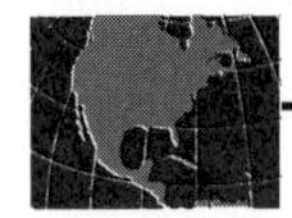

6.9 Advertising Standards

Advertising raises touchy issues of taste, legality, and social responsibility. Both legal and voluntary self-regulatory standards influence what may be advertised and what methods may be used.

Government Regulation

Section 317 of the Communications Act requires reasonably recognizable differences between radio/television commercials and programs. A station must disclose the source of anything it puts on the air for which it receives payment, whether in money or some other "valuable consideration."

This *sponsor identification rule* attempts to prevent deception by disguised propaganda from unidentified sources. Of course, anonymity is the last thing commercial advertisers desire. But propagandists who use *editorial advertising* (sometimes called *advertorials*) may not always be so anxious to reveal their true identity; nor do those who make under-the-table payments, to disc jockeys or others for on-the-air favors, wish to be identified. Outright deception in advertising comes under Federal Trade Commission jurisdiction.

Contrary to popular opinion, neither Congress nor the FCC ever set a maximum number of commercial minutes per hour of programming—by broadcast or on cable—other than those that limited advertising content in children's shows (see Section 13.9).

Program-Length Commercials

The FCC once prohibited program-length commercials—productions that interweave program and noncommercial material so closely that the program as a whole promotes the sponsor's product or service. The FCC lifted the ban on program-length commercials in 1981 for radio and in 1984 for television. They are still illegal when directed toward a child audience (see Section 13.9).

A flood of program-length commercials, by then known also as *infomercials*, ensued, often touting merchandise of questionable value. Ads for nostrums, kitchen gadgets, astrology charts, and the like overran both broadcast and cable. Despite the aura of sleaze surrounding infomercials, however, some began to achieve respectability. General Motors, for example, used the form to introduce its new line of Saturn cars. And Bell Atlantic experimented in 1992 with what it called a *sitcommercial*, an infomercial masquerading as a situation comedy. By the mid-1990s, some LPTV stations devoted their entire program schedules to infomercials.

Taboo Products

Some perfectly legal products and services that appear in print and on billboards never appear in electronic mass media. This double standard confirms that special restraints are imposed on broadcasting and, to a lesser degree, on cable because they come directly to the home, accessible to all.

Nevertheless, canons of acceptability constantly evolve. Not until the 1980s, for example, did formerly unthinkable ads—such as those for contraceptives and those showing brassieres worn by live models—begin to appear.

Congress banned the broadcasting of cigarette ads in 1971 and later extended the ban to include all tobacco products. Up until 1996, the most conspicuous example of self-imposed advertising abstinence involved the industry's refusal to accept hard liquor ads. Broadcasters were concerned that carriage of such ads would provoke opponents of all alcoholic beverages to begin an assault on beer and wine ads. In 1996, however, several television stations began accepting advertising from Seagrams for two of that company's hard liquor products. By the end of the year, over 20 stations and at least one cable network were airing Seagram's commercials. As feared by many in the industry, critics of such advertising, both inside and outside the government, called for the ads to be banned, much as cigarette ads had been 25 years earlier. Congress, the Federal Trade Commission, and the FCC began considering the issue—at least informally—at the beginning of 1997. The FCC subsequently declined to initiate a formal inquiry, leaving that task to the FTC.

Self-Regulation

The ban on liquor ads is an example of the voluntary self-regulation once codified by the National Association of Broadcasters (NAB). Its radio and television codes, though full of exceptions and qualifications, set nominal limits on commercial time. For example, NAB allowed radio 18 minutes per hour; and it allowed network-television affiliates 9.5 minutes in prime time, 16 minutes at other times.

In 1984 the Justice Department charged that the NAB standards, even though voluntary, violated antitrust laws by urging limits that reduced competition. The NAB promptly disbanded its Code Office, apparently relieved to be rid of a thankless task.

Meanwhile the television networks, and some group broadcasters, had begun dismantling their individual program and advertising codes. The networks had separate departments, variously called Continuity Acceptance, Broadcast Standards, and

Program Practices. In the late 1980s ABC and CBS sharply cut back the number of employees assigned to standards and practices departments. Advertisers, agencies, and others expressed fears that these moves might result in an overall lowering of standards. Adverse public reaction, they felt, could be followed by attempts at governmental intervention.

Unethical Practices

Aside from issues of advertising length and content, four specific types of unethical advertising practices in broadcasting have proved particularly troublesome. In the past they triggered both FCC and congressional action.

A conflict of interest occurs when a station or one of its employees uses or promotes on the air something in which the station or employee has an undisclosed financial interest. Called *plugola*, this practice usually results in an indirect payoff. A disc jockey who gives unpaid publicity to her or his personal sideline business is an example.

Direct payments to the person responsible for inserting plugs usually constitute *payola*. It typically takes the form of under-the-table payoffs by recording-company representatives to disc jockeys and others responsible for putting music on the air.

Local cooperative advertising sometimes tempts stations into *double billing*. Manufacturers who share with their local dealers the cost of local advertising of their products must rely on those dealers to handle cooperative advertising. Dealer and station may conspire to send the manufacturer a bill for advertising higher than the one the dealer actually paid. Station and dealer then split the excess payment.

Clipping occurs when affiliates cut away from network programs prematurely, usually to insert commercials of their own. Clipping constitutes fraud because networks compensate affiliates for carrying programs in their entirety with all commercials intact.

In keeping with deregulation policy then in vogue, in 1986 the FCC redefined billing frauds as civil or criminal matters, not FCC violations. It left the networks to solve clipping problems on their own by bringing suit against offending stations—a cumbersome procedure at best. The Commission did say, however, that it would consider false-billing charges when judging a licensee's character during licensing proceedings.

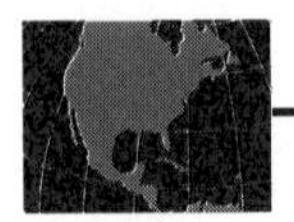

6.10 Subscription-Fee Revenue

In sharp contrast to broadcasting, cable television, wireless cable, and DBS systems rely on subscription fees for most of their revenues. In the 1990s such systems averaged about $35 total monthly revenue per subscriber. Of that amount, only about two dollars per subscriber per month came from advertising.

Cable Fee Regulation

If a graph could chart the history of cable television rate regulation, it would resemble a very wild roller coaster. Regulation of cable rates was at the discretion of local franchising authorities during cable's early history. When the FCC established its comprehensive cable regulatory scheme in 1972, it *required* local governments to regulate basic cable rates, only to rescind that requirement several years later. At the height of the deregulatory movement in the mid-1980s, Congress effectively deregulated cable rates for the vast majority of systems. This led to a period of rapidly rising prices and increasing consumer complaints.

Finally, in 1992 Congress enacted legislation that required the FCC to assume supervisory responsibility over most cable rates. The Commission responded in early 1993 and again in 1994 with a series of rulings that not only limited most cable subscriber fees but also required some cable systems to reduce their prices and even make refunds. By 1996, the regulatory pendulum had begun to reverse direction, and Congress moved to end much cable television rate regulation, effective in 1999. (See Section 12.8 for rate regulation details.)

Cable Service Tiers

Some cable systems charge a single monthly rate for their service. But most divide their product into several levels of program service, called *tiers*, with a separate fee for each level.

Most modern systems offer a *basic service* that includes local television stations, one or more distant superstations, and some advertiser-supported cable networks. The monthly fee for this basic package can vary from a few dollars to $25 or more. Some systems break their basic service into two or more tiers. For example, they may pull several of the more popular ad-supported networks (such as MTV and ESPN) out of the basic package and offer them separately as an *extended* or *expanded* basic service at extra cost.

At first in anticipation of, and later in reaction to, a provision in the 1992 Cable Act calling for governmental rate regulation of the basic program tier, some systems restructured their service. Many now offer a *broadcast basic* or *lifeline* tier limited to local stations and public access, educational, and government (PEG) channels. They then offer ad-supported networks and superstations only on one or more expanded tiers. Some systems also offer *a la carte* program options, charging subscribers a per-channel fee for individual ad-supported program services, or for groups of services that are *bundled* together.

The next level of service includes *pay-cable*, or *premium*, channels, such as HBO, Showtime, and Cinemax. In the 1990s more than 40 million homes—about three-fourths of all cable households—subscribed to pay cable. Usually subscribers pay a separate fee for each pay service they select. Such fees range from about $2 to $20 or more per service per month, averaging about $10.

Some cable operators require that their customers subscribe not only to the basic but also to an expanded-basic tier before they can buy any premium services. The 1992 Cable Act's *antibuythrough* provision prohibits this practice, although systems were given ten years to comply with the new rule.

Pay-per-View Programs

Cable systems with *addressable* converters dedicate one or more of their channels to *pay-per-view* (PPV) programs. A one-time PPV charge allows viewers to see a single program, either a movie or a special event such as a boxing match or a rock concert. By 1997 more than 30 million homes had access to PPV programs, for which they typically paid about $4 per movie and as much as $50 for each special event.

PPV programs come in one of two ways. First, individual cable systems—so-called *standalones*—negotiate directly with producers for PPV movies and events. Much more commonly, national program services—such as Request TV and Viewer's Choice—acquire PPV rights to programs and distribute them to cable systems under an arrangement that splits revenues between the program service and the cable system.

As video compression produced greater cable channel capacity, more and more systems devoted multiple channels to PPV. Some also began to experiment with two-way, interactive *video-on-demand* (VOD), a PPV technology that would compete with neighborhood VCR rental operations by permitting subscribers to order their choice of movies at the touch of a button.

Satellite-to-Home Services

In 1994 two high-power *direct broadcast satellite* (DBS) services began active marketing to potential subscribers. Hughes DirecTV offers multiple cable program services, including more than 50 pay-per-view channels, for between $35 and $40 per month, and five sports packages ranging from $8 a month for regional sports networks to $149 for a season of NBA basketball games.

United States Satellite Broadcasting (USSB) offered several different packages of service, beginning with an $8-per-month basic service that included seven advertiser-supported cable networks, and ranging up to $40 a month to get all the basics plus premium services such as Showtime and HBO.

RCA and Sony dealers market the equipment necessary to receive DirecTV and/or USSB, calling the package a Digital Satellite System (DSS). Initial cost to viewers for an 18-inch TVRO, set-top converter, and remote control ranged between $700 and $900 (plus optional professional installation charges). By 1997, pressure from new DBS competitors, notably EchoStar's Dish Network, forced retailers to lower the cost of a basic system to as low as $100 in some markets.

Another satellite-to-home service, PrimeStar, has for years been one of several companies that supply programs to homeowners who have paid $2,000 or so for their own large-diameter TVRO. PrimeStar charged about $150 for a decoder and $30 a month to view a variety of program channels. In 1994 PrimeStar increased its satellite power, thus reducing required TVRO size to 39 inches. The company now rents all necessary equipment, although the package requires professional installation at a cost of some $300—and customers also pay for programming. PrimeStar program packages, including equipment rental, average $40 to $45 per month.

Because they were nationally distributed, none of these DBS systems could include signals of local TV stations. Customers either had to switch between TVRO and rooftop antenna, or had to subscribe to basic cable to fill this void. This may change by the late 1990s as Rupert Murdoch's planned DBS system, ASkyB sought to provide local broadcast signals as part of its DBS package. (For more on DBS, see Section 3.3.)

6.11 *Personnel*

The number of people employed in an industry usually gives some indication of its importance. According to that yardstick, electronic media have relatively little importance. However, their social significance lends them far greater weight than their small workforce suggests.

Employment Levels

Exhibit 6.j provides employment data for the non-network units of the broadcast and cable industries. Overall, when networks and other related operations are included, these industries employ some 600,000 people full-time. By way of comparison, one lone manufacturing corporation, General Motors, employs nearly 800,000 people.

Many specialized creative firms support media, producing materials ranging from station-identification jingles to prime-time entertainment series. Such firms offer more opportunities for creative work than do media themselves—performing, writing, directing, designing, and so on. Other media-related jobs are found in nonprogram areas. These include jobs in advertising agencies, sales representative firms, program-syndicating organizations, news agencies, common-carrier companies, and audience-research organizations.

Aside from the major television networks, most broadcast and cable organizations have small staffs. The number of full-time employees at radio stations ranges from fewer than five for the smallest markets to about 60 for the largest, with the average being 15. Television stations have between 20 and 300 employees. A typical network affiliate employs about 90 and an independent station about 60 full-time people.

Cable systems average about 30 full-time employees but range from family-run systems in small communities (with perhaps five or six employees) to large-city systems with staffs of well over 100. Cable MSO headquarters units average about 55 full-time employees.

Salary Levels

The huge salaries reported in *People* magazine and on *Entertainment Tonight* go to top performing and creative talent and executives who work mostly at network headquarters and the production centers of New York and Hollywood.

Exhibit 6.j

Broadcast and Cable Employment

Medium	Full-time Employees (thousands)	Percent Women	Percent Minorities	Top Four Job Categories (thousands)	Percent Women	Percent Minorities
Broadcast	153	40.7	19.7	132	34.2	17.5
Cable	116	41.9	27.1	50	30.1	20.3

The top four job categories include officials and managers, professionals, technicians, and sales workers. These statistics represent broadcast and cable employment in 1995, based on reports from only those non-network units that have five or more full-time employees. The FCC, which collected this data, does not have 1995 figures for the DBS and wireless cable businesses. Total broadcast employment had been declining in recent years (its 1987 total, for example, was 176,000). Broadcast employment reached a low of 144,500 in 1993. Employment levels have increased since then. Cable television employment levels also continue to increase, up from 84,000 in 1987.

Source: FCC, *1995 Broadcast and Cable Employment Report*, Washington, D.C., 1996.

Average salaries for jobs at most stations and cable systems rank as moderate at best. Typically, those working in sales earn the highest income at broadcasting stations. They are the ones who, quite simply, bring revenue directly to the enterprise so that bills—and salaries—can be paid. At the department-head level, general sales managers usually make the most money and traffic managers the least. Program managers fall somewhere in between. Exhibit 6.k offers a more detailed salary analysis.

Typically, television stations pay higher salaries than do radio stations or cable systems for comparable positions, although in the late 1990s the salaries of radio general managers matched and then exceeded the salaries of their TV counterparts. Presumably this is a result of local radio consolidation; local radio managers may be running up to eight stations per market. Basic laws of supply and demand come into play throughout the compensation structure; because fewer people want—and are technically qualified—to be TV weathercasters than want to be sportscasters, for example, salaries for the former group tend to exceed those of the latter. Salaries for radio talent reflect the relative importance of various dayparts.

Employment of Women

The FCC enforces Equal Employment Opportunity (EEO) Act standards for broadcast stations, cable systems, wireless cable systems, DBS operators, and headquarters operations (see Section 12.5). These standards require owners to file an annual report classifying employees according to job categories, gender, and minority status.

As in all businesses, women in media often run up against corporate "glass walls" that limit their lateral movement to jobs that will provide the experience necessary for promotion to higher positions. Even when they have the appropriate qualifications, women frequently hit a "glass ceiling"—an invisible barrier that prevents their movement to the top. Exhibit 6.l, details the careers of three women who—to some extent, at least—broke through those barriers.

Exhibit 6.k

Salary Levels

In the beginning . . .
The following compares entry-level annual salaries for new college graduates in various media or related fields:

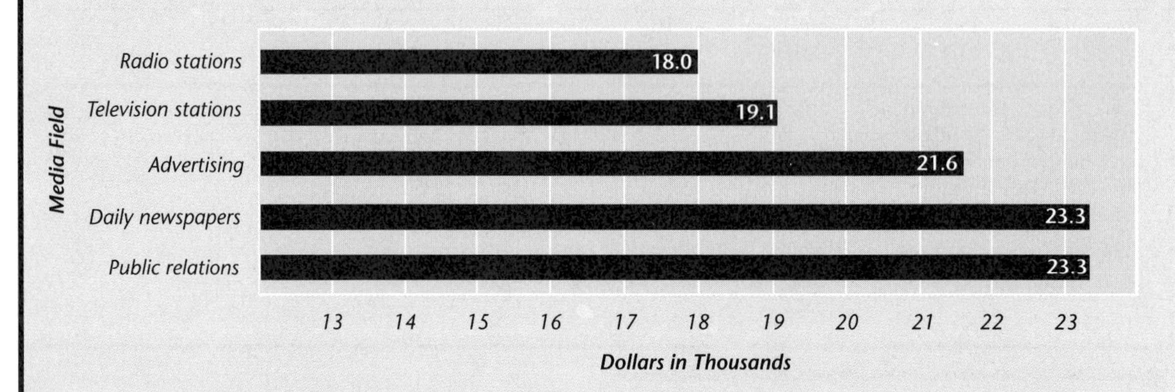

These represent nationwide *median* salaries in 1995. Some graduates found jobs at higher pay—some lower.

And on the average . . .
As they moved from entry-level to top-level positions, electronic media employees, *on the average*, earned the following annual salaries along the way. (Some salaries do not appear here because broadcast stations and cable systems use somewhat different job titles.)

In its annual employment study of all broadcast stations with five or more employees, the FCC reported that in 1995 women occupied about 41 percent of all jobs, up from 32 percent in 1979. Perhaps more important, women represented 34 percent of employees classified as officials and managers (up from 23 percent in 1979) and more than 53 percent of those classified as sales workers (up from 31 percent in 1979).

For cable systems with five or more employees in 1995, the FCC reports that women constituted nearly 42 percent of all employees and about 30 percent of officials and managers. The potentially lucrative area of sales saw major improvement: Women held more than 47 percent of those jobs, up from only 30 percent in 1983.

Minority Employment

EEO rules also require stations to report on their efforts to upgrade employment opportunities of minority-group members—defined by the FCC for this purpose as Aleutians, American Indians, Asians, Blacks, Hispanics, and Pacific Islanders. Although their overall progress has been slower than that of women, many have made outstanding

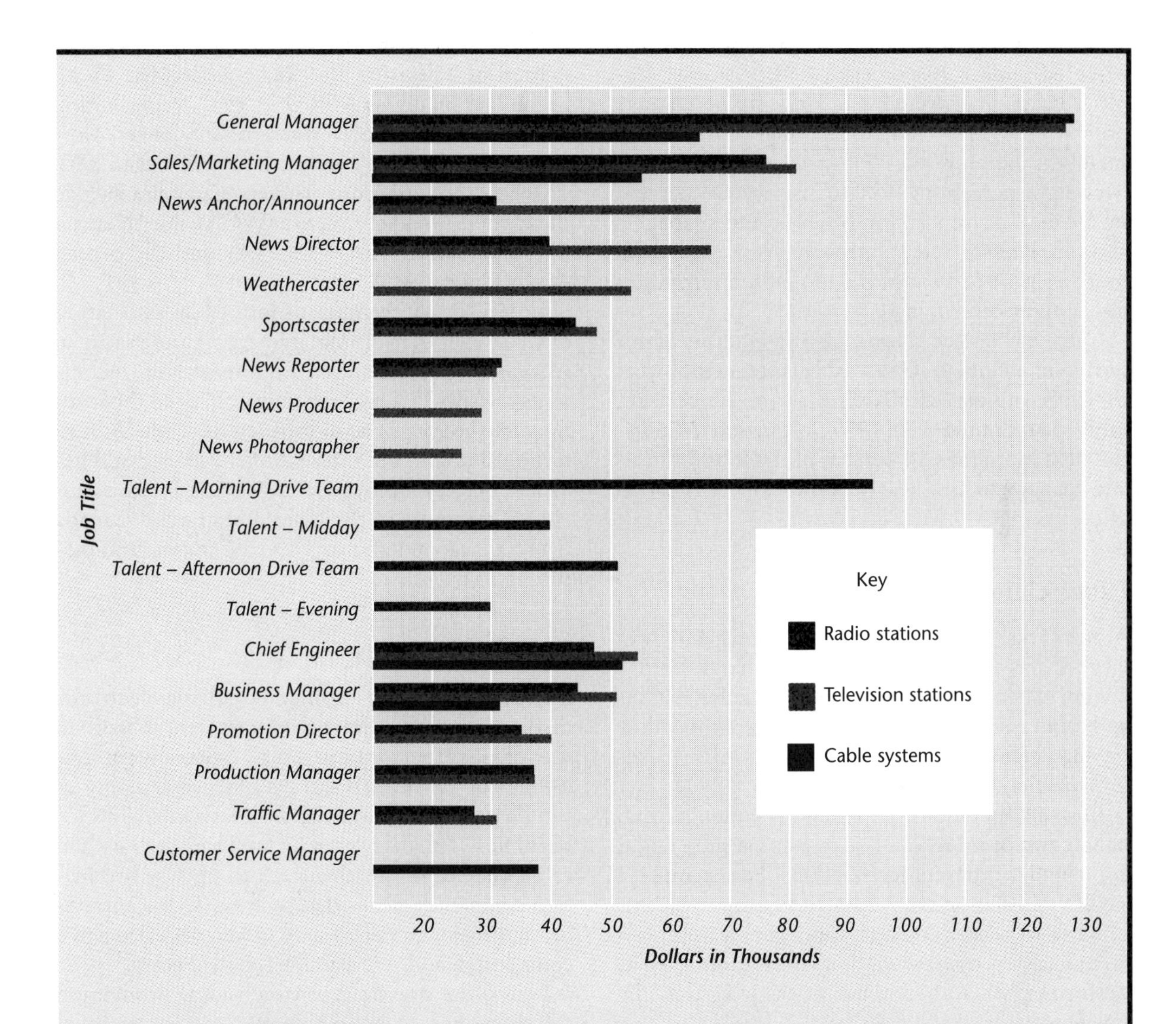

These are only *averages*. Salaries paid by organizations in large cities tend to run higher than those in smaller communities. And because averages typically rise each year, the above data prove most useful in comparing compensation levels for the various positions, rather than in studying specific salaries themselves.

Sources: Based on data in *Annual Survey of Journalism & Mass Communications Graduates*, The Ohio State University School of Journalism (Columbus, OH, 1996); *Television Employee Compensation and Fringe Benefit Report*, National Association of Broadcasters/Broadcast Financial Management Association (Washington, DC, 1995); *Radio Station Salaries*, National Association of Broadcasters (Washington, DC, 1996); and *Cable Television Salary Survey*, Jim Young and Associates (Weatherford, TX, 1994); with updating adjustments by the authors.

contributions to broadcasting and cable. Exhibit 6.m describes some of them.

The FCC reported that in 1995 minorities represented almost 20 percent of all broadcast employees, an improvement of only six percentage points over 1979. The percentage of minority members holding sales jobs increased from just over eight percent in 1979 to just over 28 percent in 1995. The report for officials and managers showed a lesser rate of growth: Minorities held more than 14 percent of the top jobs, up from just under eight percent in 1979.

Cable has outpaced broadcasting in terms of minority employment. Of all cable system employees in 1995, minorities made up about 27 percent, more than double their 1979 proportion. Minorities held more than 14 percent of top jobs (officials and managers) and nearly 28 percent of those in sales.

Labor Unions

At networks, national production centers, and most large-market network-television affiliates, unionization generally prevails. Not so at smaller stations and cable systems. The fragmentation of the industry into many units, each with small staffs, makes unionization impractical.

In small operations, staff members often have to handle two or more jobs that in a union shop might come under different jurisdictions. For example, a small radio station cannot afford to assign two employees to record an interview, paying one as a technician to operate equipment and another as performer, when the job can as easily be done by one person.

Electronic mass media unions draw on types of personnel first unionized in older industries—electrical work, music, motion pictures, stage, and newspapers. Thus the American Federation of Musicians represents every kind of professional musician, from pianists in bars to drummers on television.

People who work in broadcasting and cable can be grouped into two broad categories, the creative/performing group and the craft/technical group. Unions divide along similar lines; those representing the former usually avoid the word *union*, calling themselves *guilds*, *associations*, or *federations*.

The first pure broadcasting union, American Federation of Television and Radio Artists (AFTRA), began (originally as AFRA) in 1937, representing that universal radio performer, the announcer. Most creative/performing unions, however, began with stage and motion picture workers. Examples include the Writers Guild of America (WGA), the American Guild of Variety Artists (AGVA), and the Screen Actors Guild (SAG).

In 1953 NBC technicians formed an association of their own that ultimately became the National Association of Broadcast Engineers and Technicians (NABET). Later the union changed the word *Engineers* to *Employees* to broaden its scope. A third technical union, the International Alliance of Theatrical Stage Employees and Moving Picture Machine Operators of the United States and Canada (IATSE), expanded into TV from the motion picture industry.

Job Opportunities

Surveys of students enrolled in college electronic media programs indicate that most want to work in the creative/performing area, especially in on-camera or on-mike positions. The oversupply of candidates makes these the least accessible jobs for beginners. The delegation by broadcast stations and cable systems of creative work to outside production companies means that such work concentrates in a few major centers, where newcomers face fierce competition and where unions control entry.

News, the one field in which local production still flourishes, offers an exception to the scarcity of creative jobs at stations and systems. Nearly all broadcast stations and some cable systems employ news specialists.

Sales offers another employment area likely to expand. All commercial networks and stations, as well as a growing number of cable systems, employ salespeople. Ambitious job seekers should also note that top managerial positions have historically been filled by sales personnel, although in recent years several general managers have come from news.

Exhibit 6.1

A Woman's Place Is in the Newsroom

The path for women to equal job opportunity in broadcasting has been bumpy and long, but some progress has been made.

In 1948 ABC hired Pauline Frederick as the first woman network news correspondent. She remained the sole female hard-news network reporter for the next 12 years. In her early years, when assigned to interview the wives of presidential candidates at national political conventions, she was also required to apply their on-camera makeup. Later she became famous for her coverage of the UN, first as a correspondent for ABC, then with NBC. In 1976 Miss Frederick became the first woman to moderate a presidential debate, between Gerald Ford and Jimmy Carter. She died in 1990 at the age of 84.

In 1976 ABC hired Barbara Walters as the first woman anchor on a weekday evening network newscast. She shared the anchor desk with Harry Reasoner. Her salary: $1 million a year ($500,000 for anchoring and $500,000 for producing and hosting four entertainment specials a year), plus perquisites (a private office decorated to her taste, a private secretary, a makeup consultant, and a wardrobe person). Walters had established her reputation as co-host of NBC's *Today* show and as a successful, if sometimes controversial, interviewer of famous personalities as diverse as Princess Grace of Monaco, Fred Astaire, Ingrid Bergman, and Fidel Castro. Her tenure as anchor ended in 1977, and she moved on to more celebrity interviews and to ABC's news magazine *20/20*.

Following her graduation from Indiana University, Jane Pauley began her broadcast career in 1972 at WISH-TV in Indianapolis as a reporter and anchor. She then moved to WMAQ-TV in Chicago, where she was the first woman regularly to co-anchor a week-night news broadcast in that city. In 1976 Pauley joined the NBC network as co-anchor of the *Today* show, a position she held for 13 years. Since 1992, Pauley has co-anchored the newsmagazine *Dateline NBC*. In addition to these duties, beginning in 1996 Pauley hosts *Time and Again* on the cable network MSNBC, a daily program that chronicles past events using actual NBC news footage.

Sources: Photos from AP/Wide World (Frederick); © 1995 Capital Cities/ABC, Inc. (Walters); Corbis-Bettman (Pauley).

Pauline Frederick

Barbara Walters

Jane Pauley

Exhibit 6.m

Outstanding African Americans in Electronic Media

Although African Americans have not yet gained full equality in the media industry, many have risen to positions of prominence in broadcasting and cable. The following are but a few.

Don Barden—president of Barden Communications, the largest black-owned cable operator
Karen Barnes—vice president of Fox Children's Network
Don Cornwell—chairman, president, and controlling stockholder of Granite Broadcasting, television station group owner
Ragan Henry—owner of 10 AM and 14 FM stations
Robert Johnson—founder and president of the Black Entertainment Television cable network
Michael Moye—creator and executive producer of *Married . . . With Children*
Richard Parsons—president of media giant Time Warner
Johnathan Rogers—president, CBS Television Stations
Bernard Shaw—News anchor, CNN
Carol Simpson—News anchor, ABC News

Source: Reprinted with permission of *Broadcasting & Cable*, September 9, 1991, © 1991 by Cahners Publishing Company, with updates by authors.

Advertising agencies and national sales representative firms offer entry-level employment opportunities. Nevertheless, personnel directors frequently complain that college-educated job applicants fail to understand or appreciate the financial basics of the industry.

As technologies proliferate and competition intensifies, so does the need for creative and effective promotion. Stations, systems, and networks have to promote themselves and their products to potential audiences and advertisers. Cable networks use promotion extensively to persuade cable systems to buy their offerings.

About two-thirds of all jobs in cable are technical, but marketing, market research, and advertising also have high priority. The need for creative people will also increase as more cable systems provide public-access and local-origination programming.

Corporate Video

Applicants who look only to broadcasting and cable for entry-level jobs unnecessarily limit their employment options. Virtually every large organization that has contact with the public uses electronic mass media in one form or another. Opportunities for production and writing jobs exist in business and in religious, educational, and health organizations, foundations, government agencies, the armed services, and specialized production companies.

Many such organizations make extensive in-house use of satellite-fed, closed-circuit television. In the 1990s about 80 private business television (BTV) networks sent programs to more than 30,000 downlink sites in the United States. Most BTV organizations direct their productions to their own employees for product introductions, training, and corporate communications. The investment firm Merrill Lynch, for example, transmits three or more product-driven programs each week to its sales force in 480 offices around the country.

This rapidly growing field of corporate video applies broadcast techniques to job skills training, management development, sales presentations, and public relations. Such nonbroadcast uses of television require trained personnel for production, direction, writing, studio operations, program planning, and similar functions. They also offer the following advantages over traditional broadcast positions:

- Many broadcast stations are reducing staff sizes at the same time that corporate video organizations are enlarging theirs.

- Jobs often are available for students right out of school—even in major markets.
- Beginning salaries, benefits packages, and working hours are better.
- Employees enjoy greater job stability.
- Job security does not depend on ratings.

6.12 *Investment, Profit, and Loss*

The electronic media industries, though not as capital intensive as many manufacturing giants such as the automobile business, nonetheless require high capital investments to construct new facilities or acquire and maintain existing ones.

Capital Investments

Radio-station construction costs range from $50,000 for a simple, small-market AM or FM outlet to several million dollars for a sophisticated station in a major market. Full-power, major-market television facilities may cost tens of millions. Upgrading operations with new technologies such as SNG vans, helicopters, and newsroom computer systems call for additional capital outlays in the hundreds of thousands.

On a per-home-served basis, a cable system costs even more to construct because each home must be physically connected to the headend. Owners now pay anywhere from $650 to $1200 per home when building new systems. To *upgrade* a system—that is, to increase its channel capacity or convert from coaxial cable to optical fiber—costs between $10,000 and $15,000 per mile.

In terms of wireless delivery of multi-channel video, wireless cable and DBS systems demonstrate the range of costs to begin operations. Industry experts estimate that start-up costs for DBS systems begin at $500 million. By contrast, a 33-channel terrestrial wireless cable system costs between $500,000 and $2.5 million, depending on the power of the transmitter.

HDTV

HDTV's sharper, wider pictures will require eventual replacement of most broadcast and cable video equipment, including the home television set. Industry experts can make only rough estimates of what all this will cost. The first local television stations to convert to HDTV may pay $1.5 million to $2 million just for the equipment necessary to retransmit a network signal. To produce and transmit local programming may cost another $10 million to $12 million. Some broadcasters—who oppose the new technology and thus may not be completely objective—project that the industry may pay as much as $10 *billion* for a technology that, in all likelihood, will neither increase a station's or a network's audience nor expand its advertising revenue.

Unlike other media, electronic mass media rely on the general public—the consumer—to supply the largest part of the industry's basic capital equipment, broadcast receivers. For consumers to be able to watch HDTV in their homes on the new, wide-screen television sets (for which they may initially pay anywhere from $3,500 to as much as $6,000), whether the programs come over the air or through cable or from a (now HDTV) VCR, they will collectively pay between a conservatively estimated total of $80 *billion* and a high of $200 *billion* or more.

Ownership Turnover

Broadcasting and to a large extent cable have reached the point where investment in new facilities has come to a halt. Now the more likely route to ownership is through purchase of existing facilities.

Such factors as market size and location, format (in the case of radio), network affiliation (in the case of television), and competitive position in the market all influence asking/offering prices for stations. Through the first half of the 1990s, television stations, FM stations, and AM/FM combinations typically sold for about two-and-one half times the station's annual gross revenue, or eight to 10 times its *cash flow* (operating revenue minus operating expenses). For a standalone AM station, the price was closer to one-and-a-half times gross revenue or five times cash flow.

The Telecommunications Reform Act of 1996 changed all that. Because the national ownership limits for radio, and to a lesser extent television, were rescinded, station trading in 1996 exceeded the previous year's sales by almost 205 percent. Deals for radio groups and television stations brought prices of up to 16 times cash flow—unprecedented levels. Exhibit 6.n details the money spent on broadcast acquisitions in 1996 in comparison with 1995.

Cable systems typically sell for 10 to 12 times cash flow. Another common rule of thumb for estimating the price of a cable system assigns a dollar value to each subscribing household. In the 1990s that figure averaged between $1700 and $2000 per subscriber. Like the other formulas, this per-subscriber method yields only very rough estimates. It ignores such factors as number of homes passed by the system (which may include potential subscribers), age and channel capacity of the system, and franchise requirements.

Profits and Losses

As they moved through the 1980s and into the early 1990s, most electronic media experienced increasing financial difficulties. The nation's overall economic problems, combined with growing and ever more diverse media players, typically resulted at best in lower profits and, at worst, in losses or even bankruptcies.

For some the new decade proved more kind. In 1991, for example, ABC and Fox were the only broadcast television networks to show a profit; both CBS and NBC lost money. Two years later, after cutting costs and stabilizing their audience shares, all four were in the black. With the demise of the of fin/syn rules (see Section 8.2), and as advertising expenditures increased, the network outlook appeared positive. In 1996, for example, NBC reported pre-tax earnings of $1 billion on revenues exceeding $5 billion, a gain of 20 percent from the previous year. Nearly half of these earnings came from the NBC television network.

At the station level profits continue to rise as well. Affiliates of ABC, CBS, and NBC posted average pre-tax profits of almost $5.6 million in 1995 compared with $3.5 million only three years earlier. The average Fox affiliate earned $4.5 million in profits in 1995, and the average "independent" station weighed in with pre-tax profits of $4.2 million. Remember that these are average figures per class of station. Some major market network affiliates made as much as $75 million in 1995, while other stations lost money during the year.

On the whole, radio fortunes look positive. In 1994 radio advertising revenues exceeded $10 billion for the first time, and that figure increased to $11.3 billion in 1995. Local advertising revenues increased in 1996 by 7 percent and national revenue was up 8 percent.

Networks, on the whole, benefited from increased advertising expenditures, as did major-market stations and groups. Nationwide most FM stations and AM/FM combos operate in the black. The outlook for standalone AMs, however, especially the few remaining daytime-only operations, is grim.

Cable, on the other hand, fared less well. By the mid-1990s the cable audience had plateaued at about two-thirds of U.S. television households, and new competitors—notably DBS systems and the telephone companies—were beginning to siphon subscribers away from the cable operators. FCC-imposed cuts in subscriber fees in 1994 reduced cable system revenues and profits, and Wall Street lost some of its confidence in the industry, as evidenced by the fact that cable stocks lost 10 to 40 percent of their value in 1996. Still, things were not all bad. The industry recognized its problems and began addressing them by decreasing its debt, streamlining its administration, and consolidating its holdings. The 1996 Telecommunications Act cut

Exhibit 6.n

Dollars Flow with Dereg

Note that "deals" in the chart may include more than one station in the sale. Market analysts predicted that 1997 would see just as much action in station sales as 1996.

Station Type	1996	1995	Percent Change
TV	$10.488 billion/99 deals	$4.74 billion/112 deals	121.26%/-11.61%
AM-FM combos	$12.034 billion/345 deals	$2.79 billion/213 deals	331.33%/61.97%
FM	$2.628 billion/417 deals	$685.68 million/329 deals	283.27%/26.75%
AM	$212.82 million/254 deals	$106.76 million/195 deals	99.34%/30.26%
Total	**$25.362 billion/1,115 deals**	**$8.32 billion/849 deals**	**204.83%/31.33%**

Source: *Broadcasting & Cable.*

back on cable rate regulation to some degree, cable advertising revenues were increasing at a rate of over 10 percent per year, and cable network ratings in 1996 were up 10 to 20 percent over the previous year. Moreover, even with some continued rate regulation, cable operators managed to raise their subscriber rates by an average of 7.8 percent in 1996. Many in the cable industry predict that the last few years of the 1990s will produce an invigorated industry ready to meet the competitive challenges of the 21st century.

The outlook for the DBS and wireless cable businesses appears mixed. By the end of 1996, total DBS subscribership had reached between 3.5 and 4 million nationwide, lower than most industry predictions but not alarming enough to discourage new entrants into the business. Wireless cable, once seen as a formidable competitor to cable systems, experienced a loss of confidence among investors, notably the telephone companies. By the end of 1996 there were fewer than one million wireless cable subscribers nationwide, although some predicted that number would rise to 4 million by the end of the century.

6.13 Critique: Bottom-Line Mentality

From the public-interest standpoint, stations, cable systems, and networks need to earn profits. When they operate at a loss, their public-service programs tend to suffer first. Moreover, the lowering of standards by money-losing firms can be contagious: Their rivals tend to match lower standards to compete in the marketplace.

But obsessive concern for short-term profits has its dangers, too. Federal deregulation and permissive interpretation of antitrust laws have encouraged media acquisitions, mergers, takeovers, and consolidations. These transactions focus so single-mindedly on profit that they produce a *bottom-line mentality*—executive preoccupation with profit-and-loss statements to the exclusion of all else.

Cable television, not explicitly required to operate in the public interest, has made little or no effort to modify profit-driven goals beyond such services as C-SPAN and Cable in the Classroom. Broadcasting, which does have a public-interest mandate, and which FCC rules formerly kept mostly in the hands of professional broadcasters, fell increasingly under the control of conglomerate officials with no broadcasting background.

Trafficking in Stations

The FCC's antitrafficking rule was designed to prevent station trading at the expense of public service. It required a broadcast station licensee to operate for a minimum of three years before selling it. Deletion of this rule in 1982 enabled new licensees to make quick entrances and exits into and out of the broadcasting business for the sake of virtually overnight profits.

First-time broadcast buyers began to specialize in *leveraged buyouts* of television stations. These deals involve buying up stock to gain a controlling interest in target firms. To do so, buyers incur huge debts that have to be repaid out of profits, leaving little money for quality programming and often requiring personnel layoffs. Increasingly stations turned to quick and easy sources of income—program-length commercials, titillating shows exploiting sex and violence, tabloid pseudo-news shows, and the like.

Network Changes

The mighty networks were not immune. When new management took over at ABC, CBS, and NBC, cutbacks in operating budgets caused the layoff of thousands of employees and the early retirement of others. This new austerity took its heaviest toll in network news as veteran reporters were fired and entire news bureaus closed.

Making an unprecedented public criticism of his own network, Dan Rather, *CBS Evening News* anchor and managing editor, wrote a piece for the *New York Times* headlined "From Murrow to Mediocrity?" Rather pointed out that CBS Chairperson Laurence Tisch "told us when he arrived that he wanted us to be the best. We want nothing more than to fulfill that mandate. Ironically, he has now made the task seem something between difficult and impossible." Rather added,

> news is a business, but it is also a public trust . . . We have been asked to cut costs and work more efficiently and we have accepted that challenge. What we cannot accept is the notion that the bottom line counts more than meeting our responsibilities to the public. Anyone who says network news cannot be profitable doesn't know what he is talking about. But anyone who says it must always make money is misguided and irresponsible. (Rather, 1987)

Four years later, as network belt-tightening continued, Rather's CBS colleague Mike Wallace observed that Tisch had disappointed broadcast veterans, "probably mostly by the manner in which it was done. It was done too coolly." Wallace continued,

> When you've worked in an outfit 25 to 35 years and you've seen guys actually lay down their lives on the line in the course of doing their job and they are tossed aside, you are led to believe that "thus far and no further will be these cuts," and suddenly it's a lot further. You begin to feel a sense of disappointment and disillusionment. (Bier, 1991)

Toward the end of the 1990s network news personnel seemed to have come to terms with the new reality. Whoever owns the networks, the owners expect the news divisions to be profitable while maintaining high journalistic standards. News personnel are learning to do more with less and still get the story to the public.

Meeting the Competition

For the electronic media companies discussed in this chapter to succeed in the next millennium, they will have to respond aggressively to changes in technology and new competition. By the late 1990s, most television broadcasters realized that the movement to digital broadcasting, while involving some costs, promised some advantages as well. In addition to broadcasting in high definition, television stations might well broadcast several different services at once, finally becoming multichannel operators. Broadcasters are also exploring high-speed data delivery and the integration of traditional television and computer services. Television manufacturers have already recognized this convergence, as is evidenced by the introduction in 1996 of WebTV, a service that allows viewers to access the Internet on their television sets. Radio broadcasters will be able to compete with the sound quality of CDs as digital audio broadcasting becomes a reality.

Traditional cable operators face a host of new competitors, most notably DBS systems and the telephone companies. The cable industry has responded by aggressively rebuilding its physical facilities to accommodate more channels and digital services. The development of cable modems will allow cable subscribers to access the Internet at speeds far greater than those available via telephone lines. Some cable systems have decided to meet the telco competition head-on by offering interactive voice and data services. Wireless cable operators are planning a move to digital delivery in order to increase channel capacity, and the next generation of DBS operators may be offering local broadcast signals as part of their service package.

One thing is certain: The technology and the business of the electronic mass media will continue to change rapidly. Only by keeping up with such change and adapting to it can existing media companies hope to survive.

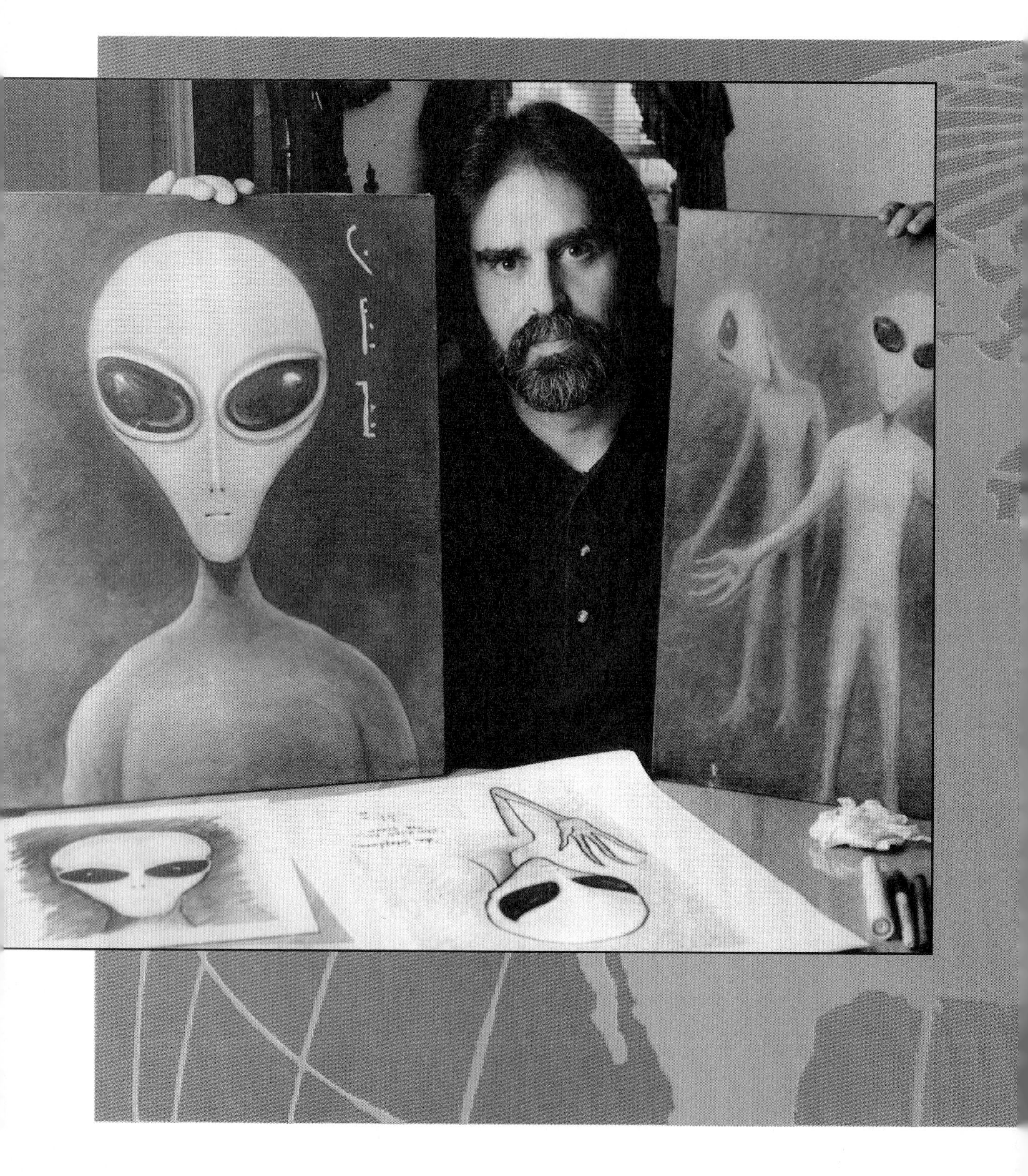

Chapter 7

Noncommercial Services

CHAPTER CONTENTS

Most of this book deals with commercial electronic media. But profit motives alone cannot be counted on to fulfill all the national cultural, educational, and informational needs that electronic media could ideally serve. Hence arose the concept of *noncommercial broadcasting*—motivated by public-service goals rather than by profit.

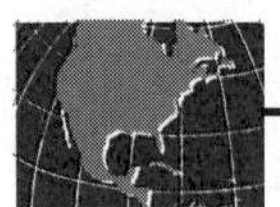

7.1 *From Educational Radio to "Public" Broadcasting*

Today's *public broadcasting* began as a more narrowly defined service devoted explicitly to education. When radio boomed in the 1920s, educational institutions joined in the rush. Most of these pioneer not-for-profit stations operated on a shoestring for only a few hours a week.

AM Radio

With broadcasting's growing financial success in the late 1920s, commercial interests began to covet AM channels tied up by educational licenses. Some schools surrendered their licenses in return for promises of airtime for educational programs on commercial stations—promises that faded with the rising value of commercial time.

Many stations that held on found themselves confined to low power, inconvenient hours (often daytime only, of limited value for adult education), and constantly changing frequency assignments. In 1927, 98 noncommercial AM stations were in operation, but by 1945 they had dwindled to about 25.

Reserved Channels

The decline in educational AM stations confirmed what critics had said from the first: Educational interests should not be expected to compete with commercial interests for broadcast channels, and the government should have set aside a certain number of AM channels exclusively for educational use.

When the Federal Communications Commission came into being in 1934, it reviewed the proposal to reserve AM channels. But the Commission accepted commercial owners' assurances that they would give ample free time for education. As stations that gave up their licenses for promises of time had already discovered, such promises never seemed to work out in practice.

The FCC finally endorsed the *reserved-channels principle* when it had an opportunity to allocate a brand-new group of channels in 1941—those for FM radio. In its final FM-allotment plan in 1945, the Commission reserved the lower portion of the FM band exclusively for educational use—20 channels (of the 100 total) running from 88 to 92 MHz.

The first noncommercial educational FM stations depended almost totally on local programs. They had neither a well-developed system of program exchange nor a network to help fill out station schedules. Their experience, of course, repeated that of commercial stations in the early 1920s—local resources alone simply could not provide adequate programming. Networking was as essential for noncommercial as for commercial broadcasting.

The arrival of television in the late 1940s stimulated a new campaign for reserved channels. It was considerably more intense than previous radio campaigns, because television attracted more attention. Educators and foundations foresaw far greater possibilities for educational television than they had for educational radio.

During the FCC's 1948-to-1952 freeze on new television-station applications, some commercial interests made a concerted effort to block the educators' campaign. But a wide spectrum of educational and cultural groups combined into a counterlobby: the Joint Committee on Educational Television (JCET). As part of its pro-reservations exhibits JCET prepared a content analysis of existing television programming (fewer than a hundred commercial television stations were then on the air). Analysis showed conclusively that commercial stations and networks did little to put television at the disposal of education and culture. JCET sounded a warning: If the forthcoming channel allotments failed to reserve channels for educational use, a once-in-a-lifetime opportunity would be lost.

The FCC took notice. Its *Sixth Report and Order*, which ended the freeze in 1952, provided for 242 reserved educational-television (ETV) channel allotments, 80 VHF and 162 UHF. Many more were added in later years, bringing total allocations to about 600, of which some 240 remain unused.

ETV Constituency

Educational radio leaders had kept the faith over many lean years, largely ignored in the seats of power. During the fight to achieve ETV channel reservations, however, the noncommercial broadcasting constituency grew larger and more diversified. Television's glamour attracted national educational, cultural, and consumer groups that previously had taken little notice of educational radio broadcasters.

Various enthusiasts developed conflicting views about the form educational television should take. Some held "educational television" to mean a broadly inclusive cultural and information service. Others took it to mean a new and improved audiovisual device, primarily important to schools and formal adult education. Some, following the model of commercial broadcasting, favored a strong national network and a concern for audience building. Others, expressly rejecting the commercial model, focused on localism and service for more limited, specialized audiences. Some wanted to stress high culture and intellectual stimulation. Others wanted to emphasize programs of interest to ethnic minorities, children, and the poor. These differences would plague public television when it appeared in the late 1960s.

A New Beginning

For its first dozen years, ETV developed slowly, long on promise but short on performance. Exhibit 7.a shows its growth rate. The few stations on the air depended on underfunded local productions and filmed programs of limited quality and interest. Their broadcast day lasted about half the length of commercial station schedules.

In 1959 many stations formed a program cooperative, National Educational Television (NET). It provided a few hours a week of shared programs, delivered to the cooperating stations by mail. Though a step in the right direction, NET still fell far short of meeting the enormous program needs of the noncommercial service.

In the mid-1960s a nonprofit foundation stepped into this bleak picture, hoping to transform it with a dynamic vision. The Carnegie Foundation felt ETV needed well-articulated national goals, top-notch public relations, and leadership at the federal level. To generate highly visible recommendations for achieving these goals, the foundation set up the Carnegie Commission on Educational Television (CCET). The Commission comprised nationally known figures from higher education, media, business, politics, and the arts. In its watershed 1967 report, *Public Television: A Program for Action*, the Carnegie Commission proposed that Congress establish a national "corporation for public television."

The CCET deliberately used the word *public* rather than *educational* to disassociate its proposals from what many had come to regard as the "somber and static image" projected by the existing ETV services. It also chose the word *public* to emphasize its recommendation for an inclusive service embracing not only formal instruction and classroom television but also a broad cultural/informational service intended for the public.

Six months after the 1967 Carnegie Commission report came out, its basic recommendations were enacted into law as part of President Lyndon Johnson's Great Society legislative program. The system it created survives to this day.

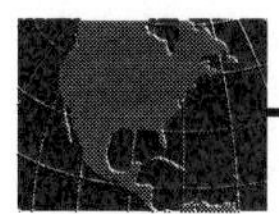

7.2 *National Organizations*

The transformation of ETV into a national public-television system succeeded in bringing it to public attention. In solving some problems, however,

Exhibit 7.a

Growth in Number of Noncommercial Broadcasting Stations (1925–1997)

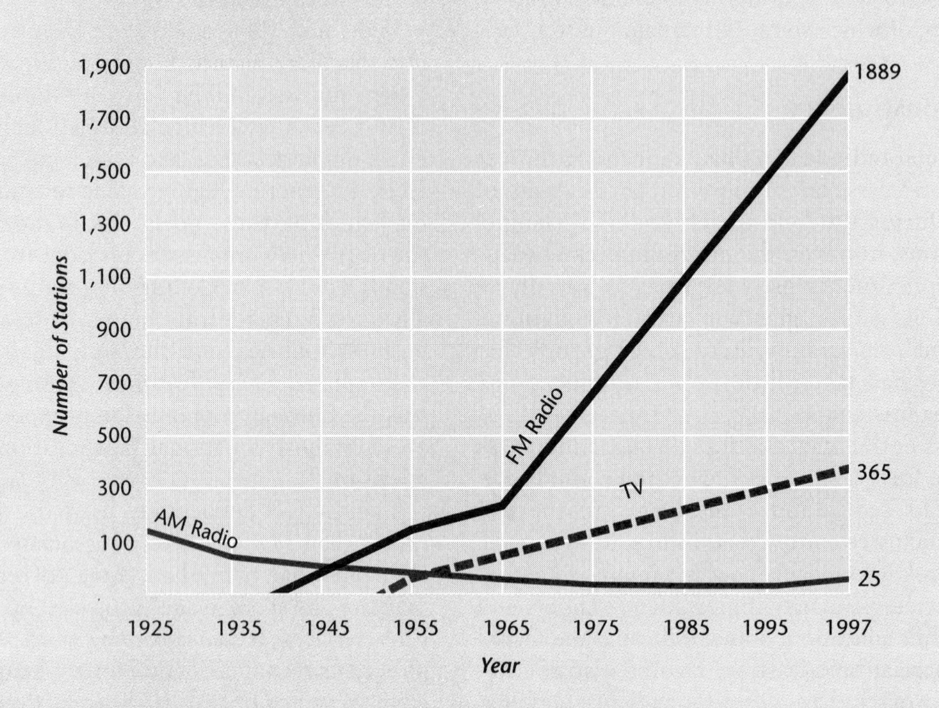

After a steady decline in the number of educational stations until 1945, noncommercial radio growth moved to the new FM service and later to television. AM figures are estimates because no official count exists.

Source: Data from FCC.

the Carnegie Commission created others. Public-television stations were often highly individualistic. They made uneasy, sometimes even acrimonious, partners with their federal allies in Washington.

Corporation for Public Broadcasting

Congress set the Carnegie Commission's keystone recommendation in place with the Public Broadcasting Act of 1967 ("broadcasting" because Congress added radio at the last moment). The new law became part of the 1934 Communications Act. It created the Corporation for Public Broadcasting (CPB), a quasi-governmental organization whose key role is to funnel money into projects supportive of the nation's public broadcasters.

The Carnegie Commission had recommended that the president of the United States appoint only half of the CPB board. Congress, however, gave all the appointive power to the president. Congress also declined to legislate the long-term federal funding that the Carnegie Commission had recommended. These two departures from the plan left CPB at the mercy of presidential politics. Unfortunate consequences followed, as detailed in Exhibit 7.b

The Public Broadcasting Act gives CPB such tasks as the following:

- making grants and contracts for obtaining and producing high-quality programs from diverse sources
- helping to set up network interconnection
- encouraging development of new public broadcasting stations
- conducting research and training

CPB itself may not own or operate stations. Congress emphasized that the stations should retain local autonomy, remaining free to select and schedule programs according to local needs. The last thing Congress wanted was to create a centralized, federal broadcaster.

The Act directs CPB to do its job "in ways that will most effectively assure the maximum freedom . . . from interference with or control of program content or other activities." Nevertheless, some stations regarded the very existence of a federal agency dispensing federal funds as a threat. The issue of local versus federal control has continually stirred up controversy between CPB and the stations.

Public Broadcasting Service

CPB launched its network in 1969–1970, calling it the Public Broadcasting Service (PBS). PBS was deliberately fashioned to differ considerably from the commercial network model. Over the years, however, these differences have diminished. PBS operates the interconnection facilities and, until very recently, increased its domination of the all-important task of program selection.

A weak national network had in fact been a Carnegie report recommendation. Though the report stressed the vital importance of network interconnection, it also warned against tight centralization of program control. Public broadcasting was expected to have "a strong component of local and regional programming" to "provide the opportunity and the means for local choice to be exercised upon the programs made available from central programming sources" (CCET, 1967:33).

As time proved, it was unrealistic to limit PBS's programming role, even though doing so was desirable to avoid the overcentralization typical of commercial networks. The Carnegie Commission failed to anticipate the practical problems of asking PBS to lay out a smorgasbord of programs from which affiliates would pick and choose at will (public broadcasters tend to avoid the term *affiliate* because of its commercial connotations, but it is used here for convenience).

Commercial experience had shown that a network needs the strong identity that can come only from a uniform national program schedule—identical key network programs available to everybody in the nation at the same time of day. But station licensees—zealous to protect their local identity, operating under varied ownership and funding patterns, and profoundly suspicious of political interference from Washington—resisted that type of network. They sided with "their" organization, PBS, against CPB in the battles over program philosophy that followed.

As a means of continuing service while the philosophical debate continued, PBS program selection from 1974 until 1990 occurred through a complicated mechanism called the *Station Program Cooperative* (SPC). Member stations (whose executives actually voted for which programs they would carry and help finance) praised SPC for its democratic participation. But SPC was also criticized for limiting program innovation, as stations nearly always voted and paid for programs successful in the past, rarely investing scarce dollars in new ideas.

Late in 1989, under pressure from a Congress concerned about SPC's decentralized and argument-ridden decision making, PBS stations concluded that the democratic SPC system had outlived its usefulness. They dissolved SPC and centralized all programming decisions under one PBS officer—much like the traditional commercial network model.

When former FCC Commissioner Ervin Duggan became PBS president in 1994, he designed a new three-division management organization. The new

Exhibit 7.b

Political Manipulation of CPB

Incidents during three conservative presidential administrations provide textbook illustrations of the difficulty of insulating a broadcast service from politics when it depends on government for substantial economic support. Section 398 of the Communications Act, added by the 1967 public broadcasting law, tries to prevent political influence by expressly forbidding any "direction, supervision, or control" over noncommercial broadcasting by officials of the U.S. government. This legal detail did not stop the Nixon administration in the 1970s nor the Reagan and Bush administrations a decade later from manipulating CPB for their own ends.

When the Public Broadcasting Service (PBS) network began beefing up its news and public-affairs programming around 1970 by hiring ex-commercial-network personnel, the White House became concerned. Regarding the PBS network as far too liberal, the administration objected to public television's concern with national affairs when, according to the administration's interpretation, it should be focusing on *local* needs. In 1973 the administration sent a more direct message when President Nixon vetoed a two-year funding measure for CPB. Long-range federal funding legislation did not finally pass until 1975.

By 1981, when the Reagan administration took office, the CPB board, never high on the priority list of political appointees, had become fertile ground for political gamesmanship. Reagan appointed several hard-right conservatives to the CPB board. One Reagan CPB appointee advocated making a content study to determine whether public television programming leaned too far to the left. The study never materialized, but the legally mandated political impartiality of the CPB had been seriously undermined.

That was evident during Bush's term (1989–1993), when CPB found itself under the combined pressure of an administration trying to trim federal funding and a Congress reviewing specific broadcasts while considering funding legislation. Late in 1991 funding for CPB for 1994–1996 was delayed for months in a controversy over both the amount of money proposed and two documentaries: *Tongues Untied,* about gay black life (1991), and *Color Adjustment,* about the role of blacks on television (1992). In the end, Congress did pass a record authorization for the three-year period, and President Bush, in the midst of an election campaign, signed it. But all sides agreed that the constant tension over funding and programming was anything but over.

Indeed, soon after the Republican party took control of Congress in 1994, House Speaker Newt Gingrich (R-Ga.) issued a call to "zero out" federal funding for CPB. He accused the Corporation of "eating taxpayers' money" and attacked its officials as "rich upper-class people" (*New York Times,* 20 January 1995:C3). Gingrich later backed off somewhat, saying that Republicans were committed to good television, but that some things might have to change at CPB. Gingrich argues for more audience support and has actively helped one of the public TV stations in his congressional district. Federal funding continues to decline (to $250 million for 1998) while debate about whether to terminate federal support entirely continues.

structure returned more choice to individual stations, allowing them to buy less from the national program service and to select more single programs or series from a new unit, PBS Syndication Services. The revised plan was intended in part to address what cost-conscious members of Congress and others called the "overlap" problem—several PBS stations that serve the same metropolitan area spending money on essentially the same program.

PBS Operations

Though more centralized, PBS still differs from commercial networks in that it carries only programs that others produce. Member stations (affiliates) contract with PBS, agreeing to pay dues determined by their respective budget and market sizes. Rather than the network paying for use of their time (long the case with commercial television network affiliates), public stations pay PBS for programs—an arrangement closer to a cable system's relationship with the many cable networks, or to program syndication in commercial broadcasting.

In the late 1990s PBS serves virtually all public-television stations on the air. Its 35-member board consists of professional station managers and members of the general public drawn from station boards. PBS staff totaled about 300 in Alexandria, Virginia (near Washington, D.C.), New York, and Los Angeles.

Satellite Interconnection

PBS pioneered use of satellites for network relays. Using satellites instead of AT&T's wire and microwave facilities for relaying public-television programs was first proposed by the Ford Foundation in 1966. A decade later, with government funding support, PBS announced a plan to interconnect its member stations by means of transponders on a domestic communications satellite.

Benefits claimed for the system included cost savings, better-quality reception, ability to relay signals both east and west as well as variously within given regions, and transmission of several signals at the same time to allow stations to choose among more program options. Satellite facilities also offered fund-raising possibilities through the lease to others of excess transponder capacity.

Stations each had to contribute about $25,000 (a hefty sum for most) toward the cost of TVROs. During 1978 public-television stations disconnected themselves from terrestrial links and began using satellite interconnection; indeed, PBS was the first national broadcast service to do so (even before most cable networks had begun operation).

National Public Radio

In 1970 CPB set up National Public Radio (NPR) both to interconnect stations (like PBS) and to produce some programs (unlike PBS). NPR provides its member stations—which do not have to carry any set amount of network programming—with about 22 percent of their daily schedules. NPR also coordinates satellite interconnection with 24 uplinks and more than 400 downlinks. Satellite distribution (the world's first satellite-delivered radio network) began in 1980 with four audio channels, which by 1997 had increased to 24.

CPB radio program funds go directly to NPR member stations, which support the network by paying a flat fee and then subscribing to morning and afternoon news services, NPR's musical and cultural programs, and Public Radio International (formerly American Public Radio, see Section 7.7) programs. Stations pay anywhere from $25,000 to more than $300,000 annually for NPR's program service, depending on their revenue, providing more than half of NPR's operating income.

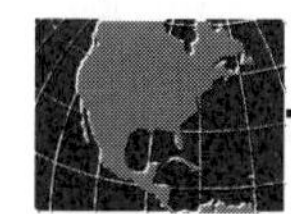

7.3 Public Stations

Both public-radio and public-television stations vary enormously in size, resources, goals, and philosophy. Though all are licensed by the FCC as "noncommercial educational" stations, some flirt with commercialism and many play no discernible formal educational role.

Television

In general, four classes of owners hold the licenses to public-television stations. The four groups differ and even conflict in their concepts of noncommercial broadcasting.

States and municipalities hold about 40 percent of the licenses, which totaled 365 in 1997. Many belong to state educational networks, usually programmed by the station located at the state capital or other key origination points. Many southern states in particular chose this means of capitalizing on educational television.

Colleges and universities own about 25 percent of the stations. They often complement educational radio stations of long standing and sometimes closely tie to college curricula.

Public-school boards hold licenses for about 3 percent. Their stations naturally focus on in-school instructional programs, many produced by and for the local school system. The low numbers of such stations reflect the failure of public television to live up to its early promise as a major adjunct of formal education. In fact, as school budgets became tighter, several school stations left the air or transferred to other licensees.

Community foundations control a third of the outlets. The licensees are nonprofit foundations set up especially to operate noncommercial stations. They recruit support from all sectors—schools, colleges, art and cultural organizations, foundations, businesses, and the general public. Usually free of obligations to local tax sources, they tend to be politically more independent than stations that depend for revenue on local or state tax dollars. Among the best-known community stations, because they produce much of what appears on PBS, are WETA in Washington, D.C., KQED in San Francisco, KCET in Los Angeles, and WGBH in Boston (see Exhibit 7.c).

NPR and Non-NPR Radio Stations

CPB decided to build its national noncommercial radio network—NPR—around a cadre of professionally competent, full-service stations (referred to as "CPB-qualified"). These stations had to meet a number of minimum standards involving such factors as power, facilities, budgets, number of employees, and so on. Today's standards for what are now called "CPB-supported" stations are much less rigorous: minimum hours of operation (12 or 18, depending on the level of financial support sought) and program schedules that do not advocate a religious or political philosophy or consist primarily of in-school or in-service educational instruction.

Stations that carry NPR programming do not necessarily have to be CPB-supported, although nearly all of them are. In the late 1990s, more than 550 noncommercial stations were "NPR members" (NPR does not use the term *affiliate*). A much larger group of more than 1,000 noncommercial stations either do not qualify for NPR membership or do not wish to join the network.

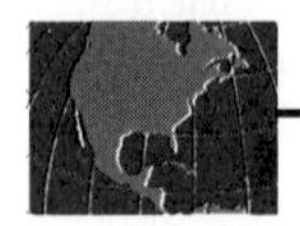

7.4 Economics

Ask any noncommercial broadcaster to name his or her most serious problem and the answer will almost always be: "not enough money."

Compared with those in many other countries, America's noncommercial system, on a per-person basis, is woefully underfunded. For every tax dollar allocated to public broadcasting in America, the governments of Britain and Canada fund their BBC and CBC services, respectively, with about $25, and Japan's NHK public network receives about $15 in government funding per person served.

Exhibit 7.d shows (excluding non-NPR stations) the diversity of public broadcasting's funding sources. But diversity has a price: Each source brings different obligations with its funding, and each has its biases.

Government Support

Local and state governments supported educational FM from its inception in the 1940s, and many states expanded into educational television over ensuing decades. By the mid-1960s state tax funds provided about half of all public broadcasting income. In the face of rising budget shortages, however, both state and local support had declined substantially by the 1990s.

At the outset the federal government gave no financial assistance at all. Federal legislative recognition first came with the Educational Television Facilities Act of 1962. It granted up to a million dollars of federal money to stations in each state, subject to their matching federal dollars with money from other sources. Extended and revised, this act continues to assist equipment and facilities purchases for public broadcasting. This program is

Exhibit 7.c

WGBH—PBS "Superstation"

Boston's WGBH-TV, perhaps best known of all Public Broadcasting Service member stations, produces about a third of PBS's prime-time schedule. It forms part of the WGBH Educational Foundation, which includes two TV stations, a radio station, a production house, and an access division that pioneered such technologies as television captions for hearing-impaired viewers and the Descriptive Video Service for visually impaired audiences.

WGBH began operations in 1951 with an educational FM station licensed to a group of Boston's prestigious universities and public organizations. Its first television broadcast appeared in 1955.

Drawing on Boston's many educational and cultural activities, WGBH-TV quickly became a major program source for other stations. It introduced public TV's first "how-to" show—and one of educational TV's first stars—with Julia Child's *The French Chef* in 1963, and went on to produce today's popular *This Old House.* It gave PBS one of its most successful and longest-running series, *Masterpiece Theatre* (underwritten from the start by Mobil), by coproducing or acquiring the best of British drama and framing it with commentary (for the series's first 22 years by Alistair Cooke, now by Russell Baker). WGBH also is the source of such mainstays as *Evening at Pops*, shown here with guest artist Jason Alexander (1970), *NOVA* (1972), *Mystery* (1980), *Frontline* (1983), and *The American Experience* (1988).

With an annual budget of about $130 million (far above that of most other public television stations), the Foundation faces the same fundraising challenges as any public broadcaster—although often with greater entrepreneurship. WGBH has been a partner since 1986 in the *Signals* mail-order catalog "for friends and fans of public television." In 1991, the first of a chain of Learningsmith stores opened—"a general store for the curious mind," which sells educational games, software, and public broadcast-related products, and in which WGBH is a minority participant.

Still, even WGBH is not immune to economic realities. In 1991, during a major downturn in the nation's economy, the station had to abandon its 15-year-old 10:00 P.M. local newscast as part of an overall reexamination of revenues and expenses.

WGBH remains active in utilizing changing technologies. Its Special Telecommunications Services unit is currently researching applications of multimedia and CD-ROM technology for in-school use of WGBH productions. In collaboration with CPB, WGBH runs the National Center for Accessible Media (NCAM). NCAM develops strategies and technologies to make media accessible to people with disabilities, minority language users, and people with low literacy skills.

Sources: photo by Michael Lutch for WGBH; logo © WGBH Boston.

Exhibit 7.d

Public Broadcasting Revenue by Source (1995)

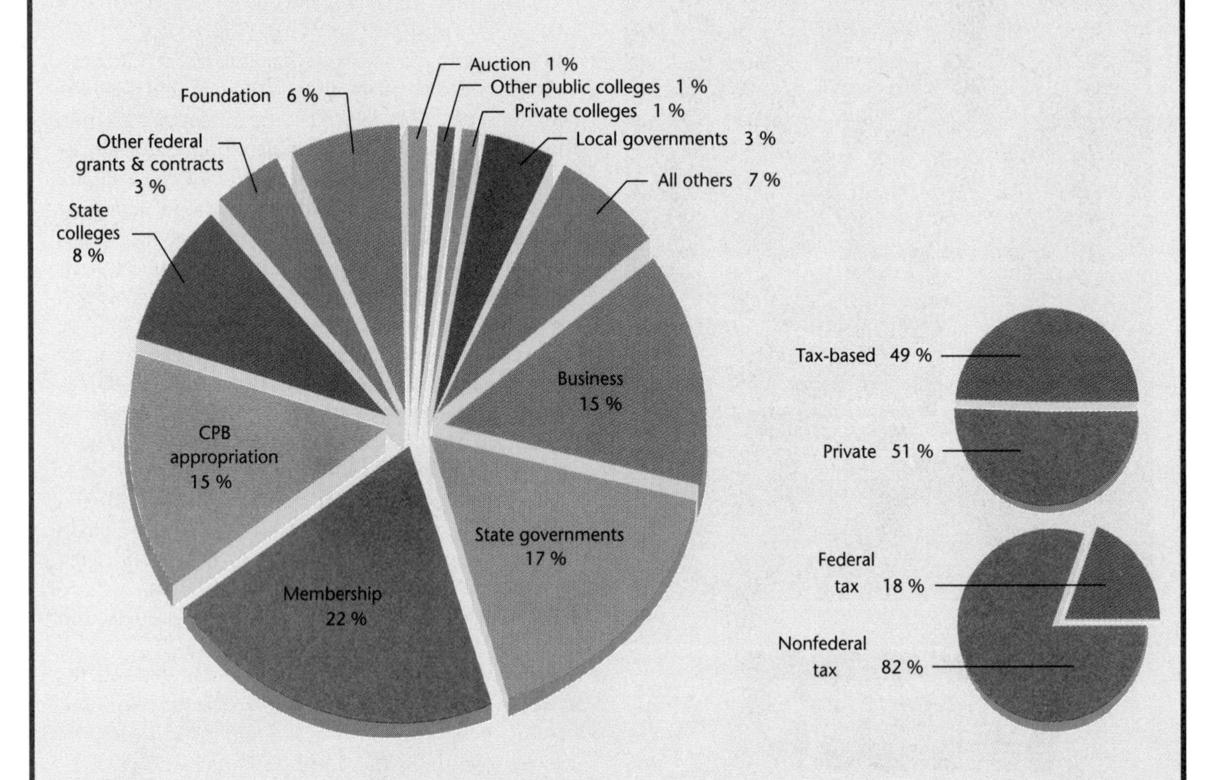

Beginning in the 1980s public broadcasting has depended increasingly on *non*tax sources of revenue. At the start of that decade, tax-based sources provided more than two-thirds of all revenue; in the 1990s tax support represents slightly less than half and has remained constant at that level over recent years. Note that public station "members," businesses, and state governments provide the largest segments of public broadcasting support.

Source: CPB, *Public Broadcasting Revenue, Fiscal Year 1995* (August 1996).

administered by the National Telecommunications and Information Agency (NTIA), an arm of the U.S. Department of Commerce. As with CPB, federal funding of NTIA also is under attack and there is considerable argument over whether this funding should continue.

Long-Range Funding

Attempts to persuade Congress to fund CPB for longer than the usual federal one-year budget cycle had failed until 1975 when the legislature finally authorized funds for a three-year period.

Although similar multiyear appropriations have followed since, Congress at times has acted to *rescind* appropriations to which it had already agreed, and there have even been attempts to halt federal funding entirely (see Exhibit 7.b). This uncertain commitment by Congress—and several state governments as well—has made it difficult for public broadcasting to plan, let alone pay, for the future.

Foundation Grants

Foundations provide substantial support for noncommercial broadcasting. Without the backing of the Ford Foundation, early educational television might not have survived its first decade. Carnegie Foundation money paid for two important studies that helped shape public broadcasting.

In 1981 the largest single gift ever made to public broadcasting came from Walter Annenberg, then publisher of *TV Guide.* He donated $150 million, spread over 15 years, to fund a project to create innovative college-level courses and programs. The Annenberg/CPB project changed focus in the early 1990s, in part because of changes in tax laws, to support programming for primary and secondary schools rather than university-level audiences.

Corporate Underwriting

A limited form of sponsorship called *underwriting* enables program producers to secure funds from businesses to cover at least some production costs for specific projects.

FCC regulations allow companies brief identifying announcements at the beginning and end of such programs. After 1981 the FCC also allowed corporations to display logos or trademarks. Some critics worry that excessive corporate underwriting might influence the content of noncommercial programming, a fear similarly raised regarding advertisers and commercial broadcasting.

Commercial Experiment

Seeking alternative or additional revenue sources for public broadcasting, Congress in 1981 established a Temporary Commission on Alternative Financing for Public Telecommunications (TCAF) to supervise an experiment with commercials as a means of support.

Following the test, TCAF suggested that most public-television stations would not carry ads because of labor union contract requirements, local economic considerations, or concerns about advertising's impact on the character of and other financial support for public broadcasting (TCAF, 1983). The experiment ended in mid-1983 and outright commercials remained taboo.

Creeping Commercialism?

Influenced by TCAF recommendations, the FCC in 1984 authorized *enhanced underwriting*, leading stations to carry what then PBS President Bruce Christensen called "almost commercials" (Smith, 1985). The FCC tried to balance station and public needs: stations may sell announcements up to 30 seconds in length and mention specific consumer products but may sell no more than two-and-a-half minutes between individual programs.

By the late 1990s, the line between "enhanced underwriting" and commercial spots has become increasingly blurred. Critics deplore this increasing commercialization of public television. After all, public broadcasting's primary justification is that it furnishes *alternatives*—programs not likely to appeal to advertisers but nevertheless useful and desirable. To the extent that pleasing advertisers motivates *both* services, the argument for public broadcasting is weakened. Still, the notion resurfaces from time to time. When a Republican-dominated Congress threatened in 1995 to eliminate CPB funding, full commercialization of public broadcasting was suggested as a possible replacement funding source.

Local Contributions

Federal appropriations (and often national foundation grants) trickle down only when stations match federal grants. The desperate scramble to match federal money forces local fund raisers to push membership/subscription drives to the saturation point. Polls indicate that many viewers detest marathon fund-raising drives (derisively called *begathons*). On-air auctions of donated goods and services (much of them from commercial sources) promote the givers so blatantly as to amount to program-length commercials. However, as Exhibit 7.d demonstrates, these efforts are vital to the survival of noncommercial stations: Fully 22 percent of public broadcasting's revenue comes from individual viewers and listeners.

Other devices considered (and in some cases actually used) to raise funds locally, and to some extent nationally, include the following:

- selling commercial rights to market merchandise associated with programs. (The Children's Television Workshop defrays a large part of *Sesame Street*'s production costs from such sources.)
- selling newly produced programs to commercial television or pay cable for initial showing before release to public television
- renting station facilities (usually studios) to commercial producers
- selling books, videotapes, and other items to viewers
- trading a reserved educational channel for a less desirable commercial channel for a price. (Such channel reclassification requires FCC approval.)
- acquiring assured tax-based revenue by charging commercial broadcasters a tax on profits, a receiver licensing fee, or a spectrum-use fee
- selling access to the FM subcarrier or TV vertical blanking interval for commercial use

In sum, public broadcasting executives have to serve too many masters and spend an inordinate amount of time on fund raising. The year-to-year uncertainties of congressional appropriations, corporate underwriting, and local membership drives make it impossible for public broadcasters to plan rationally for long-term development. As a result, in the mid- and late 1990s some prominent legislators began calling for the establishment of a public broadcasting trust fund, the proceeds of which could provide predictable long-term funding. Interestingly, this is exactly what the Carnegie Commission suggested in its initial report in 1967. As of mid-1997 these proposals had not been formally addressed by Congress.

7.5 TV Program Sources

Although noncommercial broadcasting's basic purpose is to supply an alternative to commercial services, program overlap does occur. Noncommercial stations often show feature films and syndicated series obtained from the same distributors that commercial stations use. On the other hand, noncommercial outlets do more local production and experimental programming than do their commercial counterparts.

Stations as Network Producers

A handful of major-market producer stations act as primary producers for PBS. In the 1990s they provided more than a third of national programming, with another 10 percent or so coming from lesser public-television stations. Major producer stations, such as WGBH-Boston (see Exhibit 7.c), have long histories of creative innovation in public television. They tend to develop specializations: WETA-Washington, D.C., for news and public affairs, for example, and WGBH for science, documentary, and drama presentations.

Foreign Sources

By the 1990s nearly a fifth of all PBS series contained some overseas material, while others came straight from such sources as British commercial and BBC television series. Best known are the long-running *Masterpiece Theatre* and *Mystery* series, both co-produced by WGBH.

Independent and Syndicated Production

Under pressure from producers who claimed PBS largely ignored their output and abilities, Congress in 1988 required CPB to fund an Independent Production Service, separate from CPB, to encourage more independent program sources.

Easily the best known of these is the nonprofit Children's Television Workshop. It won international fame for *Sesame Street* and subsequent children's series. The *Sesame Street* series began in 1969 with initial funding from government and foundation grants. By the late 1980s ancillary commercial ventures, such as merchandising items using the program name and characters, defrayed well over half of its budget.

Local Production

The typical noncommercial television station produces and uses more local programming than do most commercial stations. Locally produced programs consist mainly of news and public affairs, along with some educational/instructional material telecast during daytime hours for in-school use.

However, local production is often one of the first things to go when budgets are cut, as was common during the economically difficult early 1990s. Accordingly, many stations cut local production by half from its mid-1970s level, now averaging slightly more than 5 percent of the total program schedule.

Cable Alternatives

Just as cable has increasingly taken on many functions of traditional broadcasting in entertainment, so has it done in education and public affairs. C-SPAN, in addition to its extensive public-affairs programming (see Sections 3.2, 9.5, and 11.5), offers C-SPAN in the Classroom, a free educational service that allows teachers to tape material for instructional use.

By the end of the 1990s about 75 percent of all schools in the country had been supplied with free cable service, most by local cable systems as a part of their franchises. Cable in the Classroom, a free service of the commercial cable industry, provides some national coordination. The service provides some 500 hours of commercial-free programming per month, supplied largely by 31 cable-delivered networks. A 1996 survey of some 800 elementary and secondary school teachers found that 70 percent used Cable in the Classroom programming at least once during the last school year, and 25 percent indicated they used it up to four times per week.

Cable television also provides noncommercial alternatives through its access channels—channels set aside for public, educational, and governmental use. Many communities have vital cable access operations, providing one or more channels of local entertainment, educational, and government programming. Public access channels provide a forum for residents to showcase their artistic (though in some cases that is certainly open to argument) talents on television, a proposition too expensive for most people to pursue on a broadcast station. Some cities use their government access channels to transmit council meetings, school board meetings, and various other local government activities, providing their citizens with greater access to local decision-making.

DBS and the Internet

In the Cable Act of 1992, Congress imposed a requirement on all DBS operators that they reserve from 4 to 7 percent of their channel capacity, as determined by the FCC, "exclusively for noncommercial programming of an educational or informational nature." Following lengthy litigation over the new law, a federal court in 1996 upheld the regulation and the FCC in 1997 began considering ways to implement the legislation.

Increasingly in the 1990s the Internet is providing significant amounts of educational "programming" to a global audience. Many universities offer courses and even entire degree programs on-line. With the push to reach nontraditional students through "distance education," the Internet is unique in its ability to deliver interactive instructional programming to widely dispersed audiences.

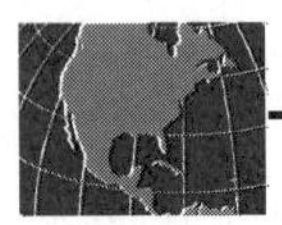

7.6 Noncommercial TV Programs

PBS began in 1970 with several programs either still on the air (*Washington Week in Review*) or well remembered (*The French Chef*, which featured Julia Child demonstrating that good French cooking was within everyone's reach). Many initial programs focused on news and public affairs.

News and Public Affairs

In contrast to commercial networks, PBS found its affiliated stations receptive to long-format (more than half-hour) newscasts.

The NewsHour with Jim Lehrer (and its immediate predecessor, *The MacNeil/Lehrer NewsHour*) has consistently been acclaimed as one of TV's best news and information programs, featuring in-depth interviews and analyses of current news stories. *NewsHour* provides a significant alternative to the largely news headline newscasts of the commercial TV networks.

PBS began the series in 1976 as a half-hour news program, expanding it to an hour in 1983. In 1994 Robert MacNeil announced he would leave the program in October 1995, with Lehrer continuing as sole anchor. Despite critical praise, *NewsHour* has failed to build a large following, even by noncommercial standards. Still, its loyal audience has kept the program on the air for over 20 years. In 1996 PBS introduced *NewsHour Online*, a continuously updated news service accessible on the PBS World Wide Web home page.

PBS provides a steady diet of other public-affairs programs as well, including *Washington Week in Review*, *Nightly Business Report*, and *Frontline*. Bill Moyers specials are irregularly scheduled but get both wide coverage and high praise for their in-depth analysis of and sensitivity for social problems.

When in 1992 the commercial networks abandoned full-time coverage of the quadrennial political conventions, PBS (and cable services) stepped into the breach, offering full prime-time coverage—combining Robert MacNeil and Jim Lehrer with NBC's top correspondents, including anchor Tom Brokaw, when NBC was not airing convention news itself. Such coverage was repeated for the 1996 conventions.

Another aspect of PBS news and public-affairs programs appeared in 1992 as well: PBS-distributed programs were now legally *required* (by CPB long-range funding legislation) to offer objective and balanced content. If an expert review panel finds questions of balance or fairness in a program, CPB may support costs of producing balancing programs. As a part of this "Open to the Public" process, CPB announced more ways for its audience to "talk back" on fairness and other issues—a toll-free telephone number, quarterly meetings of the CPB Board with public groups away from Washington, and the like.

In anticipation of the 1996 election, PBS introduced "The Democracy Project," which it described as a "laboratory for citizen-oriented news and public affairs programs." The Democracy Project featured a series of programs on elections, government, and public participation. A classroom component of the Democracy Project—activities and special events for elementary schools—focused on engaging youths in civic education and the democratic process.

History

Occasionally documentaries attract viewers who do not normally tune in on the noncommercial service. This category was exemplified by *The Civil War*, an 11-hour 1991 series, which won widespread praise and viewership—and helped sell thousands of copies of a handsome "companion" book and videotape copies of the series, proceeds from which helped meet the program's production costs (see Exhibit 7.e). A similar, though smaller, reaction greeted *The West*, an eight-part documentary that aired in 1996.

Public television's provision of such programming goes back more than two decades. The multipart *Civilization*, a review of world history and cultural development featuring Sir Kenneth Clark, ran in PBS's first season and helped to introduce the "companion book" marketing idea.

Vietnam: A Television History, co-produced by Boston's WGBH and French television, became the highest-rated documentary of 1983 and won several awards. The documentary's less-than-positive assessment of America's role brought howls of protest from conservative groups—and a controversial decision by PBS to allow several critics to air their differences during an hour or so of airtime with the program's producers.

The American Experience has served as host vehicle for a series of historical documentaries on American history and key historical figures—with such varied programs as those on the building of the Brooklyn Bridge, the impact of rigged television quiz shows, immigration, and several studies of recent American presidents, including a well-received two-part study of *The Kennedys.*

Performing Arts

All of the performing arts—drama, dance, music—find a home in public television. Among public television evening programs, *Masterpiece Theatre* is probably best known. The Sunday evening hour-long staple provides British-made drama, ranging from short series built around historical events, famous novels, or contemporary views of the human condition, to such lengthy costume favorites as *Upstairs, Downstairs*, on life in a fine Edwardian house. Another popular program since its first broadcast in 1980 is *Mystery*, hosted first by Vincent Price and then by Diana Rigg. Viewers of this program watch as their favorite sleuths—Miss Marple, Hercule Poirot, Inspector Morse, Sherlock Holmes, Rumpole of the Bailey—solve crimes and other mysteries in often humorous and intellectually stimulating ways.

Music and dance appear regularly on three PBS programs that began in the mid-1970s and are still going strong two decades later: *Great Performances* (1974), *Austin City Limits* (1975), and *Live from Lincoln Center* (1976). A more recent series, *Evening at Pops*, features the Boston Pops Orchestra.

Science and Nature

One of the first programs dealing with science and nature aired in the 1970s. *The Ascent of Man* featured philosopher Jacob Bronowski's view of developing mankind in a multipart series and yet another useful companion volume.

Since then PBS has offered an eight-part series on *The Brain*, as well as the multipart *Cosmos*, with famous astronomer Carl Sagan introducing many Americans to the wonders of astronomy. The long-running series *Nature* as well as some of the most popular public-television documentaries, such as the *National Geographic Specials* and *National Audubon Society Specials*, proved that Americans could also produce and enjoy top-notch nature films.

Over the last two decades, *NOVA*, the longest-running nationally broadcast science series, has presented on PBS some 400 different hour-long science programs, each devoted to a single topic. Each week the program draws some eight million viewers, the largest audience of any PBS weekly series.

Children

Public broadcasting has a clear mandate to provide constructive, imaginative children's programs. Accordingly, many public-television stations fill a large portion of their daytime hours with in-school programs (paid for by school districts) and schedule additional special programs for children both early in the morning and again in the afternoon.

Since 1967 Pittsburgh station WQED's *Mister Rogers' Neighborhood* has taught social skills to very young preschoolers, focusing on their values, feelings, and fears with gentle conversation and songs

Exhibit 7.e

Ken Burns and The Civil War

tailored to their cognitive needs. It is second in viewing popularity only to the famous *Sesame Street.*

Children's Television Workshop (CTW) brought all the technical resources of television—as well as all the capabilities of educational research—to bear with its 1969 premier of *Sesame Street.* A wonderfully original group of large-scale puppets, the Muppets, became the program's hallmark. Research showed that children who watched *Sesame Street* learned to read more quickly than did other children. No series on either commercial or public television has ever been given as much scheduled airtime as has *Sesame Street.* By the 1990s it had taken on classic status as the preeminent program for preschoolers.

Building on *Sesame Street*'s success, CTW later branched out with *The Electric Company* for older children. In 1980 CTW began a daily science program, *3-2-1 Contact*, and later *Square One TV*, a math series for young children. CTW also began *Encyclopedia*, a children's series for HBO, and in

For five consecutive nights in September 1991, some 14 million viewers watched *The Civl War* unfold, making it the most-watched series in the history of public television. It earned an average rating of nine and was seen by four times the normal PBS audience.

The fruit of a five-year effort, and of some $3.2 million in support from CPB, Washington, D.C.'s WETA, General Electric (the only corporate underwriter), and the National Endowment for the Humanities, *The Civil War* was the creation of filmmaker Ken Burns. Burns had done earlier documentary programs for PBS. But this was different—an 11-hour series that brought the 1861–1865 struggle to life as most viewers had never seen or thought about it before. Combining creative use of period music, live shots of battlefields, as well as old photos over which cameras seemed to roam lovingly—and interspersed with quotes from experts—*The Civil War* took a neutral view of the war's causes. (Viewers in both North and South later complained about bias—suggesting the neutral approach had been successful.)

Syndicated newspaper columnist George Will summed up the series and its maker: "Our Iliad has found its Homer: He has made accessible and vivid for everyone the pain and poetry and meaning of the event that is the hinge of our history." Concluded producer Burns:

> There's no way that *The Civil War* could have been produced anywhere but on public television. Public television not only tolerates but encourages the kind of experimentation and risk-taking that we take as independent producers. And on public television I don't lose the viewer's attention every six to eight minutes for a commercial. The ability to hold that attention is critical to me as an artist. Public television is also the only kind of television where the creator actually retains control. (CPB, 1991: 44–45)

The series proved financially valuable, too. Viewer contributions to local stations rose sharply, thanks to money pleas during breaks in the program. The "companion book" sold 750,000 copies, and related video and audio recordings also sold well. Burns followed his *The Civil War* success with a 90-minute program on early radio called *Empire of the Air*, which traced the lives of Lee de Forest, Edwin Armstrong, and David Sarnoff, a multipart documentary series tracing the history of baseball, and, with co-producer Stephen Ives, an eight-part series descriptively titled *The West*.

Source: Photo by Ken Regan/Camera 5.

1991 launched a $20 million multimedia literacy project, *Ghostwriter*. The following year it joined with the United Nation's UNICEF and Mexico's Televisa to create a Spanish-language version of its old standby, *Sesame Street*.

At the beginning of 1997, the national programming feed of PBS distributed 18 different children's television shows to PBS affiliates across the country. These programs target children from preschool age through their teens and include such popular favorites as *Bill Nye the Science Guy*, *Barney and Friends*, *Reading Rainbow*, and *Wishbone*, which features the exploits of a small dog who introduces children to the great classics of world literature.

Education and Instruction

Since 1977 CPB has funded more than 20 instructional television (ITV) series for use in classrooms, providing texts and other support for each series.

A 1991 CPB survey showed that some three-quarters of the country's teachers had used instructional programs (some on videocassettes) in the previous school year, up from 54 percent in a survey eight years earlier (CPB, 1992: 11).

Beginning in 1981, PBS offered the Adult Learning Service (ALS), a cooperative effort with local stations that provides college-credit television courses for viewers at home. Many of these courses reach colleges directly by means of the Adult Learning Satellite Service, which began in 1988 and serves some 2,000 institutions of higher learning. In parallel fashion, PBS coordinates distribution of programs for elementary and secondary schools as well.

In carrying out their classroom instruction mission, public-television stations draw on several large libraries of instructional materials from companies that act as syndicators. Notable examples include the Agency for Instructional Television (AIT) in Bloomington, Indiana, innovative producers of interactive videodisc programs for children; the Great Plains National Instructional Television Library (GPN) of Lincoln, Nebraska, producers of well-known classroom series for children; and the Annenberg/CPB Project, producers of several prime-time adult-learning series.

Not all of the educational and information programming from PBS is intended solely for the classroom. Many of the most popular PBS television series feature how-to themes. In the late 1990s programs such as *Baking with Julia* (Child), *This Old House*, and *The Victory Garden* delivered entertaining instructional content to "do-it-yourselfers" across the country.

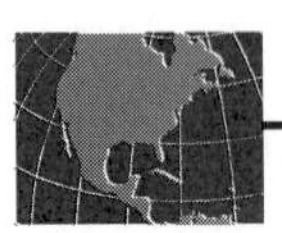

7.7 *Noncommercial Radio Programs*

Typical public-radio stations affiliated with National Public Radio (NPR) and/or Public Radio International (PRI) fill about two-fifths of their weekly total airtime with programs from these networks and other syndicated sources.

National Public Radio

NPR's first continuing success was its weekday late-afternoon news-and-feature program, *All Things Considered*, which began in 1971. The program quickly won every radio program award in sight and became a listening addiction for its loyal fans. NPR took the same approach to news on *Morning Edition*, which premiered in 1979, and *Weekend Edition*, which debuted in 1985. In 1993–1994, NPR first offered around-the-clock hourly newscasts, continuing an 18-hour-a-day schedule of newscasts on weekends.

In addition to news programs, NPR distributes a series of talk, cultural, and informational programs to its affiliates. One particularly popular talk show is the Peabody Award-winning *Fresh Air* with Terry Gross. This weekday magazine of contemporary arts and issues features in-depth interviews with prominent cultural and entertainment figures plus criticism and comment. *Performance Today* takes listeners behind the scenes to learn more about composers and performers through interviews and features. Perhaps the most unlikely NPR offering—though one of the most popular and financially successful for stations that carry it—is *Car Talk*, featuring the Magliozzi brothers, whose answers to called-in car-maintenance questions are spiced with jokes, word play, and a general sense of fun.

Public Radio International

A second national public-radio service, Public Radio International (PRI), began as American Public Radio (APR) in 1983. Formed by Minnesota Public Radio, it first served chiefly as a distribution channel for the popular *A Prairie Home Companion*, featuring Garrison Keillor (see Exhibit 7.f). By the early 1990s it was the largest distributor of public-radio programming. Effective July 1994 it changed its name to reflect its increasing global emphasis in news and other programming.

Exhibit 7.f

The Mythical World of Garrison Keillor

Every Saturday evening, increasing numbers of public radio listeners across the country tune in to the soft and zany homespun humor of Garrison Keillor, creator and host of *A Prairie Home Companion.* Produced by Minnesota Public Radio and distributed by Public Radio International, the program regularly airs on more than 370 stations to a dedicated (fanatical?) listenership of about 2 million.

The program's appeal centers on Keillor and a cast that includes "The Guy's All-Star Shoe Band" and a sound-effects expert, Tom Keith. The live variety show runs for two hours and features comedy sketches, music heard nowhere else, guests, and Keillor's monologue on the news from all those people in mythical "Lake Wobegon."

To a considerable degree, the characters and style of Lake Wobegon grow out of Keillor's own background. A Minnesota native who grew up in a small town, Keillor had aspirations—and some success—as a writer early in life. He worked as both writer and staff announcer for the campus radio station while earning a degree in English at the University of Minnesota. He wrote occasionally for *The New Yorker* after college, and it was while doing a story on radio's *Grand Ole Opry* that the notion of what would become *A Prairie Home Companion* came to him in 1974. The show first aired in Minnesota that year, and went national in 1980.

Keillor explains why he likes radio:

> The beauty of radio is its comparative simplicity. We count on this. It means I can work on a show, be inventing things, right up until airtime and not be so beset with theater or transportation problems that I can't pay attention to the real job at hand.

Television, he feels, is just "the Wal-Mart of the mind. Radio is infinitely sexier." This attitude comes through in his novel *WLT: A Radio Romance* (1991). His characters are people who work at a Minneapolis station (the call letters stand for "with lettuce and tomato") in radio's golden age, an era of creativity he tries to emulate every Saturday night.

Source: photo from Gamma Liaison.

Unlike NPR, PRI provides programs to only one station per market. Also unlike NPR, it does not produce programs but, rather, acquires them from member stations. In addition to classical and other music programs, PRI offers *Monitor Radio*, a news-and-feature program of the *Christian Science Monitor* newspaper, and the half-hour weeknight *Marketplace* business news program. It also brings BBC World Service programming to American listeners.

Local Public Programming

The primary appeal of public stations is that they offer content usually heard nowhere else—jazz, opera, local public affairs, folk music, and the like. Of the on-air hours of more than 500 public-radio stations surveyed in 1992, 74 percent were devoted to music (34 percent classical, 14 percent jazz—and 9 percent rock!), 17 percent to news and public events, 4 percent to call-in and other public-affairs programs, and about 2 percent to non-English programs (Giovannoni, 1992: 37). More recently, a growing number of stations, especially in larger radio markets with multiple noncommercial licensees, have shifted their formats from music to all-talk.

Public stations tend to differ most from their commercial counterparts in that fully half vary their content considerably each day rather than consistently hewing to a single format. They come close to the "middle-of-the-road" or "something for everybody" formats that traditional radio stations provided in the 1940s and 1950s.

Community Radio

Several hundred noncommercial FM stations and a growing number of low-power AM stations (and a few cable-only FM services) provide one or more types of community radio. (College stations operating on student fees could be considered a type of community radio.) In the 1960s, reacting against rigidly formatted radio and dissenting from establishment values, some small stations began the *underground* or *free-form* radio format.

Such radio is now lumped under the *progressive* format, which mixes live and recorded music and talk, usually at the whim of the presenter. Community stations often feature, if not progressive music, a classical, jazz, or diversified format. Small, low-salaried staffs supplemented by volunteers mean low overhead, enabling such outlets to present formats that would not be commercially viable.

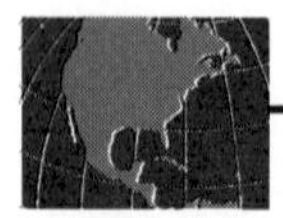

7.8 *Changing Roles*

With their small but generally upscale audiences, noncommercial stations are concerned with survival in the 1990s in at least two areas—financial support and competition from other services. But they face an even more basic dilemma as well: Can they successfully continue to justify their very existence?

New Options

A common argument heard these days is that development of alternative media, especially cable, has weakened the major rationale for public television. Programs for children, good drama, science and history, and other cultural programming—once the nearly exclusive domain of public broadcasting—now appear on several cable networks delivered via DBS as well as cable systems.

The counterargument, of course, is that these newer distribution technologies are subscription-based and many people who might otherwise want these services cannot afford them, while over 98 percent of the U.S. population has access to at least one over-the-air public television signal. Further, audience research shows that five times as many people watch PBS prime-time programs as watch The Discovery Channel, and seven times more see PBS than Arts and Entertainment, two of the most-often-cited channels offering PBS-like programming.

Still, it seems clear that cable, DBS, and Internet competition—and increasingly widespread use of VCRs and CD-ROMs for educational material—makes public broadcasters more aware of the need to specify clearly what role they will play amid a wider variety of viewer choices. But it remains just as clear that, for the foreseeable future, newer media cannot provide noncommercial-system-type services to all who want them.

Critical Views

In addition to citing an increasing variety of media options providing educational and cultural fare, critics in recent years have broadened their attack on the current system of noncommercial broadcasting on several related but distinct fronts:

- The funding shortage always cited by public broadcasters is largely a matter of evasion: Lack of sufficient audience appeal (and thus support) lies at the root of their problem.
- Public participation in public television is in too many cases a sham because professional managers make most decisions, just as they do in commercial operations.

- Public television has little public appeal. Public television's vaunted fine arts and high culture merely serve privileged groups that are easily able to afford other sources of such material without resorting to publicly supported broadcast channels.
- Conservatives have long targeted public broadcasting as being too liberal in news and documentary programs, presenting a one-sided view of American life—this despite regular appearances by such alternative spokespersons as William F. Buckley and John McLaughlin.

Outlook

Some of these criticisms reflect the deregulatory philosophy that dominated discussions of U.S. media policy during the conservative political ascendancy of the 1980s. Similar attacks on public broadcasting surfaced in other countries, also stimulated by a market-oriented, *laissez-faire* approach to media regulation. Even so widely esteemed a service as the BBC came under attack as elitist and lacking in the fiscal responsibility that, according to deregulatory theory, only "discipline of the market" can impose.

In Europe, as in the United States, these attacks raise a basic question: Should all broadcasting be regarded strictly as an *economic* undertaking, with programs treated as ordinary consumer goods? Or should at least some part of any national electronic media system be regarded as a cultural undertaking, with programs treated as a significant aspect of national life? Must everything depend on that slogan of free marketers, "consumer choice"? Not everyone chooses to attend great museums, galleries, and libraries that public funds support—often in locations not far from festering slums. Yet few would advocate dismantling all such cultural treasures and diverting their government grants to public housing.

Nevertheless, market-oriented thinking and the deregulatory policies that ride its crest cannot be ignored. Leaders of national public-service broadcasting organizations readily understood and agreed with the need for in-depth self-assessments in the early 1990s. CPB worked within a constantly updated five-year plan. NPR undertook a review of its role and options.

The whole system received a strong boost from these words in cable legislation of 1992:

> The Federal Government has a substantial interest in making all nonduplicative local public television services available on cable systems because . . . public television provides educational and informational programming to the Nation's citizens, thereby advancing the Government's compelling interest in educating its citizens . . . [and because] public television is a local community institution. (Cable Act of 1992, Section 2(a)(8))

As discussed earlier in this chapter, the funding for public broadcasting increasingly is shifting to non-federal sources as federal tax dollars now represent only 15 percent of total public broadcasting revenues. Many stations report that fund-raising appeals based in part on reduced federal funding are effective. Further, additional noncommercial sources of programming are available and are slowly increasing. While the future of public broadcasting might not necessarily be considered bright, these developments suggest that noncommercial services are important to the public and that the public, when faced with a diminution of those services, will respond with increased financial support in an attempt to retain them.

Chapter 8

Programs and Programming Basics

CHAPTER CONTENTS

Ask anyone for a personal opinion about programs, and, quite properly, every listener/viewer proves to be an expert on the subject. This chapter, however, deals not with personal preferences but with anticipating group preferences. The collective term *programming* refers to practices used in selecting, scheduling, and evaluating programs. Whether viewed from the perspective of broadcast or cable, the overriding purpose of commercial programming is either to attract advertisers by delivering audiences of sufficient size and appropriate composition or to attract audiences who pay subscription fees—or both.

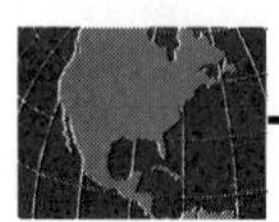

8.1 Program Costs

Programs must be both affordable and relevant to the programming goals of particular schedules. Programmers never end their search for what they refer to as usable *product*—a term, derived from the motion picture industry, that gives a hint of the programmer's neutral point of view as to program quality.

Audience Targeting

As the numbers of stations, cable systems, and networks grew, many program services gave up aiming at the mass audience that was the original target of broadcasting. The terms *narrowcasting* and *niche services* came into vogue, suggesting a specific alternative goal. Cable networks such as CNN, MTV, Nickelodeon, and Home Shopping deliberately limit their appeal to specific audience segments. Radio stations adopt rigid music formats (a particular radio format may repel most listeners but nevertheless prove irresistible to a narrow audience segment).

Advertisers influenced the trend toward narrowcasting. They are not inclined to pay to reach people who have no money to spend on their products or no interest in buying them. The need to target people willing and able to pay the bills, whether as advertisers or as subscribers, has led to *audience targeting* and segmentation throughout electronic media.

Even though noncommercial broadcasting and pay cable sell no advertising, they also target specific audience groups. They try to attract audience segments most likely to support them by paying subscription and membership fees. Public television favors programs that appeal to middle-to-upper-income, well-educated families. Pay cable tends to select movies that attract women and families with children. Pay-per-view (PPV) programs on cable and direct broadcast satellites focus heavily on movies, often those with adult content, and events, such as heavy-weight boxing matches.

The major broadcast television networks target the largest groups of viewers. Some cable networks target the same broad audience; others program for more narrowly defined groups. Each service defines audiences in terms of *demographics* (age range and gender) or *psychographics* (lifestyle and interests). Targeting women 18 to 34 years is a demographic goal; targeting sports fans is a psychographic goal.

Radio has further refined the process of targeting by using *segmentation*, defining extremely narrow subsets of the potential radio audience. Radio usually segments audiences in both demographic and psychographic terms: teenagers-who-want-to-hear-only-hit-songs or 25-to-44-year-old-adults-who-prefer-the-music-of-the-1970s, for example. Audience segmentation makes attracting an economically desirable audience of adequate size more difficult. Many media owners feel it essential they control multiple program channels if they are to prosper or even survive.

Parsimony Principle

Product that can attract audiences of adequate size and desirable composition costs so much that filling all channels continuously with brand-new—not to mention "good"—programs would be economically impossible.

To cope with high program costs, writers, producers, and programmers resort to a variety of strategies based on what might be called the *parsimony principle.* This rule dictates that program materials must be used as sparingly as possible, repeated as often as possible, and shared as widely as possible.

Sparing use of material means, for example, stretching a dramatic plot over many episodes instead of burning it up in a single performance—the technique of the soap opera. Often material collected for a news feature is released in a series of minifeatures. An expensive live sports event is not programmed as a single item: it is stretched to the utmost, with a pre-game show, half-time interviews, and a post-game wrap-up. The most dramatic example, of course, is the Super Bowl. Pre-game and post-game shows often fill more air time than the contest itself.

Repeated use is illustrated by the standardization of openings, closings, and transitions in daily newscasts, quiz shows, weekly dramatic series, and the like. Cartoons, movies, and television series replay endlessly over time. Some cable networks repeat the same movie at several different times on the same day. Successful talk shows such as *Montel Williams* and *Jenny Jones* may also double-run (air twice the same day) on broadcast stations.

Shared use is best illustrated by networks, which enable the same programs to appear on hundreds of different stations and thousands of different cable systems. Such sharing by stations and systems spreads high costs among many users, bringing prices down to a reasonable level for each. The sharing does not stop there, because network entertainment programs reappear in syndication (the mechanics of which are explained later in this chapter); this enables further sharing worldwide among stations, cable systems, and networks. Stations in many markets also share their newscasts with local cable systems or reuse source materials to produce another version of the newscast for cable distribution or broadcast on another (usually independent) station.

From Prime Time to Any Time

The best-known and most expensive entertainment programs appear during the network portion of *prime time*, between 8:00 and 11:00 P.M. EST and PST, 7:00 and 10:00 P.M. CST and MST. Of the $5 billion or so that ABC, CBS, and NBC TV networks spend each year on programs, about $3 billion goes for prime-time shows alone. Budgets on that scale put enormous pressure on those who conceive, develop, and schedule the programs. Networks spend up to $200 million each year just on *pilots* for new entertainment series. Only a quarter of them survive long enough to go on the air as first episodes in series. Exhibit 8.a depicts the high attrition rate in the progress from program concept to on-the-air product.

Deficit Financing

Outside companies have traditionally produced many network programs, then licensed networks to show them a limited number of times (usually twice). In the early 1990s the license fees that producers charged commercial networks for a one-hour prime-time drama or action/adventure program averaged about $1 million per episode. Costs can be somewhat higher, however. To attract a younger audience CBS lured *Family Matters* away from ABC by paying $1.7 million per episode, about $200,000 more than ABC had paid for the show.

High as such payments may seem, they rarely cover the full cost of production. NBC, for example, pays Warner Brothers a license fee of about $1 million per episode of *ER*, but each episode costs an estimated $1.2 million to produce. Producers of prime-time entertainment series count on subsequent resale fees to bring in the profits. This maneuver, known as *deficit financing*, capitalizes on the peculiar dynamics of the program market. Initial showing by the broadcast networks enhances the future resale value of a series. Exposure on ABC, CBS, NBC, Fox, The WB, or UPN confers recognition and a ratings track record.

After reverting to their producers, such shows become known as *off-network* programs. Many go on to earn handsome profits when licensed for exhibition by broadcast stations, cable systems, cable networks, and foreign broadcast systems. In fact, the staggering profits that can be made from the distribution of entertainment programs have

Exhibit 8.a

Prime-Time Series: Concept to Air

Networks | **Producers**

1
About 1,000 program concepts

2
Initial interest $

3
Deals for about 300 sample scripts and treatments

4
Further interest $

5
About 100 pilot programs produced

6
Approval $ and scheduling

7
6 to 11 episodes produced (for each of about 35 series: 210 to 385 episodes)

8
Renewal $ or cancellation

9
Production of more episodes for successful series

New program concepts usually originate with producers. Once a network shows an initial interest, it advances funds ($) stage by stage. Concepts drop out at each stage. In the hypothetical year depicted, nearly 29 program concepts were discarded for each one actually produced. The numbers listed here typify program development from many concepts to relatively few completed episodes.

caused some of the major movie studios in Hollywood to pay more attention to distributing television programs than to producing them.

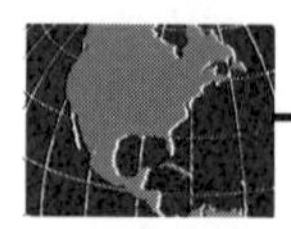

8.2 Syndication

The distribution method just described, known as *syndication*, offers both a supplement and an alternative to networks as a mechanism for financing centrally produced, high-cost product. Syndication occurs worldwide wherever broadcasting or cablecasting exists.

The FCC defines a *syndicated* television program as

> any program sold, licensed, distributed, or offered to television stations in more than one market within the United States for noninterconnected [that is, nonnetwork] television broadcast exhibition, but not including live presentations. (47 CFR 76.5[p])

In practice, programmers classify as syndicated all *nonlocal* programs not currently licensed to a network, including movies, and even some live presentations.

There are two types of syndicated programming. *Off-network syndication* refers to programs that have appeared (or may still be running) on a broadcast network; *first-run syndication* refers to programs originally designed for syndication and never seen on a broadcast network. *Seinfeld*, *Mad About You*, and *Home Improvement* are examples of off-network programs in syndication. *Wheel of Fortune* and *Jeopardy!* are first-run syndication programs. While a successful network run is important in creating a strong market for the show in syndication, first-run programs such as *Oprah* and *Entertainment Tonight* appear in most of the nation's 200-plus markets. This reach rivals that provided by the national broadcast networks.

How Syndication Works

Rather than *selling* programs, the syndicator *licenses* the "buyer" (actually the lessee) to use a syndicated product (program or series). The license permits a limited number of *runs* or *plays* of a program, or each *episode* or *title* of a program series, within a limited time period. The buyer pays a fee to the syndicator—in cash (also over time) or in free advertising time (*barter*, described later in this section) or in a combination of the two. After the license period expires, the programs return to the syndicator, who can recycle them over and over again.

Syndicated distribution resembles network distribution insofar as the same programs go to many outlets. But syndication differs as to both financial arrangements and timing. After a station leases a syndicated program, it usually may schedule the program at any time. Network affiliates, however, contract to broadcast programs in exchange for compensation and normally carry them at network-stipulated times.

Syndicators deliver their programs by various means. A station might receive one syndicated program through a satellite feed and either broadcast it at the feed time or videotape it for later playback. The same station might receive another program by *bicycle*—a system whereby the syndicator ships perhaps two weeks' worth of program tapes to station A, which airs the shows and then forwards them to station B, and so on.

Major television syndication firms often operate as units within vertically integrated companies—those that engage in several related activities. Disney, for example, has cable-network as well as studio and broadcasting interests. King World Productions ranks as the top syndicator, in terms of hit shows, with *Wheel of Fortune*, *Jeopardy!* and *Oprah*. All three consistently rank among the four or five most-popular syndicated programs on TV.

Long-lived syndicated television series run season after season for decades, replayed scores of times. *I Love Lucy* (1951–1956), the quintessential off-network syndicated series, dates back to precolor days and has been syndicated in virtually every country in the world. At times as many as five *Lucy* episodes have been available on the same day in a single American city.

Syndication firms also provide *packages* of programs dealing with a particular theme or sharing a common characteristic. A syndicator may bundle a collection of westerns, crime dramas, science fiction, or "action" movies and offer them for one price. Often syndicators offer programs featuring a particular performer, such as Clint Eastwood.

Television syndicators showcase new productions at annual meetings of the National Association of Television Program Executives (NATPE) and at other national and international program trade fairs. The elaborate NATPE convention sometimes draws criticism for being both too expensive and unnecessary. Some question its value following the widespread merging of program buyers and sellers under consolidated ownership, but its survival does not appear immediately threatened.

The Nielsen rating service documents the track records of syndicated programs already on the market, issuing special reports on the size and composition of the audiences attracted by current syndicated series.

Syndication Exclusivity (Syndex)

Stations may obtain from distributors syndication exclusivity—the sole right to show a product within the buyer's own broadcast market for the term of the syndication deal. Cable program services, especially superstations, often distribute the same programs, bringing them into markets where broadcast stations have paid extra for local exclusive rights to the same programs. That duplication, of course, divides the audience and diminishes the programs' value to the broadcast station.

Responding to this program-duplication dilemma, in 1988 the FCC reintroduced *syndication exclusivity (syndex) rules* it had rescinded eight years earlier in the first wave of deregulation. Syndex rules empower broadcast television stations to force cable systems in their coverage areas to delete duplicate syndicated programs to which the stations hold exclusive rights. Cable systems must block out an offending superstation or other program service, substituting another program or a slide that explains the deletion.

In some cases a superstation's satellite carrier handles this task by inserting alternative programs before downlinking the station's signal to cable systems. When it operated as a superstation, Turner Broadcasting's WTBS resolved the issue by acquiring only those shows that had no syndex problem and thus ran a "blackout-proof" schedule. But syndex conflicts involving superstations may resolve themselves as superstations cease operations or evolve into basic cable channels. The fees cable operators must pay the U.S. Copyright Office to distribute the signal of a distant broadcast station place superstations at an economic disadvantage compared to basic cable networks in the competitive 1990s. New York superstation WWOR reverted to strictly local status at the close of 1996 and in mid-1997, following Turner's lead, Tribune Broadcasting announced its WGN-TV superstation also may convert to operation as a basic cable network.

Syndex rules do not apply to two affiliates of the same network carried by a cable system or to very small cable systems, nor are all syndication contracts exclusive. In fact, obtaining exclusivity increases the price of syndicated programs; sometimes stations prefer foregoing exclusivity for the sake of a lower price.

Financial Interest and Syndication (Fin/Syn)

ABC, CBS, and NBC currently own relatively few off-network syndicated programs—the result of an FCC decision and a U.S. District Court antitrust consent decree. Designed to prevent network domination of program production and distribution, the *financial interest and syndication (fin/syn) rules* severely limited network freedom to participate in production and ownership of prime-time programs or in their domestic syndication.

The FCC modified the fin/syn rules slightly in 1991 and effectively repealed them in 1993. The financial interest portion of the rules ended immediately, permitting ABC, CBS, and NBC to increase their involvement in program production. The Commission delayed repeal of the syndication portion, however, until two years after dismissal of the court's consent decree. As part of its 1993 action, the FCC exempted the Fox network from the rules entirely.

Later in 1993 a federal judge in Los Angeles lifted the consent decree, removing the last barrier to full participation by ABC, CBS, and NBC in program ownership and syndication, although the Commission's two-year delay of its repeal of the syndication rule remained in effect. As expected, the networks moved rapidly to take advantage of their new opportunities. Elimination of the fin/syn rules helped smooth the way for the melding of networks, production companies, and syndication firms.

Prime-Time Access

Another FCC regulation, the *prime-time access rule* (PTAR), gave syndicators a boost. Before 1971 the three major commercial television networks filled nearly all the best evening hours of their affiliates' schedules. This network monopoly left little opportunity for producers to sell programs aimed at the national market but not good enough (or lucky enough) to be selected for network exhibition. On affiliated stations, only the fringe hours (late afternoon and immediately following prime time) remained open for syndicated material. Of course, independent stations have prime time available, but in the early 1970s they could not afford to pay for recently produced, high-quality syndicated shows (later some could, partly because of PTAR help).

In part to encourage local programs and to enlarge the market for producers of new programs, and in part to diminish the network hold on the highest audience hours, the FCC adopted PTAR, effective in 1971. It limited network entertainment programs to no more than three of the four prime-time hours—the evening hours when the television audience reaches maximum size and hence when stations can pay the most for non-network programs. As defined by the FCC, *prime time* consists of the four evening hours between 7:00 P.M. and

11:00 P.M. Eastern and Pacific time (one hour earlier in Central and Mountain time zones, with variations during daylight-saving time).

In practice, the networks had already abandoned the 7:00 to 7:30 P.M. slot to their affiliates. PTAR therefore gave the affiliates only the additional prime-time half-hour between 7:30 and 8:00 P.M.

With PTAR, the entire 7:00 to 8:00 P.M. hour became known as *access time.* Affiliates in the top-50 markets could fill access time with either locally produced programs or nationally syndicated *first-run* programs, but they could not schedule either regular network feeds or former network programs.

After a quarter-century, however, PTAR had outlived its reason for being. Television, with its multiple outlets, was no longer a medium of scarcity, and the Big-Three networks had lost much of their dominance. In July 1995 the FCC scrapped all aspects of the prime-time access rule, effective in 1996. Although there was dispute about some effects of this action, most observers agreed that affiliates would not allow their networks to retake any part of the highly valuable access hour.

Barter Syndication

Because program costs are high, stations sometimes run short of ready cash with which to buy syndicated programs. Distributors and stations therefore work out deals for trading advertising time for programs. The practice, known as *barter syndication,* has become well established.

At first barterers mostly offered only once-a-week programs of little interest to most stations. Generally these were *straight* (or *full*) *barter* deals, meaning the producer/syndicator sold all the commercial time nationally and stations received a free program but no commercial time to sell locally and, hence, no opportunity for profit. Later nearly all first-run access programs, including such hit series as *Wild Kingdom*, *Hee Haw*, and *The Lawrence Welk Show*, were sold as *partial barter*, meaning the station retained some minutes for local advertising sales.

Then syndicators discovered that former network hits could command both cash payments and advertising time. In the late 1980s most barter deals for off-net sitcoms were of this *cash/barter* or *barter-plus-cash* type. By the 1990s, however, as more and more stations became cash poor, syndicators had returned to partial barter arrangements. *Designing Women* was one of the first programs offered on the newly revived barter–no-cash basis.

Barter continued with newer first-run access shows, notably hit game shows such as *Wheel of Fortune* and *Jeopardy!* Soon barter included most daytime talk shows—*Oprah* and *Donahue*—as well as access magazines such as *Entertainment Tonight.* The original *Cosby Show,* for years NBC's top-rated show, sold on a cash/barter basis and shattered all revenue records for a syndicated program. Licensed mostly to network affiliates as a lead-in to early evening news, it earned more money than any other syndicated series, before or since (see Exhibit 8.b).

Barter occasionally presents stations with a dilemma. A station that carries a barter program typically agrees that, if the program is moved to another, less desirable time period, the station will continue to air the syndicator's client's commercials in the original time period. This could mean that the replacement show, if it too is on a barter deal, fills with syndicators' commercials, leaving little if any time for station sales. Stations may also find that some of the most popular barter shows are too "expensive." That is, the value of commercial time given to a syndicator may be far more than the station would have been willing to pay in cash for the right to carry the program and sell all the commercial time itself.

Radio Syndication

Syndication and barter syndication also operate in radio. Syndicators use satellites to relay news, sports, and entertainment material to stations. Programmers distinguish between *syndicated formats* and *syndicated features.* A station might buy the use of a ready-made syndicated country-music format, for example, supplementing it with syndicated news and

Exhibit 8.b

Off-Network Syndicated Program Revenues

Series	Distributor	Estimated Revenue per Episode
The Cosby Show	Viacom	$4.8 million
Who's the Boss?	Columbia	2.5 million
Cheers	Paramount	1.7 million
Magnum, P.I.	MCA	1.6 million
Webster	Paramount	1.6 million
Roseanne	Viacom	1.5 million
The Golden Girls	Disney	1.4 million
Family Ties	Paramount	1.4 million
*M*A*S*H*	Fox	1.1 million

Some successful programs continue in syndication virtually forever. And each new contract or contract renewal means still more income for the series. The above estimates indicate per-episode revenues as of the early 1990s for some of the more popular off-network programs. 20th Century Fox originally concluded that *M*A*S*H* would not perform well on local stations and, consequently, set outrageously low prices for what turned out to be one of the highest-rated—though relatively low-revenue-per-episode-earning—programs in syndication history. In contrast, Viacom offered *The Cosby Show* at outrageously high prices, and stations, expecting results even better than those achieved by *M*A*S*H,* came up with the money. Unfortunately, however, *Cosby* did not meet local rating expectations, and industry observers do not expect to see any future series reach comparable per-episode revenues.

entertainment features from other sources. Thus stations can create unique programming mixes from commonly available syndicated elements. Radio feature material is often bartered, whereas syndicated formats are usually cash deals.

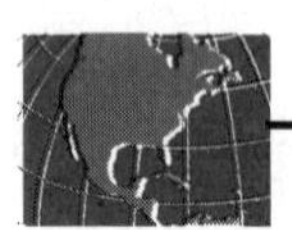

8.3 Program Types

As the previous sections indicate, programs can be classified according to their method of distribution as *local*, *network*, and *syndicated.* This section deals with other classification methods based on

- *content*, in terms of a broad two-way division into entertainment and information;
- *scheduling*, in terms of frequency ("strip scheduling") and time of day ("prime-time soap opera");
- *format*, in terms either of individual programs ("the talk format") or of entire services ("the all-talk radio format"); and
- *genre*, usually in terms of content ("the sitcom genre") but sometimes in terms of target audience ("the children's genre").

Entertainment vs. Information

Most programs qualify primarily as entertainment. However, information programs receive special at-

Exhibit 8.c

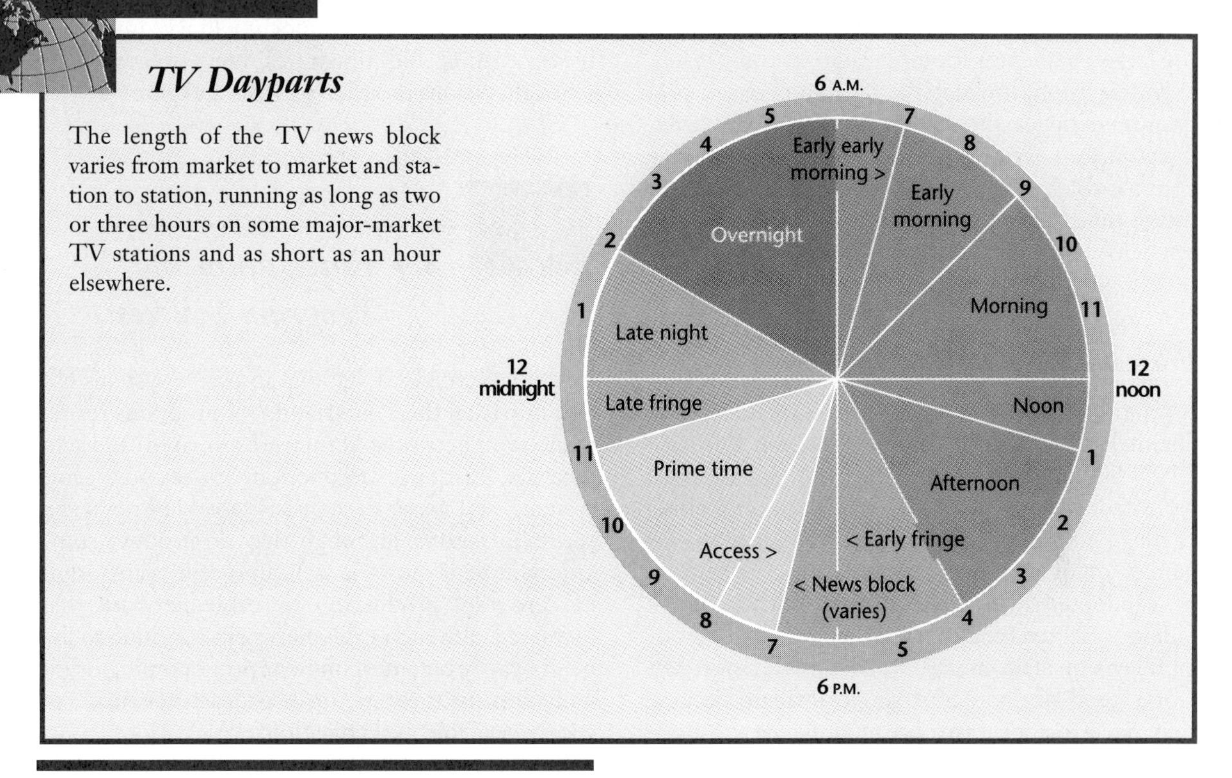

TV Dayparts

The length of the TV news block varies from market to market and station to station, running as long as two or three hours on some major-market TV stations and as short as an hour elsewhere.

tention and deference because many of them enhance electronic media's social importance. A hybrid class of programs sometimes called *reality shows*, became trendy in the late 1980s and expanded in the 1990s. *Hard Copy* and *Inside Edition* are representative examples. They earned the name *infotainment* because they take on the aura of information programs but aim primarily at entertaining. *Infomercials*, long-form commercials that often resemble talk shows, also mix information (or sales pitches masquerading as information) and entertainment. Legitimate information programs have two main subtypes, news and public affairs.

Some people argue that sports programs should be classed as information rather than as entertainment. However, programmers regard sports as primarily entertainment. Network sports departments differ in purpose, style, and types of personnel from news departments.

Dayparts

Two programs of the same type that are scheduled in two different times of day take on different colorations. Prime-time programs have distinctive qualities, as do other programs associated with specific schedule positions. Thus a daytime soap opera differs from a prime-time soap opera, and prime-time sports shows usually differ from weekend sports shows.

For scheduling purposes, programmers break the 24-hour day into blocks they call *dayparts.* Radio programmers generally divide the day into *morning drive*, *midday*, *afternoon drive*, *night*, and *overnight* segments. Morning and afternoon drive periods, with commuters traveling to and from work, have the largest audiences for most radio stations.

In television, broadcast and cable network programmers divide the day as shown in Exhibit 8.c. *Network prime time*, the most important segment, commands the largest audiences. *Access time*, the prime-time hour preceding network prime time, often delivers the most revenue for network affiliates.

Formats

A program may have a "quiz-show format." A cable channel may have a "home-shopping format." CNN has an all-news format, ESPN an all-sports format, and so on. The term *format* can refer either to the organization of a single program or to the organization of an entire service. Most radio stations have adopted distinctive formats, such as AOR (album-oriented rock), classical music, or all news.

Television stations and broadcast television networks tend not to adopt single formats because they need to appeal to a broad audience. As a single-channel service, a station cannot afford to narrow its audience to followers of one particular format. Cable can afford such specialization, however, because it is a multiple-channel service.

Genres

In the context of programs, *genre* (pronounced "zhahn′-ruh") usually denotes a type of content. Familiar entertainment genres include the situation comedy, the game show, the western, and the soap opera. Sometimes target audience rather than content defines genre. Children's shows, for example, can take many different forms—not only cartoons and action/adventure drama but also news, discussion, quiz, and comedy.

Programmers identify a program by genre as a shorthand way of conveying a great deal of information about its probable length, seriousness, subject matter, visual approach, production method, and audience appeal. Hybrid programs sometimes use combined names such as *dramedy* (a drama/comedy hybrid) and *docudrama* (a documentary/drama hybrid). Exhibit 8.d illustrates how programmers schedule various program formats and genres.

8.4 Entertainment Program Sources

Individual broadcast stations and cable systems obtain nearly all their entertainment programs ready-made from networks (if they are affiliated) and syndicators. In-house production by stations, cable systems, and networks consists mostly of news and sports programs, although they do produce some entertainment shows as well. In recent years, original program production by cable networks has increased, although cable networks continue to depend heavily on entertainment programming originally seen in theaters, on broadcast networks, or widely available in syndication.

Before the mid-1990s, nearly all original entertainment programming on networks was provided by *outside* producers falling into five creative groups:

- about a half-dozen major Hollywood film studios;
- a few additional studios below the "major" rank;
- about a dozen major independent producers;
- a host of smaller and more specialized independent producers;
- and some foreign syndicators distributing in the United States.

See Exhibit 8.e for examples of TV program distributors and their products.

Elimination of the fin/syn rules (see Section 8.2) and increased co-ownership of production studios, cable networks, and broadcast networks means there will be more in-house—or at least within-conglomerate—production and distribution of entertainment programming in the future. Indeed, one reason Warner Brothers and Paramount

Exhibit 8.d

Genres, Formats, and Dayparts

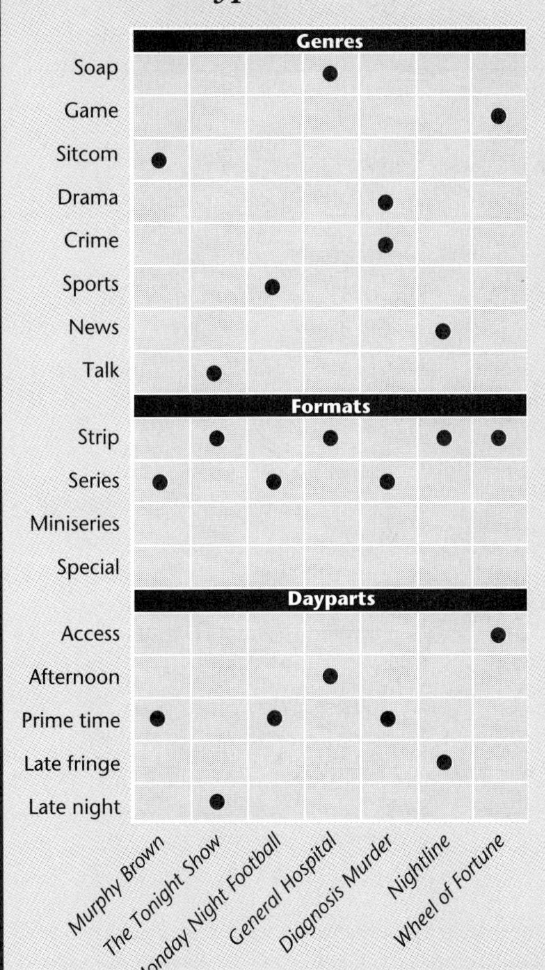

Particular television shows can be classified by their content *(genre)*, the way they are usually scheduled on the networks *(format)*, and the time of day they are typically scheduled on the networks *(daypart)*. In syndicated reruns, the methods of scheduling and daypart often differ from the original network run.

launched broadcast networks is that they provide additional distribution vehicles for the studios' programming. It would be a mistake, however, to assume a network carries only those programs produced by a co-owned company. It is not unusual for a producer's product to air on a network that competes with the producer's co-owned network. This occurs in part because of contractual obligations made before recent mergers and acquisitions, but also because of the constant demand for good programming at the lowest cost, regardless of the source.

Major Studios

The major Hollywood studios have familiar movie-studio names, such as Paramount and 20th Century Fox. Best known to the public for feature films made originally for theatrical release, they also produce a less expensive type of film, made-for-TV movies, as well as program series made especially for broadcast and cable television networks.

By definition, theatrical feature films normally reach audiences first through motion picture theaters. Thereafter distributors release them successively to pay cable and the other electronic media at intervals governed by windows of availability, as shown in Exhibit 8.f. Despite the exhibition priority enjoyed by movie theaters, movie studios make more money from television, cable, videocassette, and foreign rights than from American theatrical exhibition. Feature films are a staple item in broadcast and cable network schedules and even in some individual broadcast station schedules. They have the advantage of filling large blocks of time with material that has strong audience appeal, especially for the young female subaudience that advertisers often target.

Independent Producers

Independent producers range in size from large firms that usually have several series in production simultaneously, to smaller producers with only a

Exhibit 8.e

Major Providers of TV Programming and Program Examples

Company	Original Broadcast Network Programs	Programs for First-Run Syndication & Cable Networks
Disney/Buena Vista/ Touchstone	*Home Improvement; Ellen; Boy Meets World; Smart Guy*	*Regis & Kathy Lee; Debt; Siskel & Ebert*
Carsey-Werner	*Cybill; Third Rock from the Sun; Men Behaving Badly; Cosby; A Different World*	*You Bet Your Life*
Columbia/Tri-Star	*The Nanny; Party of Five; Early Edition; Mad About You; The Days of Our Lives; Sleepwalkers*	*Wheel of Fortune; Jeopardy!; The Dating/Newlywed Hour; Ricki Lake*
CBS Entertainment/ Eyemark Entertainment	*Touched by an Angel; Dr. Quinn, Medicine Woman; Promised Land; Caroline in the City*	*Martha Stewart Living; Bob Vila's Home Again; The George Michael Sports Machine; Psi Factor*
Viacom/Paramount	*Frasier; Diagnosis Murder; Fired Up; Family Ties; J.A.G.; Star Trek: Voyager; Sister, Sister; The Sentinel; Sabrina the Teenage Witch*	*Entertainment Tonight; The Montel Williams Show; This Morning's Business; Hard Copy*
20th Century Fox	*The X-Files; The Simpsons; Chicago Hope; Buffy the Vampire Slayer; The Practice; Millennium; Ally McBeal*	*Student Bodies; Real Stories of the Highway Patrol; Access Hollywood; Cops*
Universal	*TimeCop; Players; Nash Bridges; Law & Order*	*Xena: Warrior Princess; Hercules: The Legendary Journeys; My Secret Identity*
Warner Brothers	*ER; Suddenly Susan; The Drew Carey Show; Murphy Brown; Family Matters; Friends; The Wayans; Veronica's Closet*	*The Rosie O'Donnell Show; EXTRA; Jenny Jones; Babylon 5; The Maureen O'Boyle Show*

During the 1990s companies were bought and sold, merged, and otherwise organized into complex media enterprises. Production and distribution of programming reflects this complexity. Credits for *The Sentinel,* for example, show Pet Fly Productions "in association with Paramount." This list gives examples of such organizations and programs produced or distributed by them.

single series under contract at any one time. In the 1970s independent producers created some of the most innovative television series, such as *All in the Family* and *The Mary Tyler Moore Show.* By the 1980s and on into the 1990s, independent producers such as Spelling, Goldberg, Lear, Tandem, MTM, and Cannell produced an enormous number of prime-time series.

Exhibit 8.f

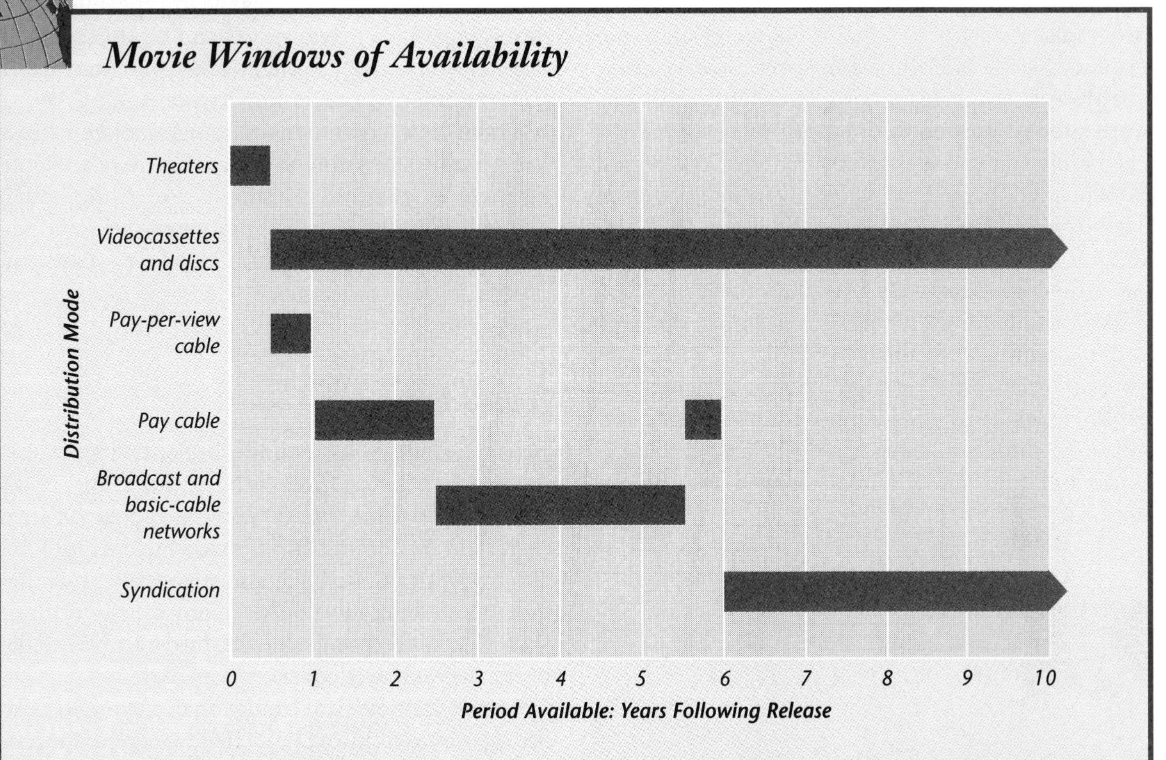

Occasionally, exceptions to this sequence occur. Sometimes distributors withhold highly popular movies from broadcast and cable for many years; sometimes a broadcast network pays extra to step in ahead of the pay-cable window; some unsuccessful films move directly to syndication; and some movies never appear in theaters or television at all, but go directly to video stores. With increasing vertical integration among producers and distributors, the ability to move a product into the most economically advantageous window has improved.

Made-for-Cable Programs

The growth of national cable program services created an insatiable demand for more product. The leading pay-cable network, HBO, alone consumes more than 200 movies a year.

To increase their program supply, to gain control over content and costs, and to create a distinctive image, cable services co-finance movies with producers. This investment gives them *exclusivity*, the right to show the product first on cable. A few movies co-financed by cable, however, such as *On Golden Pond* and *Sophie's Choice*, had box-office success in theaters. HBO's parent corporation is Time-Warner Inc., a magazine publisher and cable MSO. By the 1980s it had become Hollywood's largest financial backer of movies.

Each year, ad-supported as well as pay-cable services increase their investments in production activities. By the mid-1990s basic and pay-cable networks spent about $4 billion annually for original programming. They develop made-for-cable dramatic and comedy series and variety programs for first-run cable use. About one-third of Showtime's schedule and nearly half of the Disney Channel's programs consist of original cable material. Some products, such as HBO's *The Larry Sanders Show*, qualify as true cable-only programs (although they were later syndicated to broadcast television); others merely add episodes to existing television series, as happened in the case of *The Days and Nights of Molly Dodd* on the Lifetime network.

Although made-for-cable programs may not recover their approximately $600,000 (sitcom) to $1.1 million (hour-long drama) cost per episode, they help to counteract the image of cable as a mere parasite on broadcast television. They also make good cable's promise to enhance program diversity and help one cable service distinguish itself from another and gain subscribers seeking new program options.

8.5 News Sources

A markedly different atmosphere surrounds the creation of news programs. By assuming the serious function of conveying news and information, broadcasters, at least initially, took on a more important role than that of merely entertaining. Admittedly, some broadcasters cared little about being part of a great free-speech tradition if doing so interfered with another great American tradition—that of making money through free enterprise. But in time, news programs changed from money losers into moneymakers, ensuring the survival of broadcasting's special role as bearer of information-as-entertainment.

News Agencies

Long before broadcasting and cable, newspapers established news-collecting agencies, the news field's equivalent of program syndicators. Today more than a hundred news services are available to U.S. electronic mass media. Many specialize. For example, ESPN offers sports and CNBC presents business news; HBO offers television stations its Entertainment Satellite Report.

Major news agencies such as Associated Press, Worldwide Television News, and Reuters are international in scope. They supply text, audio, video, and still pictures to subscribing networks and stations for incorporation into their newscasts and other programs.

Network News

Broadcast networks retain tight control over most news and public-affairs production. ABC, CBS, and NBC operate news divisions separate from their entertainment divisions, each employing about 1,000 people. Each also supports its own foreign news bureaus in the major world capitals, staffed by correspondents with in-depth knowledge of the regions they cover.

Broadcast networks traditionally avoided news footage or documentaries from outside sources. This policy began to erode in the 1980s when, to cut costs, networks began accepting news footage from affiliates and from *pool feeds*—cooperative arrangements whereby media rotate responsibility for coverage of news events, or at least share costs.

Networks also cut back their news department operating budgets, reduced staffs, and consolidated or closed some of their news bureaus, both overseas and in the United States, relying more than ever on affiliates and on outside organizations such as Reuters and Worldwide Television News (WTN). In contrast, Cable News Network (CNN) expanded its presence abroad in the 1990s by opening bureaus in Asia, Africa, and Latin America. In 1997 CNN opened a bureau in Havana, the first U.S. network to establish a news operation in Castro's Cuba since 1969. CNN's success in challenging the major networks resulted in further cable specialization, with entire channels devoted to such news genres as finance, sports, weather, and coverage of court trials.

Local News

National and international news is the natural province of networks, local news the province of stations. But stations have increasingly expanded their coverage beyond local and regional news. They have many sources to call on beyond those of their local staffs and regional news services.

Both television and radio affiliates can obtain what amounts to syndicated television news services from their own networks. Provided by the networks' news divisions, these services feed hard news and features over regular network relay facilities during hours when these facilities carry no scheduled network programs. Affiliates can record these feeds, selecting items for later insertion in local newscasts. They can obtain the right to record regular network news programs as sources of stories for insertion in local programs.

Affiliates and independents alike can draw on established news agencies, AP, Reuters, and others. With the advent of electronic news gathering (ENG) and satellite news gathering (SNG), local television news teams could fill longer newscasts and provide more on-the-scene coverage of both local and distant events. And the proliferation of the home camcorder has provided yet another news source—the amateur.

CNN, though originally a news network only for cable, now acts also as a syndicated news source for television stations, affiliates as well as independents. CNN exchanges news stories with stations and supplies them with *Headline News* in 5- or 30-minute blocks, or continuously overnight. Another station source, Conus Communications, coordinates satellite news gathering by more than 100 stations, facilitates exchange of news footage and stories, and, with Viacom, produces its own 24-hour news service, All News Channel.

Cable systems select from a bewildering array of sources to meet their news and sports needs: CNN, CNBC, ESPN, ESPN2, ESPNews, MSNBC, Fox News Channel, CNN/SI, CNNfn, plus regional networks supplying news and sports of mostly local interest. All News Channel supplies material to some regional cable news networks (New England Cable News, for example), which, in turn, supply newscasts for cable systems.

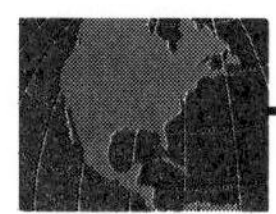

8.6 Sports Program Sources

Networks

Except for professional football games, broadcast networks rely increasingly on production companies that specialize in televising sports. Networks simply pay the cost of national rights to the events and of producing live games, and their own announcers' salaries and travel costs.

Mega-events such as the World Series, the Olympic Games, the Super Bowl, and regular-season NFL football, remain exceptions to this trend. Cable networks and individual broadcast stations also employ sports production companies. ESPN produces most of the games it carries.

Sports Sponsorship

Advertisers gave up most program sponsorship long ago, but the practice has revived for some lesser sports events. Sponsorship means that the advertiser obtains the rights to broadcast or cablecast the event and controls the program.

The network carrying a sponsored sports event supplies only play-by-play and color announcers. Sponsors hire nearby television stations or sports production houses to do the rest of the production work. They also often participate in lining up celebrity guests, promoting the program, and even selling tickets. They sometimes also recover some of their expenses by selling spots to other advertisers and peddling subsidiary coverage rights to the event to radio stations and cable systems.

Both networks and advertisers benefit from this division of labor. Sponsors retain control of costs and ensure maximum promotional value; networks gain hundreds of program hours of minor sports events too risky financially for them to cover by themselves.

8.7 *Network Schedule Strategies*

Whatever their program sources, stations and networks strive to structure their selections into coherent schedules. Effective scheduling requires, among other things, coordinating program types and production tempos to complement typical audience activities.

For example, during the busy early-morning period, listeners/viewers get ready for their day's activities at work, at school, in the home, or at play. This period calls for light, up-tempo treatment of news, weather, traffic reports, and short entertainment features. A more relaxing tempo and longer program units suit the less-structured evening period.

Audience Flow

Schedulers try to draw audience members away from rival channels and to prevent rival channels from enticing away members of their own audience. These efforts focus on controlling *audience flow*—the movement of viewers or listeners from one program to another. Flow occurs mostly at the junctions between programs or, on radio, after one block of songs ends and before the next begins. Audience flow includes both *flowthrough* (on the same stations or channel) and *outflow* or *inflow* (to or from competing programs).

Audience research can measure the extent and direction of audience flow. Such data give programmers guidance on how to adjust schedules and program types to retain audience members and gain new ones at the expense of opposing channels.

However, remote-control channel selectors and VCRs, both now in more than three-fourths of all households, have made tracking television-audience flow more difficult. A special vocabulary has evolved to describe how some audience members use these devices: they *surf* or *graze* through the available channels, *zap* the sound of unwanted commercials, *jump* between pairs of channels, *flip* around to see what's happening on nearby channels, and *zip* through the boring parts of prerecorded cassettes.

Newer technologies such as *picture-in-picture* allow viewers to divide their attention among several competing on-screen channels simultaneously. Stations distributing data, such as supplementary Web pages embedded in their broadcast signal, also provide viewers with more, potentially distracting, information on-screen along with the primary program content. Not all viewers take advantage of this new freedom, however, and programmers continue to employ scheduling strategies designed to influence audience flow.

The unique continuousness of electronic mass media differentiates them from other media, whose products reach consumers only at intervals, coming in individually packaged physical units—a videocassette, a compact disc, an edition of a newspaper or a book, or an issue of a magazine. Only broadcasting, DBS, and cable television afford the consumer an instantaneous choice among several continuously flowing experiences.

Specific Strategies

Taking into account the lineups on competing channels, programmers adjust their own program schedules, fine-tuning them to take advantage of opponents' weak points. Following are some typical scheduling strategies used by networks (broadcast and cable) and by individual systems and stations to exploit audience flow.

- *Counterprogramming* seeks to attract the audience toward one's own station or network by offering programs different from those of the competition. For example, an independent station might schedule situation comedies against evening news programs on the network affiliates in its market.
- *Block programming* tries to maintain audience flowthrough by scheduling programs with similar appeal next to each other—for example, by filling an entire evening with family-comedy programs.
- *Stripping* tries to create viewing habits by scheduling episodes of a series at the same time every weekday.
- A strong *lead-in* attempts to attract the maximum initial audience by starting a daypart with a particularly strong program in the hope of retaining the audience for subsequent programs.
- *Leading-out* with a strong program may similarly attract viewers to the program that precedes it.
- *Hot switching* fine-tunes the above lead-in strategy through the use of "seamless," commercial-free transitions from one show to another. Running crunched-together credits at warp speed while viewers are enticed to stay tuned by promotions occupying two-thirds of the screen further enhances a seamless transition. An NBC study that found almost a third of the audience stopped watching when the closing credits appeared prompted this practice.
- A *hammock* tries to establish an audience for a new program, or to recover the audience for a show slipping in popularity, by scheduling the program in question *between* two strong programs. Flowthrough from the previous (lead-in) program should enhance the initial audience for the hammocked program, and viewers may stick with the weak (or unfamiliar) show in anticipation of the strong (lead-out) program that follows it.
- *Bridging* attempts to weaken the drawing power of a competing show by scheduling a one-hour (or longer) program that overlaps the start time of the competing show.
- *Repetition*, a pay-cable strategy, makes it convenient for viewers to catch a program, such as a movie, because repeat showings are scattered throughout the schedule.
- *Stunting* seeks to keep the opposing networks off-balance in the short term by such tactics as making abrupt schedule changes, opening a new series with an extra-long episode, and interrupting regular programming frequently with heavily promoted specials. Networks resort to stunting especially during the February, May, and November ratings periods. Unfortunately, stunting may not only confuse and frustrate competing networks but also viewers when a favorite program is suddenly replaced or moved in the schedule.

Two general theories tend sometimes to get in the way of these scheduling strategies. One suggests that no matter what is offered, viewers will watch the *least objectionable program* (LOP) in the time period. The other, *appointment television*, holds that successful programs may be scheduled anywhere because serious-minded fans will follow wherever they go.

Network Prime-Time Arena

The main arena of broadcast network rivalry is the weekly 22 hours of network prime time—the three hours from 8:00 P.M. to 11:00 P.M. each night of the week in the Eastern and Pacific time zones (an hour earlier in the Central and Mountain time zones), plus an extra hour from 7:00 P.M. to 8:00 P.M. on Sundays (EST/PST).

Prime time gives major networks access to the largest and most varied audiences of any medium. It therefore demands programs with the broadest appeal. Because networks pay so much for prime-time programs and because they need scheduling maneuverability, they normally run prime-time series once a week. NBC's *Dateline*, a prime-time news magazine running four nights a week, is a remarkable exception to this general rule. If they scheduled all of prime time with daily half-hour series, each network would have only six weeknight programs to schedule, providing too few chances for exploiting scheduling strategies.

In other dayparts, however, broadcast networks strip most of their shows—scheduling the same series at the same time, Monday through Friday. Network morning talk programs, afternoon soap operas, and evening newscasts, for example, occur at the same times each weekday.

Cable Network Strategies

In the early 1980s most cable networks showcased their best product in dayparts *other* than prime time. They were satisfied with reaching demographic subgroups in prime time. By the mid-1980s, however, the largest cable program services, such as ESPN, TBS, and USA Network, had begun to compete head-on with broadcast networks for the prime-time mass audience.

Advertiser-supported cable networks commonly adopt habit-forming strategies to build loyal audiences. The broad-appeal cable services strip their programs across the board both in daytime and in prime time. USA Network and Lifetime, for example, stripped costly off-network series such as *Murder, She Wrote* in prime time, hoping to draw viewers away from broadcast stations and to build a cable-watching habit among first-time viewers.

Pay-cable networks rely heavily on *repetition*, scheduling repeat showings of their movies and variety shows in various time periods, cumulatively building audiences for each program. Each movie plays at varying start times on different days. A movie may be recycled as many as a dozen times a month. For this reason, and also because cable program guides come out once a month and cable companies need to encourage monthly subscription renewals, pay-cable networks plan their schedules in monthly cycles.

Pay cable also uses the bridging strategy, scheduling across the start times of other programs. HBO and Showtime movies, for example, usually start at 8:00 P.M., bridging the 9:00 P.M. station break. This is the period with the largest number of people watching television. Sometimes pay-cable networks try to get the jump on the broadcast networks by starting their movies earlier in the evenings (at 7:00 or 7:30 P.M.). This strategy works best when broadcast schedules have been disrupted by late-running sports or political programs.

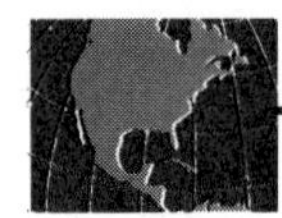

8.8 Local Schedule Strategies

TV Stations

ABC, CBS, and NBC fill about 70 percent of their affiliates' schedules. The newer networks, especially The WB and UPN, provide less. During time not filled by network programming, affiliates apply the strategies previously described to maximize their audience. For affiliates the programmer's most important decisions concern the choice of programs for the early fringe and access dayparts.

Programmers at independent stations have charge of their total schedule. Their chief stratagem, counterprogramming, capitalizes on the inflexibility of the affiliate's schedule because of its prior commitment to network programs. For example, an independent station can schedule sports events at times when affiliates carry major network shows. Networks can afford to devote prime time to only a few top-rated sports events of national interest. Independents, however, can schedule sports events of local interest, even during prime time.

Both network affiliates and independent stations utilize stripping on weekdays. Saturday and Sunday programs are scheduled only weekly. Monday-through-Friday stripping of syndicated and local programs has three advantages for stations. First, daily same-time scheduling encourages the audience to form regular viewing habits, such as the 6:00 P.M. news habit. Second, a single promotional spot can publicize an entire week's schedule for a given time slot. Finally, purchasing many episodes of a syndicated series in a single transaction earns quantity discounts from syndicators.

The practice of stripping off-network programs led to an enormous demand for television series with many episodes already "in the can." Stripping ideally requires 130 episodes for a half-year run. Off-network series that have generated such large numbers of episodes have, by definition, earned good ratings on a network over more than one season. They therefore command the highest prices among syndicated series. Popular long-running shows such as *Murphy Brown*, with more than 200 episodes, are ideal off-network products for stripping in syndication.

In recent years, competition made network schedules more volatile. The networks nervously canceled shows at the first sign of weakening ratings. Runs became shorter, building up too few episodes for strip scheduling over the long period needed for best results. This in turn meant increasing scarcity of, and thus intensified competition for, successful off-network syndicated series.

Besides syndicated entertainment programming, local stations increasingly turn to paid programming such as infomercials and home shopping to fill time (often late night and Sunday mornings) when viewing levels are typically low. These programs provide a reasonable profit for the station, even if they attract a small audience or none at all.

Cable Systems

Technical considerations play a major role in determining cable system program line ups. Despite having multiple channels, most systems cannot carry all of the 150-plus program services available to them, although video compression and fiber optics promise eventually to expand channel capacity almost without limit.

Cable systems usually include signals of most television stations operating in their service area. Subscribers find convenient, and stations strongly advocate, *on-channel* carriage—that is, for example, a station broadcasting on channel 2 appearing as well on the cable system's channel 2. Occasionally this proves impossible, especially in cases where a strong station signal would cause *ghosting* (double images) if carried by the system on its own channel.

Economic considerations also have an effect. Most cable networks charge systems a fee to carry their programs. Were systems to carry them all they would have to charge subscribers an unacceptably high monthly rate. Understandably, a system's MSO often dictates that its own systems carry those networks in which the MSO has an ownership interest. Similarly, a cable network may offer its programming at reduced prices to those MSOs that agree to carry it on all of their systems. Systems also favor ad-supported networks that offer the most time for local advertising. As cable systems seek new revenue sources and technology increases channel capacity, cable operators are dedicating more channels to pay-per-view services.

Some cable systems group channels together by program genre, a practice referred to as *clustering*. They place pay-cable networks—as well as other services that target similar audiences—on adjacent channels. Cable operators also consider program content and the tastes of their subscribers. Many, for example, refuse to carry Playboy; some have even canceled MTV.

Radio Stations

Radio stations use counterprogramming, stripping, and blocking strategies even more than television stations do. Most radio stations schedule their program elements, whether songs or news items, in hourly rotations, creating 60-minute cycles. As the day progresses the hourly pattern is altered by daypart to match changing audience activities. Exhibit 8.g shows an hourly plan for a Top-40 format. Radio stations pay special attention to their programming during drive-time hours, the periods when stations reach the largest audiences and commercials command top dollar.

Long before technology brought increased diversity to TV programming, competing local radio stations battled each other for small segments of the audience. With about 12,000 radio stations on the air, the number of possible variations in programming boggles the mind.

8.9 Program Promotion

Having the best programs in the world means little if audiences don't know about them. Promotion therefore ranks as a major aspect of programming strategies.

On-Air

Broadcast stations and broadcast/cable networks consider on-air promotional spots the most cost-efficient way to advertise their programs. Breaks between and often within programs usually contain promos for upcoming shows. Credits at the end of programs often include audio-only *voice-over* announcements urging viewers to stay tuned to what follows. Stations and broadcast networks also air *teasers*—brief mentions of upcoming news stories—during the hour preceding newscasts. Networks and affiliated stations wrestle to claim precious seconds in which to promote their respective programming. So fierce has this contest become that popular shows may sometimes run 21, or even 20, minutes rather than the traditional 22 to provide more time for promotional (and commercial) announcements.

Television stations and cable systems obtain from program distributors, or produce themselves, promos for syndicated shows. Announcements that promote an entire series are referred to as *generic*, while those that highlight one episode in a series are called *specific* or *episodic*.

Pay-cable networks schedule elaborately produced *billboards* of upcoming programs as filler between the end of one show and the start of the next. To encourage viewers to sample their wares, some cable networks run complete programs on cable's tv! channel. Cable systems and DBS often dedicate one or more channels entirely to program listings. Some form of on-screen "navigation device" is increasingly important to viewers as they try to find programming of special interest among dozens of available channels. New technologies promise viewers the ability to see program titles on-screen as they graze channels, to search for programs by theme or genre, to call up a program by title, to scan a seven-day program guide organized by channel and time, and even to program their VCRs for desired programs or program types.

Other Media

To reach nonviewers, networks and stations advertise in newspapers and magazines, on radio, on outdoor billboards, and in other media. Radio stations promote television programs, and television stations carry spots promoting radio stations—sometimes as paid advertising and sometimes as *trade-outs*, whereby media exchange airtime with no cash involved.

The World Wide Web is an important new promotional medium that developed rapidly in the mid-1990s. Persons responsible for designing Web sites use anything they think will generate more interest or viewership.

Still other promotion forms abound. Radio and TV stations sponsor rock concerts and ice shows

Exhibit 8.g

Radio Station Format Clock

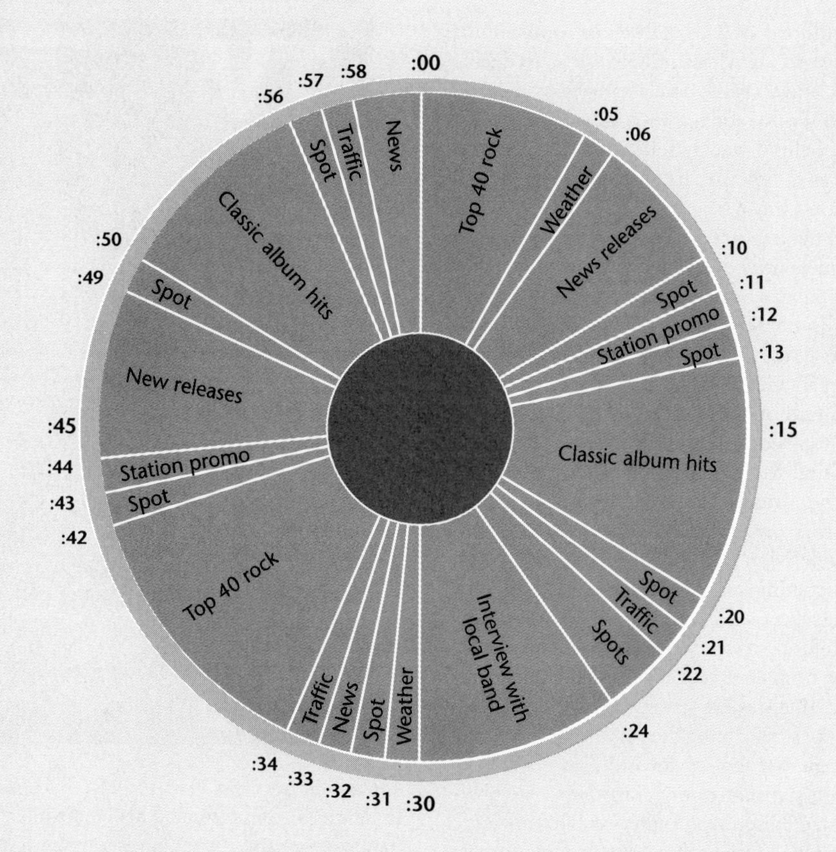

In the beginning, radio programmers planned their schedules hour by hour, drawing a circle around a 45 rpm record and dividing the circle into segments to indicate where program elements would air. Many continue to use this concept, employing different clocks to adapt to audience activities in various dayparts by incorporating changes in both content and tempo of presentation. The example given here represents a typical afternoon drive-time clock for a commercial station.

Exhibit 8.h

TV Guide

Every week millions of U.S. television households turn to *TV Guide* to see what's on television, to decide what to watch, and to read articles and gossip about the television and cable industries. *TV Guide* comes in a distant third, behind weekly television supplements and daily listings in newspapers, as the source of program information most frequently cited by viewers. But these local publications cannot match *TV Guide's* nationwide readership.

Walter Annenberg (later a U.S. ambassador to Great Britain) combined three local weekly television program guides to create the first edition of *TV Guide* in 1953. He began with 10 regional editions and a circulation of 1.5 million subscribers. Media baron Rupert Murdoch took control of *TV Guide* in 1988 when he paid $3 billion for its parent company, Triangle Publications; he already owned half of *TV Guide's* Australian counterpart, *TV Week.*

Each week, television stations throughout the country send in upcoming program schedules, and each week, magazine editors call stations to update program information. Refusing to list program content based on press releases from networks and other program providers, magazine staffers screen every program advance, or read the script, or, at the very least, talk to the show's producers, writers, or talent. The magazine's computers store summaries of more than a quarter-million syndicated episodes and some 36,000 movies.

Dozens of multichannel cable systems may operate within a region covered by a single *TV Guide* edition. Because systems often carry different channel lineups, the magazine lists cable programs not by system channel but by program service. Still, more cable networks exist than could be listed in a magazine of its size. *TV Guide* currently solves this problem by including only the most-watched services. Rarely, for example, does it offer program information for C-SPAN, CNBC, or The Learning Channel.

Moving into the 21st century, *TV Guide* now offers its services "on screen." This electronic version of the magazine promises not only to permit home viewers to check program listings, but also to remind them when a favorite program is on, to tune directly to the show, to record it on their VCR, to search for programs by theme or by title, and even to lock out those shows they don't want their children to see.

Source: Reprinted with permission from News America Publications, Inc., publisher of *TV Guide* magazine. Copyright Vol. 45, No. 37, September 13, 1997, News America Publications, Inc.

and give away T-shirts and bumper stickers. Television networks send news anchors abroad to report from the locales of major events. Daily newspapers and Sunday supplements devote considerable space to broadcast and television listings. Cable systems mail customized program guides to their subscribers; some are free, whereas others cost money. Home satellite dish owners turn to magazines such as *Satellite Orbit* to find out which transponders carry which programs.

The most widely recognized printed source for program information, *TV Guide*, publishes about a hundred different editions weekly within the continental United States, as well as many localized editions in other countries. It also stands at the forefront of efforts to produce electronic program guides. Exhibit 8.h tells more about this uniquely successful publication. Even *TV Guide* cannot provide full details on all available programs for viewers with access to dozens of channels. But that would be asking too much of any print-based technology pacing the dynamic electronic media in the late 1990s.

Chapter 9

Programs: Network, Syndicated, Local

CHAPTER CONTENTS

Programs readily fall into three broad categories—network, syndicated, and locally produced. Subcategories are based on paired differences: prime time versus other times, radio versus television, broadcast television versus cable television, first-run syndication versus off-net syndication, network affiliates versus independents, and entertainment versus information content. Each subcategory has its own characteristic audience; its own production, distribution, and delivery methods; and its own type of content.

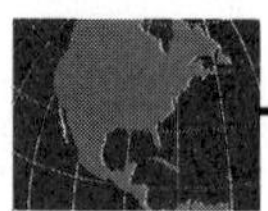

9.1 *Network TV Prime-Time Entertainment*

The best-known and most popular programs appear on television broadcast networks in prime time. They consist mostly of light entertainment—comedy and drama. Prime-time entertainment network programs come as weekly series, with a sprinkling of movies and occasional one-time specials. A series can run for an indefinite number of episodes. Those designed typically with three to eight episodes are known as *miniseries.*

Movies include both theatrical feature films and made-for-TV movies. After being shown in movie theaters, feature films are licensed for release to networks and other outlets in sequence, as shown in Exhibit 8.f. But it is important to remember that the windows of availability illustrated in Exhibit 8.f are not inflexible. Made-for-television movies come in both 90-minute and two-hour lengths. Typically, though not universally, they have lower budgets than theatrical films and are adapted to the limitations of the small television screen.

Audience Share

During prime time, the broadcast networks vie for huge audiences—larger than any in the previous history of entertainment. The percentage of viewers watching the broadcast networks, however, has been declining for two decades. In the 1970s the top-rated network typically attracted a third or more of the prime-time TV audience. By the mid-1990s a network could be number one if it captured about a fifth of the total TV audience during prime-time. NBC's *ER* lost around 10 percent of its Nielsen rating in a single year (1995–1996) but retained its status as the most successful prime-time program.

Despite a decline in their combined audience share, the broadcast television networks still draw the most massive audiences of any media. The rise and fall of their prime-time programs makes headline news. And after (sometimes even during) their contractual network runs, those same programs often travel throughout the world as syndicated off-network shows.

Situation Comedies

Among television prime-time series, situation comedies (sitcoms) are the staple product. Sitcoms such as *Seinfeld* and *Friends* topped the charts in the 1990s. Hit sitcoms attract huge audiences and tend to earn higher ratings than do hour-long dramas.

Writers of situation comedies create a group of engaging characters who find themselves in a particular situation—often a family setting. Plots spring from the way characters react to a specific new source of tension injected into the situation week by week. The characters have marked traits—habits, attitudes, and mannerisms—that soon become familiar to the audience. *The Nanny*, *Cybill*, and *Third Rock from the Sun* are sitcoms that lean heavily on the unique personae of principal characters for their success.

Early TV sitcoms, such as *Ozzie and Harriet* and *Father Knows Best*, portrayed only stereotypical modular family units with a handsome, middle-class father, loving mother/homemaker, and adorable kids, and emphasized idealized "wholesome" values. *The Honeymooners* was a little different—temperamental working-class husband, no kids—but was still rather conventional. *All in the Family* broke new ground in the 1970s by dealing with generational conflict, civil rights, racial and religious prejudice, homosexuality, and gender issues.

In the 1980s and 1990s some sitcoms moved out of the living room, and the definition of "family" changed. Hits such as *M*A*S*H*, *Cheers*, and *Murphy Brown* revolved around character interaction in the workplace, but surrogate family relationships and support networks remained basic to the plots.

Some 1990s sitcoms increasingly probed the boundaries of acceptable home viewing. *Empty Nest* featured a motherless family and *Seinfeld* offered a variety of offbeat plots ranging from an attempt to author a television show about "absolutely nothing" to a contest to determine which of the four leading characters could survive the longest without masturbating. *Ellen* created controversy when the lead character, and the actress portraying her, revealed that she was gay. But *Home Improvement, Dave's World*, and other 1990s sitcoms successfully continued patterns established decades earlier with only minor changes.

A variant, sometimes called the *slobcom*, depicted less-than-ideal family situations. *Married . . . With Children* featured a frustrated housewife, a husband who failed as a provider, and outrageously undisciplined children. *Roseanne*, especially during its most popular years, featured an aggressively earthy, blue-collar family, truly different from *Ozzie and Harriet*.

In the early 1980s producers began injecting comedy elements, formerly the exclusive province of sitcoms, into action/adventure shows. Series such as *Moonlighting* included lighthearted comic scenes. Critics referred to some of these shows as *dramedies*—blends of drama and comedy. The late-1980s shows *thirtysomething* and *Beauty and the Beast* added nostalgia and fantasy to the mix. In the 1990s programs such as *Xena: Warrior Princess* combined fantasy, action, adventure, and tongue-in-cheek humor. *Twin Peaks* and *Northern Exposure* added the word *quirky* to the vocabulary of millions of viewers. *The X-Files* uniquely blended several themes to generate ratings and attract a cult following worldwide (see Exhibit 9.a).

Spinoffs and Clones

A successful program with a new angle instantly begets brazen imitations. Programmers speak of such copies as spinoffs and clones. A *spinoff* stars secondary characters from a previous hit on the same network, placing the characters in a new situation. *Touched by an Angel*, for example, spun off *Promised Land*.

A *clone* closely imitates an already popular program, often on another network, changing only the stars and details of plot and setting. Examples include sitcoms featuring stand-up comics (*Grace Under Fire* and *Ellen*) that hoped to duplicate the popularity of *Seinfeld* and *Home Improvement*. (Indeed, this latter form of clone runs in cycles, dating back at least as far as *Make Room for Daddy*, with Danny Thomas, through *Cosby* and *Roseanne*.)

Crime Shows

Police, courtroom, and detective dramas such as *Dragnet* and *Perry Mason* peaked in the ratings around 1960. In the mid-1980s a new breed of more authentic crime show captured top ratings. *Hill Street Blues* and *Cagney and Lacey* began a trend toward crime dramas dealing with tough social issues, using multilayered plots involving many characters.

NYPD Blue tested the limits of acceptable program content for broadcast networks by using limited nudity and strong language. Although the series enjoyed some success with viewers and received numerous awards and critical praise, many advertisers withheld their support. In contrast, *Murder, She Wrote* featured sanitized homicide mysteries, usually set in upper-class surroundings, with a well-bred, mature mystery writer as the unlikely detective.

Medical Dramas

For years programs in hospital settings were staple fare on network prime-time television. Shows such as *Ben Casey* and *Dr. Kildare*, exploring the lives of

Exhibit 9.a

The Truth Is Out There

Following its debut on Fox in fall, 1993, *The X-Files* developed into both an unqualified network ratings success and a cult program with fanatical fans worldwide. Chris Carter's mod-Gothic melodrama combines two popular themes, a belief that contemporary science cannot explain everything and a nagging concern that government officials do not always level with the public. The show owes much of its success to the main characters, F.B.I. agents Fox "Spooky" Mulder (David Duchovny) and Dana Scully (Gillian Anderson), who together investigate bizarre top-secret cases involving the paranormal. Mulder is an intense, brooding individual who believes in the paranormal; Scully is a physician trained in forensic medicine, and a skeptic. The show also benefits from top-notch production values and its music. The show is shot in Vancouver, Canada, where the fog and mist contribute to an enveloping atmosphere of dread. As critic James Wolcott says, "Even the sunlight looks a little ill." (*The New Yorker*, 6 January 1997:76)

The X-Files reaches an estimated 10 million American TV households and has a large following internationally. It is the focus of intense interest among fans (who call themselves X-Philes) on the World Wide Web. The series inspired a bewildering array of promotional merchandise, and at *X-Files* conventions fans interact with people associated with the program, sometimes even Chris Carter himself.

After more than four years on Fox, the show, some critics suggest, may be losing the unique appeal that made it successful initially. But the show won a 1997 Golden Globe award as best drama and was nominated for 12 Emmy awards, including outstanding drama series. Duchovny and Anderson received Emmy nominations for outstanding lead actor and actress in a dramatic series. The ratings remain good, and Chris Carter continues looking for unusual news reports, believing there are still good story ideas "out there" for future episodes.

Source: © 1997 Fox Broadcasting Company.

both patients and practitioners, drew millions of viewers every week. Then, bowing to the cyclical nature of audience taste, they disappeared. After some successes, such as *St. Elsewhere*, the genre returned in the mid 1990s with *Chicago Hope* and *ER*, the latter often finding itself at the top of the Nielsen ratings. *Dr. Quinn: Medicine Woman* was an interesting variation on the genre in that it combined several unusual elements: a *woman* practicing a *profession* in the *late 1800s* in the *Old West*.

Movies

By the 1970s the television networks were paying increasingly astronomical prices for licenses to exhibit hit feature films. A decade later ABC set a new record by paying $15 million for the right to air

Ghostbusters, even though it had already played for two years in theaters and on pay cable. By the 1990s, VCRs and pay cable had devalued network showings of movies to the point that the average network licensing fee dropped to about $3 million per showing.

Nevertheless, both broadcast and cable networks still pay huge fees for the movie blockbusters. NBC's $50 million payment in 1995 for multiple runs of *Jurassic Park* set a new record. These costly films rarely earn back their full rental fees in advertising revenue but can be useful for clobbering opposing networks in the ratings.

Typically a broadcast network's rental fee for a single showing of a major theatrical feature would more than pay for producing a brand-new, modest budget, made-for-TV-movie. In recent seasons, the top made-for-TV movies commanded higher audience shares than most televised theatrical movies. By 1990 two-thirds of network movies were "made-fors." They also sometimes served as pilots for prospective prime-time network series. For plots, producers often turn to actual, usually sensational events. This practice reached new heights—or exploitative depths—in 1993 when ABC, CBS, and NBC each aired a made-for-television movie based on the true story of the attempt by a teenage girl, dubbed the "Long Island Lolita," to murder her adult lover's wife. By the following year, these networks had begun to back away from this strategy, turning more to what some called "uplifting, feel-good" movies. Fox, on the other hand, stayed on the crime-and-tragedy path with, among others, its O.J. Simpson docudrama. Waiting until the jury had been selected and sequestered, Fox aired the Simpson movie during the February 1995 ratings sweep, where, to the network's certain disappointment, it ran fourth in its time period.

NBC fared better in 1997, easily winning its time period with the first prime-time network broadcast of Steven Spielberg's acclaimed black-and-white docudrama, *Schindler's List*. Sensitive to the program's serious content, NBC aired it largely unedited and without commercials, thanks to underwriting by Ford. It received the then-new parental advisory system's most restrictive TV-M (mature audience) rating based on its violence, full-frontal nudity, and profanity. There were some complaints, but most people praised NBC's handling of the broadcast, including politicians often critical of TV programming.

Beginning in the mid-1980s, the major pay-cable networks (such as HBO and Showtime), as well as some other cable services (USA and TNT), also plunged into the financing of *made-for-cable* movies. Although pay-per-view grows in availability, pay-cable still relies on theatrical feature films as its bread-and-butter entertainment.

Miniseries

Roots, a 14-hour adaptation of a bestseller about the evolving role of blacks in American life, ran for eight successive nights in 1977. It started a trend toward miniseries. At the time, experts doubted the drawing power of the series because of the subject matter and its ability to sustain viewership for so many hours, but *Roots* took them by surprise. The audience increased for each episode, breaking all records on the eighth night. A western miniseries, *Lonesome Dove*, scored a similar surprise success in the late 1980s. Miniseries proved able to compete well against pay-cable movies and to attract new viewers, especially upscale professionals, who may otherwise watch little entertainment television.

Miniseries such as *Roots* have not been successful in the 1990s. The hectic lifestyles of viewers and the multiple channels competing for attention make it difficult for miniseries covering a dozen or more evenings to be successful, the VCR notwithstanding. Miniseries today typically are limited to fewer than a half-dozen episodes.

Music and Variety

Except for variety shows and a few programs such as *American Bandstand*, which in the 1950s became one of television's first hits, and *Your Hit Parade*, which aired from 1950 to 1974, television paid little attention to popular music. That changed in 1981 with the formation of the Music Television (MTV) cable network (see Exhibit 9.b). MTV quickly became a 24-hour rock-video powerhouse

Exhibit 9.b

MTV

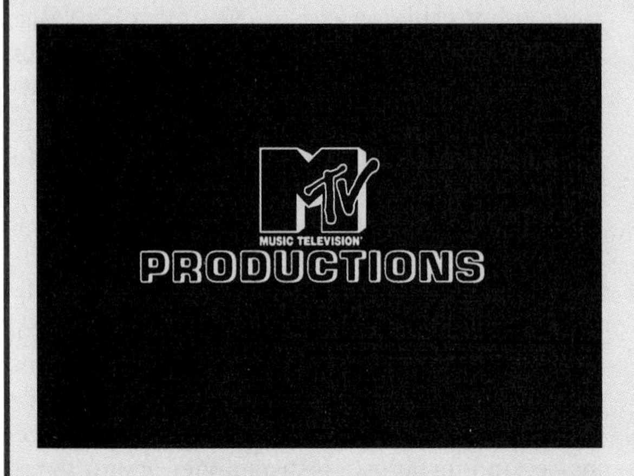

MTV ushered the music-television era into the electronic-media world on August 1, 1981, at 12:01 A.M. Its very first video clip featured an obscure British band known as the Buggles.

Says Tom Freston, head of MTV Networks, "From the very beginning, we made a lot of hay out of the fact that MTV was meant to alienate a lot of people. It was *meant* to drive a 55-year-old person crazy" (Williams, 1989). MTV may well have driven the older generation crazy with worries that, like Elvis's pelvis in the 1950s, MTV would seduce their children with its hypnotic sexuality. But it does delight the young. Today, MTV is available in more than 60 million U.S. cable homes.

MTV's music videos—essentially commercials for songs—rescued the record industry, whose sales in 1981 had dropped 30 percent from their 1978 peak. Over the ensuing decade, sales hit new highs, doubling by 1991. With its fast edits, bizarre camera angles, and exaggerated colors, the MTV "look" has also influenced nearly all areas of the entertainment world. Commercials, TV shows, even movies have adopted the format.

In 1985 a sister channel, VH-1, was born, intended for an audience somewhat older than the 12- to 24-year-olds targeted by MTV. In 1996 another channel, M2: Music Television, was added. The network has also expanded to include game shows, newscasts, stand-up comedy, documentaries, a soap opera, and an action-adventure series.

MTV's animated series, *Beavis and Butt-head*, came under attack in 1993 when a woman blamed the program for a fire, set by her five-year-old son, which killed his two-year-old sister. She said her son began playing with matches after watching the cartoon, in which the characters often talked about or played with fire. Although MTV denied any responsibility, it agreed to delete from the show all references to fire. The controversy did no damage to the commercial viability of the show's characters. Besides remaining on MTV, they starred in *Beavis and Butt-head Do America*, a 1996 feature-length movie. MTV moved the show from its original 7:00 P.M. slot to 10:30 P.M., but later moved it back to 8 P.M.

About 85 percent of MTV airtime remains devoted to music. And not just for Americans. Versions of the network are now available in Europe, Asia, Central and South America, Australia, Russia, Japan—the list goes on and on.

Source: Photofest.

that targets teens and young adults aged 12 to 24. A co-owned network, Video Hits One, programs to attract 25- to 34-year-olds.

In 1996 Viacom/MTV launched yet another channel devoted to music videos, M2: Music Television. There are more than a dozen video music channels distributed nationally featuring many kinds of music. For example, Country Music Television (CMT) provides country-western music videos and music-based features, as does Great American Country (GAC). BET's Cable Jazz Channel programs a broad variety of innovative jazz productions, films, and documentaries. HTV is a 24-hour, all-Latin, all-music service in Spanish.

Originally intended to promote record sales, music videos became a television genre in their own right. Performers act out song lyrics, interpret them, or otherwise create imaginative visual images for songs. As promotional tools, videos came free of charge to stations and networks. But MTV changed the ground rules in 1984 by *paying* for exclusive rights to Michael Jackson's much-publicized *Thriller* video. MTV now contracts for exclusive early *windows* (periods of availability) for some videos. These strategies essentially demolished music videos as a source of free program material.

MTV's success had wide ramifications. As one commentator put it,

> the fast pace and kaleidoscopic style of music video has ricocheted across popular culture, changing the way people listen to music and leaving its frenetic mark on movies, television, fashion, advertising and even TV news. (Pareles, 1989)

Variety programs once played a role both in viewers' habits and in the world of music. Hosted by such stars as Ed Sullivan (who introduced America to Elvis and the Beatles), Judy Garland, Frank Sinatra, Dean Martin, Perry Como, and Sonny and Cher, these programs once brought hours of music into the home.

But their popularity faded as production costs increased and as musical tastes changed. By the 1990s they had disappeared as series from network television, reappearing only occasionally as specials.

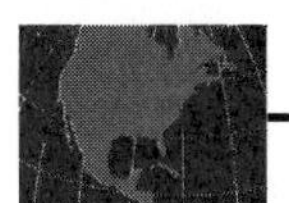

9.2 Non-Prime-Time Network TV Entertainment

Although the broadcast networks put their best creative efforts into prime-time programs, daytime programs yield a higher profit margin. The huge production costs and the smaller number of commercial minutes in prime-time shows make them less efficient revenue earners.

Daytime programs air about 5 to 7 more commercial minutes an hour than do prime-time programs—as many as 16 commercial minutes each hour. Constant switching of soap-opera plot lines from the doings of one character to those of another accommodates interruptions that would be intolerable in most dramas. Daytime talk shows do not provide convenient breaks and sometimes their hosts reveal frustration when they are required to interrupt their programs to insert commercials.

Dayparts

Network non-prime-time breaks down into the dayparts depicted in Exhibit 8.c. A characteristic type of program tends to occupy each daypart:

- Networks fill the early-morning daypart with newscasts and talk, weekends with sports, all discussed later in this chapter.
- Soaps, games, and talk shows dominate daytime network television just as they once dominated daytime network radio.
- Late-night fare consists mostly of talk and comedy/variety programs.

Soap Operas

In the early days of radio, soap companies often sponsored daytime serial dramas whose broad histrionics earned them the nickname "soap operas." This genre is a classic case of parsimonious use of program resources. Notorious for their snail-like pace, soaps use every means imaginable to drag out each episode's story. Most minimize scenery costs, relying heavily on head shots of actors in emotion-laden, one-on-one dialogue.

Contemporary soaps have responded to changing public tastes. They include story lines based on once-forbidden subjects such as drug addiction, social diseases, and family violence. Women and members of minority groups began appearing in more varied roles by the mid-1970s. In 1989 NBC began *Generations*, a trailblazing if unsuccessful soap featuring close relations between a black family and a white family.

Changes in audience composition encouraged soap-opera writers to deal with controversial topics and new social roles. First *General Hospital* and then *The Young and the Restless* stimulated a faddish interest on the part of younger viewers, including males, during the 1980s. ABC launched the promotion *My Time for Me* during the summer of 1988 to lure young viewers during the school vacation.

Responding to criticism that they excessively emphasized sexual relations between married and unmarried characters, soaps adopted a more "socially responsible" approach in the 1990s. Characters continue their sexual activity, but condom use, once taboo on TV, appears with increasing frequency. Alternative lifestyles also are treated with greater sensitivity in story lines.

Even a modestly successful soap generates profits amounting to tens of millions annually, thanks in large measure to relatively low production cost. Not surprisingly, the economic importance of soaps to daytime programming continues to inspire efforts to launch new shows, but with little success. Between 1987 and 1997 only one new soap launched successfully, CBS's *The Bold and the Beautiful*. During the same period two NBC soaps (*Generations* and *Santa Barbara*) and three ABC soaps (*Ryan's Hope*, *The City*, and *Loving*) disappeared. But the search for another *Guiding Light* goes on. Rumors persist that Disney/ABC is planning a Soap Channel for cable featuring syndicated soap operas and related programming.

Soap operas have highly successful counterparts on ethnic and foreign-language outlets. Two competing U.S. Spanish-language cable/broadcast networks, Univision and Telemundo, feature imported *telenovelas*. These enormously popular soap operas reach tens of millions of viewers throughout the Hispanic world. In contrast to American soaps, which endure for decades (see Exhibit 9.c), telenovelas burn themselves out in a few months. Though primarily entertainment, they often also carry educational messages, typically promoting socially approved conduct and family values.

Game Shows

Another parsimonious format, the audience participation game show, became a staple of network radio more than a half century ago. One of the cheapest formats, game shows cost little in time, talent, effort, and money once a winning formula has been devised and a winning emcee selected. Talent expenses are limited to the host's salary and, in some cases, fees for show-business personalities, who usually work at minimum union scale because of the publicity value of game-show appearances. Taping five or more episodes in a single day reduces production costs further.

Opportunities for enhanced commercial content also contribute to the profitability of game shows. The giveaway format justifies supplementing the normal daytime limit of 16 minutes or so of advertising spots with *plugs*—short paid-for announcements on behalf of advertisers who donate prizes and other services such as transportation and wardrobe items. The prizes come free of charge from advertisers, who write them off as business expenses. Game shows were among TV's most popular programs in the 1950s and remained on the broadcast networks well into the 1980s. But by the mid-1990s game shows had practically disappeared from broadcast network schedules. In the second half of the decade Fox experimented with *Big Deal!* (a spinoff on the old *Let's Make a Deal* format) and The WB announced plans to launch a wacky show called *In the Dark* that has contestants perform mundane tasks in total darkness.

Seldom seen on broadcast networks today, game shows remain popular on cable networks. In 1997 the Family Channel, Lifetime, Nickelodeon, and MTV each scheduled at least two game shows. The Family Channel ran a half dozen. Even Odyssey and MSNBC ran game shows. Underlining the popularity of the genre is cable's Game Show Network (GSN) where game shows run 24 hours a day. GSN fills its schedule mostly with syndicated

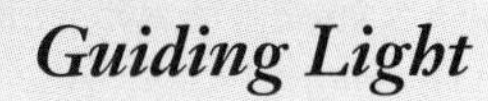

Exhibit 9.c

Guiding Light

Guiding Light, television's longest-running soap opera, celebrated its 45th anniversary in June 1997. The program, which began on January 25, 1937, as a 15-minute radio serial, made the transition to television on June 30, 1952. Its longevity attests to the enduring popularity of the soap opera genre.

Source: Photos courtesy CBS, Inc.

product that had its debut elsewhere, but about 20 percent of the programming is original.

Game shows such as *Wheel of Fortune* and *Jeopardy!* remain among the most successful programs in first-run syndication. *Wheel* benefits both from the suspense element of the wheel's unpredictable stopping places and from the winning talent combination of emcee Pat Sajack and his assistant, Vanna White. She won celebrity status for her unique letter-turner "talent." The producers later made her task even less demanding by replacing the puzzle board's cards with multiple high-tech computerized touch screens, but Vanna remains popular. As she herself candidly admits, her primary role is simply to look beautiful.

Most games run the half-hour length favored for syndication to enable flexible scheduling in all dayparts. CBS runs *The Price is Right* in the late morning, a traditional slot for game shows on the broadcast networks. Cable networks often schedule game shows in the early afternoon, opposite soaps on the broadcast nets, although they can show up in all dayparts. Network affiliates typically purchase first-run game shows to schedule in the hour before prime-time network programming begins.

Part of a game show's appeal is that it encourages viewers to play along with the contestants. The popularity of interactive games available on CD-ROM has not been lost on producers, and they are working to enhance viewer loyalty and involvement by providing ways viewers can participate in a version of the game. Game show producers, of course, are not alone in seeking audience participation. The Internet is a popular vehicle for such activities, and many Web sites promoting programming provide varying degrees of interactivity.

Magazine Shows

In the 1950s programmers began extending network television into hitherto unprogrammed early-morning and late-evening hours, a radical move at the time. For those innovative shows, NBC developed the *magazine format*—a medley of short features bound together by a personable host or group of hosts. *Today*, NBC's pioneer early-morning magazine show, started in 1952. Continuing into the 1990s, the *Today* show shared the 7:00 A.M. to 9:00 A.M. morning spotlight with CBS's *This Morning* (in various guises) and ABC's *Good Morning America*.

Talk Shows

Although related to magazine shows, the talk show more closely resembles an essay than a magazine. It emphasizes the talker's personality, which colors the interviews and other segments of the show. NBC's *Tonight* (later called *The Tonight Show*), a late talk-show companion to the morning magazine show, started in 1954 as a showcase for the comic talents of Steve Allen. After a series of other hosts, including Jack Paar, Johnny Carson took over in 1962.

For 30 years Carson reigned as the "King of Late Night." He hosted some 5,000 shows, talked with more than 22,000 guests, and ran about 100,000 commercials. Carson's reign finally came to a close in 1992. NBC estimated that a record 55 million people watched his final show. The lead-in gave David Letterman the highest-ever rating for his *Late Night* program, which followed.

Letterman, who moved to CBS in 1993 to compete directly against the program, had aspired to move up to *The Tonight Show*. But it was Jay Leno who succeeded into the host chair. Leno had first appeared as a *Tonight Show* guest in 1977, becoming one of several guest hosts in 1986 and exclusive guest host in 1987. Although they had been friends, Letterman and Leno became competitors for late-night viewers—and the millions of dollars in network revenue they represented.

Cable television also takes advantage of the relatively low production cost and flexibility of talk shows. Lifetime, for example, specializes in talk programs, carrying a dozen or so on health, consumer services, and the like. In addition to *Larry King Live*, CNN regularly schedules talk shows on money management, as well as news interviews and discussions. Religious networks rely heavily on inspirational talk programs, and sports channels on sports talk segments and interview shows.

Home Shopping

Not all viewers find commercials boring, at least not those that appear on home shopping networks. These broadcast/cable operations market consumer items such as clothing, jewelry, home appliances, and novelty ware, claiming that bulk purchasing and low overhead cost enable drastically reduced prices compared with in-store prices. A cable system carrying a shopping network receives a percentage of each sale in its service area.

Such nonstop commercialism represents a radical break with established broadcasting tradition. The pre-deregulation FCC enforced an arm's-length relationship between programs and advertising by penalizing stations for program-length commercials—programs that so interwove ad and program content that audiences could not tell one from the other. The FCC reasoned that full-time advertising violated the public interest by displacing normal program functions.

9.3 Network TV Sports

For true fans, sports are ideal television and radio subjects: real-life events that occur on predictable schedules, yet are filled with suspense.

Professional football, basketball, and major-league baseball attract by far the largest audience. The numbers of viewers for other sports fall off rapidly. Super Bowl football and World Series baseball rank among the all-time hit programs because their excitement and spectacle appeal to a broad audience.

The Olympic Games are also a major TV sports attraction. Atlanta's 1996 Summer Olympics, for example, generated high ratings for NBC. Unfortunately, some of that success was attributable to additional viewer interest created when a bomb exploded in the Olympic Park.

Network television strategists value sports programs because they appeal to middle-class males, an audience not well reached by most other programs. The ability of sports to capture such elusive consumers justifies charging higher-than-normal advertising rates for commercials on sports programs.

Evolution of Network Sports

By 1990 combined broadcast/cable rights payments for on-the-spot sports events exceeded a billion dollars annually. Yet both professional and college football ratings, long a sure thing, had begun to decline. Sports fans increasingly divided their attention among dozens of televised sports events.

Broadcast networks responded by purchasing stock in cable networks. In 1984 ABC acquired control of ESPN (originally Entertainment and Sports Program Network). The 24-hour all-sports network furnishes full-length coverage ABC cannot provide as well as enhanced bidding power for rights to sports events, a significant advantage. ESPN2 launched in 1993, providing even greater sports coverage, and three years later ESPNews appeared. ESPN and ESPN2 focus mostly on covering sports events, while ESPNews is a studio-based service providing scores, interviews, and features related to sports.

Following ABC's lead, NBC in 1989 moved into basic cable to create supplementary sports and news outlets. In 1990 NBC's SportsChannel America introduced high school sports to national television. In 1997 NBC assumed a 25 percent interest in Rainbow Programming Holdings, Inc., owner of several SportsChannel regional networks and also Bravo, American Movie Classics, and the Independent Film Channel. CBS was slower to become involved in cable. CBS's participation in cable, including regional sports networks Home Team Sports and Midwest Sports Channel, picked up after its purchase by Westinghouse.

In 1996 Time Warner-Turner launched CNN/SI, a joint effort of Turner's CNN and Time Warner's *Sports Illustrated* magazine. CNN/SI competes directly with ESPNews, serving as a source of sports-related information and features rather than live coverage. Other sports cable networks were developing, some of them, such as The Golf Channel, devoted to a single sport or sports genre.

During the 1990s newcomer Fox emerged as a major player in sports programming, both on-air and on-cable. In 1993, Fox made a spectacular entry into sports broadcasting by outbidding CBS for four years of National Football League (NFL) games. The Fox bid topped that of CBS by $400 million, showing the seriousness of Fox's desire to become a power in sports programming. When a six-year deal signed in 1994 between NBC and ABC for major league baseball fell through, thanks largely to a baseball strike, Fox stepped in. A dramatic come-from-behind win by the New York Yankees in the 1996 World Series gave Fox its first-ever weekly win against ABC, NBC, and CBS during prime time. The following January Fox attracted another huge sports audience with Super Bowl XXXI coverage. Between these two major sporting events Fox launched the Fox Sports Net, based on regional cable sports networks covering much of the country.

The proliferation of sports channels on cable and the desire of broadcast networks to increase sports coverage attests to the importance of the genre. *Monday Night Football* continued to pull top prime-time ratings for ABC through the 1996 season, and cable network TNT finished first in prime time among basic-cable networks in 1996 by running a heavy schedule of NFL and National Basketball Association (NBA) games. NBC announced early in 1997 it would add weekly coverage of Women's National Basketball Association games to its schedule. CBS, still reeling from its loss to Fox in 1993, vowed to bid aggressively for NFL games when the broadcast contract expired following the 1997 season. By 1997, major league baseball was out of its earlier slump and enjoying multi-year broadcast, cable, and World Wide Web rights contracts worth several billion dollars.

Broadcast network viewership of most sports programming declined during the 1990s, but sports remain a critically important genre. Audience demographics make sports programming a "must buy" for many advertisers. In spite of the huge prices networks command for commercials in mega-events such as the Super Bowl, sports programs do not always pay for themselves with ad revenue. Some estimate that total ad revenues from NFL games fell about half a billion dollars short of the 1993 price Fox paid the NFL to edge out CBS for broadcast rights. However, sports programming remains the key to attracting hard-to-reach adult male viewers and is critical to keeping affiliates happy, providing lead-ins to entertainment programming, and establishing a "brand-name" image for the network. In the convoluted logic of sports programming, a network sports division awash in red ink can be considered extremely successful.

Pay-Cable and PPV Sports

Increases in rights charges for sports events forced the gradual migration of many outstanding sports events to pay cable, which can afford the high costs. Despite its relatively small subscribership, HBO, for example, receives stable, predictable revenue that handily covers the costs of sports rights.

Events such as championship boxing and professional wrestling have strong appeal for small but intensely loyal and willing-to-pay audiences. Such events can be profitably scheduled on national pay-per-view (PPV) television. A 1997 heavyweight rematch between Evander Holyfield and Mike Tyson grossed more than $90 million from almost 2 million PPV subscribers. PPV programs are seen mostly by means of addressable cable technology in homes, bars, and hotels. The average cable home has access to a half-dozen PPV channels.

Scheduling and Buying Sports

The seasonal nature of sports and the limited control that stations and networks have over their timing create scheduling complications. ABC made a daring innovation in 1970 when it started scheduling *Monday Night Football* in network prime time. This scheduling risked devoting a long stretch of extremely valuable time to a single program with only selective audience appeal. And once the football season ended, replacement programs had to be found.

With the ABC Monday night exception, major broadcast networks usually keep sports out of prime time. Regular-season sports generally get lower ratings than do prime-time entertainment programs but capture the largest audiences on weekends.

In the late 1980s ESPN also carried a weekend package of late-season NFL games and college games, generally those the commercial broadcast networks did not want. By 1991 TNT had picked up the rights to pre-season and some early-season games. Gaining the rights to these games enhanced both ESPN's and TNT's stature with advertisers and audiences as sports powers.

In 1993 broadcast TV networks, despite having suffered both audience and revenue declines, entered into unprecedented arrangements for coverage of major sporting events. First NBC signed a four-year deal for NBA basketball at $750 million *plus* an agreement to share advertising revenues with the league once the network reaches a certain sales level (about $1 billion). Next ABC and NBC announced a six-year contract for major-league baseball with no rights fees at all, but with an immediate sharing of advertising revenue from televised games. This deal dissolved after the 1995 season when disappointing revenue triggered an escape clause in the contract.

Cable followed the revenue-sharing trend later in 1993 when Turner Broadcasting System renewed its exclusive NBA basketball cable rights for four years (through 1997–1998) for $350 million, a 27 percent increase over the previous contract, plus agreement to split advertising revenues 50-50 after sales exceed that $350 million. The escalating cost of sports programming prompt networks to join forces in bidding for rights. Broadcast networks often pair with cable partners in working out package deals involving shared costs of production, sales, and promotion.

Cable sports networks have dramatically improved sports coverage of not only professional "big-event" sports, but also nonprofessional and "minor" sports. Viewers now see a variety of sports once largely ignored by the broadcast networks. Team sports such as volleyball, rugby, and soccer appear frequently. Individual event sports such as track and gymnastics, along with tennis, golf, swimming, diving, and skiing receive unprecedented coverage. More automobile, motorcycle, and boat races are available. Even events such as dog and cat shows, that many would question as being sports, receive coverage.

Coverage of women's sports also improved in the 1990s. Women's sports attracted loyal viewers of both genders. Female viewers also found significantly more sports programming that appealed to them rather than to the traditional male audience. Ice skating, a sport that attracts a large female audience, appeared so often that popular professional skaters such as Katarina Witt began doing commercials.

Issues in Sports Broadcasting

Television has increased the popularity of sports and opened new opportunities for professional sports careers. The downside is that the enormous fees exacted for television rights have brought up troubling issues:

- When coverage of sports events moves to cable, and especially when it moves to pay-per-view, it becomes unavailable to millions of viewers.
- Astronomical player salaries, induced by television stardom and revenues, invite press and fan criticism.
- The millions of dollars in television rights fees that go to colleges have heightened the temptation to commercialize recruiting, to tolerate low graduation rates among players, and to manipulate college and NCAA rules.
- Television influences event scheduling and even how some games are played. Stations and networks sometimes try to overcome the unpredictability of sports events by staging them expressly for television or by presenting videotaped coverage of a presumably live event—practices that are innocent enough when the events are clearly labeled, less innocent when the manipulation is concealed.

- Over the protests of fans, East Coast games are played under lights to maximize Pacific-time audiences. Also controversial is the delay of college basketball start times to 9:30 P.M. so that networks can carry double-header games on weekends. Such late starts mean that players and student fans arrive home as late as 2 or 3 A.M.
- Referees call arbitrary time-outs about every 10 minutes to accommodate advertising spots within football and basketball games. Such artificial breaks interrupt a team's momentum and undermine coaching strategies.
- Teams and sports associations often insist on controlling the hiring of play-by-play and color announcers. Critics regard such control as an illegal surrender of broadcaster responsibility for programs.
- Critics deplore beer sponsorship of televised college games, which inappropriately links the consumption of alcoholic beverages with the enjoyment of sports by underage college students.

9.4 Children's Programs

Children's programming encompasses virtually all genres. Though cartoons dominate, most adult program types have their counterparts in children's programs.

In the early 1990s cable provided about 60 percent of total television hours devoted to children; Fox and independent stations, about 17 percent; and ABC, CBS, and NBC, only 3 percent. (Chapter 7 includes a discussion of children's programs on PBS, which supplies the remaining 20 percent.)

By the mid-1990s broadcasting's share of the children's programming audience had declined, falling to less than half the total number of children viewing. In 1997 broadcasters sometimes attracted only about a quarter of the audience watching children's programs.

Broadcasting

Congress boosted children's programming in 1990 when it passed legislation requiring that broadcasters air an unspecified amount of programming designed specifically to meet children's needs. In 1996 the FCC clarified the obligation by mandating a minimum of three hours of children's programs weekly between 7:00 A.M. and 10:00 P.M. Shows meeting the FCC's requirements are flagged on-screen with an E/I icon. Stations not meeting the requirement must explain their failure at license renewal time, as discussed in Section 13.9.

For decades, ABC, CBS, and NBC scheduled several hours of children's programming for their affiliates every week, consisting mostly of Saturday morning animated cartoons. They were joined by the Fox Children's Network, which by 1994 had become the highest-rated supplier of children's programs on weekdays as well as Saturdays.

In 1992 NBC began a phaseout of its Saturday morning animated-cartoon line up, moving instead to live-action programs that the network hoped would attract a somewhat older audience—so-called teens and 'tweens—as well as expanding its *Today* program to seven days a week.

Four years later CBS, suffering massive erosion of its Saturday morning audience, also considered quitting the kid business and focusing on teen viewers. Ultimately CBS added four new shows to the Saturday morning schedule, but all were live-action programs rather than animated fare. Right in the middle of Saturday morning the network scheduled a two-hour block of *CBS News Saturday Morning*, followed by *The Sports Illustrated for Kids Show*. ABC, in spite of its ownership by family-oriented Disney, was down to fewer than a dozen hours of children's programming weekly by 1997. The WB, in contrast, doubled the number of hours devoted to kids, from nine to 19, and Fox added seven more children's shows to its already kids-friendly schedule. UPN took a more conservative approach, scaling back to one hour its original plan to launch a new two-hour weekday program targeted at teens.

Pressure remains on the networks to help affiliates meet the FCC-imposed three-hour minimum of educational and informational programming for children. But reaching the kids is increasingly difficult because of competition from CD-ROM games, computer services such as America Online, and, most seriously, the cable networks.

Children's programs earn a relatively small percentage of broadcast television's total advertising revenues, but most run at times that might otherwise go unsold. Toy, candy, and cereal manufacturers support most commercial children's programs.

Cable

Traditionally a strong competitor for young viewers, in the mid-1990s Nickelodeon became the principal channel for kids. By 1997 Nickelodeon accounted for almost 60 percent of TV viewing by kids aged two to 11. Nickelodeon targets younger children in the daytime. After 8:00 P.M. (Eastern) the network runs Nick at Nite, featuring a heavy schedule of classic off-network sitcoms such as *Happy Days*, *Bewitched*, and *Mary Tyler Moore*. Besides sitcoms, cartoons, and game shows, Nickelodeon also originates high-quality programs that avoid violence as entertainment and feature a broad range of role models. In 1996 Nickelodeon began another 24-hour channel, Nick at Nite's TV Land, featuring shows such as *Ed Sullivan*, *Sonny and Cher*, *That Girl*, *Green Acres*, and *Gunsmoke*.

In 1992 Ted Turner launched his new, 24-hour-a-day, all-Cartoon Network. Many of its programs come from the Hanna-Barbera library, which includes old television series such as *The Jetsons* and *The Flintstones.* Turner also owns about 800 MGM half-hours, including *Tom and Jerry.* The merger of Turner with Time Warner in 1996 also improved Turner's access to the well-stocked Warner Brothers cartoon library, further assuring the Cartoon Network of programming well into the future. USA Network also schedules programs, mostly syndicated cartoons, for children weekdays and Sunday morning.

In 1997 News Corporation CEO Rupert Murdoch acquired a 50 percent interest in International Family Entertainment Inc., owner of The Family Channel. Murdoch announced he planned to fill the daytime schedule (12 hours) with the kinds of cartoons and live-action children's programs featured on his Fox broadcast network during the afternoon.

Several cable channels offer children's programs that, while entertaining, have an educational or informational objective. The Learning Channel, an advertising-supported Discovery Communications, Inc. channel, programs commercial-free, nonviolent shows weekdays for preschoolers. A service devoted entirely to children, the Children's Cable Network, started in 1995. Children's Cable Network is also advertising-supported and offers nonviolent educational programs, such as *Dusty's Treehouse*, *The Metric Series*, and *Giggle Snort Hotel*, 30 hours weekly on leased access channels. In 1996 Discovery Communications, Inc. dramatically increased the number of children's programs on basic cable by launching two new advertising-supported networks: Animal Planet, featuring animal-related programs designed specifically for children, and Discovery Kids, a 24-hour programming service targeted at children aged two to 14. Discovery Kids provides documentaries dealing with science, history, technology, and the environment as well as other nonfiction programs.

Long a premium cable channel, The Disney Channel began migrating to basic cable during the mid-1990s but remained a commercial-free service. The Disney Channel programs for children in the daytime and for the family in the evenings. In the 1990s Disney expanded its operation to offer Disney Afternoon, a two-hour block of syndicated programs for broadcast stations.

The lack of available time slots during peak children's viewing periods caused Disney to shorten the package of rotating cartoons to 90 minutes in 1997. Also in 1997 Disney announced plans to challenge Nickelodeon by launching a new cable network targeting young children during the day, and teens and young adults at night. Tentatively called the ABZ Channel, Disney announced plans to begin operations early in 1998, but potential new competition from The Family Channel caused some observers to question whether Disney would move ahead with ABZ. The Children's Television Workshop, famous for its award-winning programs on PBS, had plans for Kid City, another cable network for children.

Children's Television Issues

Children have such easy access to television, they consume so much of it, and it exerts such a powerful hold on their attention that society has a special stake in the quality of programs made especially for them.

Most foreign countries regulate children's programs in considerable detail; some forbid advertising to children altogether. The FCC, however, has imposed somewhat less regulation (Section 13.9). Advocates of greater control raise such issues as these:

- the inappropriateness of much television viewed by children in terms of their needs and vulnerabilities
- the negative impact of violent and aggressive program content
- the absence of a wide range of suitable role models on television
- exploitation of children by advertisers, especially by those that encourage eating candy and sugar-coated cereals
- the shortage of age-specific programs, especially for very young children

Congress reflected concern about these issues in the Telecommunications Act of 1996 by requiring that manufacturers install a V-chip (*V* for violence) in new television receivers that parents can use to block programs containing objectionable content. The V-chip will sense codes transmitted by the broadcast station or cable network that reflect ratings assigned to each program. The law gave manufacturers two years to begin installing the chip and gave program distributors a year (until February 1997) to devise their own ratings system or have one developed by an FCC advisory committee. The broadcast and cable industry reluctantly chose to develop their own system rather than have one imposed on them, as discussed in Exhibit 9.d.

Many TV critics and members of the Clinton administration, especially Vice President Al Gore, hailed the V-chip legislation, and its associated program rating codes, as an important step in providing parents the tools needed to exercise greater control over their children's viewing. Skeptics questioned whether the legislation would achieve its goals. Content creators and distributors viewed the new requirements as infringing their First Amendment rights.

Later in 1996 the FCC adopted rules mandating that broadcast stations air a minimum of three hours of educational children's programming weekly. This new regulation and the V-chip legislation are efforts to address a growing concern that much TV content has a harmful effect on children.

9.5 Network TV News and Public-Affairs Programs

The strict network separation of news and public-affairs programs from entertainment began to erode as competition for audience attention increased, beginning in the 1970s and continuing into the 1990s. Networks reacted by squeezing more value from their heavy investment in their news division—more frequent news and news magazine programs and more syndication of network news product to affiliate and foreign organizations. At the same time, and deplored by critics, show-biz culture invaded news, making some news offerings increasingly superficial.

Exhibit 9.d

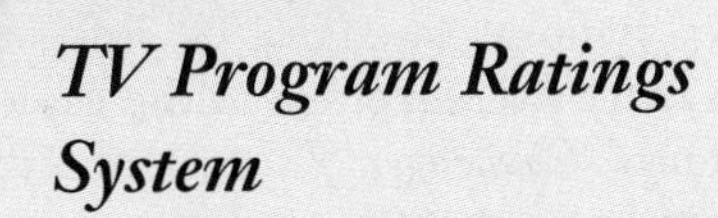

TV Program Ratings System

The following programs are designed solely for children:

TV-Y Appropriate for all children. The themes and elements in this program are specifically designed for a very young audience, including children ages 2–6. This program is not expected to frighten younger children.

TV-Y7 Designed for older children, age 7 and up. This program is appropriate for children who have acquired the developmental skills needed to distinguish between make-believe and reality. Programs in which fantasy violence may be more intense or more combative than in other programs in this category will be designated **TV-Y7-FV**.

The following programs are designed for all members of the audience:

TV-G General Audience: This program is appropriate for all ages. It contains little or no violence, no strong language, and little or no sexual language or situations. Most parents would find this program suitable for viewing by children of all ages.

TV-PG Parental Guidance Suggested: This program contains material that parents may find unsuitable for younger children. The theme itself may call for parental guidance and/or the program contains one or more of the following: moderate violence (V), some sexual situations (S), infrequent coarse language (L), or some suggestive dialog (D).

TV-14 Parents Strongly Cautioned: This program contains material that many parents would find unsuitable for children under the age of 14. The program contains one or more of the following: intense violence (V), intense sexual situations (S), strong coarse language (L), or intensely suggestive dialog (D).

TV-M Mature Audiences Only: This program is specifically designed to be viewed by adults and therefore may be unsuitable for children under 17. This program contains one or more of the following: graphic violence (V), explicit sexual activity (S), or crude indecent language (L).

Late in 1996, under political pressure to do so, the television industry adopted this six-category age-based rating system for programs. The system allowed programs to be rated both as series and as individual episodes. The ratings applied only to entertainment programs, not news. Infotainment shows, such as *Inside Edition*, received ratings unless they involved hard news stories or interviews with recognized news figures such as the First Lady. Program guides began publishing the ratings and they appeared superimposed on-air at the beginning of programs for 15 seconds.

Critics immediately objected to the system. Most, including outspoken Congressman Edward Markey (D-Mass.), said the ratings lacked the specificity needed by parents, arguing instead for a system that ranked each program according to the levels of violence, sex, or language it contained. Objections increased when most prime-time shows received a TV-PG rating. Critics said TV-PG was an especially vague category that set neither age limits nor provided guidance regarding the level or specific type of objectionable material in the program. Television industry leaders, saying it was a work in progress, asked critics to give the system a fair chance before adopting a government-designed alternative. But the outcry was too broad and too shrill to ignore.

After only seven months the cable and broadcast industry, except for NBC, agreed to add content warnings for violence (V), sexual content (S), strong language (L), suggestive dialogue (D), and fantasy violence (FV) to the existing rating categories. Additionally, it was agreed that the ratings symbols would be made larger on the screen for easier reading, but the time on screen would remain the same as before, 15 seconds. The new system went into effect October 1, 1997.

Broadcast News

Back in 1963 CBS and NBC expanded their prime-time news programs from 15 minutes to a half-hour (actually only about 22 minutes, after time for commercials and openings and closings is subtracted). The text of an entire half-hour network newscast fills less than a single page of a full-size newspaper. Nevertheless, moving to the half-hour format played a role in elevating network television to the status of the country's most widely accepted news source.

A CBS movement to expand the evening news to a full hour was defeated in 1975 by affiliates who wanted to retain both their own highly profitable local newscasts and the 7:00 to 8:00 P.M. access hour for revenue-producing syndicated fare. CBS, unwilling to invade network prime time with news at the expense of entertainment programs, gave up the plan.

In 1979 ABC started an experimental late-night network news program, *Nightline*, featuring Ted Koppel. The experiment became a permanent half-hour at 11:30 P.M. Each evening Koppel concentrates on one or two current news stories.

In 1992 Fox announced plans to establish a national news operation. The following year, however, the network backed away from its national news ambitions, although it did expand its local news presence on its owned-and-operated stations.

In 1995 Fox concluded a deal with the international news service, Reuters, to provide news feeds for its stations. In the mid-1990s stations with strong local news operations became Fox affiliates and in 1997 Fox announced it was making local news programming a condition for continued network affiliation. Fox promised to help by forming a Fox News production unit that would provide about 80 percent of the content needed for a local newscast.

Cable News Networks

Ted Turner launched Cable News Network (CNN) as a fourth major television news service in 1980. With a 24-hour schedule to fill, CNN can supply in-depth reportage as well as continuous coverage of breaking news stories.

Early doubts about CNN as a serious news service began to disappear when in 1986 it was the only network to carry live coverage of the explosion of the space shuttle *Challenger*. Those doubts dissolved completely in January 1991 as government officials and even ABC, CBS, and NBC turned to CNN for its coverage of the war in the Persian Gulf. In 1992, after having received abysmal ratings in 1984 and 1988, the major networks conceded to CNN, PBS, and C-SPAN gavel-to-gavel coverage of the Democratic and Republican conventions, limiting their own to a few hours each day.

Between news roundups, CNN schedules interviews (*Larry King Live* is the best known), features on managing money, stock market analyses, sports news, and public-affairs discussions. Today more than 70 million cable and DBS television subscribers have access to CNN worldwide.

Turner's companion news service, CNN Headline News, provides news headlines and frequent updates in continuous half-hour cycles. It resembles all-news radio with pictures. More than half of all cable systems devote an entire channel to Headline News in addition to a channel for CNN.

Other news channels joined CNN on cable in the mid-1990s, including CNN International (CNNI) in 1995. Available in Europe since 1985, CNNI is a 24-hour global news and information channel providing three hours of programming daily in Spanish for U.S. subscribers. In 1996 News Corp. launched the Fox News Channel and NBC collaborated with Microsoft to start MSNBC, a cable network with an exceptionally strong Internet presence. Heated public clashes between CNN's Ted Turner and News Corp.'s Rupert Murdoch regarding cable system carriage made the launch of Fox News Channel especially memorable.

CBS launched Eye on People in 1997 with only about 2 million subscribers. CBS does not bill Eye on People as a news network but rather as an entertainment and information network focusing on people and personalities. The network does not, for example, go live with news coverage. The network leans heavily on CBS News, with anchors such as Paula Zahn, Charles Osgood, and even Dan Rather hosting programs. Eye on People also features a spinoff of *60 Minutes*, the broadcast network's most prestigious news and public affairs program. CBS says it expects the network subscriber base to expand rapidly but does not expect to see profits for at least five years.

Informational Networks

Many cable networks provide informational programming. NBC's Consumer News and Business Channel (CNBC) attempted to become a major cable player when it acquired its prime competitor, Financial News Network (FNN), in 1991. In addition to its cable programs, it offers CNBC Radio through the Unistar network.

Drawing upon its considerable news resources, CNN began a competing financial news network, CNNfn, in 1995. Also launched in 1995, Bloomberg Information Television provides financial and other news featuring a multi-screen format with a newscaster in the upper corner and text filling the remainder of the screen. Court TV features live coverage of courtroom trials from around the country with analysis and commentary by experienced legal journalists. Court TV's ratings skyrocketed during the O. J. Simpson criminal trial.

The Weather Channel was among the first (1982), and remains one of the most "pure," information channels. Many others offering healthy doses of information along with entertainment have either joined it or announced launch dates. They include The Learning Channel; The Silent Channel (for the hearing impaired); The Travel Channel; The Enrichment Channel (self-help, healing); Hobby Craft Network; Fitness Interactive; Home Improvement Network; The Love Network (education regarding relationships, self-esteem, and self-improvement); and the Recovery Network (information for people dealing with alcohol and other chemical dependencies).

Public-Affairs Programs

News and public-affairs programs tend to overlap, but the FCC makes a distinction, defining "public affairs" as

> local, state, regional, national or international issues or problems, including, but not limited to, talks, commentaries, discussions, speeches, editorials, political programs, documentaries, minidocumentaries, panels, roundtables and vignettes, and extended coverage (whether live or recorded) of public events or proceedings, such as local council meetings, congressional hearings and the like.

The Commission stressed public-affairs programs because of the traditional view that broadcasters in a democracy have a special obligation to serve the needs of citizen-voters.

Most commercial broadcast television networks and most large stations maintain at least one weekly public-affairs discussion series, sometimes also a news documentary series, and often minidocumentaries within newscasts (especially during sweeps periods).

The most striking development in network public-affairs programs in the past two decades has been the rise of *60 Minutes*, the weekly CBS magazine-format documentary series (Exhibit 9.e). It ranks as one of television's most-watched programs. That success violates all conventional wisdom, which had said that documentaries repelled the mass audience. One reason previous documentaries had low ratings, aside from not being as well done as *60 Minutes*, is that the networks usually scheduled them in unfavorable time slots and denied them the long-term stability they needed to build an audience.

After years of wandering through the CBS schedule, *60 Minutes* finally arrived at a good hour—a byproduct of the prime-time access rule, which left the Sunday 7 to 8 P.M. time slot open for nonentertainment network programs. Another factor may have been the CBS counterprogramming strategy of scheduling *60 Minutes* against children's programs and movies. In the late 1980s CBS started a second public-affairs series, *48 Hours*, which turns cameras on a single topic, such as a hospital or an election.

Exhibit 9.e

60 Minutes

A stellar team of correspondents, originally consisting of Mike Wallace, Harry Reasoner, and Morley Safer, contributes to the *60 Minutes* success story. As a *New York Times* commentator put it:

> Their gray or graying hair, their pouched and careworn countenances, the stigmata of countless jet flights, imminent deadlines, and perhaps an occasional relaxing martini, provide a welcome contrast to the Ken and Barbie dolls of television news whose journalistic skills are apt to be exhausted after they have parroted a snippet of wire service copy and asked someone whose home has just been wrecked by an earthquake, How do you feel? (Buckley, 1978)

Later, Dan Rather joined the team; later still, Ed Bradley and (for a time) Diane Sawyer improved its ethnic/gender balance. Rather left the show in the mid-1980s to anchor *CBS Evening News*, the most prestigious news position a network can offer. In 1989 Sawyer moved to ABC, which gave her a multimillion-dollar contract, and in 1991 Harry Reasoner retired. Today, Leslie Stahl and Steve Kroft, together with commentator Andy Rooney, round out the team.

The magazine format allows the program to treat a great variety of subjects in segments of varying lengths. With executive producer Don Hewitt and a staff of some 70 producers, editors, and reporters, the *60 Minutes* team develops about 120 segments annually, each of which, on the average, requires six to 10 weeks to produce. The *ambush interview*, a Mike Wallace specialty, adds drama to investigative reports. Presenting his victims on camera with damning evidence of wrongdoing, Wallace grills them unmercifully. The victims' evasions, lies, and brazen attempts to bluff their way out of their predicament—often captured through extreme-close-up lenses—fascinate some viewers, while others question these tactics. Tabloid TV programs have imitated the Wallace-type confrontation to the point of turning it into a cliché.

The top-rated Sunday evening newsmagazine, which has aired since 1968, now stars (clockwise from upper left) Steve Kroft, Ed Bradley, Morley Safer, Mike Wallace, Leslie Stahl, and Andy Rooney.

60 Minutes is generally acknowledged to be the most successful news program—in terms of both ratings and revenues—in television history. It may well be the most successful of *any* program, news or otherwise. It is the only broadcast to have finished in the top-ten Nielsen ratings for more than 15 consecutive seasons, and the only program to be rated number one in each of three different decades. On the revenue side: it generates a profit for CBS of between $60 million and $70 million every year, year after year after year after year . . . .

Source: Photo from CBS, Inc.

ABC's *20/20*, a magazine-format show similar to *60 Minutes*, survives as a respected and successful prime-time program. It stars Barbara Walters and Hugh Downs, both of whom began their careers at NBC. ABC later added two additional hour-long prime-time public affairs programs, *Prime Time Live* and *Turning Point.*

NBC had difficulty in developing a prime-time news magazine—but not for want of trying. It had attempted 18 times to make the format work, beginning with *First Tuesday* in 1971, when finally 1992's *Dateline,* with former *Today* anchor Jane Pauley and former *20/20* correspondent Stone Phillips, attracted enough viewers to survive, eventually expanding to four nights a week.

Fox did not enter the arena until 1993 when it launched its first prime-time news magazine, *Front Page.* In May 1994 Fox canceled the show and aborted others—variously named *Assignment* and *Full Disclosure*—before they even went on the air.

Networks favor prime-time news magazines because, compared with other entertainment programs, they are relatively inexpensive. They often use and reuse resources already in place in network news divisions. However, many of them compete for the same stories and for the same audience. By late 1994 ratings for most news magazines had begun to slip, and networks found themselves canceling or combining several of these offerings.

ABC, CBS, Fox, and NBC schedule a public-affairs question-and-answer session with newsworthy figures on Sundays, usually around midday. NBC's *Meet the Press* started in 1947 and is the oldest continuously scheduled program on network television. *Meet the Press* made history again in 1997 when it became the first live network television program broadcast in digital high definition. CBS launched *Face the Nation* in 1954, and ABC began *Issues and Answers* in 1960, later replacing it with *This Week with David Brinkley.* Brinkley retired in 1996 and the title was shortened to *This Week.* In the mid-1990s Fox joined the older networks with *Fox News Sunday.*

Sunday Morning provided a vehicle for CBS's much-loved Charles Kuralt. It drew only low ratings, but the time slot suited Kuralt's leisurely, reflective, low-key style. Many considered the program to be one of the very best on television. After 37 years with CBS, 15 of them as *Sunday Morning* anchor, Kuralt retired in 1994. Charles Osgood, who followed him onto the anchor stool, said "Although I am succeeding Charles Kuralt, I know that I cannot replace him. Nobody could" (*Miami Herald,* 2 April 1994: A2).

Public Affairs on Cable

Cable developed a unique public-affairs vehicle specifically to cultivate a positive image for its industry. C-SPAN (Cable Satellite Public Affairs Network), a nonprofit corporation, originates a full 24-hour public-affairs service. Together with its companion C-SPAN II, it offers live coverage of congressional floor sessions, hearings, political conventions, and other informative programs.

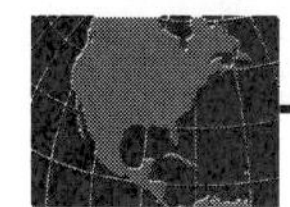

9.6 Syndicated TV Programs

The mechanics of syndication and how stations and cable television use syndicated programs are discussed in Section 8.2. Here we focus on the syndicated programs themselves, divided into three categories: off-network programs, first-run programs, and theatrical films (movies). Some additional syndicated material comes from religious and noncommercial sources.

Off-Network Syndication

After having completed their contractual network runs (usually two showings), network entertainment programs revert to their owners, who then make them available for licensing to broadcast stations or cable networks as off-network syndicated programs.

Sometimes early episodes of long-running network series go into syndication while new episodes are still being produced for the network. Off-network programs are the largest syndicated category. They are highly prized because their previous exposure on networks confers recognizability and a ratings track record. An off-network series newly released in syndication typically earns a rating about ten points lower than it earned in its initial network appearance. On cable networks, ratings for such programs usually fall still lower.

Most stations schedule half-hour syndicated series because they attract a desirable demographic group (typically women 18 to 49 years) and permit scheduling flexibility. Hour-long programs, on the other hand, do not attract many new viewers during their second 30 minutes.

Sitcoms attract both children and adult women. Stations usually put their most popular, male-appealing syndicated program in the slot leading into local news. Sixty-minute action/adventure series have had some success in late afternoon because of their male demographics. Action/adventure shows are also used on weekends to lure and retain hard-to-reach male viewers before, between, and following sports programs.

Cable networks have found that some hour-long dramatic series appeal to cable viewers in access-time, prime-time, and late-fringe-time periods. Lifetime, for example, successfully stripped *L.A. Law*, and USA stripped *Murder, She Wrote*—both in prime time.

First-Run Syndication

Shows that never appeared on a network—*first-run* programs—are brand new when first sold to stations. They are produced especially for the syndication market. Casual viewers can hardly distinguish the best first-run syndicated programs from network programs. For example, the daily syndicated magazine show *Entertainment Tonight* (described in Exhibit 9.f) is so timely and slickly produced it could easily be taken for a network program.

By the mid-1990s producers were offering even high-priced dramatic programs in first-run syndication. Examples include *Hercules* and *Star Trek: Deep Space 9*. Sometimes programs discarded by networks later turn up in first-run syndication (*Baywatch*, for example, which eventually became the most-watched program worldwide).

Producers of first-run syndicated programs most often employ low-budget genres such as quizzes and games. The reigning king of first-run programs, *Wheel of Fortune*, typically gets higher ratings than any competition, including off-network shows.

Syndicators and stations also favor first-run talk shows. Most find homes in morning or afternoon time slots. *Oprah Winfrey* ranks at the top of the syndicated-talk-program ratings ladder.

A much-discussed first-run syndicated program genre emerged in the late 1980s. Known variously as *reality shows*, *trash TV*, and *tabloid TV*, these shows have in common a pseudo-journalistic approach to real-life topics, usually items currently or recently in the news. Such programs cater to morbid interest in the sensational and the bizarre and exploit scandals, sex, and violence. Their blending of entertainment and journalism were symptomatic of a broad trend toward treating news from an entertainment perspective. *A Current Affair* started the genre in 1988 and remained in production through the mid-1990s. Its success inspired clones such as *Inside Edition*, *Hard Copy*, *American Journal*, and *Extra*. Other reality-based programs include *Rescue 911*, *Cops*, and *Real Stories of the Highway Patrol*.

Movies

In addition to off-network and first-run programs, television syndicators offer movies, singly or in packages. Distributors often sell "all or none" packages—a questionable practice resembling block booking, a practice legally banned in the motion picture industry. The buyer has to accept second-rate movies to get the more desirable films in the group of films sold as a package. Broadcast network affiliates usually strip movies in late-night and Saturday afternoon slots. Independent stations and cable networks are more likely to schedule them in prime time and on Sundays.

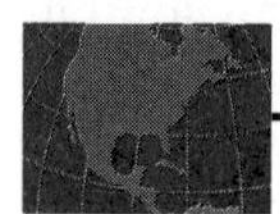

9.7 Radio Network and Syndicated Programs

Competition from television devastated the original radio networks as full-service program suppli-

Exhibit 9.f

Entertainment Tonight

Entertainment Tonight (ET) revolutionized the syndication business by proving that expensive, original, non-network programs—other than game shows—could be profitable for stations as well as for producers and syndicators. The economics of such an effort are all the more unusual in that each episode is an original—with no reruns to help spread the cost.

Introduced in the late 1970s, the program took off in 1981 when Group W began delivering the topical show by satellite. Mary Hart joined ET in 1982 (with special lighting to illuminate her legs beneath the plexiglass desk), and John Tesh arrived in 1986. Tesh later left the show to develop his career as a musician and composer. Bob Goen now shares anchor duties with Mary Hart.

Entertainment Tonight provides a classic example of the constant innovation essential to keep a series from going stale. Initially, the program capitalized on the audience's appetite for gossip, personality exploitation, and show-biz fluff. Later the producer countered the lightweight image by introducing brief "think-pieces," some hard news, and more in-depth stories. By the 1990s, in order to compete with a glut of tabloid shows, the program had begun to turn more and more to sexually oriented items.

Among the best elements of the show are the contributions of Leonard Maltin, who became a regular contributor in 1982 and obviously knows and loves "the business." His consistently perceptive pieces are as rare on popular commercial television as is an episode of *Entertainment Tonight* that uses neither the word *exclusive* nor the phrase *behind-the-scenes*.

ers, but they survived by reducing their role to that of supplements to radio music formats adopted by their affiliates. In the 1970s new networks began to offer their own music formats, made possible by inexpensive satellite relays.

Statistical Research Inc., which twice each year conducts its radio network listener survey called RADAR (Radio's All-Dimension Audience Research—see Chapter 10), defines a network as a program service that has continuity of programming; written, contractual agreements with its affiliates; the capability of an instant feed to all affiliates; and a clearance system so it can determine which affiliates carry which programs. This does not keep any number of station alliances from calling themselves "networks." RADAR reports refer only to those few (about a dozen) satisfying the definition devised by Statistical Research Inc. Tracking network radio is difficult because "radio networks" form and dissolve frequently to better target a desired audience.

Network News

Radio networks associated with television networks get the benefit of daily newscasts voiced by well-known television news personalities. Some television journalists anchor radio news bulletins such as morning-drive reports; others prerecord news stories for later inclusion in scheduled radio newscasts. Network radio's most popular anchor/commentator is radio veteran Paul Harvey. Famous for his distinctive phrasing and use of pauses, Harvey, who began his broadcasting career in the early 1940s, reaches an estimated audience of more than five million people with his daily newscasts. Besides his newscast (which he describes as "news and commentary"), Harvey also produces a radio feature.

Talk Shows and Sports

The Mutual Broadcasting System pioneered all-night talk on network radio with *The Larry King*

Show, predecessor of the CNN television program, *Larry King Live*. For years it remained the most popular program in its genre. In the 1990s talk radio turned strongly conservative. Best known of the conservative talkers is Rush Limbaugh. The cliché that the audience for radio talk shows is old, white and male is generally accurate. Surveys suggest it also may be better educated, more affluent, and more politically active than the population as a whole. Some people think Limbaugh and other conservative hosts helped Republicans take control of the House of Representatives in the 1994 election. Their political influence seemed diminished by 1996, although Limbaugh continues to draw a devoted audience.

Several radio talk show hosts have also tried working in television. In 1994 King dropped his syndicated radio show, electing instead to simulcast on Westwood One the audio portion of his CNN television program. Both Limbaugh and Paul Harvey migrated to television, but their styles are more effective on radio.

The sports component of radio networks is sufficiently important to affiliates for CBS to pay $50 million for radio baseball rights for the early 1990s, nearly doubling the previous contract payment. The importance of radio sports became even more evident with the advent in 1992 of ESPN Radio, one of the services offered by ABC Radio Networks.

Music Networks

The largest music networks, such as Westwood One and Unistar, each has more than a thousand radio affiliates. *American Top 40 with Casey Kasem*, perhaps the best-known show on music radio, shifted to Westwood One in 1989 after running on ABC for nearly 20 years. Kasem's countdown technique, incorporating brief stories about artists and their music, and now in multiple formats to accommodate a fragmented Top-40 audience, has become a radio standard.

Commercial radio networking and syndication are likely to appear indistinguishable to listeners, but each has its own characteristic type of content, delivery means, advertising procedures, and payment practices. The major networks supply news, sports, and specials, or all-talk or all-music, relaying programs to affiliates by satellite, accompanied by national advertising. Syndicators concentrate on popular music formats and features. Radio networks often pay compensation to major-market affiliates, but radio syndicators usually charge for their programs. Music format syndicators allow stations to change formats rapidly as competitive pressures shift in local radio markets.

9.8 Locally Produced TV Programs

Both broadcasting and cable disappointed many who saw them as playing a strong role in giving opportunities for producing local programs. The economics of program production, however, inevitably favored program centralization—networking and syndication.

Local News Production

The one program category to escape the centralizing tendency is local/regional news. Audience interest in local television newscasts escalated during the 1970s, converting news from a loss leader into a profit center. It is not unusual for a network affiliate to make up to half its annual profit from advertising during local newscasts. It helps if the affiliate is the top-rated news station in the market, but even a modestly successful news operation attracts sponsors who would not otherwise advertise.

Multimillion-dollar budgets for local television news departments became commonplace in large markets, enabling stations to invest in high-tech equipment such as that shown in Exhibit 9.g.

Major-market stations developed their own investigative reporting and documentary units. Satellite technology enlarged the reach of local stations even further, minimizing time and distance constraints. By the mid-1980s large-market stations

Exhibit 9.g

Electronic News-Gathering Equipment

The unwieldy remote-production trailer in the background contrasts with the compact, lightweight ENG (electronic news-gathering) van in the foreground. The machine gun–like object atop the van is a microwave antenna for relaying pictures back to the studio.

Many metropolitan stations use helicopters to cover news events when ground-based units cannot do so or when an aerial perspective is more desirable. A dramatic example was the helicopter coverage of police "chasing" O. J. Simpson's white Bronco on the Los Angeles freeway in 1994.

Sources: Upper photo courtesy WTVJ-TV, Miami, FL; lower photo © Tom Martin, Aspen Photography/ The Stock Market.

routinely dispatched local news teams to distant places to get local angles on national news events. They send, or *backhaul,* live stories back to the home base via satellite.

Local news departments sometimes capitalize on the availability of home camcorders to turn amateur photographers, at least temporarily, into *stringers* (self-employed professionals who sell individual stories to radio and TV stations). Stations have used amateur footage of everything from earthquakes and tornados to train wrecks, and, perhaps best known, scenes of the Rodney King beating by police officers whose subsequent acquittal on assault charges set off the 1992 riots in Los Angeles. News professionals use such donated video cautiously, knowing the donor may have less than honorable motives. Most amateur video seen on-air involves natural disasters or provides information independently verified by other sources.

Responding to criticism that local news followed an "if-it-bleeds-it-leads" philosophy, some stations have adopted a policy of not using sensational or emotionally charged video in their early evening newscasts. Some viewers praise the broadcasters for their sensitivity while others complain the news is artificially sanitized. One station promoting its "family *centered*" newscast received angry calls objecting to the "family *censored*" news.

News Origination by Affiliates

Most network affiliates originate an early-evening newscast and a late-fringe newscast. Affiliates' local evening news shows either lead into network news or both precede and follow it to form a "sandwich." In the Eastern and Pacific time zones, network evening news usually starts at 6:30, preceded by local newscasts ranging from a half-hour to two hours in length.

Late newscasts typically appear at 11:00 P.M. (Eastern). For years they lasted a half-hour. By the 1990s, however, they had expanded to 35 minutes—a network concession to affiliate desires for more local advertising time and to the network's own efforts to improve clearances of their programs that followed.

Fox stations often schedule their late newscast an hour earlier than affiliates of other networks. Modern lifestyles make staying up past 11:00 P.M. for the news unappealing, especially when many people can get national news, sports, and local weather anytime from cable or DBS. Fox's early news is a more successful programming strategy on the East and West coasts than in the nation's midsection, where most network stations air late news at 10:00 P.M. local time.

Many affiliates also schedule a half-hour or hour of noon news. About 80 percent of affiliates originate early-morning half-hour newscasts or magazine/talk shows, preceding the network morning programs. The network shows provide slots into which affiliates can insert local-news segments. Some stations broadcast brief *updates* or *news capsules* during breaks in prime-time programs, many of which actually provide little news, serving mostly as promotion for their late newscasts. Some also offer "news-all-night," a mix of network overnight news programs, CNN Headline and other alternative news services, and repeats of the affiliate's own late newscasts.

To increase their reach, some stations produce special newscasts for, or rerun their regular programs on, independent stations or cable systems. They also offer the audio portion of their newscasts to local radio stations. Some TV stations produce weather reports for local insertion into cable networks such as The Weather Channel. Others provide local radio stations frequent weather updates voiced by their highly promoted on-air meteorologists, expensive talent many radio stations cannot afford.

News Origination by Independents

Relatively few independent stations ("indies") produce an early-evening newscast. They usually schedule their most popular syndicated entertainment shows against local news programs on affiliates.

Many produce a late local newscast, usually at 10:00 P.M. (Eastern and Pacific times) to counter-program network entertainment on competing stations. Independents also carry national news programs provided by such services as CNN, scheduling them immediately before or after their late local newscasts.

A handful of indies have devoted enormous resources to local news. In the 1990s Miami's WSVN-TV at times considered itself an independent station, at others an affiliate of Fox and of CNN. Whatever its status, it had the highest-rated 10:00 P.M. newscast in the nation. Not only did its late newscast typically achieve ratings higher than those on some affiliates at 11:00, but also its 6:00 P.M. local newscast often ranked second in the time period.

Station-Produced Non-News TV Programs

Aside from news, broadcast stations produce relatively few programs. Affiliates generally schedule local talk and public-affairs shows on weekday or Sunday mornings. During election years or at times when important local issues arise, they may increase local production, but they normally avoid scheduling local shows in valuable time periods such as early fringe, access, or prime time. Affiliates that produce regularly scheduled local non-news/public-affairs shows usually choose a magazine/talk program for women, scheduled on weekday mornings. Non-network magazine shows, though actually syndicated, may give the appearance of local productions. In fact syndicated formats sometimes acquire local color by providing for live or recorded inserts by local personalities.

Local Cable Production

About a third of the approximately 11,600 cable systems in the United States originate local programs other than automated services such as time, weather, and channel guides. Those that do mostly offer either commercial *local-origination* (LO) channels, controlled and programmed by the cable operators themselves, or noncommercial *public-access* channels, programmed by private citizens and nonprofit institutions such as schools and municipal governments.

Although some occasionally originate coverage of local sporting events, cable systems rarely undertake the expense required to produce regularly scheduled half-hour newscasts. Many do, however, produce local news inserts in CNN's *Headline News*.

An increasing number of large-market cable systems devote one channel full-time to regional news services. Pioneering *News 12 Long Island*, launched in 1986, provides 24-hour news service to more than 70 cities on New York's Long Island. *News 12 Connecticut*, *News 12 New Jersey*, and *News 12 Westchester* serve more than a million subscribers. *New England Cable News* serves about 2 million subscribers in a six-state area, and *News Channel 8* reaches about a million homes in the Washington, D.C., market. *New York News 1* provides news about New York City to some 1.5 million subscribers in the area. *ChicagoLand Television News* serves about the same number in an eight-county region of Illinois and Indiana. *NorthWest Cable News* (Seattle) has almost 2 million subscribers; *BAYTV* (San Francisco) has more than a million; *Orange County* (California) *NewsChannel* and *Pittsburgh Cable News Channel* each have half a million. These services, and others launched during the 1990s, indicate the increasing interest in regional cable news.

Local-origination channels are less common than access channels. The latter focus mostly on informational and cultural programs. Much of cable-access programming consists of public affairs, typically school board meetings, city and county council sessions, hearings on community issues, and discussion or documentary programs on political, environmental, and educational matters. But, as described in Exhibit 9.h, access channels have increasingly begun to offer somewhat more lively fare. Public access channels have sometimes become the focus of heated debate when that "lively fare" deals with controversial issues or handles subjects in ways some feel do not meet community standards of decency.

9.9 Local Radio Programs

Most locally programmed radio stations use modified block scheduling, changing format slightly to suit each daypart. For example, many stations emphasize news and weather/traffic information during drive times, shifting to music at other times.

Music Formats

Most parsimonious of all broadcast formats, the disc-jockey show relies on recorded music, reducing local production costs to the lowest possible level. The format exploits the ability of a DJ to build up a loyal following and to comment on the rapidly changing popular music scene.

Until about 20 years ago, at least some DJs could select pretty much whatever music they wanted. Today stations that still employ live DJs typically set up strict *playlists* of songs from which the host must select. A form of expression once unique to radio, the DJ format has migrated to television through video jockeys (VJs) on MTV and other program services.

Formats based broadly on rock music predominate, as Exhibit 9.i shows. Radio draws more fine distinctions among types of rock than among other musical genres, producing what some refer to as *format fragmentation*. Rock formats include *adult contemporary* (AC, a broad array of popular music and golden oldies), *contemporary hit radio* (CHR) or *Top-40* (playlists restricted to about a hundred of the most recent hits), *classic rock* (familiar songs from popular albums of the 1960s, 1970s, and increasingly the 1980s), *oldies* (hits from the 1960s and 1970s), and *urban contemporary* or simply *urban* (a mix of rhythm and blues with jazz favoring black artists). Among these subformats, many of which themselves are subdivided, adult contemporary has been the most successful. The *alternative*, or *progressive*, format featuring avant-garde music not

Exhibit 9.h

The Wacky World of Public-Access Television

Saturday Night Live's outrageous *Wayne's World,* starring Mike Meyers (left) and Dana Carvey, parodied—or in some cases emulated—public-access shows on real-life cable systems.

"You, too, can be a star!" Such is the promise of some public-access channels on local cable systems around the country.

It all started in the 1970s when, as one inducement to city fathers to grant them an operating franchise, cable operators committed to opening their facilities to virtually all comers. Many such channels carry meetings and activities of local organizations, public forums on environmental issues, and local election debates.

In the 1990s, when *Wayne's World* had progressed from a segment on NBC's *Saturday Night Live* to a hit motion picture, the access concept skyrocketed. One such program, *The Late Mr. Pete Show,* began as an access show in Los Angeles and moved to USA Network. Today, cable subscribers can see most anything on their local system's access channel—so long as it is advertiser-free and not obscene. Among the offerings:

- environmentalists modeling shoes recycled from old tires
- would-be talk-show hosts with an endless parade of nervous guests and belly dancers
- Catholics, Jews, Mormons, atheists, pagans, Peruvians, Sikhs, Democrats, Republicans, right-to-lifers, and Indian gurus, and
- hordes of *Wayne's World* wannabes, hoping for their big break.

Whether anybody watches is an entirely different matter.

Source: Photo from the National Broadcasting Company, Inc.

played by conventional commercial outlets became popular on many college stations in the early 1990s. Exhibit 9.i shows some commercial interest in the format, but most stations featuring the alternative/progressive format are noncommercial. In the mid-1990s a niche format called *rap,* or *gangsta rap,* received criticism disproportionate to its popularity because of controversial messages it sometimes conveyed regarding violence, drugs, sex, and the value of human life. Concern about the social impact of rap escalated following the almost back-to-back drive-by shooting deaths of two top rap performers, Tupac Shakur and The Notorious B.I.G (Christopher Wallace).

In the aggregate, more stations play *country music* than any other single type except rock, if all rock formats are added together. Though Exhibit 9.i shows country as a single format, it has followed rock's trend of subdividing, splitting into urban, traditional, crossover, oldies, and other specialties. After years as a mostly AM format, country music today is equally at home on FM. Adult contemporary has been successful on both AM and FM stations. *Religious/gospel* radio, one of the top-three formats listed in the exhibit, shows a steady shift to the AM band, forced out of FM by the more profitable rock music formats.

Variety/diverse (or *eclectic*), as a format, refers to the *full-service* station that schedules a mix of news, talk, and music. Similarly, some foreign-language, black, and other ethnically oriented stations mix music and informational programming as a commercial format, targeting specialized groups.

Beautiful music, a broad-appeal format that uses mostly unobtrusive vocal or instrumental music, has now largely been replaced by *easy-* (or EZ-) *listening*. Broadcast in most large radio markets, it saturates waiting rooms, elevators, department stores, and other public spaces. Syndicated easy-listening, commercial-free music services such as Muzak go to subscribers by satellite or FM subcarriers. More than 100 CD-quality music formats are provided to cable and DBS subscribers by Digital Music Express (DMX) and Music Choice.

Classical music appears mostly on noncommercial FM stations, but a few commercial FM and AM stations have adopted the format in the largest markets.

Information Formats

The radio formats listed in Exhibit 9.i include four predominantly information types: *talk*, *news*, *educational*, and *agricultural*. Newly developed AM formats, such as all-business and all-sports, attempt to capture fresh audiences for AM radio, but only about a fourth of radio's audiences listen to AM stations. Most under-45 listeners tune to FM, but in 1996 the top-rated morning and/or afternoon drive-time broadcasts in about half the nation's 10 largest markets were still on AM stations.

Talk radio occurs primarily on commercial AM stations. It combines call-in and interview programs with feature material and local news. Talk content varies between the extremes of sexual innuendo and serious political or social commentary. In major markets, shock radio deliberately aims at outraging conservative listeners by violating common taboos and desecrating sacred cows. Shock radio's contempt for adult authority and social tradition tends to attract listeners younger than the usual talk radio audience. Howard Stern, probably the best-known practitioner of the shock-radio format, has had repeated run-ins with the FCC (see Section 13.5). Stern's tussles with the FCC have (so far) resulted in indecency fines amounting to about $2 million. This may seem expensive, but Stern's problems with the FCC have probably helped his "bad-boy" image with fans. Besides his radio show, Stern has promoted books and videotapes, and has appeared in a movie, *Private Parts*. Segments of Stern's radio program appear on E!, the cable network. Another popular radio talk show host, "I-Man" Don Imus, can be seen doing his radio program on MSNBC.

Most talk programs focus on controversial issues, using guest-expert interviews and call-in questions. Authors on book tours, an essential promotional ritual for mass-marketing new books, also provide a constant stream of interviewees. The two-way telephone call-in show attracts an older and generally conservative group of listeners—people who have both time and militant convictions that incline them to engage in discussions with talk-show hosts. Program directors have to be alert lest a small but highly vocal group of repeat callers, often advocates of extremist views, kill listener and advertiser interest in telephone-talk shows. Of course, not all talk shows focus on controversial issues. Some help callers solve problems with their health, careers, marriages, money, even their automobiles and pets.

During the 1980s talk radio emerged as an important public forum in the black community, even though only a few black-oriented stations have adopted the format. *Black talk* stations provide a window through which candidates for public office,

Exhibit 9.i

Radio Station Formats by Popularity and Station Type

	Popularity in Terms of:		Station Types			
Format	No. of Stations	Percentage of Stations	AMs	FMs	Commercial	Noncommercial
Country/Western	2,823	17.1	1,198	1,625	2,789	34
Adult Contemporary	2,061	12.5	551	1,510	1,971	90
Religion/Gospel	1,739	10.5	942	797	1,214	525
Oldies/Classic Rock	1,425	8.6	497	928	1,367	58
News/Talk	1,021	6.2	880	141	921	100
News	712	4.3	323	389	378	334
Rock/AOR	685	4.1	48	637	512	172
CHR/Top-40	560	3.4	58	502	480	80
Talk	538	3.3	434	104	493	45
Variety/Diverse	505	3.1	106	399	122	383
Classical	473	2.9	17	456	53	420
Spanish	447	2.7	302	145	413	34
Sports	423	2.6	371	52	420	3
MOR	401	2.4	320	81	381	20
Jazz	399	2.4	20	379	71	328
Progressive/ Alternative	375	2.3	15	360	82	293
Urban Contemporary	325	2.0	103	222	274	51
Big Band/Nostalgia	284	1.7	231	53	269	15
Educational	252	1.5	18	234	10	242
Beautiful Music/ Easy Listening	233	1.4	89	144	201	32
Black	172	1.0	120	52	141	31
Agricultural	100	.6	73	27	100	0
All Others	582	3.5	302	280	417	165

Included in the "All Others" category are such disparate formats as blues and bluegrass, disco and drama, folk and foreign language, New Age, polka and reggae.

Source: Based on data in *Broadcasting & Cable Yearbook* (1996): B 604–605.

community organizations, and reporters can obtain a unique perspective on black public opinion. Politicians such as Harold Washington, Chicago's first black mayor, and the Reverend Jesse Jackson, found that black talk radio gave them political input that they could not get through mainstream media.

The *all-news* format costs a lot to produce yet earns only low ratings compared with successful music formats. All-news stations count on holding listeners' attention for only about 20 minutes at a time, long enough for listeners to arm themselves with the latest headlines, the time of day, weather tips, and advice about driving conditions. To succeed, this revolving-door programming needs a large audience reservoir that only major markets can supply.

9.10 World Wide Web Programming

Programming on the Web is in its infancy. Program providers are moving toward offering content, but the Web in the late 1990s remains a poor medium for transmission of audio and video compared to broadcasting and cable. It is likely to remain so until most people can access the Web with something other than analog modems connected to conventional phone lines. Even if the user's connection were instantly and dramatically improved, the bandwidth demands of audio and video would require significant equipment upgrades by Internet service providers and the entire Internet infrastructure would need strengthening. Upgrading the infrastructure is easier than eliminating the bottleneck between the ISP and the home user, however, and until this restriction is eliminated or some truly revolutionary breakthrough in technology occurs, the Web will remain mostly a supplement to conventional program delivery systems.

Audio and Video Streaming

The technological restraints on the Web have not prevented people from trying to use it to deliver real-time audio and video. Several companies provide software that allows content providers to stream audio and video over the Web for reception by persons with the necessary hardware, software, and Internet connection. The audio quality, at best, duplicates that of AM radio. The video is restricted to a small area of the screen and falls far short of full motion.

Web content often duplicates what the provider already offers through other distribution systems. For example, Audionet claims to broadcast the live programming of more than 160 radio stations. It is unlikely many people able to hear a station over the air will choose to listen on the Web. Putting the signal on the Web does, however, provide the station a potential audience among other listeners wanting access to a unique form of programming offered by the station that may not be available from local broadcasters. Putting the signal on the Web also allows the station to promote itself as a "worldwide" outlet.

The greater bandwidth requirements of video (which also needs bandwidth for its associated audio) makes video streaming less viable. Content, such as news, which involves less motion and requires lower image resolution, works best. Streamed audio and video of music concerts also have attracted interest, but the video quality provides more a sense of the event rather than much detailed information. For devoted music fans, of course, the Web concerts may provide content not otherwise available.

Supplemental and Promotional Content

Most program producers and distributors use the Web to supplement content provided through other channels and to promote their other delivery systems. In this respect they differ little from many other businesses in their use of the Web.

Content producers sometimes have more material than they use. They constantly make editorial decisions about what to delete, trying to please the largest audience or to fit the content into a restrictive time frame. The Web lets producers put more of their content out there and allows people to select what they want based on *their* interests and time availability. Someone who catches the Surgeon General commenting in a CNN report about a new AIDS treatment, for example, can later use the Web to hear the entire interview from which that comment was selected, or read a text transcript of it. The interview might even contain links to physicians providing the new treatment. A rock concert on MTV probably will not include an interview with the producer of the event discussing site selection, booking of performers, and other pre-production planning details. Such a long interview would break the program's flow and cause many viewers to start flipping channels. Putting the interview on MTV's Web site, however, does not tie up valuable time on the cable channel and allows those with an interest to get "inside" information.

The fact that the Web does not tie up the broadcast or cable channel allows producers to provide long-form programming that would not otherwise be available. There are limits to how many people can simultaneously access information on a Web site, but in a sense the Web provides the programmer many new "channels" for content distribution.

Developing Rapport and Talking Back

Besides providing users the ability to select the content they want when they want it, the Web encourages them to interact with it. On-air contests, slogans, and promotions seek to create a sense of "belonging" among audience members, to make them feel "we're your" station, network, or channel. Web site content often extends the opportunities for interaction by offering contest clues or earlier access to promotional materials.

The Web site also offers users an opportunity to learn more about the personalities they see and hear on the main distribution channel. Most radio station Web sites, for example, provide pictures of their on-air staff, along with biographical information. The Web site is also a helpful promotional tool. TV stations may include pictures and a detailed explanation of how their "exclusive Doppler radar" makes the station's weathercast superior to other stations in the market. Although not program content, these Web site materials enhance and promote programming.

Stations and networks also use their Web sites to get audience reaction to the programming offered on their main distribution channels. Radio stations test new musical selections by asking Web site visitors to listen to short excerpts and rate them. Or they may ask listeners to say what they like most or least about the station. Deciding how much weight to give these responses in making managerial decisions is a difficult matter, because the respondents are self-selected and may not represent the typical audience member. But most stations want as much feedback as possible and the Web site is another source. Managers may also use the Web sites of distant stations with streaming audio and/or video to sample programming in other markets.

Future Programming

Whether the Web will remain a supplementary content vehicle or become a program delivery channel like broadcasting and cable will depend on how quickly, if ever, the technological and economic restraints are eliminated. Meanwhile, programmers jump in response to each new advancement in technology and shift in the marketplace. As the vice president of new media for Lifetime Television, Brian Donlon, says regarding Web programming, "It's a brave new world and there are no rules. You just keep putting material up there and see if it works." (*Broadcasting & Cable*, 10 February 1997:24)

9.11 Program Critique

This and the preceding chapter have examined broadcast and cable programs largely from the industry viewpoint—as vehicles to carry commercial messages or to entice paying subscribers. Here we touch on other perspectives—those of the critics and consumers.

Diversity

A persistent complaint about commercial television, especially prime-time network programming, deplores its lack of diversity—the sameness of its program types, themes, plots, production styles, and sources. Broadcast networks risk so much on each program series that they take the safe route of copying successful shows again and again. Spinoffs and clones reduce prime-time entertainment essentially to sitcoms, crime dramas, and movies. Because mass media strive to reach the largest absolute audience possible, the broadcast networks must provide programs with the greatest appeal to the largest number of people, and they must do so most of the time.

The network's homogenizing influence also affects production styles. Programs from one production company look much like those from another. Yet programmers desperately seek novelty. This seeming paradox comes from wanting to be different but not wanting to take chances.

Popular cable networks depend on the same mass-appeal programs as broadcast networks. Although they offer many original shows, the programs are mostly variations on successful genres. Specialized cable networks do provide viewers much more of what mainstream broadcast and cable networks offer only in snippets. As cable systems and DBS continue to increase their channel capacity, and as competing delivery technologies such as open video systems and the Internet develop, there will be more networks serving niche audiences, with greater program diversity. Economics, rather than technology, will become a more critical limiting factor.

Journalism

As the public's primary news source, television has special obligations not always met in a competitive marketplace. Some observers express concern over reduced journalistic diversity resulting from the trend toward media concentration in the hands of a few huge corporations with no direct accountability to the public.

Others contemplate the impact of television trial coverage on the rights of defendants and victims. The criminal and civil trials of O. J. Simpson received unprecedented coverage. Some say trial coverage made the public more aware of judicial procedures. Others say the minute coverage of the trials with their starkly different verdicts did more to confuse, frustrate, and divide people than inform or reassure them.

The announcement of the verdict in the Simpson civil trial during President Clinton's 1997 State of the Union address posed further problems for the networks. Managers had to decide whether to stick with Clinton talking about issues affecting the entire nation or switch to LA for news that, strictly speaking, had a direct impact on only a few individuals. Managers had to decide between giving the audience what journalists consider most important and what viewers consider most interesting.

Judges also take note of media coverage. Consistent with policies of the Judicial Conference of the United States regarding federal trials, U.S. District Judge Richard Matsch banned courtroom news cameras during the 1997 trial of Timothy McVeigh in connection with the Oklahoma City bombing that killed 168 persons and injured hundreds more. Judge Matsch did allow relatives of victims in Oklahoma City unable to attend the Denver trial to watch a closed-circuit feed provided by a hidden, remote-controlled camera.

Traditional news programs have come under attack for seeking popularity through the incorporation of too much show business. More and more stations have reacted to the success of tabloid television by stressing violence and tragedy at the expense of more far-reaching but less visual stories.

The need for visually appealing stories has tempted producers to create *staged news events*. Television also is susceptible to manipulation by persons or organizations who stage dramatically visual *pseudoevents* designed to attract news coverage. TV journalists may sometimes use, but fail to identify, footage from *video news releases* (VNR) produced by outside sources such as government agencies and businesses. Sometimes broadcasters deceive viewers by not identifying apparently live coverage as being, in fact, recorded. See Section 11.2 for a more thorough discussion of these issues.

Digital enhancing offers high-tech opportunities for deception. Already in use by print media, it permits a reporter to manipulate pictures in limitless ways. For example, *New York Newsday* in February 1994 published an electronically doctored photograph showing Olympic figure skating competitors Tonya Harding and Nancy Kerrigan apparently (but not actually) side by side on the ice.

Critics deplore *checkbook journalism*—the practice of paying for interviews. The syndicated tabloid show *Inside Edition* likely set a record in 1994 when it reportedly paid Harding about $600,000 to talk about the media-sensationalized physical attack on her rival, Kerrigan. Reputable news organizations seldom pay for interviews, but they do cater to newsmakers by sending their most prestigious air personalities. CBS, for example, dispatched Connie Chung, hoping Chung's prestige would put CBS nearer the top of Harding's media interview list. Critics say the Harding-Kerrigan affair is typical of incidents with limited legitimate news value that too often become the focus of media frenzy.

Many people dislike the way journalists go about their business, particularly when technique creates more controversy than content. In 1997 a North Carolina jury slapped a $5.5 million fine against ABC for a story aired on *Prime Time Live* about alleged wrongdoing involving the handling of food products at Food Lion supermarkets. Jurors did not fault the report so much as the fact reporters gained access to Food Lion's internal operations by falsifying information on the supermarket's employment application forms. The verdict reflected public concern about invasive reporting techniques and prompted news operations to review their policies related to undercover investigative procedures.

Programs and the Public Interest

Fundamentally, the viewer/listener perspective asks whether commercial motives should be the primary factor in program choice and quality. Debate has swirled around this issue ever since broadcasting began—commerce versus culture. How should society balance the sometimes conflicting claims of these two goals?

Is it enough for media to treat programs simply as "product"—articles of trade? Some argue that programs can also be broadly cultural and contribute to the intellectual, artistic, and moral quality of national life. Seen in that perspective, programs should do more than encourage audiences to watch commercials.

"Wasteland" vs. "Toaster"

The best-known critique of the industry's program performance in public-interest terms came from an FCC chairperson appointed by the Kennedy administration, Newton Minow (1961–1963). In an address to the National Association of Broadcasters in 1961, Minow challenged station owners and managers to sit down and watch their own programs for a full broadcast day. They would, he assured them, find a "vast wasteland" of violence, repetitive formulas, irritating commercials, and sheer boredom (Minow, 1964: 52). The "vast wasteland" phrase caught on and became a permanent part of broadcasting lore.

Some 20 years later a Republican-appointed FCC chairperson, Mark Fowler, pointedly refrained from talking to the industry about program quality. In his view, the FCC had no business interfering with the workings of the marketplace. He too coined a memorable descriptive phrase when, in addresses to various broadcast industry groups, he described television as "a toaster with pictures" and the agency he headed as "a New Deal dinosaur." These dismissive phrases reflected a then-dominant theory in Washington: The economic laws of supply and demand suffice to ensure that commercial television will supply suitable programs; if programs degenerate into a vast wasteland, blame not the industry but the audience.

In 1991 Newton Minow (long since an attorney in private practice) revisited television, commenting on his famous words, "vast wasteland":

> Today that 1961 speech is remembered for two words—but not the two I intended to be remembered. The words we tried to advance were "public interest." To me, the public interest meant, and still means, that we should constantly ask: What can television do for our country?—for the common good?—for the American People? . . .
>
> If television is to change, the men and women in television will have to make it a leading institution in American life rather than merely a reactive mirror of the lowest common denominator in the marketplace. Based on the last thirty years, the record gives the television marketplace an A+ for technology, but only a C for using that technology to serve human and humane goals. (Minow, 1991:9, 12)

Despite the often exaggerated charges of some critics, most would concede that television sometimes rises to peaks of excellence—even though between the peaks lie broad valleys (vast wastelands?) of routine programs. How green the valleys are depends on the viewer's personal tastes.

Taste

A continuously available mass medium such as television cannot satisfy every taste all the time. *Most* of the time, but not *all* of the time, it must try to please *most* of the people, but not *all* of the people—an extraordinary demand. In meeting it, television has exposed the low common denominator of mass popular taste as contrasted with the more cultivated standards of high taste. As Daniel Boorstin, an authority on American cultural history, put it:

> Much of what we hear complained of as the "vulgarity," the emptiness, the sensationalism, of television is not a peculiar product of television at all. It is simply the translation of the subliterature onto the television screen . . . . Never before were the vulgar tastes so conspicuous and so accessible to the prophets of our high culture. Subculture—which is of course the dominant culture of a democratic society—is now probably no worse, and certainly no better, than it has ever been. But it is emphatically more visible. (Boorstin, 1978:19)

From the standpoint of the middle ground between the extremes of programs-as-merchandise and programs-as-culture, it is unrealistic to expect programs always to rise above the lowest common denominator. As the most democratic of media, broadcasting necessarily caters to popular tastes. That mission, however, need not preclude serving minority tastes as well. One hallmark of a democracy is that though the majority prevails, minorities still have rights. Public broadcasting exists in part to compensate for the omissions of commercial broadcasting in this regard. But even public broadcasting has had to increase its popular appeal to attract program underwriters and to broaden its subscriber base. It has so far lacked the financial support it needs to offer an adequate alternative program service.

Promoters of the marketplace philosophy argued that deregulating broadcasting and cable would automatically bring diversity and enhance quality. Deregulation did indeed give viewers more choices—but at an added price. Those who can afford, and choose to have, cable or a satellite dish can browse through scores of channels to find programs of interest. Owners of home video recorders can browse through tape inventories of video stores and also use time shifting to escape the tyranny of broadcast/cable schedules.

So far, however, too many of the new options have turned out to be merely repetitions of the old options. After all, every delivery system responds to the same marketplace imperatives that drive commercial broadcasting. Significantly, the shining exception in the public-affairs field, the program service that opens a window on government in action—C-SPAN—does not operate as a self-supporting commercial venture. The cable industry subsidizes C-SPAN as a public-relations showcase of the good things television can provide—when not constrained by the need to make money.

Popular Taste versus Bad Taste

Although commercial broadcasters have always catered to *popular taste*, they have generally refrained from catering to the appetite for downright *bad* taste in programs. In its now-abandoned Television Code, the National Association of Broadcasters emphasized the role of television as a family medium, warning that "great care must be exercised to be sure that treatment and presentation are made in good faith and not for the purpose of sensationalism or to shock or exploit the audience or appeal to prurient interests or morbid curiosity" (NAB, 1978:2).

That kind of sensitivity, along with the Television Code, has fallen victim to changing times. In fact, the Code's statement of what not to do accurately describes exactly what succeeded on the air beginning in the late 1980s. "Raunch' on a Roll," proclaimed a headline in a trade journal over a story about the rise of slobcom—sitcoms that "stretch the bounds of what's acceptable" (*Broadcasting*, 21 November 1988:27). The Fox network's *Married . . . With Children* occasioned the article. Its plots dealt with such topics as premenstrual syndrome, treated with outrageously vulgar humor. Despite some complaints, such programs drew high ratings, which in turn ensured widespread imitation.

During a single week in 1992 ABC offered episodes of the following established series, accompanied by their respective advertising lines:

- *The Commish*—Tony goes underground to bust an X-rated movie racket.
- *Life Goes On*—Jesse's nude painting of Becca makes her the talk of the town.
- *Doogie Houser, M.D.*—Doogie's caught with his pants down when he swims naked with his mom's sexy boss.
- *20/20*—How to have better sex.
- *Civil Wars*—An art photographer convinces Sydney to pose nude.

By mid-decade there were isolated examples of changes in some prime-time programming. CBS, in particular, tried "going home" to its traditionally older, more conservative viewers with "wholesome" shows such as *Chicago Hope*, *Touched by an Angel*, and its spinoff, *Promised Land. Touched by an Angel*, one of CBS's top-rated entertainment shows in 1997, has a strongly religious flavor that appeals to viewers aged 25 to 54. But, in spite of federal legislation requiring that future TV sets contain violence-blocking V-chips, violence remains central to some hour-long dramas, such as *Walker: Texas Ranger*, and to many network movies.

Although containing little or no violent content, sitcoms continue to test the boundaries with their use of language and sexual innuendo. A watchdog organization critical of TV programming reported that during the week when content-rating icons (see Exhibit 9.d) first appeared on-air, one sitcom character reminisced about having had teenage sex in a driver's-ed car; another boasted of having had sex four times in one day, including once on the hood of an automobile; another talked about having had sex with his father's fiancée; yet another joked about a man who masturbated by rubbing against a tree; and still another fantasized about eating pralines off the naked body of a single woman. The TV-G- and TV-PG-rated sitcoms also included words such as "ass," "suck," and "bastard," and one referred to a part of the male anatomy as "Mr. Sturdy." For some viewers, questionable language and sexually based humor is more objectionable than a gunfight, where the good guy usually wins.

Cable, which operates on a standard less strict than that of broadcast, "double-shoots" some programs—one version with topless scenes, for the pay-cable network and foreign syndication; another without, for ad-supported cable and domestic station use. Pay-cable "adult" services such as Playboy and Spice offer full nudity and carefully edited, but nonetheless graphic, movies.

Violence, mayhem, and sociopathic behavior, long the subjects of critical debate, continued unabated in the 1990s. One can speculate on influences that may have had a hand in bringing about this shift in program standards. Among the many possible influences might be these:

- the FCC's laissez-faire policy during the 1980s, which encouraged broadcasters to test the limits of public tolerance
- abandonment of the NAB codes, implying to some that anything goes
- heightened competition, encouraged in part by FCC policies, requiring ever more strenuous efforts to capture audience attention
- the impact of cable television, which until the 1992 Cable Act had been unconstrained either by legislated public-interest standards or by a tradition of self-restraint
- corporate mergers that replaced experienced broadcast and cable executives with cost-conscious corporate managers saddled with huge debts
- social changes in the direction of more open and permissive behavior, marked especially by violence, sex, and rebellion against conventional standards.

Program producers have a difficult task. They face constant pressure to produce a "product" generating maximum profit regardless of its artistic quality or social impact. On the other hand, the electronic media, especially the licensed media, have a legal and moral responsibility to do more than offer programs that are simply popular and make money. Indeed, they have a responsibility to provide programs that sometimes may be very unpopular.

At least theoretically, ultimate power over programming rests with the audience. Programming must compete for survival as does any other product in a consumer-driven economy, and there are political channels of control available when the marketplace fails to produce culturally acceptable programming. This does not mean the media do not have a responsibility to strive independently toward a higher standard than the lowest common denominator without being goaded to do so by economic or legislative threats. Nor does it mean the media are not often leaders, rather than followers, in establishing cultural standards of taste. It does suggest that a passive public will get no better programming than it collectively accepts without complaint.

Chapter 10

Ratings

CHAPTER CONTENTS

Commercial broadcast and cable services compete to win the popularity contest that ratings measure. If people don't watch or listen—as measured by ratings—then advertisers lose interest and programs die, to be replaced by others, or stations fail and leave the air. Electronic media want unbiased and consistent audience information. For this, broadcasters, cable services—whether delivered via traditional systems, DBS, or wireless cable—and advertisers employ independent companies to conduct most day-to-day audience research, using scientific methods for probing into human behavior and attitudes.

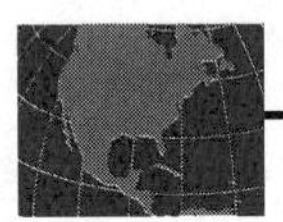

10.1 *Ratings Business*

Practically everyone knows what television ratings are in a general sense. In the long broadcast television network battle for ratings dominance, the rise and fall of prime-time programs always make news. But unlike most other industries, broadcasting and cable deliver no physical products. Program "publishing" goes on continuously, with audiences flowing at will from one program to another. Hence the need for highly specialized research.

Arbitron and Nielsen

Two ratings firms, Arbitron and Nielsen, dominate the ratings business as sources of most measurements used by electronic media and their advertisers. Other competitors—Statistical Research, Inc (SRI), for example—focus more on limited or specialized types of research—or have lasted only a few years.

By 1994 Arbitron had ended the television (and later cable) rating service it had begun in 1949, leaving Nielsen as the sole provider of audience research in those areas. In an effort to inject some competition in the television ratings business, several broadcast and cable networks and large advertisers are backing the establishment of a new ratings system being developed by SRI. The new service, tested in Philadelphia since 1994, is known as SMART: Systems for Measuring and Reporting Television. Arbitron continues to dominate radio ratings.

The two services' revenues come mainly from subscriptions by cable and broadcast networks, stations, advertising agencies, sales rep firms, program suppliers, and syndicators. Station subscription rates vary according to station revenue. Major ad agencies subscribe to both network and local-market ratings reports, some spending close to a million dollars a year.

Local Market Ratings

Nielsen gathers and publishes television data for virtually all television markets. Local reports reflect the relative position of each station among its competitors and estimate local audience size for network, syndicated, and locally produced programs.

For all but the largest markets, it costs too much to collect data for local television ratings continuously. Instead, researchers gather data in short spurts known as *rating periods.*

Nielsen Station Index (NSI), the major Nielsen local television market measure, covers 211 markets—173 by means of written diaries, the other 38 by a combination of diary and passive household meters (see Section 10.2). The company measures all markets four times a year in four-week-long *sweep weeks* (or simply *sweeps*); it measures larger markets up to three additional times per year. A sample page from a local-market report appears in Exhibit 10.a.

Arbitron is now the primary firm providing radio market ratings—Nielsen stopped measuring radio use in 1964, and Birch Radio left the business at the end of 1991. Arbitron covers more than 260 radio markets: about 95 year-round and the rest in the spring and fall. Exhibit 10.b offers a sample page from a diary used in producing radio rating reports.

Network Ratings

Broadcast television networks demand faster and more frequent reporting than do local stations. But

Exhibit 10.a

Local-Market TV Ratings Report

Nielsen uses diaries and household meters to measure 211 local market television audiences. Shown here is part of a sample page of late-night ratings and shares for network and syndicated programs in the New York market. This report includes data from four weeks and shows a variety of demographic breakdowns so that station management and advertisers can obtain ratings for, say, women 25 to 49 years of age.

Sources: Illustrations © Nielsen Media Research. *Viewers in Profile*, February 1997, Nielsen Media Report. Used by permission.

New York, NY
Metered Market Service
February 1997
SPECIAL ETHNIC TREATMENT USED IN THIS MARKET
(See Page 3)
Nielsen Station Index
Viewers in Profile
Nielsen
tv

NEW YORK, NY — WK1 1/30-2/05 WK2 2/06-2/12 WK3 2/13-2/19 WK4 2/20-2/26

MONDAY-FRIDAY 11:30PM - 12:30AM

TIME PERIOD

METRO HH RTG	METRO HH SHR	STATION	PROGRAM	DMA HH RATINGS WEEKS 1	WEEKS 2	WEEKS 3	WEEKS 4	MULTI-WEEK AVG RTG	MULTI-WEEK AVG SHR	SHARE TREND NOV 96	MAY 96	FEB 96	PERSONS 2+	PERSONS 18+	PERSONS 12-24	PERSONS 12-34	PERSONS 18-34	PERSONS 18-49	PERSONS 21-49	PERSONS 25-54	PERSONS 35+	PERSONS 35-64	PERSONS 50+	WOMEN 18+	WOMEN 12-24	WOMEN 18-34	WOMEN 18-49	WOMEN 25-49	WOMEN 25-54	WOMEN WKG	MEN 18+	MEN 18-34	MEN 18-49	MEN 21-49	MEN 25-49	MEN 25-54	TNS 12-17	CHILD 2-11	CHILD 6-11
1	2			3	4	5	6	7	8	11	12	13	15	16	17	18	19	20	21	22	23	24	25	26	27	28	29	31	32	34	35	36	37	38	39	40	41	42	43
LT LT			R.S.E. THRESHOLDS 25+% (1 S.E.) 4 WK AVG 50+%	1 LT	1 LT	1 LT	1 LT	LT LT					LT LT	LT LT	1 LT	1 LT	1 LT	1 LT	1 LT	LT LT	LT LT	1 LT	1 LT	LT LT	2 LT	1 LT	1 LT	1 LT	1 LT	1 LT	LT LT	2 LT	1 LT	1 LT	1 LT	1 LT	3 1	2 LT	3 1
			11:30PM																																				
8	16	WABC	ABC-NITELINE	8	8	8	8	8	16	18X	16	18	4	5	1	2	2	3	3	4	7	6	9	5	1	2	3	3	4	5	5	2	3	3	3	4			
<<		WBIS	AVG. ALL WKS	<<	<<	<<	<<	<<																															
<<			#SPORTS-DAILY	<<				<<																															
<<			#HEARD-ST PM R		<<	<<	<<	<<																															
4	8	WCBS	D LETTRMAN-CBS	4	4	4	5	4	9	10X	9	10	2	3	1	2	2	3	3	3	3	3	3	3	2	2	2	2	2	3	3	2	3	3	3	3	1		
<<		WLIW	AVG. ALL WKS	<<	<<	<<	<<	<<																															
<<			#ITN WORLD NEWS	<<			<<	<<																															
<<			ITN WORLD NEWS		<<	<<		<<																															
1	1	WLNY	STAR TK-GENRTN≠	1	<<	<<	1	1	1	X																													
7	15	WNBC	TONITE SHW-NBC	7	8	8	7	8	15	16X	16	15	4	5	2	3	3	4	4	4	6	6	8	6	2	3	4	4	4	5	5	3	3	4	4	4	1		
1	3	WNET	CHARLIE ROSE	2	1	1	1	1	3	4X	3	3	1	1						1	1	1	1	1				1	1		1					1			
<<		WNJN+	NITE BSNSS RPT	<<	<<	<<	<<	<<		X																													
<<		WNJU	CBS-TELNT 11PM	<<	<<	<<	<<	<<		X		1																											
4	8	WNYW	AVG. ALL WKS	3	4	4	4	4	8	7	8	8	2	2	4	3	3	2	2	2	1	2	1	3	6	4	3	3	3	2	1	2	2	2	1	1	4		
4	7		#MAD ABOUT YOU	3	4			4	7				2	2	5	4	3	2	2	2	1	1	1	2	6	4	3	3	2	2	1	2	1	1	1	1	6		
4	8		MAD ABOUT YOU			4	4	4	9				2	2	3	3	3	3	3	2	2	2	2	3	5	4	3	3	3	3	1	1	2	2	2	2	3		
7	15	WPIX	CHEERS	7	7	7	7	7	14	15X	15	15	4	5	4	5	6	5	5	5	4	4	4	4	3	5	5	5	5	5	5	7	5	5	5	5	2	1	1
7	13	WWOR	JENNY JONES R	6	5	7	6	6	12	10X	9	11	4	4	7	7	6	4	4	4	3	3	3	5	11	9	6	5	5	4	3	4	3	3	3	3	7	1	1
2	3	WXTV	NOTC UNI LATE	1	1	1	1	1	3	2X	3	2	1	1	2	2	2	1	1	1	1	1	1	1	1	2	1	1	1	1	1	2	2	1	1	1	1	1	1
50			HUT/PUT/TOTALS*	50	50	50	49	50		49	47	47	29	34	25	28	31	31	31	33	36	36	40	36	29	33	33	32	34	33	32	29	29	30	31	31	19	4	5
			11:45PM																																				
6	12	WABC	ABC-NITELINE	6	6	6	5	6	13	16X	14	15	3	4		1	1	2	2	3	5	4	6	4	1	1	2	2	3	2	3	1	2	2	2	2			
<<		WBIS	AVG. ALL WKS	<<	<<	<<	<<	<<																															
<<			#SPORTS-DAILY	<<				<<																															
<<			#HEARD-ST PM R		<<	<<	<<	<<																															
3	7	WCBS	D LETTRMAN-CBS	4	3	3	4	3	7	8X	8	9	2	2	1	1	1	2	2	2	2	3	2	2	1	1	2	2	2	2	2	2	2	2	3	3	1		
<<		WLIW	AVG. ALL WKS	<<	<<	<<	<<	<<																															
<<			#ITN WORLD NEWS	<<			<<	<<																															
<<			ITN WORLD NEWS		<<	<<		<<																															
1	1	WLNY	STAR TK-GENRTN≠	1	1	<<	<<	1	1	X																													
6	13	WNBC	TONITE SHW-NBC	6	7	6	5	6	13	14X	14	13	3	4	2	2	3	3	3	3	5	4	6	4	2	2	3	3	3	4	4	3	3	3	3	3	1		
1	3	WNET	CHARLIE ROSE	1	1	<<	1	1	2	3X	2	2		1							1	1	1	1							1								
<<		WNJN+	NITE BSNSS RPT	<<	<<	<<	<<	<<		X																													
<<		WNJU	CBS-TELNT 11PM	<<	<<	<<	<<	<<		X																													
4	8	WNYW	AVG. ALL WKS	3	3	4	5	4	8	7	8	8	2	2	4	3	3	2	2	2	1	1	1	3	6	4	3	3	3	2	1	2	2	2	1	1	4		
4	8		#MAD ABOUT YOU	3	3			3	7				2	2	4	4	3	2	2	2	1	1	1	2	6	4	3	3	2	2	1	2	1	1	1	1	5		
4	9		MAD ABOUT YOU			4	5	4	9				2	2	3	3	3	3	3	2	2	2	2	3	5	5	3	3	3	3	1	1	2	2	2	2	3		
7	15	WPIX	CHEERS	7	6	6	6	6	14	15X	15	15	3	4	3	5	6	5	5	5	3	4	3	4	3	5	4	5	5	4	4	6	5	5	5	5	2	1	1
6	13	WWOR	JENNY JONES R	5	5	7	5	5	12	10X	9	11	3	3	6	6	6	4	3	3	2	2	2	4	9	7	5	4	4	4	2	3	2	2	2	2	6	1	1
1	3	WXTV	NOTC UNI LATE	1	1	1	1	1	2	2X	3	2	1	1	2	1	2	1	1	1	1	1	1	1	1	1	1	1	1	1	1	2	1	1	1	1	1	1	1
46			HUT/PUT/TOTALS*	47	45	45	44	45		45	43	44	25	29	22	25	28	27	27	28	30	30	33	30	26	30	28	28	29	27	28	26	26	27	27	28	17	4	5
			12:00MD																																				
5	12	WABC	AVG. ALL WKS	5	6	5	5	5	13	17	14	15	2	3	1	1	2	2	2	2	3	3	5	3	1	2	2	2	3	3	2	2	2	2	1	2			
5	12		#POLIT INCT-ABC	5	6	5		5	13				2	3	1	1	2	2	2	2	4	3	5	3	1	2	2	2	3	2	3	2	2	2	2	2			
4	11		POLIT INCT-ABC				5	5	12				2	3	1	1	2	2	2	3	3	3	4	4	1	3	4	4	4	3	2	1	1	1	1	2			
<<		WBIS	AVG. ALL WKS	<<	<<	<<	<<	<<																															
<<			#PAID PROGRAM	<<	<<	<<		<<																															
<<			PAID PROGRAM				<<	<<																															
3	7	WCBS	D LETTRMAN-CBS	3	3	3	4	3	8	9X	8	10	2	2	1	1	1	2	2	2	2	2	2	2	1	1	2	2	2	2	2	2	2	2	2	2	1		
<<		WLIW	#THIS-AM-WHOLEY	<<	<<	<<	<<	<<																															
<<		WLNY	LATE SHOW	<<	<<	<<	<<	<<		X																													
6	14	WNBC	TONITE SHW-NBC	5	7	6	6	6	15	14X	14	13	3	4	2	2	3	3	3	3	4	4	6	4	2	3	3	3	3	3	3	3	3	3	3	3			
1	2	WNET	VARIOUS	<<	1	1	1	1	1	2		2																											
<<		WNJU	VARIOUS	<<	<<	<<	<<	<<																															

Exhibit 10.b

Local-Market Radio Ratings

You count in the radio ratings!

No matter how much or how little you listen, you're important!

You're one of the few people picked in your area to have the chance to tell radio stations what you listen to.

This is *your* ratings diary. Please make sure you fill it out yourself.

Here's what we mean by "listening":

"Listening" is any time you can hear a radio — whether you choose the station or not.

When you hear a radio between Thursday, (Month, Day), and Wednesday, (Month, Day), write it down — whether you're at home, in a car, at work or someplace else.

When you hear a radio, write down:

TIME

Write the time you start listening and the time you stop. If you start at one time of day and stop in another, draw a line from the time you start to the time you stop.

STATION

Write the call letters or station name. If you don't know either, write down the program name or dial setting.

Check AM or FM. AM and FM stations can have the same call letters. Make sure you check the right box.

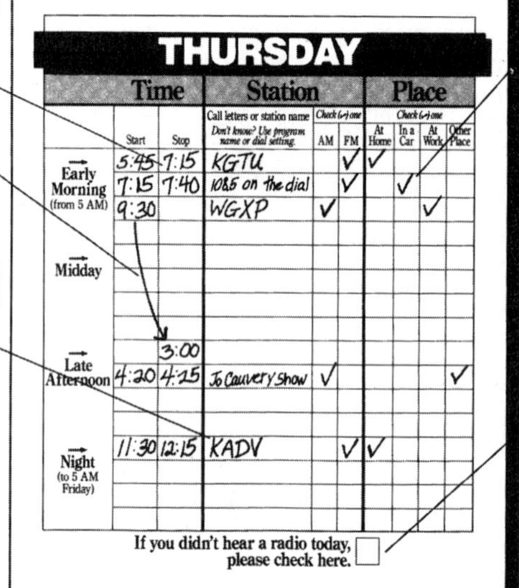

THURSDAY

	Time		Station			Place			
			Call letters or station name. Don't know? Use program name or dial setting.	Check (✓) one		Check (✓) one			
	Start	Stop		AM	FM	At Home	In a Car	At Work	Other Place
Early Morning (from 5 AM)	5:45	7:15	KGTU		✓	✓			
	7:15	7:40	1065 on the dial		✓		✓		
	9:30		WGXP	✓				✓	
Midday									
		3:00							
Late Afternoon	4:20	4:25	Jo Cauvery Show	✓					✓
Night (to 5 AM Friday)	11:30	12:15	KADV		✓	✓			

If you didn't hear a radio today, please check here. ☐

PLACE

Check where you listen:

- at home
- in a car
- at work
- other place

Write down *all* the radio you hear. Carry your diary with you starting **Thursday, (Month, Day).**

No listening? If you haven't heard a radio all day, check the box at the bottom of the page.

Questions? Call us toll-free at 1-800-638-7091. In Maryland, call collect 301-497-5100.

This instruction page from a radio diary illustrates the Arbitron system. Each individual in a household fills out a separate diary.

Source: Copyright © 1995. Used by permission of Arbitron Ratings Company.

in two respects network ratings are easier to obtain: (1) not every market need be surveyed, as a sample of network markets yields a reliable picture of national network audiences; and (2) far fewer broadcast television networks than stations compete at any one time. Nielsen issues the only network television ratings. RADAR, the only ratings for network radio, is compiled by SRI.

Nielsen Television Index (NTI) issues television network ratings in what is called a "pocket piece" (see Exhibit 10.c). Nielsen's regular network reports use a national sample of about 5,000 peoplemetered homes. Nielsen also reports on its national Hispanic sample (National Hispanic Television Index), which includes just under 1,000 peoplemetered homes.

Radio's All-Dimension Audience Research (RADAR), financed by networks that contract with Statistical Research, Inc., uses telephone recall interviews. RADAR issues reports twice a year, each covering a sample week, based on surveys of 12,000 respondents. Unlike most telephone surveys, RADAR calls the *same* people each day, seven days in a row.

Syndicated-Program Ratings

By using data already obtained for market ratings reports, Nielsen provides regular analyses on relative performance of nationally syndicated television programs.

The Nielsen *Cassandra* service provides detailed audience demographic information for both first-run and off-network syndicated programs (as well as for network and some local programs). Cassandra allows researchers to compare household demographic information for syndicated programs across markets, and to collect and measure similar data on both lead-in and lead-out programs.

Special Studies

Many supplementary reports, drawing upon data gathered in preparing regular ratings reports, are also available. Clients can order special reports tailored to their needs. For example, to help determine when to schedule its news promotion announcements, a TV station might order a special "audience flow" study to discover which of its own programs attract viewers who normally watch a competing newscast. Another might commission a study to find out how much a program appeals to specific audience subgroups.

10.2 Collecting Data

Whatever electronic medium they analyze, researchers use several methods for collecting data on which to base ratings: diaries, meters, peoplemeters, and two different kinds of telephone calls. Special studies often use other methods and combinations. Each has advantages and disadvantages.

Diaries and Passive Meters

Arbitron and Nielsen researchers use a written diary method for daily gathering of most local-market data. To obtain radio data, Arbitron sends a separate diary to each person over 12 years of age in every sample household. It asks diary keepers to write down for one week their listening times and the stations they tune to, keeping track of away-from-home as well as in-home listening.

Nielsen uses diaries for its Nielsen Station Index television ratings outside the largest markets, asking one person in each sample home to take charge of reporting all household viewing. VCR use is measured separately (see Section 10.8).

Diaries suffer a serious drawback: People often enter inaccurate information, purposely (to show they have good taste in programs) or not (many sample families fill in a diary in "catch-up" style at the end of the week rather than as listening or viewing takes place). Some also tire of making entries, suffering from *diary fatigue*.

The *passive household meter* (Nielsen's Audimeter) was first used to report ratings in 1950. Attached to each television set in a sample household, the so-called black box automatically records when the set is on and to which channel it is tuned. These meters are capable of identifying tuning on cable

Exhibit 10.c

Nielsen TV Network Ratings Report

A-8 **NATIONAL *NielsenTV* AUDIENCE ESTIMATES** — **EVE.THU. SEP.25, 1997**

TIME	7:00	7:15	7:30	7:45	8:00	8:15	8:30	8:45	9:00	9:15	9:30	9:45	10:00	10:15	10:30	10:45
HUT	48.3	51.0	53.1	55.9	59.7	61.8	62.4	64.1	66.0	67.0	66.6	67.0	66.6	64.5	61.8	59.6
ABC TV					←NOTHING SACRED→				←CRACKER→				←20/20-THURS→			
HHLD AUDIENCE% & (000)					4.4	4,330			5.0	4,930			7.5	7,340		
TA%, AVG. AUD. 1/2 HR %					*6.5*	4.4•		4.5•	*7.3*	4.6•		5.4•	*12.4*	7.3•		7.7•
SHARE AUDIENCE %					7	7•		7•	8	7•		8•	12	11•		13•
AVG. AUD. BY 1/4 HR %					4.6	4.1	4.5	4.4	4.5	4.7	5.2	5.7	6.8	7.7	7.8	7.6
CBS TV					←PROMISED LAND→				←DIAGNOSIS MURDER→				←48 HOURS→			
HHLD AUDIENCE% & (000)					11.3	11,100			10.8	10,610			6.6	6,490		
TA%, AVG. AUD. 1/2 HR %					*14.8*	10.5•		12.2•	*13.3*	10.8•		10.8•	*10.6*	6.6•		6.7•
SHARE AUDIENCE %					18	17•		19•	16	16•		16•	11	10•		11•
AVG. AUD. BY 1/4 HR %					10.3	10.7	11.8	12.5	10.9	10.7	11.0	10.7	6.6	6.6	6.9	6.4
NBC TV					FRIENDS		UNION SQUARE		SEINFELD		VERONICA'S CLOSET		←E.R.→			
HHLD AUDIENCE% & (000)					19.5	19,110	17.3	16,940	24.6	24,130	23.3	22,860	28.5	27,960		
TA%, AVG. AUD. 1/2 HR %					*22.2*		*19.9*		*27.5*		*26.4*		*34.7*	29.3•		27.7•
SHARE AUDIENCE %					32		27		37		35		45	45•		46•
AVG. AUD. BY 1/4 HR %					18.6	20.4	17.1	17.5	23.8	25.4	23.1	23.6	29.9	28.7	28.2	27.3
FOX TV					LIVING SINGLE (PAE)		BETWEEN BROTHERS		←413 HOPE ST→							
HHLD AUDIENCE% & (000)					5.1	5,020	5.2	5,080	5.4	5,290						
TA%, AVG. AUD. 1/2 HR %					*6.1*		*6.0*		*7.1*	5.3•		5.5•				
SHARE AUDIENCE %					8		8		8	8•		8•				
AVG. AUD. BY 1/4 HR %					5.0	5.3	5.1	5.3	5.2	5.4	5.5	5.5				
WB TV																
HHLD AUDIENCE% & (000)																
TA%, AVG. AUD. 1/2 HR %																
SHARE AUDIENCE %																
AVG. AUD. BY 1/4 HR %																
UPN TV																
HHLD AUDIENCE% & (000)																
TA%, AVG. AUD. 1/2 HR %																
SHARE AUDIENCE %																
AVG. AUD. BY 1/4 HR %																
INDEPENDENTS (Inc. superstations except TBS) AA%	13.0 (+F)		13.7 (+F)		6.2		6.9		6.4		6.1		10.3 (+F)		8.6 (+F)	
SHR%	26		25		10		11		10		9		16		14	
PBS AA%	1.2		1.4		2.1		2.1		1.9		2.7		2.5		2.7	
SHR%	2		3		3		3		3		4		4		4	
CABLE ORIG. (Including TBS) AA%	14.3 (+F)		17.2 (+F)		19.4		21.6		19.1		19.1		15.7 (+F)		13.5 (+F)	
SHR%	29		32		32		34		29		29		24		22	
PAY SERVICES AA%	2.1		2.2		2.3		2.5		2.7		2.6		2.4		2.1	
SHR%	4		4		4		4		4		4		4		3	

U.S. TV Households: 98.000,000

For explanation of symbols. See page B.

This ratings report for a weekday evening compares the prime-time appeal of the national broadcasting networks with independent stations, Public Broadcasting Service, cable program services—shown here as *cable orig. (including TBS)*—and pay-cable services. Shown for each broadcast network program is the average audience in TV households (of almost 98 million total in 1997), the share of audience, and the average audience rating by quarter hour. Note that for the independent stations and nonbroadcast network services, ratings show only overall average audience and share of audience, as the figures are a combination of individual market ratings, often for different programs in different cities.

Source: *National Pocketpiece*, Nielsen Media Research. Used by permission.

converters, VCRs, and DBS systems. Newer meters currently in development will be able to measure the multiple channels made possible through digital technology. Researchers refer to the household meter as *passive* because no effort is required on the part of the viewer to record its information, and to distinguish it from the newer peoplemeter. Because it provides no information as to who, if anyone, is actually watching, Nielsen supplements meter data with written diary data from some of the same households to obtain demographic data on actual viewers. Although passive household meters were phased out of network rating use in 1987 with activation of peoplemeters, they remain essential for individual market ratings in the 1990s.

Peoplemeters

Like Nielsen's Audimeter, the *peoplemeter* keeps a record of receiver use. It also offers research formerly only diaries provided by simultaneously collecting demographic data. It does this by requiring each viewer, whenever she or he watches television, to "check in" and "check out" by pushing a special handset button (see Exhibit 10.d). Data on both receiver use and viewer identity go by telephone line to a central computer containing basic household demographic data, stored earlier when the peoplemeter was installed.

Nielsen began a three-year test of a peoplemeter sample in 1983 and started its peoplemeter-based national network ratings service in 1987. Arbitron, meanwhile, had developed *ScanAmerica*, a three-pronged single-source system providing ratings, audience demographic data, and—through use of a *wand* that read universal bar codes—a measure of product purchasing. In 1992, lacking sufficient advertiser interest, Arbitron dropped use of both the wand and the name "ScanAmerica," and soon after ended all of its TV and cable ratings activities.

Introduction of peoplemeters created enormous controversy. Network ratings in all dayparts dropped dramatically while ratings for many independent stations and cable channels rose. In 1989 the networks funded an independent study that concluded peoplemeters appeared to slight such typical network heavy viewers as children and older people who might be intimidated by (or simply too impatient with) the computerlike button-pushing that peoplemeters required.

Passive Peoplemeters

Intense network and advertiser pressure to improve their methods pushed researchers to announce plans to replace peoplemeters with *passive* peoplemeters. Using computerized image-recognition, these passive devices presumably could "recognize" regular (family) viewers. Further, the devices eventually could tell when viewers were actually watching the screen or doing something else.

A number of variations are under consideration and being tested—including one that counts someone as watching television only if that person is actually facing the receiver. There were kinks—several demonstrations confused pets with people! But when perfected, the passive peoplemeter should be able electronically to record regular family members, and others such as guests, tabulating who and how many are watching and when. Although such systems might meet major objections to "active" peoplemeters, they raise other more troublesome concerns—chiefly viewer privacy; TV viewers may experience severe discomfort knowing that their television is watching back.

Coincidental Telephone Calls

Some researchers consider coincidental telephone data gathering the most accurate means of obtaining audience information. The term *coincidental* means that researchers ask respondents what they are listening to or watching during (that is, coincidental with) the time of the call. Putting questions that way eliminates memory concerns and reduces possibilities of intentional misinformation. Researchers ask whether respondents have a set turned on at that moment and, if so, what program, station, or channel the set is tuned to, plus a few demographic questions such as the number, gender, and age of those watching or listening.

Exhibit 10.d

Peoplemeter Ratings

The Nielsen national audience reporting system depends heavily on automation, from the home meter through the gathering, analyzing, and reporting processes. The peoplemeter used for national ratings consists of two parts: the hand-held unit on which each viewing member of the family and guests can "punch in," using individual keys, and the base unit on top of the receiver that stores home viewing information.

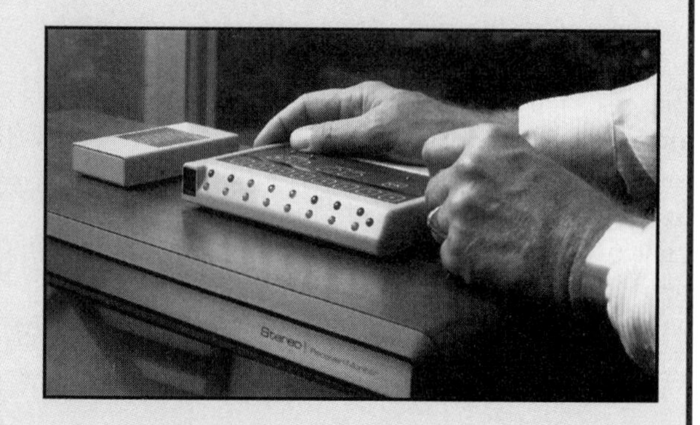

The Nielsen system begins with the individual sample household (1), whose peoplemeter base unit results are (2) "read" over a telephone line (leased by Nielsen) at 3:00 A.M. local time each morning by a

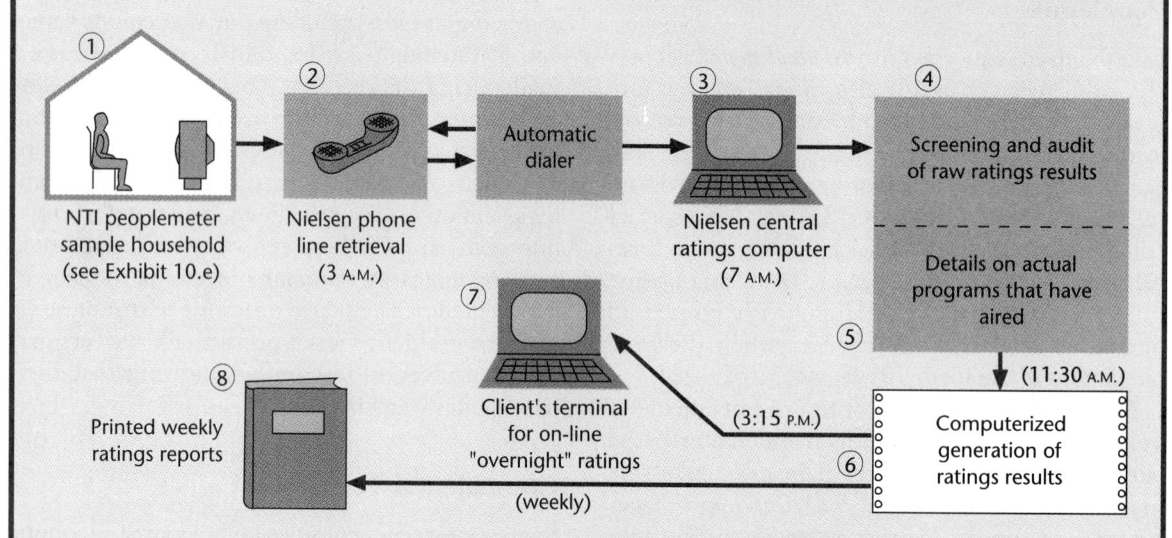

computer that sends the viewing results of the sample household into (3) the Nielsen central ratings computers. The collected raw ratings are (4) sent to the screening and audit process, where all ratings materials are assembled and (5) combined with the detailed program information constantly gathered by Nielsen. They are then sent for (6) computer generation of actual overnight ratings. By 3:15 P.M. the overnights are ready for (7) client (agencies, stations, networks) retrieval by computer terminal, and (8) weekly printed ratings reports. Nielsen diary and passive household meter markets go through a similar process to generate market-by-market reports.

Source: Data used by permission of A.C. Nielsen. Photo courtesy Nielsen Media Research.

Because coincidental methods provide only *momentary* data from each respondent ("What are you listening to or watching *now*?"), they require many calls, spaced out to cover each daypart, to build up a complete profile of listening or viewing. Properly conducted, coincidental calls require large batteries of trained callers, making the method expensive. Nor can coincidental calls cover the entire broadcast day; information on audience activity after 10:30 P.M. and before 8:00 A.M. must be gathered during more socially acceptable calling hours, when researchers ask respondents to recall programs they listened to or viewed during these nighttime and early-morning hours.

Telephone Recall

Tricks of memory make *telephone recall* less reliable than coincidental calls, but it costs less because more data can be gathered per call. RADAR, the only source of radio network ratings, uses telephone recall, collecting samples daily over a period of seven days, thus minimizing memory errors while obtaining a week's coverage.

Personal Interview

The use of in-person, door-to-door surveys has declined in recent years largely because of dangers implicit in knocking on doors in strange neighborhoods. Typically interviewers question people on the street or in shopping centers, or, for car radio listening, at stoplights. Data gathered in these ways cannot be projected to the general population, however, because the samples are not at all representative.

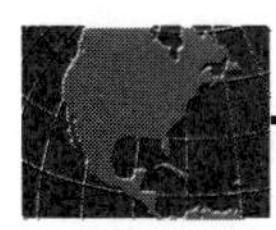

10.3 Sampling

Around-the-clock monitoring of all the private listening and viewing behavior of millions of people is obviously impossible. The task becomes possible only by *sampling*—studying some people to represent the behavior of all.

Using Samples to Simplify

Sampling simplifies three aspects of ratings research: behavior, time, and number of people.

- *Behavioral sampling.* Researchers agreed years ago on a minimum measurable behavioral response—turning on a receiver, selecting a station, and later turning the set off.
- *Time sampling.* The second simplification used in ratings takes advantage of repetitive daily and weekly cycles of most broadcast and cable programming. A sample taken every few weeks or months from this continuous program stream suffices for most purposes. Daily measurements occur only for network and major-city audiences.
- *Number of people sampled.* The most controversial ratings simplification arises from the use of only a few hundred or thousand people to represent program choices of thousands or millions of others. On a local-market level, Nielsen uses a few hundred households per market. It uses about 5,000 homes to represent almost 98 million for its national network ratings. For a very large radio market like Los Angeles (over nine million people aged 12 and older in the metro area), Arbitron will send out more than 14,000 diaries. For its fall 1996 report, 6,894 usable diaries were returned. By contrast, in a market the size of Lubbock, Texas (market 172; 12 and older population of 192,000), Arbriton received 911 usable diaries after distributing 1,940. The RADAR national radio network surveys 12,000 people.

Random Samples

In fact, small samples *can* give reasonably accurate estimates. The laws of chance, or *probability*, predict that a *randomly selected* small sample from a large population will be representative (within a predictable degree of accuracy) of the entire population. Random selection means that, ideally, *every* member of the entire population to be surveyed has an *equal* chance of being selected. Major defining characteristics of the sample will appear in about the same proportion as their distribution throughout the entire population.

Exhibit 10.e

Multistage Sampling Method

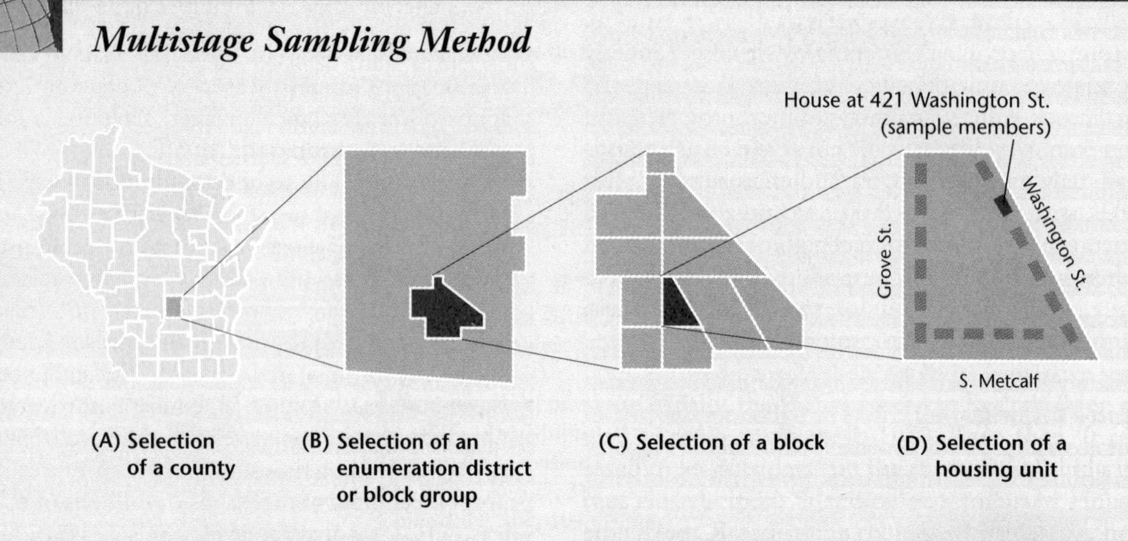

The four steps show selection of a sample member on the basis of geography: (A) random selection of a county for the sample; (B) random selection of a specific district or block within the county; (C) random selection of a specific city block; and (D) random selection of a specific housing unit within a block. Official Census Bureau data provide the sampling frames.

Source: Nielsen Media Research. Used by permission of A.C. Nielsen.

However, choosing at random is not as easy as it sounds. Drawing a sample randomly from a large human population requires some means of identifying each member by name, number, location, or some other unique distinguishing label. In practice, this usually means using either lists of people's names or maps of housing unit locations. Such listings are called *sample frames.*

Selecting a Ratings Sample

Nielsen draws its national sample of metered television households from U.S. census maps by a method known as *multistage area probability sampling. Multistage* refers to step-by-step narrowing down of selection areas, starting with counties and ending with individual housing units, as shown in Exhibit 10.e. For local-market ratings Nielsen uses special updated telephone directories. *Random digit dialing,* a computer method of generating telephone numbers without reference to directories, can solve the problem of reaching unlisted and newly installed telephones, but it increases the number of wasted (unused or business address) calls, thus increasing survey cost.

Ideally, each time a company takes a survey it should draw a brand-new sample. On the other hand, if the company uses expensive sampling and data-gathering methods, it cannot afford to discard each sample after only one use. Nielsen tries to retain each peoplemeter household in its national sample for no more than two years but allows local-market diary families to stay in samples for up to five years.

Exhibit 10.f

Nonresponse Errors

Typical nonresponse errors for each of the principal methods of ratings data collection include:

- *diaries:* refusal to accept diaries; failure to complete accepted diaries; unreadable and self-contradictory diary entries; drop-off in entries as the week progresses ("diary fatigue'); and failure to mail in completed diaries.
- *passive (household) meters:* refusal to allow installation; breakdown of receivers, meters, and associated equipment; telephone-line failures.
- *peoplemeters:* same drawbacks as passive meters, plus failure of some viewers (especially the very old and very young) to use the buttons to "check in" and "check out," having succumbed to "response fatigue."
- *telephone calls:* busy signals; no answers; disconnected telephones; refusals to talk; inability to communicate with respondents who speak foreign languages.

Sample Size

Having established a sample universe, the researcher must next decide how large a sample to choose—the larger the sample, the greater its reliability. But reliability increases approximately in proportion to the square of sample size (the sample size multiplied by itself). For example, to double reliability requires a fourfold increase in size. Thus a *point of diminishing returns* soon arrives, after which an increase in sample size yields such small gains in reliability as not to be worth the added cost.

Sources of Error

At its best, sampling yields only estimates, never absolute certainties. Thus the real question becomes how much uncertainty can be accepted in a given sampling situation.

Even when researchers carefully select their samples and sample sizes, two kinds of error can invalidate research findings. The built-in uncertainty of all measurements based on samples arises first from *sampling error.* No matter how carefully a sample is drawn, that sample will never exactly represent the entire population (or universe) being measured. With the use of statistical techniques, the *probable* amount of statistical uncertainty in ratings (that is, the amount of sampling error to be expected) can be calculated in advance.

The second possibly invalidating problem—*nonsampling error*—arises from mistakes, both intentional and inadvertent. Such errors produce *bias* in the results. Bias can come from lying by respondents as well as from honest mistakes. Researchers are probably consciously or unconsciously prejudiced. The wording of questionnaires may be misleading. And mistakes can occur in recording data and calculating results.

Sometimes reports state sample size as the number of people (or households) the researcher contacted, when the key element should be how many actually *participated.* A 45 percent response rate indicates that only 45 out of 100 homes or individuals contacted actually participated. In practice, a response rate of 100 percent never occurs. Depending on the research method and purpose, a response rate of under about 85 percent for other than day-to-day ratings research is cause for concern.

Notwithstanding a series of incentives and entreaties from the ratings services, nonresponse poses a serious limitation on ratings accuracy. Some age groups, teenagers, for example, have a particularly low response rate for radio diaries. Moreover, those who do return diaries may not represent those who do not. Exhibit 10.f lists typical *nonresponse errors.*

Overall, diary and meter methods yield a usable response rate of about 40 percent ("usable" being variously defined, but meaning "returned on time," plus other measures of how completely the diary is filled in); the telephone method comes close to 75 percent. The number of such usable responses is termed the *in-tab* sample, the sample actually used in tabulating results and producing a given ratings report.

10.4 *Determining Ratings and Shares*

Defining Markets

A crucial step in ratings research is an accurate definition of the local market to be measured. Advertising depends on a universally recognized, national system of clearly defined, nonoverlapping markets.

Arbitron's *Area of Dominant Influence* (ADI), first developed in 1965, was the most widely accepted system for defining TV markets until the company abandoned the TV rating business three decades later. Nielsen has its own version, called *Designated Market Area* (DMA). A DMA consists of one or more counties in which stations located in a central town or city are the most viewed (see Exhibit 10.g). DMAs usually extend over smaller areas in the East, where cities are closer together, than in the West. Nielsen assigns each of the more than 3,000 counties in the United States to a single DMA, updating the assignments annually. DMAs range in size from No. 1 (New York City, with about seven million television households) to No. 211 (Glendive, Montana, with fewer than 4,000).

Arbitron's ADI is still used to define local radio markets. As with Nielsen's DMA, Arbitron assigns every county in the United States to one specific ADI. In addition to measuring audience use in the ADI, Arbitron reports on two other related areas. One, the *metro* area, usually consists of one or more counties around a central city core. The metro area covers an area smaller than the ADI. The other, *total survey area* (TSA), the largest local region on which Arbitron reports, includes 98 percent of a market's listening audience, thus covering counties outside the ADI and overlapping with adjacent ADIs. Thus, although counties are assigned to only one ADI, they may be shown in more than one TSA, depending on listening in that county. Nielsen takes a similar approach with its TV ratings.

Households

Another preliminary step in ratings research is to define what will count as "one" when measuring audience size.

Television viewing has traditionally been a family activity, making the *household* a logical unit of measure, even though a majority of households now have two or more television sets, and much viewing takes place as a solo rather than family activity. A single diary or meter records viewing of each television receiver, whereas for radio, each listener has a separate diary.

Radio researchers prefer to count persons rather than households because (1) radio listening usually occurs as an individual activity, and (2) much radio listening takes place outside the home, especially in autos and workplaces.

Ratings

The term "rating" is widely misused and usually confuses two quite separate industry measures: rating and share (discussed in the next section). Specifically, a *rating* is an estimate of the number of households (or persons in the case of radio) tuned to a specific channel (station, network, or program), expressed as a percentage of *available* households (or persons). "Available" refers to those who could watch (or listen) if they wanted to, but may or may not have their receivers turned on.

Prime-time broadcast television network shows average about a 17 rating while daytime shows average 6. The most successful nonsports entertainment program of all time, the final episode of *M*A*S*H* in 1983, had a Nielsen rating of 60.2. This means that 60.2 percent of all U.S. television households watched at least part of that episode. The Super Bowl of 1982 achieved a rating of 49.1—a record for that annual football championship.

By contrast, the most popular cable television networks had prime-time ratings of 2.1 or less. TNT, the highest-rated cable network in 1996, had an average prime-time rating of 2.1, followed by USA with a 2.0. Nickelodeon and ESPN garnered ratings of 1.8 and 1.6, respectively.

Exhibit 10.g

TV Market Definition Concepts

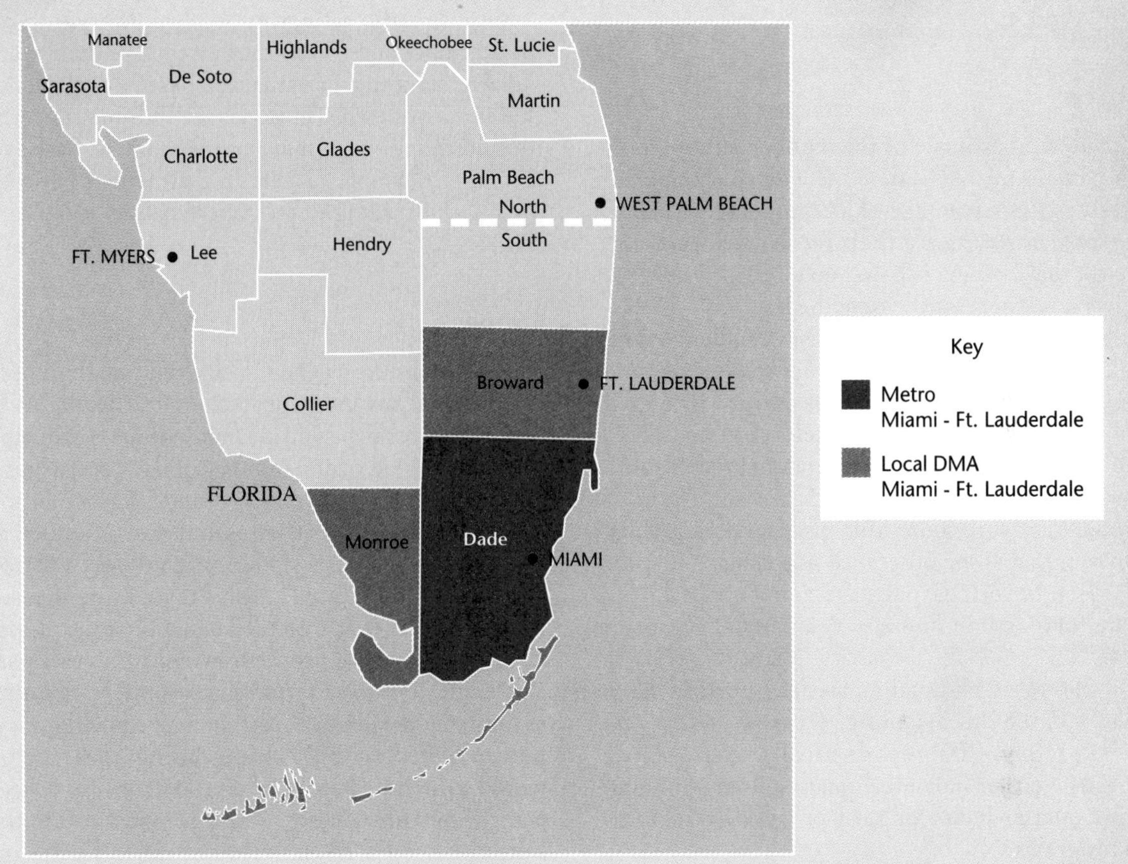

In the Miami–Ft. Lauderdale TV market, the *designated market area* of stations located in or near Miami extends to three counties—Miami's own (Dade, also shown here as the *Metro* or home county) and two adjacent counties (Broward and Monroe, which includes the Florida Keys). In practice, viewers often receive programs from stations in two or more markets. In those cases, Nielsen determines which stations are viewed most frequently in the market being defined and *designates* which counties are to be assigned to which DMAs. For example, in this illustration, when viewers in Broward County receive signals from stations in both Miami and West Palm Beach, Nielsen has determined that most of the time Broward County viewers tune to Miami stations.

No county is assigned to more than one DMA. Based on viewing patterns, however, Nielsen for some reporting purposes splits some counties into north and south (as with Palm Beach County on this map) or east and west sections. Nielsen also measures audiences in the *total survey area*, which includes all counties wherever viewership is reported. For superstations, the *total area* is virtually the entire United States.

Source: Map © 1993 The Arbitron Company. Used by permission.

Radio stations—and public-television stations—often earn ratings of less than 1, and rarely more than 2 or 3. Such low ratings make no meaningful distinctions among stations; radio therefore relies more often on *cumes* (measures of cumulative audiences) and shares.

Shares

A *share* is an estimate of the number of households (or persons for radio) tuned to a given channel, expressed as a percentage of all those households (or persons) *actually using* their receivers at that time. Recall that a rating is based on all those *owning* receivers, not necessarily *using* them.

A station's share of the television audience is calculated on the basis of *households using television* (HUT). A HUT of 55 indicates that at a given time an estimated 55 percent of all television households are actually tuned in to *some* channel receivable in that market; the remaining 45 percent are not at home, busy with something else (screening videotapes perhaps), or otherwise not using television. HUTs vary with daypart, averaging about 25 percent for daytime hours and about 60 for prime time.

Radio research usually measures persons rather than households, yielding a *persons using radio* (PUR) rating. PUR reports usually refer to blocks of time—either individual quarter-hours or cumulative quarter-hours for a day or week—rather than to programs.

Shares are figured based on either HUT or PUR data. *A station always has a larger share than rating for a given time period.* For example, top broadcast television network prime-time programs usually average ratings of about 20, but their corresponding shares are closer to 30. Exhibit 10.h explains how to calculate both ratings and shares.

Television programmers use shares in making programming decisions; salespersons usually use ratings as advertising sales tools. The reason is that shares give programmers a better estimate of their competitive position within a medium, whereas ratings more readily allow comparison of advertiser exposure on radio or television with that in other media. Also, because HUT levels vary widely during different times of the year (higher in November, lower in May), programmers use share rather than rating to compare a program's competitive performance from one time of year to another.

Because radio ratings are typically very low, radio stations more often than not judge their competitive standing in a market based on audience shares. With few exceptions (see Exhibit 10.i), the top-rated stations in major radio markets, based on listening of people 12 years old and older between 6 A.M. and midnight, average between a 6 and 7 share.

Cumes

A radio program reaches a relatively small number of people in any given quarter-hour. *Cumulatively*, however, over a period of many hours, or during the same period over a number of days, it reaches a large number of *different* listeners. A *cume* rating gives an estimate of the (cumulative) number of *unduplicated* persons a station reaches over a period of time. "Unduplicated" means that during the two or four weeks that typically make up a rating period, a person who listened several times to a particular station on different days would be counted as only one person in constructing a cume figure. A person listening only once during that period would also be counted as one person, because a cume shows how many *different* people tuned to the station during a given period of time. The terms *reach* and *circulation* usually refer to cume audience measurements. Cumes are useful in commercial broadcasting because ads are often repeated. Consequently, the larger the cume for a station, the greater chance that more listeners have been exposed to the advertiser's message.

Reporting Demographics

Rating reports detail audience composition by gender and age. These *demographic breakouts*, or simply *demographics*, divide overall ratings into such subgroups as those for men, women, and teens. Adult audience age-group categories typically consist of

Exhibit 10.h

Ratings Concepts

(A) The pie chart shows television set–use information gathered from a sample of 400 households, representing a hypothetical market of 100,000 households.

Note that program ratings are percentages based on the entire sample (including the "no-response" households). Thus Program A, with 100 households, represents a quarter (25 percent) of the total sample of 400. The formula is 100 ÷ 400 = .25; the decimal is dropped when expressing the number as a rating.

Projected to the entire population, this rating of 25 would mean an estimated audience of 25,000 households. The formula is .25 (the rating with the decimal restored) × 100,000 = 25,000.

(B) The smaller pie chart, representing 55 percent of pie A, includes only the households using television, in this case 80 + 100 + 40 households, or a total of 220 (expressed as a households-using-television or HUT rating of 55, as shown in A). Shares are computed by treating the total number of households using television, in this case 220, as 100 percent. Thus program A's 100 households divided by 220 equals about .4545 expressed in rounded numbers as a share of 45.

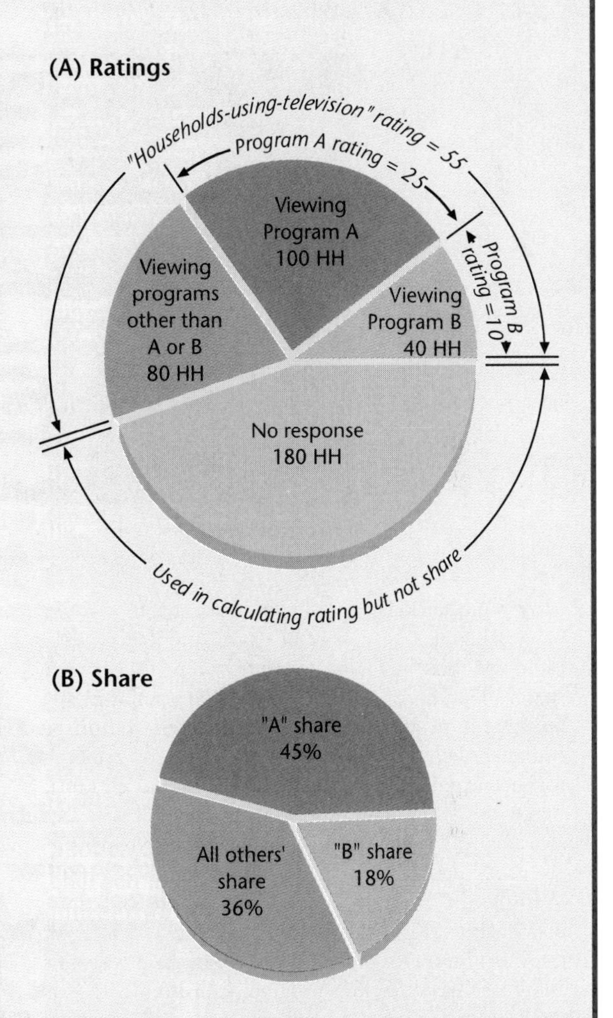

decade units (such as men 35–44) for radio and larger units for television (for example, women 18–34 or 25–49), although "persons 12+" serves as the basic category for determining, for example, which radio station in a given market is "number one."

Advertising agencies usually "buy" demographics rather than generalized audiences. Most advertisers would rather have an audience of moderate size with the right demographics for their product than a huge audience containing many members not likely to be interested.

Exhibit 10.i

WFBQ, Indianapolis: Not your average major market station

The Bob and Tom Show, featuring (from left, clockwise) Bob Kevoian, sportscaster Chick McGee, Tom Griswold, and newscaster Kristi Lee.

Most radio stations operating in major markets would be very happy with a share of 4 or 5, and the highest-rated stations in major markets rarely exceed a 7 share.* That figure, however, would be a major disappointment to WFBQ, the highest-rated station in Indianapolis, and consistently the first- or second-highest rated rock-and-roll station in the country. Arbitron's fall 1996 report indicates that WFBQ scored a 12.7 share in the Indianapolis metro area.

According to Marty Bender, program director for WFBQ, the station's success can be attributed to many factors. First is definitely *The Bob and Tom Show*, the morning drive program hosted by Bob Kevoian and Tom Griswald. This program, consistently one of the highest-rated morning shows in the country with shares between 20 and 25, features interviews with comedians and musicians, discussion of off-beat news stories, original music and advertising parodies, and a long list of wacky characters played by various members of the station staff.

Following *The Bob and Tom Show*, the station moves to its rock-and-roll format, featuring a mix of classic rock and new material. The station capitalizes on Bob and Tom's popularity by playing excerpts from the morning show throughout the day—an example of the parsimony principle discussed in Section 8.1.

In addition to its programming, Bender attributes much of the station's popularity to its community involvement. A recent charity auction sponsored by the station raised more than $65,000 for leukemia research. Bob and Tom typically produce two CDs per year, with the proceeds going to a variety of local charities. And when the military considered closing a local facility, Fort Benjamin Harrison, the station initiated a petition campaign to try to save the base. These kinds of activities increase the sense of loyalty among WFBQ's listeners.

The station is not just a ratings success; it has enjoyed significant critical acclaim as well. The station and its staff have won four Marconi awards, presented annually to stations demonstrating excellence in radio. No other station in the country has won more Marconis than WFBQ.

*Measure of all listeners aged 12 and over using radio between 6 A.M. and midnight.

Sources: *Arbitron Fall Radio Local Market Report*, 1996; *Radio and Records*, Vol. 2, 1996; *Broadcasting & Cable*, 13 Jan. 1997: 127; Photo courtesy WFBQ Radio.

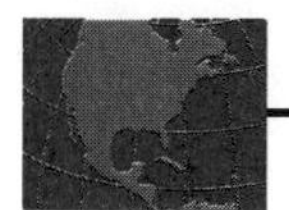

10.5 Use and Abuse of Ratings

Ratings and shares perform a vital function for operators and advertisers—which makes them subject to misuse. Blind consideration of only the numbers could lead to one station getting a disproportionate amount of advertising dollars when the spread between that station and its nearest competitor is less than the sampling error reported. Through ignorance or misrepresentation, some users present ratings as hard-and-fast measurements instead of what they are: only estimates.

Facing Complaints

In response to complaints and investigations about ratings in the 1960s, a Broadcast Rating Council was established in 1964. Eighteen years later, in 1982, it became the Electronic Media Rating Council (EMRC), serving as an independent auditing agency representing ratings users. The EMRC accredits ratings services that meet its standards and submit to annual auditing paid for by the service provider, not by the EMRC. The procedures are sufficiently involved and costly that many smaller companies don't bother to apply.

Reliability of Ratings

Reliability in research refers to the degree to which methods yield consistent results over time. For example, the drops in broadcast television network ratings in initial peoplemeter surveys raised questions about ratings reliability, because those results clearly were *not* consistent with earlier findings. Consistent reports of lower network ratings since the peoplemeter's introduction suggest that the method is generally reliable, though not necessarily valid.

Validity of Ratings

Validity in research refers to the degree to which findings actually measure what they purport to measure. Ratings purport to measure the *entire* broadcast audience, but in practice they can account for only the broad middle-range majority of that audience. Thus ratings tend to underrepresent the very rich, the very poor, the very young, and ethnic minorities—and are less valid for that limitation.

Television's use of households rather than individuals as its ratings measurement unit also affects validity. About a quarter of today's households consist of lone individuals whose lifestyle (and therefore broadcast use) patterns often differ from those of multiperson households.

Further, measuring only households excludes many other venues for watching television. Not included in the ratings data is viewing from bars, hotels and motels, hospitals, and college dormitories. Until recently, comparatively little was known about how much viewing was done away from home. A 1993 Nielsen study found that about 4 percent of adult weekly viewing takes place away from home, a figure that, given the recent proliferation and popularity of sports bars, is surely rising. The ever-increasing amount of programming being delivered via the Internet could also significantly affect away-from-home viewing rates.

Hyping and Tampering

A widespread industry practice known as *hyping* (or, sometimes, *hypoing*) can also bias ratings. Hyping refers to deliberate attempts by stations or networks to influence ratings by scheduling special programs and promotional efforts during ratings sweeps. For example, both radio and television stations sometimes hype with listener contests.

Both the FTC and the FCC have investigated hyping, but with little effect—networks and nearly all stations do it to some degree. The arrival of 52-weeks-a-year meter ratings for television has reduced the incidence of hyping somewhat, although many stations continue the practice during sweep weeks. In any case, Arbitron and Nielsen note in ratings reports any exceptionally blatant activities so that advertisers can take them into account in market analyses.

Less commonly, ratings can be vulnerable to *tampering;* someone who can influence viewing habits of even a few households in a small sample can have a substantial impact on resulting ratings. Rating companies therefore keep sample household identities a closely held secret. Still, a few cases of outright manipulation of viewers have become public, causing ratings services to junk reports for some programs, and even entire station or market reports for a given ratings period.

Qualitative Ratings

As long as ratings have dominated programming strategies, critics have complained that ratings encourage mediocrity by emphasizing sheer size to the exclusion of *qualitative* program aspects. Time and again, programs that are seemingly of above-average quality receive enthusiastic reviews and audiences, but fail to meet the rigid minimum-share requirements for commercial survival. Critics question whether programs that are merely accepted by large audiences should automatically win out over programs that attract smaller but intensely interested audiences. Quantitative ratings dictate this kind of judgment—the system favors "least objectionable" majority programs over possible alternatives.

The multi-channel environment of the late 1990s has changed this situation somewhat, though not entirely. Prime-time network programs today survive with ratings and shares that would have resulted in sure cancellation even five years ago. And because of the proliferation of cable networks eager to find programs, programs that are forced off the broadcast networks may find a place on cable. Numbers are still the name of the game for the networks, but shows do not have to be as popular as in years past to survive commercially.

About the only qualitative ratings in which U.S. commercial broadcasters have shown interest are those conducted by Market Evaluations, Inc. Using questionnaires completed by a random national sample of 1,000 families, this firm regularly estimates image (familiarity and likability—not necessarily the same thing) ratings of major performers and the popularity of specific programs among various demographic groups.

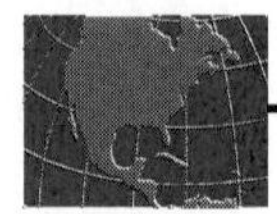

10.6 Broadcast Audiences

Over decades of intensive ratings research, a vast amount of knowledge about electronic media listening and viewing habits has been accumulated—this is surely the most analyzed media activity in history.

Set Penetration

The most basic statistic about broadcast audiences is *set penetration* or *saturation*—the percentage of all homes that have broadcast receivers. In the United States, radio and television penetration has long since peaked at more than 98 percent. Indeed, most homes have several radios and more than one television set. In short, for practical purposes the entire U.S. population constitutes the potential broadcast audience.

Set Use

HUT measurements tell us the average percentage of television households actually using sets at different times of the day.

Television viewing climbs throughout the day from a low of about 12 percent of households at 7:00 A.M. to a high of about 70 percent in the top prime-time hour of 9:00 to 10:00 P.M. (Eastern time). Audience levels for television change somewhat with the seasons: Viewing peaks in January–February and bottoms out in June, reflecting some influence of weather and leisure activity on audience availability.

On the other hand, radio listening has a flatter profile than television. Radio listening typically reaches its highest peak during morning drive-time hours, recedes through the mid-day shifts, and climbs again during the afternoon drive-time period. Listening after 6:00 P.M. continually declines through the evening, as more listeners begin watching television news and then prime-time programming. Unlike television, radio listening has very little seasonal variance.

TV as Habit

Long-term trends aside, people tend to turn on their television sets day after day in the same overall numbers, with no apparent regard for the particular programs that may be scheduled. Expressed in terms made famous by Marshall McLuhan, the *medium* matters more than the *message*. Paul Klein, former CBS programming chief, proposed a similar theory, that of the *least objectionable program* (LOP). He theorized that people stay with the same station until they are driven to another station by an objectionable program. But even if they find *all* programs objectionable, they will stay tuned to the least objectionable one rather than turn off the set.

Impact of Remote Controls

A corollary of LOP was *tuning inertia*. Whether because of viewer loyalty to a station or network, or simple laziness before widespread use of remote controls, viewers tended to leave sets tuned to the same station. Tuning inertia still strongly affects radio audiences; in large markets with as many as 40 stations from which to choose, listeners tend to confine their tuning to only two or three favorites.

By the late 1980s, however, remote control devices (RCDs) for television—and the increased number of cable channels—began to modify old patterns. Whereas fewer than 30 percent of households had remote controls in 1985, by the mid-1990s nearly 90 percent did—and could thus very easily change channels. One result has been *restless viewers*, or *grazers* — those who use a remote control to switch rapidly among several channels. These patterns of use may be already changing. Recent research indicates that many RCD users are becoming much more sophisticated and deliberate. One study found that the most common use of the RCD was punching in a specific channel number, a direct substitute for on-set tuning (Eastman and Newton, 1995).

Time Spent

The total amount of time that people devote to television serves as a broad measure of its audience impact. This statistic arouses the most widespread concern among critics: Any activity that takes up more time than sleeping, working, or going to school—as watching television does for many viewers—surely, they reason, has profound social implications.

In the 1982–1983 television season, average daily viewing per household was 6 hours and 55 minutes (or, more precisely, at least the TV set was turned on for that length of time). Fifteen years later that average had climbed to 7 hours and 16 minutes. This total combined viewing by all household members. Exhibit 10.j details other average viewing levels by various demographic groups. According to Arbitron's most recent data, radio use by all persons aged 12 and over averages 22 hours and 51 minutes per week, down slightly from previous years.

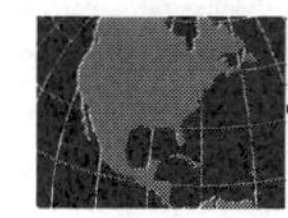

10.7 Cable Audiences

Researching cable audiences somewhat parallels the patterns familiar in broadcast research, in terms of both methods and findings.

Cable Research

Cable television determines its potential audience on the basis of basic-cable subscribers, although one might argue that the number of *homes passed* by cable feeder lines—which includes those who choose not to subscribe—is a better indicator of cable's real potential. Though *subscriber* numbers tell nothing

Exhibit 10.j

TV Viewing Trends

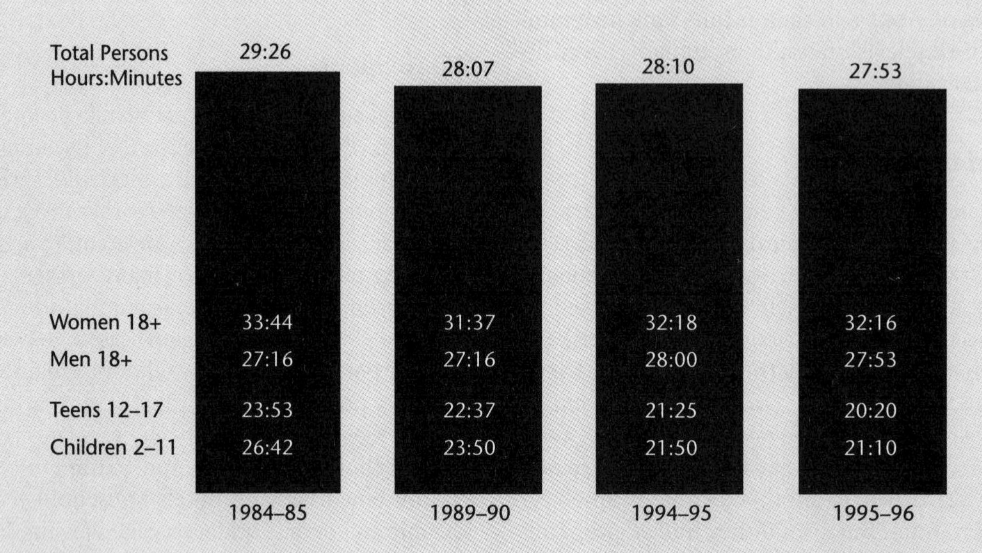

During the 1995–96 broadcast year, the amount of time persons spent viewing declined slightly from that of the previous year. The largest shift was among teens, followed by children.

Source: Television Audience 1996, Nielsen Media Research. Used by permission.

about actual cable *use*, they *did* show that by 1997, cable was present in about two-thirds of U.S. television households.

Because of cable's many channels (with more being added all the time), its total audience subdivides into many small groups, posing severe measurement difficulties. Further, some cable tuning converter boxes make it difficult for Nielsen to attach passive household meters. Cable systems that provide addressability (the ability to control and record access to each subscribing household from the headend) can resolve this problem, but only as the number of systems with that capability continues to expand.

Cable Ratings

Nielsen introduced audience reports for a few national cable services in 1979, gradually increasing the amount of information provided on basic and pay networks. Based on Nielsen's national peoplemeter sample, Nielsen Homevideo Index (NHI) issues quarterly *Cable National Audience Demographic Reports* and *Nielsen Cable Audience Reports*, which provide cable household viewing data in all dayparts for the largest national cable and broadcast networks.

However, *local* cable audiences are far more difficult to measure than national audiences. Political

Average Hours of Viewing per TV Household per Week

Year	Hrs:Mins	Year	Hrs:Mins
1971	42:04	1984	49:58
1972	42:46	1985	50:00
1973	43:49	1986	50:16
1974	43:41	1987	48:22
1975	43:24	1988	49:04
1976	43:29	1989	49:19
1977	43:37	1990	48:29
1978	43:41	1991	48:40
1979	45:05	1992	49:35
1980	46:06	1993	50:24
1981	47:07	1994	50:50
1982	47:44	1995	50:42
1983	48:31	1996	50:44

Prior to 1993, 12 months ending August each year; 1993–96 dates follow broadcast seasons (mid-September to mid-September). Data prior to 1987 based on NTI Audimeter sample.

boundaries define cable franchise areas, often comprising only a single county or even just part of one, whereas broadcast signals, by ignoring such artificial lines, create larger markets that are more cost-effective to measure. Nielsen includes local cable viewing only in markets where cable audiences reach minimum television reporting levels, usually a share percentage of three or more.

In 1992 Nielsen previewed a new local cable ratings book designed to show market-level viewing of major national cable services and in 1993 announced a new computer-readable service that would more equitably compare local, cable, broadcast, and syndicated program viewing patterns. Meanwhile, several cable systems had turned to telephone coincidentals to measure their audiences.

Uses of Cable

By the mid-1980s research had begun to show clear evidence of national cable network influence on audience viewing patterns. Among households subscribing to cable in 1995, basic-cable networks collectively had a viewing share of 43 percent, compared with 19 percent in 1985. Pay-cable hovered at an 8 percent share, down from 10 percent in 1985. Broadcast networks' share of viewing in those same households had declined from 56 percent to 40 percent, across all dayparts (NCTA, Spring 1997). Use of television increases in households with more than one set, and as more than one set is connected to cable television.

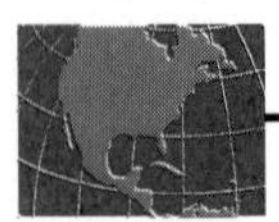

10.8 VCR Audiences

Videocassette recorder use-patterns call for research approaches quite different from those for broadcasting or cable. Yet as VCRs become increasingly common (they were in more than three-quarters of TV homes by the mid-1990s, with over 20% of TV households owning two or more), understanding their many uses becomes vital.

VCR Research

Time shifting enables VCR owners to control when they watch broadcast or cable programs. VCRs also encourage skipping commercials by *zipping* (the avoidance of commercials during VCR playback) or *zapping* (the deletion of commercials during recording). A recorded program (and possibly its advertisements) may be seen several times and by different viewers, or the ads may be seen only once if at all. To trace all of these varied patterns, Nielsen publishes a quarterly *VCR Usage Study* based on special VCR diaries filled out monthly by sample households. The time broadcast material is recorded is considered the time of viewing for purposes of determining ratings—playbacks are not counted. Thus, even if no one ever watches the tape (a common phenomenon), the material is still counted as having been viewed.

Uses of VCRs

VCRs are almost twice as likely to be used to play tapes than to record them. Although they are still used mostly to supplement real-time broadcast and cable program viewing, VCRs do give viewers more control over their time in at least four ways:

- to time shift broadcast or cable programs for later viewing (sometimes more quickly, as in sports events, by fast-forwarding through dull spots and commercials) but not for retention after that viewing;
- to develop a permanent home video library;
- to view purchased or rented prerecorded tapes; and
- to watch tapes made with home camcorders.

Most playback occurs during television prime-time hours. Exhibit 10.k offers more data on VCR usage.

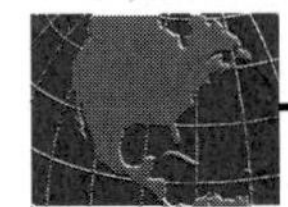

10.9 Measuring Internet Use

As the Internet, particularly the World Wide Web, becomes a more important advertising medium (see Section 6.8), more interest is placed on discovering how people use the Web. A number of companies now attempt to measure Internet use. According to one of these companies, PC Meter, at the beginning of 1997 there were between 13 and 15 million households capable of Internet access. Consequently, a ratings point translates to 130,000 to 150,000 users. Web site providers and advertisers on those sites want to know how many people access their sites, but also for how long. PC Meter uses a sample of 10,000 computer-equipped households to calculate its Internet ratings.

According to the December 1996 report from PC Meter, 58.8 percent of all households equipped to access the Internet contacted at least one news/information/entertainment Web site during the month. The average page request per site was 2.11, with viewers spending an average of 1.18 minutes per page. The average usage of the Internet among all users totaled 35.35 minutes per use. Exhibit 10.l shows the Internet ratings of the most popular sites at the end of 1996.

Exhibit 10.k

Patterns of VCR Use

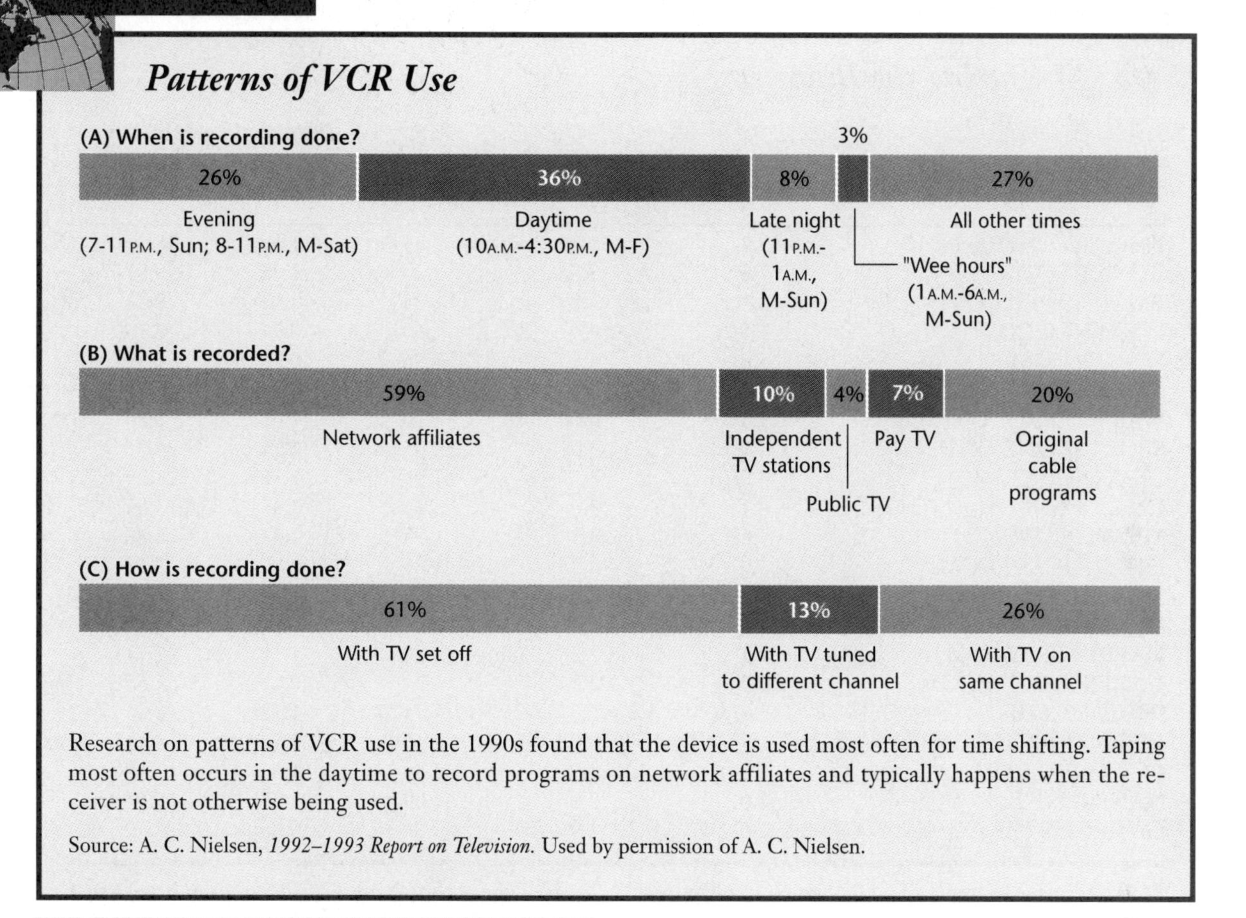

Research on patterns of VCR use in the 1990s found that the device is used most often for time shifting. Taping most often occurs in the daytime to record programs on network affiliates and typically happens when the receiver is not otherwise being used.

Source: A. C. Nielsen, *1992–1993 Report on Television.* Used by permission of A. C. Nielsen.

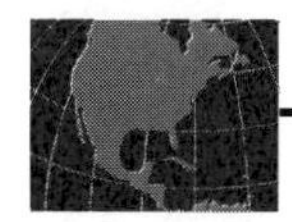

10.10 Other Applied Research

Nonratings research tries to find out what people like and dislike, what interests or bores them, what they recognize and remember, and what they overlook and forget. To study such subjective reactions, investigators usually use *attitudinal* research methods, which reveal not so much people's actions (such as set use) as their reactions—their *reasons* for action, as revealed in their attitudes toward programming.

Focus Groups

Commercial attitudinal research often makes no attempt to construct probability samples, because it usually does not try to make estimates generalizable to whole populations. Instead, investigators choose respondents informally. They assemble

Exhibit 10.1

Measuring the Internet

	Nov. '96	Dec. '96	Page requests per viewing	Minutes per page request	Minutes of usage per person
News/info./entertainment	59.2	58.8	2.11	1.18	35.35
ZDNET.COM	7.8	8.0	2.20	1.25	9.07
PATHFINDER.COM	6.7	6.8	1.85	1.02	8.33
WEATHER.COM	3.5	4.9	1.68	0.91	8.19
DISNEY.COM	5.8	4.8	1.86	0.90	11.59
C/NET.COM	5.2	4.5	2.06	1.24	8.72
WARNERBROS ONLINE	3.3	4.0	2.20	1.08	8.54
SPORTSZONE.COM	4.0	3.9	2.05	1.35	22.76
USATODAY.COM	4.0	3.7	1.99	1.38	15.39
CNN.COM	4.1	3.6	2.17	1.47	13.43
MSNBC.COM	3.8	3.5	1.93	1.41	8.88
INTELLICAST.COM	3.0	3.4	1.85	0.90	7.18
SPORTSLINE.COM	2.9	2.9	2.59	1.54	16.19
NFL.COM	2.1	2.7	2.04	1.50	15.84
HAPPYPUPPY.COM	2.1	2.2	2.29	1.19	7.13
UNITEDMEDIA.COM	1.7	2.2	1.71	0.81	7.66
IGUIDE.COM	2.1	2.0	1.75	1.12	8.43
WARNERBROS.COM	1.3	2.0	2.18	1.15	8.33
PCWORLD.COM	1.8	2.0	2.36	1.23	11.16
NANDO.NET	2.0	2.0	1.80	1.06	5.67
GAMECENTER.COM	1.6	1.9	1.89	0.86	7.39

PC Meter uses a sample of 10,000 computer-equipped households to measure the 13 to 15 million homes that access the Internet. Internet content providers and advertisers are interested in both the number of people visiting their site and the length of the visit.

small panels, called *focus groups*, and gain insights about people's motivations through informal discussion-interview sessions.

Program concept research, for example, tries out ideas for potential new shows. A focus group's reactions to a one-page program description can help programmers decide whether to develop an idea further, to change details, or to drop it entirely. Advertisers often test concepts for commercials before making final commitments to full production. These tests may use simple graphic storyboards, or they may employ *photomatics*, videotaped versions of original storyboards with camera effects and audio added to make them look and sound something like full-scale commercials.

To test new or changed programs, producers often show pilot versions to focus groups. People watch, give their reactions, and sometimes discuss reasons for their attitudes with a session director. Frequently producers, writers, and others study reactions by watching these discussions through one-way mirrors or by screening videotapes of the session.

Program Analysis

Minute-by-minute reactions to a program can be studied using a *program analyzer*—a device first developed in the 1940s and now thoroughly computerized—that enables test-group members to express favorable, neutral, or unfavorable reactions by pushing buttons at regular intervals on cue. The machine automatically sums up the entire test group's reactions, furnishing a graphic profile. A follow-up discussion can then probe for reasons why audience interest changed at given moments in a script, as revealed by peaks and valleys in the graph.

Theater vs. In-Home Testing

The movie industry has long used theater previews to gauge audience response. Several firms specialize in staging similar theater previews (sometimes called *auditorium testing*) of television programs and commercials. Investigators sometimes test advertisements under a pretext of testing programs, with commercials seeming to appear only incidentally. Viewer reactions come out in questionnaires or discussions.

Physiological Testing

Most of the methods described thus far depend on self-analysis by panel members. In an attempt to eliminate subjectivity and to monitor responses more subtly, researchers have identified a number of involuntary physical reactions that give clues to audience thinking. Reactions measured for this purpose include changes in brain waves, eye movements, eye pupil dilation, breathing rates, pulse rates, voice quality, perspiration, and sitting position (the "squirm test").

For example, a number of researchers have capitalized on the two-sided nature of the human brain. Each side has its own specialized functions and reacts to different stimuli. Reasoning ability seems to be located in the brain's left side, and emotions in the right. It follows that product ads with emotional appeal should, if correctly designed, stimulate primarily the right side of the brain more than would commercials that appeal mainly to logic. Commercials shown to viewers wired for brain-wave recording can be tested to determine whether the messages draw the desired brain-wave responses.

Audience Response

Telephone calls and letters from listeners or viewers about programs, performers, and commercials provide additional audience information. Indeed, early broadcasters relied entirely on voluntary listener mail for audience information. But people who write or call a station are not a good representative sample of the entire audience. Research has shown that letter writers differ significantly from the general population in terms of race, education, income, type of job, age, and marital status—all differences important to advertisers and programmers. Further, letter-writing campaigns for or against a given point of view, product, or service can give misleading impressions about general audience reaction. For these reasons, stations take phone calls less seriously than systematically gathered data. Likewise, unsolicited e-mail comments concerning a company's Web page may not be considered as carefully as responses to a systematic survey.

Chapter 11

Effects

CHAPTER CONTENTS

The time we spend with electronic media—typically more each day than with any other leisure activity—underlines our need to understand their varied impacts. This chapter reports some of what research has discovered about the many ways electronic media have affected our lives—and raises questions for which we don't yet have complete answers. Researchers now strongly feel that the background audiences bring to their use of electronic media is more important to understanding media "effects" than any study of media content.

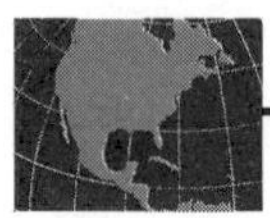

11.1 Developing Effects Research

People have long assumed that spending so much time with electronic media simply had to have *some* impact, good or bad. Such assumptions take on practical significance when policy decisions are based on supposed or real media impact. For that reason, and because of scientific curiosity, researchers have spent considerable effort to determine media effects.

Early Findings

Early media research focused on the impact of *mass propaganda* efforts used during World War I (1914–1918), before broadcasting arrived. After the war, the extent of deception by propagandists on all sides showed how thousands had been manipulated. The postwar advent of radio broadcasting—along with emergence of the politically threatening new communist regime in Russia—made fear of increased propaganda manipulation more alarming. This led to more research on the social and psychological dynamics of propaganda.

The 1920s' concept of media impact saw messages as so many bullets of information (or misinformation) aimed at passive groups. Because researchers assumed that messages penetrated and caused specific reactions, the concept became known as the *bullet* or *hypodermic-injection* theory.

By the late 1930s researchers had realized the bullet theory was oversimplified in treating message receivers as mere passive targets. New and more sophisticated studies discovered that audiences react to messages as individuals. Media effects therefore depended on many variables within individual audience members.

Researchers labeled such factors, many not directly observable, *intervening variables* because they come between messages and effects. They vary a message's impact because of each person's previously acquired attitudes, traits, experiences, social situation, and education. Intervening variables explain why an identical message often has a different impact on different people.

One important intervening variable studied in the 1940s was the personal influence of *opinion leaders*, as contrasted to media's *im*personal influence. One theory suggested that media influence often passes through leaders to followers rather than to all individuals directly. For most people, opinion leaders play a greater role in various decisions than does direct influence of radio, newspapers, magazines, or books. Other studies confirmed and refined this *two-step flow* hypothesis of media influence (Katz and Lazarsfield, 1955), which had great impact on research for two decades.

Selective Effects

Yet even opinion leaders are not influenced in direct proportion to the amount of persuasive media content they receive. People pay attention to messages that fit their established opinions and ignore those that don't. Researchers discovered that because of this *selective exposure*, media tend to *reinforce* existing views, rather than converting people to new ones. Existing mindsets color how people perceive media they select. Those who select the same message may interpret it in different ways. Though the stimulus is constant, the response varies. Called *selective perception*, this variable shows that people interpret messages based on their opinions and attitudes rather than receiv-

Exhibit 11.a

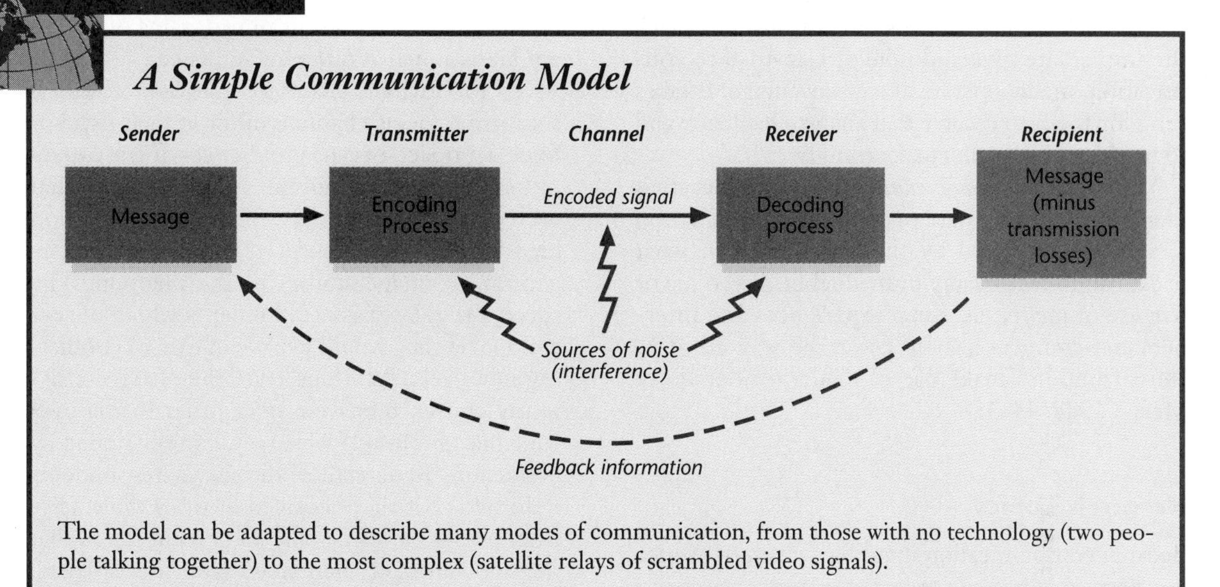

The model can be adapted to describe many modes of communication, from those with no technology (two people talking together) to the most complex (satellite relays of scrambled video signals).

ing them passively (as the old bullet theory had supposed).

Selective perception also accounts for a *boomerang effect*. Experiments have shown that those with strong beliefs tend to misinterpret messages with which they disagree—they distort evidence or retain only those elements that reinforce their existing attitudes, rather than allowing the message to change even slightly their present thinking. Thus propaganda can boomerang, producing exactly the opposite of the intended effect (Cooper and Jahoda, 1947). As a classic example, many television viewers misread the antibigotry message of the "Archie Bunker" character in *All in the Family* in the 1970s. They saw an endorsement of the very prejudices that producer Norman Lear intended to satirize.

This refusal to forsake an existing mindset suggests that people have a kind of internal gyroscope that tends to maintain a consistent set of attitudes, opinions, and perceptions. Psychologists developed several versions of a concept known generally as *congruence theory* to account for this tendency, which is an important intervening variable in determining communication effects. Such theories hold that a person's internal state of mind is usually in balance or congruence. A message that contradicts established opinions causes dissonance, lack of congruity, or imbalance. An effort to restore balance (conscious or unconscious) follows. It might take the form of rejecting the message (saying the source is "unreliable," for example), distorting the message to make it fit the existing mindset, or, least likely, adjusting the balance by accepting the new idea—conversion to a new point of view.

These concepts all came out of the social sciences. A very different way of looking at communication came from engineering with publication of *The Mathematical Theory of Communication* (Shannon and Weaver, 1949). Here, information is seen as a transmission system, using such familiar engineering terms as channel capacity, noise, encoding, and decoding (see Exhibit 11.a). *Information theory* contributed valuable insights to media research, including the concept of *feedback*

(Wiener, 1950)—whereby communicators modify their messages in response to information that comes back from audiences. But because they lack the immediate give and take of face-to-face conversation, media operate at a disadvantage. It takes time for them to receive and analyze feedback and to modify their product accordingly.

More recent *accommodation theory* argues that media are but one part of a person's information system as evidenced by the importance of what background factors any individual brings to his or her use of media. Personal experiences and interpersonal contacts heavily color the way all of us learn from and make use of media (Anderson & Meyer, 1988: 44–45).

Research Topics

Because communication involves a chain of events, it is best explained as a *process*. A communicator initiates a process that produces an end result—if indeed definable effects occur.

Study of the process can be directed to any one, or any combination, of five different elements: (1) *originators* of messages, (2) *contents* of messages, (3) *channels* through which messages travel, (4) *audiences* that receive messages, and (5) *effects* of messages. A pioneer communications researcher, Harold Lasswell, summarized these stages by saying that the objects of research could be identified by posing the question: "*Who* says *what* in which *channel* to *whom* with what *effect*?" (adapted from Smith et al., 1946:121).

- Researchers study the "who" of communication to learn about sources and shapers of media content—those who act as *gatekeepers*. FCC regulation and station or network "clearance" offices are types of gatekeeping, as is the selection, placement, and editing of all content presented. Gatekeeping studies examine how controls operate, where gates in information flow occur, and what effects they have on content when it reaches its destination.
- "What" studies usually consist of content analyses of electronic media programming and thus concern us only insofar as they indicate potential impact—as with measures of the amount and types of violence present in programs of different kinds aimed at different audiences.
- As to the "which channel" question, research shows that media channels differ in their psychological impact because audiences form expectations for each medium and interpret what each delivers accordingly. Since 1959 television broadcasters have sponsored biennial Roper Organization image studies of the medium. The surveys regularly ask respondents which of several media they would believe in case of conflicting news reports. Since 1961 they have consistently chosen television over other media by a wide margin. In a 1997 survey, commissioned by Newseum, 50 percent of the survey respondents indicated that they received most of their news from television, while only 24 percent stated that they got most of their news from newspapers. When asked whether they trust all or most of what news professionals say, 53 percent said they trusted local TV news anchors, 45 percent indicated they trusted network TV news anchors, and only 31 percent stated they trusted newspaper reporters (Newseum, 1997: http://www.newseum.org).
- Ratings are an example of the "to whom" factor of Lasswell's model inasmuch as they detail media use with *media exposure* or *time-spent* data. Breakdowns of audiences into demographic subcategories give further details about the "to whom" of broadcasting. But in researching the "to whom" question, scholars use more detailed personal and social-group indicators to study composition of audiences than does ratings research. In particular, the child audience has been extensively analyzed, using such variables as race, intelligence, social class, home environment, and personality type. These analyses relate audience characteristics to effects. Researchers ask such questions as: What types of children will be most likely to believe what they see on television—and what types will imitate what they see?
- Finally, Lasswell's "with what effect?" culminates his question because communicators, content,

channels, and audiences all help determine the ultimate outcomes of communicating. Joseph Klapper summarized effects theories as of 1960 in his influential *The Effects of Mass Communication.* Summarizing more than a thousand research reports, Klapper concluded that ordinarily communication "does not serve as a necessary and sufficient cause of audience effects, but rather functions among and through a nexus of mediating factors and influences." Media largely reinforce existing perceptions and beliefs. They may persuade people to buy a product, but not to change a political allegiance or religion. The broadcasting industry welcomed this conservative conclusion, known as the *law of minimal effects*, because it gave apparent scientific sanction to the industry's argument that programs or advertisements could not be blamed for causing antisocial behavior.

In the 1970s, opinion began to shift away from minimal effects concepts, largely because of intensive research on violence (discussed in Section 11.7). Researchers now even avoid talking about *effects* as such. The very word implies an oversimplification of what is now understood to be an extremely complex process. Without denying that specific media content might under specific conditions have specific effects on some specific people, researchers prefer to speak in terms of *associating* certain inputs with certain outputs. They avoid reliance on a simple, straight-line cause-effect relationship. Exhibit 11.b summarizes the stages of research development, from the simplistic cause-effect concept of early studies to our present interest in looking more deeply into the many variables involved in impact.

The People Problem

In designing studies on media impact, researchers face the frustrating problem that impact is usually made up of *subjective responses* hard to determine by direct observation and measuring instruments.

Most impact research today relies in whole or in part on questioning people about their subjective experiences—how they think and feel about something—rather than on observing their reactions. Self-reporting, of course, is not altogether reliable. People are sometimes unwilling—or unable—to tell the truth about their inner experiences, or they may be forgetful or unaware of their own subconscious motivations.

Methods of Research

Investigators usually select one of five major types of research:

- *Sample surveys:* The research strategy most familiar to the general public is the sample survey used in opinion polls and audience ratings reports. Such surveys can estimate characteristics of entire populations through use of very small random samples. Sample surveys tell us nothing, however, about the *causes* of tuning—a major weakness of the survey strategy.
- *Content analysis:* Classification of programs into various categories constitutes one form of content analysis. On a more sophisticated level, content analysis categorizes, enumerates, and interprets message content. Media researchers have used content analysis to study advertising copy, censors' actions, television specials, violent acts in programs, and the portrayal of minorities in television dramas.
- *Laboratory experiments* allow researchers to control some experimental factors precisely while excluding others. Both subjectively reported and objectively observed data can be derived from such experiments. Lab experiments do, however, put people in artificially simplified situations unlike the complex ways in which most of us actually use media.
- *Field studies* record behavior in "the real world" without unduly intruding or otherwise influencing participants. In studying the impact of violent programming, for example, field-study researchers can watch children in their normal home or school environment rather than in a lab setting. However, these experiments require a great deal of time and effort to arrange, and

Exhibit 11.b

Development of Media Effects Research

Time Period	Prevailing Viewpoint	Empirical Basis
1. (1920s–1930s)	Mass media have strong effects	Observation of apparent success of propaganda campaigns Experiments show immediate attitude change after exposure to messages Evidence of selective perception—persons ignore messages contrary to existing predispositions
2. (1940s–1950s)	Mass media largely reinforce existing predispositions, and thus outcomes are likely to be the same in their absence	Evidence of personal influence—persons are more influenced by others than the mass media Evidence of little influence on voting No relationship observed between exposure to mass-media violence and delinquent behavior among the young
3. (1950s–1960s)	Mass media have effects independent of other influences, which would not occur in the absence of the particular mass-media stimuli under scrutiny	Evidence that selective perception is only partially operative Evidence that media set the context and identify the persons, events, and issues toward which existing predispositions affect attitudes and behavior Evidence that television violence increases aggressiveness in the young
4. (1970s–1980s)	Process behind effects so far studied may be more general, suggesting new areas for research	Research finds that under some circumstances TV may influence attitudes and behavior *other than* aggressiveness Many agenda-setting studies
5. (1980s–1990s)	Individual differences in audience background and needs are found central in defining media impact	Qualitative studies (including ethnographies) of media as one of many social resources

Source: First four segments adapted from George Comstock et al., *Television and Human Behavior*. Copyright 1978. Used by permission of RAND Corporation.

some of their methods can be just as questionable as those used in laboratories.

- *Ethnographic studies:* Here the researcher becomes part of the (usually small) group being studied and this *participant observation* over extended time periods allows a realistic "window" on media consumption patterns to develop, along with some sense of subtle media effects. Ethnographic studies are expensive because of their intensive and time-consuming nature and are not readily generalizable to larger groups in society because of their narrow study-group base.

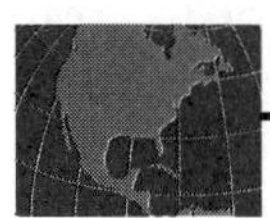

11.2 News

Because most Americans depend primarily on television for news, TV journalism clearly plays an important role. Most of us see the world beyond our own lives pretty much the way cable and television news present it to us.

Gatekeeping

Media can report only a tiny fraction of everything that happens in a day. On its way to becoming neatly packaged pieces on the TV screen, raw reporting of events passes through many editorial *gatekeepers.* News directors, news producers, photographers, and reporters open and close gates deliberately, deciding which events to cover in which places and how stories should be written, edited, and positioned in the news presentation. Some gatekeeping occurs inadvertently, depending on accessibility of news events or availability of transportation or relay facilities.

Some gatekeeping has an institutional or an ideological bias—or both. Institutional biases can develop from news organization priorities ("If it bleeds, it leads"); ideological biases grow out of individual political, social, economic, or religious beliefs ("We don't reveal the names of rape victims").

Television demands pictures. This tends to bias the medium toward covering events that can be visualized. Impacts of this visual bias include a preference for airing stories that have dramatic pictures (fires, accidents, disasters) and a forced effort to illustrate nonvisual stories with often irrelevant stock shots (as when file pictures of bidding on the floor of the stock exchange illustrate a story on financial trends).

Agenda Setting

Gatekeeping focuses our attention on selected events, persons, and issues that temporarily dominate the news. The list changes frequently as old items drop out and new ones claim attention.

No more vivid example of this process exists than the 1994–1995 coverage of the arrest and prosecution of football and media star O. J. Simpson for the murder of his ex-wife and her male friend—hours of live helicopter-based shots of Simpson's Ford Bronco on Los Angeles freeways and of the ensuing courtroom drama, all of which preempted regular programming from basketball playoffs to prime-time programs to newscasts to soap operas.

Researchers term this overall process of selection and ranking *agenda setting,* one of the primary ways in which media shape our perception of the world. In short, although media may not tell us *what* to think, they do tell us what to think *about.*

A related impact is *status conferral.* An event's very appearance on the air gives it importance. Well-known anchors and correspondents lend glamour and significance to the events and people they cover personally. If the story were not important, would Dan Rather, Tom Brokaw, Peter Jennings, or Bernard Shaw be covering it? Conversely, can an event really matter if television chooses *not* to be there?

Covering Controversy

Similar agenda setting is evident in coverage of the AIDS epidemic. Constant television coverage of the disease's spread after 1980, health agency efforts to counter that spread, case studies of individuals who are infected and become spokespersons to warn others (especially celebrities such as basketball star Magic Johnson)—all serve to focus more attention on AIDS and the search for its cure.

Concern about the environment in the past few years has been encouraged by television coverage of the bad news (burning tropical forests, disappearing animal species, air and water pollution) and the good (planting of new forests, fining polluting companies, enacting tougher laws).

Coverage of the abortion controversy features highly vocal groups and speakers on both sides of this emotional, political, religious, and personal issue. Here again, television news largely *reflects* divisions evident in society at large. The camera and microphone merely record events—though in so doing, they keep a controversy at the top of the current agenda.

Media News Staging

Television's need for images creates the temptation to enhance the pictorial content of news stories artificially. Even when news crews make no move to provoke reactions, the very presence of cameras in tense situations can escalate or sensationalize ongoing action.

As a practical matter, some artificiality is accepted as normal in news coverage. In televised interviews, for example, a camera usually focuses on the interviewee. "Reverse angle" shots of the interviewer are usually taken afterward and spliced into the interview to facilitate editing and to provide a visual give-and-take. This tactic allows a single camera to cover interviews.

Deliberate staging by reporters, however, is a far more serious violation of normal news practice. All too often we hear of a station or network carrying as news something set up by a news crew showing a drug bust, illegal betting, or the like, without any indication of the staging process. Early in 1993 NBC News got into trouble over what turned out to be a staged and partly faked story concerning General Motors truck safety. The head of the News Division was forced out and several others were fired. But the damage to NBC's credibility was still considerable.

More recent and troublesome is the ability to change photographs, film, or videotape digitally to show things that never happened. This kind of staging may merely modify a sky to make it more dramatic or may totally alter a real event to suggest something quite different.

Pseudoevents

Outright staging of events by the *subjects* of news occurs when press agents or public-relations counselors seek to plant stories in media or to create happenings to attract media coverage.

Daniel Boorstin (1964) coined the term *pseudoevent* to describe these contrived happenings. He analyzed the many forms they take, such as press conferences, trial balloons, photo opportunities, news leaks of confidential information, and background briefings "not for attribution." Free video news releases tempt stations short on photographic material. These brief items supplied by business and government PR offices are often visually interesting, with the real message subtly buried.

Court Trials

The former long-time ban on cameras and microphones in virtually all courtrooms arose from assumptions that broadcast coverage would affect behavior of participants (witnesses, lawyers, and defendants) and that these effects would be detrimental to the judicial process. Later, experience showed that once the novelty of being photographed wore off, subjects of coverage betrayed little reaction.

Three factors account for this minimal impact. Equipment became smaller and less obtrusive (able to operate with existing light levels, for example). News crews became more professional and sensitive to the need to avoid disruption. Finally, society became more tolerant, even expectant, of electronic media access to official activities. (For more on cameras in the courtroom, see Section 13.4).

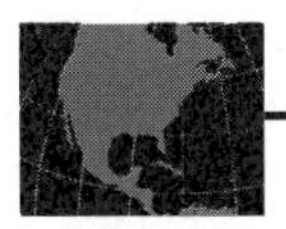

11.3 World Events

During World War II (1939–1945), broadcast radio's first war, network news and entertainment played a highly supportive role in building both civilian and military morale.

Vietnam

Television's first war coverage came with the Vietnam conflict (the Korean War of 1950–1953 occurred during the formative years of television news, when live coverage from such a remote distance was impossible). Television coverage for the 1965–1975 decade made Vietnam a "living room war," in the words of *New Yorker* critic Michael Arlen (1969).

In total, this longest war in U.S. history played in living rooms for 15 years. CBS sent its first combat news team to Vietnam in 1961, and news photography of the final evacuation of Saigon, showing desperate pro-American Vietnamese being beaten back as helicopters lifted off the landing pad atop the U.S. embassy, came in 1975.

At first, Vietnam coverage tended to be "sanitized," stressing U.S. efficiency and military might, and playing down the gore and suffering of actual combat. Although little direct military censorship was imposed, as had been the case in previous wars, in this undeclared war generals and presidents had public relations uppermost in their minds. Thus control of access to events could serve the same end as more blatant censorship.

A more violent phase of news coverage came as a result of the early 1968 Tet offensive, which brought fighting to the very doors of the Saigon hotels where correspondents stayed. Broadcast news from a host of network and local-station reporters turned increasingly to combat realities, and Vietnam became a real war in American living rooms, not the sanitized version of military public relations.

Following the Tet offensive by the North Vietnamese, several events combined to turn U.S. opinion against the war. Television's leading journalist, CBS's Walter Cronkite, gave a sharply negative assessment of the war after a visit to Vietnam. Loss of Cronkite's support for the war was said to have solidified President Lyndon Johnson's decision not to run for reelection. At about the same time, vivid photographic images from the battlefields became icons of American disillusionment.

More Wars

Television's effect on public perception of the Vietnam conflict raised troublesome questions about future war reporting. Could a nation at war afford to allow television freely to bring home the horror of combat night after night?

A 1983 terrorist bombing that killed some 240 Americans in a U.S. Marine barracks in Beirut seemed a case in point. Television coverage of the bombing's aftermath undermined public support for the "peacekeeping" role of the Marines in Lebanon, making the venture politically untenable. American troops pulled out shortly thereafter.

On the other hand, would the public support a war that was made temporarily invisible by rigid military censorship? To the dismay of journalists, a U.S. invasion of the tiny Caribbean island of Grenada in 1984 suggested that it might. Alleging security concerns, the military barred all media access to the initial assault. For the first 48 hours the world knew what happened only from military press releases. Subsequent press disclosures of military bumbling in the invasion and apparent official misrepresentation of circumstances leading to the attack came too late to dispel entirely the aura of success and righteousness surrounding the Grenada "rescue" of American civilians allegedly endangered by the volatile political situation on the island.

The public largely supported the action and applauded the administration's decision to limit media access—and impact. Strong media criticism of this abrupt departure from prior practice, however, led to agreement that in future actions the military would set up a news pool to cover events from the outset.

In the 1980s, Reagan administration efforts to bring down the Sandinista regime in Nicaragua and related guerrilla fighting in several Central American countries served as testing grounds for television journalism's post-Grenada maturity. Controversy surrounded coverage of these conflicts. New Right politicians claimed that television weakened efforts to gain public support for "freedom fighters"; others, remembering Vietnam, said that aggressive television coverage might serve to keep American soldiers out of another prolonged, undeclared conflict.

When President Bush sent troops to Panama to help remove President Manuel Noriega from power, the media were critical of the press pools set up by the military. According to many correspondents covering that action, media personnel once again were not allowed to accompany the combat troops and were admitted to battle areas only after most of the fighting had ceased.

Persian Gulf Crisis

Lessons learned in all these events heavily influenced television's coverage of and impact on the 1990–1991 Gulf War, when the United States led a successful UN effort to throw Iraqi troops out of occupied Kuwait (see Exhibit 11.c). The Pentagon invoked strong censorship and pooling of reporters, making it almost impossible for media to report from troop sites during the months-long build-up unless a military public-affairs officer was present (SGAC, 1991).

An intense air war began in mid-January 1991 and was seen live on American East Coast television screens in prime time as CNN cameras in Baghdad focused on falling bombs and anti-aircraft fire. Reporters Peter Arnett and Bernard Shaw continued live *voice*casts from Baghdad by means of a high-tech satellite telephone.

When reporters questioned Pentagon officials the next day as to the impact of initial bombing runs, officers made clear that much of what they knew had come from the same CNN reports seen by everyone else. That pattern was evident in the first few days of the air war, when (as shown by research) 44 percent of those turning to television watched CNN coverage while 41 percent turned to one of the three traditional broadcast networks (Birch Scarborough, 1991: 9). Later analyses often attributed CNN's "arrival" as an accepted news source to its stellar role in reporting all aspects of the Gulf conflict.

Even more than had been the case with Vietnam, Operation "Desert Storm"—thanks to time-zone differences and satellite relays, as well as around-the-clock reporting—was a "living room war" in American prime time. But it was a far more *controlled* television war than Vietnam had been, the result of tight Pentagon limits on media access to troops and events—and of censorship, often directed more at the possible political impact of negative coverage than at any military security need.

Only months after the war was over did disquieting television reports about Iraqi civilian losses, allied soldiers killed by "friendly fire," and attempted cover-ups of other military mistakes become widely reported—but by then few were paying attention. Pentagon and administration officials called the Gulf War press pools a model for future conflict coverage, while reporters argued the opposite—that censorship had prevented their usual multifaceted job of reporting.

Most public-opinion surveys found that viewers were generally satisfied with what they had seen and believed military controls to have been justified. The controversy arose again when television images of starving women and children pressured the United States to send troops to Somalia to help stabilize that country. Military personnel attempting to disembark on the beaches of Somalia under cover of darkness were met by the bright lights of the television cameras, recording their every action. Public opinion turned against U.S. participation after several of the troops were killed and their bodies were shown being abused by their killers. It seemed likely the contest of wills over the "proper" role of news media in wartime would continue into the *next* war.

Terrorism

News media problems took a vicious turn when terrorist organizations began committing crimes to

Exhibit 11.c

A Tale of Two Television Wars

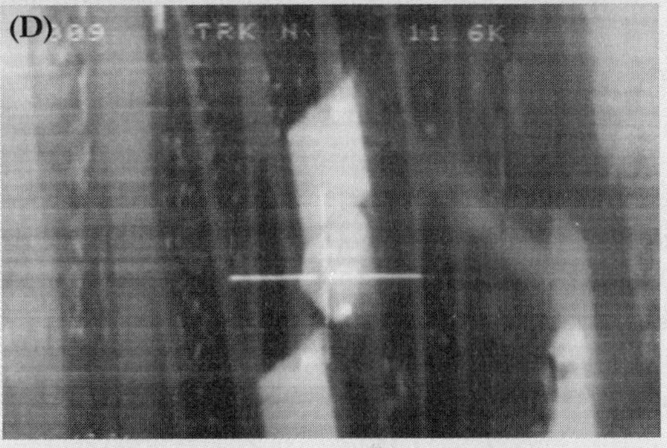

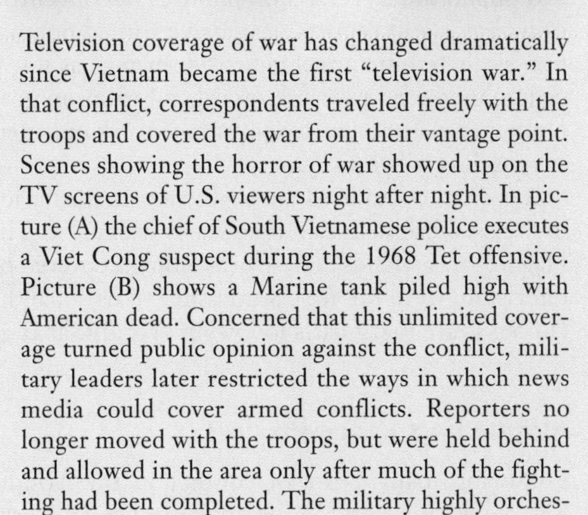

Television coverage of war has changed dramatically since Vietnam became the first "television war." In that conflict, correspondents traveled freely with the troops and covered the war from their vantage point. Scenes showing the horror of war showed up on the TV screens of U.S. viewers night after night. In picture (A) the chief of South Vietnamese police executes a Viet Cong suspect during the 1968 Tet offensive. Picture (B) shows a Marine tank piled high with American dead. Concerned that this unlimited coverage turned public opinion against the conflict, military leaders later restricted the ways in which news media could cover armed conflicts. Reporters no longer moved with the troops, but were held behind and allowed in the area only after much of the fighting had been completed. The military highly orchestrated the ways in which television viewers experienced the Persian Gulf conflict. News media were limited to traveling in groups escorted by military personnel, and the military itself controlled much of the news available to the media. Here (C) General Colin Powell explains the campaign during a news conference. Picture (D) provides an infrared night view from a Stealth F-117 bomber of a "smart" bomb homing in on its target. Such pictures were regular features of the Persian Gulf war coverage. Very little of the death and destruction caused by the conflict was ever seen by the TV audience. Critics of the coverage suggested it sanitized the war; military leaders and other supporters of the coverage opined that the coverage maintained important public support for the operations.

Sources: (A) AP/Wide World; (B) UPI/Corbis-Bettmann; (C) UPI/Corbis-Bettmann; (D) Reuters/Corbis-Bettmann.

gain news coverage. In the 1980s small and desperate political or religious groups seeking world attention increasingly employed violence against usually innocent third parties. Such *publicity crimes* paradoxically transform pseudoevents into real events.

Bombings of airport terminals and subway stations, hijackings, and kidnappings, the most common kinds of terrorist stories, pose difficult ethical dilemmas for all news media. The very act of reporting a publicity crime transforms media into accomplices, and the eagerness with which the public awaits news about it makes all of us inadvertent accomplices as well.

Observers complain that television's massive coverage encourages future terrorists by providing them with an international forum. Criticism focuses especially on coverage of news conferences staged and controlled by terrorists, whereby media appear captive to terrorist manipulation. Cable and broadcast news people often counter that they are merely accommodating the public's insatiable demand for more coverage. And surely these criticisms do not apply to terrorist acts such as the Oklahoma City or Olympic Park bombings, where no terrorist group came forward to claim "credit" for the actions.

Revolution in Eastern Europe

Television covered the dramatic changes in Eastern Europe and the USSR in and after 1989. Indeed, television's impact seemed to speed collapse of the Soviet system.

The highlight—really an icon for a revolution—was the televised scenes, many at night under harsh lights, of Berliners tearing down the hated wall that had divided their city for nearly three decades. The hard-line East German regime collapsed in days. In the following weeks, similar upheavals were reported in Poland, Hungary, Czechoslovakia, Bulgaria, and Romania.

In mid-1991 television highlighted the tension of what turned out to be a short-lived but frightening coup attempt in Moscow, when hard-liners tried to revive the former rigid communist regime that had prevailed before Mikhail Gorbachev's more open government. For two days television showed tanks moving in the streets of Moscow and highlighted the brave front put up by Russian leader Boris Yeltsin, leading those who challenged the coup plotters. The weakened Gorbachev soon stepped down, as reaction to the coup hastened the end of the nearly 75-year-old USSR. The world watched in amazement as television showed the red hammer-and-sickle flag being lowered over the Kremlin on New Year's Eve 1991, to be replaced with the Russian flag—a scene previously imaginable only in spy novels.

But just as lingering and even more haunting in the months that followed was television's coverage of rising ethnic unrest in former communist-dominated countries. As had often happened in centuries past, Balkan political divisions broke into bitter civil war in what was once Yugoslavia. Viewers were treated in mid-1992 to oddly juxtaposed pictures of a war-torn Sarajevo and its devastated population (the city had been the site of the 1986 Winter Olympic games), and prosperous Barcelona hosting the 1992 games.

Night after night, television news showed the destruction of once-handsome cities and their hapless populations. Television pictures of concentration camps and reports of massive forced population shifts (ethnic or religious "cleansing" as it was called) increased pressure on other European governments to intervene to stop what struck outsiders as senseless tribal warfare.

Atrocities committed during the war in Bosnia led to the establishment of a war crimes tribunal in The Hague, Netherlands. These proceedings, covered by television, were the first war crimes trials since the famous Nurenburg trials following World War II.

Middle East Peace Process

Following many years of conflict in the Middle East, television covered the beginning of a new era in 1993 as Israel signed a peace accord with the Palestine Liberation Organization, followed shortly

Exhibit 11.d

Television Covers a Troubled Peace Process

Although most of the peace process went on out of sight of the public, Palestine Liberation Organization President Yassir Arafat shook hands with Israeli Prime Minister Yitzhak Rabin after signing a peace accord in Washington, D.C., in front of a delighted President Bill Clinton—and a global television audience. Television coverage then followed the peace process as Palestinian police took local control of areas once patrolled by the Israeli soldiers. Within a year television also recorded the assassination of Prime Minister Rabin by an Israeli student upset at his peace initiatives.

Source: Reuters/Corbis-Bettmann.

by a peace treaty in 1994 between Israel and Jordan. (See Exhibit 11.d.) Thereafter, newscasts covered the peace process as Palestinian populated areas in Gaza and the West Bank town of Jericho prepared for Palestinian self-rule, which began in 1994. Television cameras also captured the 1995 assassination of Israeli Prime Minister Yitzhak Rabin following a peace rally in Tel Aviv.

Media attention then turned to the Israeli elections between Rabin's successor Shimon Peres, a major architect and supporter of the peace process, and opposition leader Benjamin Netanyahu, a frequent critic of many aspects of the peace accords. Since the election of Netanyahu in 1996, and his government's decision to continue building new settlements in the West Bank, violence between the

Israelis and Palestinians has increased. Once again, as before the peace accords, nightly newscasts frequently carry scenes of demonstrations, terrorist attacks, and families on both sides of the conflict grieving the deaths of the victims.

11.4 Politics

Early on, politicians recognized that radio could have significant impact on elections. So, in writing laws controlling broadcasting in 1927 and again in 1934, they saw to it that they would have access to radio (and, later, to television) when running for election. But beyond this, electronic media have acted as facilitators and reporters of democratic processes.

Election Campaigns

Radio exerted an impact on political campaigning almost from the start. Radio speeches by Calvin Coolidge, whose low-key delivery suited the microphone, may have been a factor in his 1924 reelection. Radio became especially important to Democratic candidates because it gave them a chance to appeal directly to voters, bypassing Republican-controlled newspapers. Franklin Roosevelt used radio masterfully in his four presidential campaigns, starting in 1932.

Television brought hard-sell advertising techniques to the presidential campaign of 1952, when a specialist in such commercials, Rosser Reeves, designed television spots for Dwight Eisenhower. By 1968, when Richard Nixon won the presidential election, television had become—and remains—the most important factor in national political campaigning.

After the 1960s parties redesigned their conventions to become more effective television presentations. Convention managers began to time events to the second, with set minutes allowed for "spontaneous" demonstrations. By 1988 networks had begun to reconsider just how much news value remained, given conventions' declining appeal to viewers. The 1992 conventions were the first the major networks did not fully cover (C-SPAN, and to a lesser extent CNN, did), limiting themselves to an hour or two each evening.

Research and experience make clear that candidates using more television do not always prevail. One reason may be that television visualizes campaigns in two very different ways—with paid partisan advertising that candidates control, but also with objective news and public-affairs coverage. Bona fide news programs about candidates have a credibility that 30-second advertising spots and candidate-controlled appearances can never obtain. Politicians understand that news coverage provides more credibility than political ads. Thus, candidates actively try to get as much favorable news coverage as possible during their campaigns.

Televised Candidate Debates

Starting in 1960, televised debates between presidential candidates began to steal the spotlight from conventions and other candidate appearances. The first debates, four 1960 confrontations between candidates John F. Kennedy and Richard M. Nixon, were made possible by a special act of Congress exempting the televised event from the "equal opportunities" law (see Section 13.7). Often called "Great Debates," these carefully choreographed face-offs probably decided that close race. Exhaustive research suggests that Kennedy seemed more precise and visually crisp to television viewers, while Nixon sounded more persuasive to radio listeners (Rubin, 1967).

The two 1988 televised "debates" (a panel of journalists questioned both candidates who rarely addressed each other directly) between President George Bush and Democratic nominee Michael Dukakis seemed to have little overall impact on that election's outcome. Bush came across as "lean and mean," whereas Dukakis seemed wooden and cold.

In 1992 Bush faced Democrat Bill Clinton, who was seasoned after months of hard primary campaigns studded with televised debates among Democratic candidates. A third candidate—inde-

Exhibit 11.e

Talking Politics

The 1992 presidential election campaign saw widespread use of a new campaign device—the topical talk show. In a pioneering example, independent candidate Ross Perot appeared with CNN's Larry King to announce his original candidacy, to explain his withdrawal six months later, and again to discuss his reentry in the campaign. After being denied the right to participate in the 1996 presidential debates, Perot once again used talk shows and long-form political advertising to take his message directly to the electorate.

Source: Corbis-Bettmann.

pendent Ross Perot—took part in the debates, adding often witty commentary on and sharp criticism of both major parties.

In 1996, the independent commission overseeing the presidential debates refused to allow third-party candidate Perot to participate, leaving Democratic candidate Bill Clinton and Republican candidate Bob Dole as the only debaters. The commission, made up of Republicans and Democrats, ostensibly excluded Perot because he did not garner enough support in the 1992 election to warrant inclusion. For a discussion of the legality of excluding candidates from television debates, see Section 13.7.

Talk Show Syndrome

During a February 1992 appearance on CNN's Larry King call-in show, Texas billionaire Ross Perot began his grassroots independent candidacy by announcing his willingness to be drafted if volunteers placed his name on all 50 state ballots. Later, in July, Perot returned to King's show to explain why he had abruptly pulled out of the race before even formally declaring his candidacy (see Exhibit 11.e). Then, in a yo-yo campaign, he came back yet again to explain his *re*entry just a month before election day.

Every candidate appeared on local and national call-in talk shows during the primaries, giving viewers an unprecedented chance to pose questions. Bush, Clinton, Perot, Quayle, and Gore all appeared on network morning programs as well as other talk shows as the campaign reached its height in October. Democratic candidate Bill Clinton even played the saxophone on the late-night *Arsenio Hall Show*, and George Bush tried to appeal to young voters by appearing on MTV.

This use of talk shows by candidates continued in the 1996 election, though not to the extent seen in 1992. The 1992 election may very well have been the highwater mark for the political use of broadcast talk shows.

Election-Night Predictions

Early television reports of voting trends in eastern states have been blamed for lower voter turnout in the West, where polls remain open. Members of Congress have been concerned that early predictions of winners and losers—or candidate concessions—in national races adversely impact western state and local elections.

The networks agreed in 1985 not to air predictions of any state's vote until after polls in that state had closed. They largely adhered to that promise in reporting results of the national elections of and after 1986.

In 1996, most network news organizations announced that President Clinton had been reelected hours before many polls had closed on the West Coast. This did not technically violate the pledge they had made earlier, because their projections were based only on results in states whose polls had already closed. The debate continues over whether such announcements affect races in states where the polls remain open.

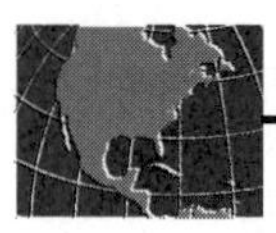

11.5 *Government*

Aside from coverage of seemingly endless election campaigns, television has dramatically affected the governing of America.

Presidential Television

A newly elected president enjoys a huge advantage over political opposition in obtaining media coverage. The chief executive has endless opportunities to manufacture pseudoevents to support his policies or to divert attention from his failures. No matter how blatant this exploitation may seem, editors rarely dare ignore presidential events. Virtually everything the head of state says or does has news value.

For greatest impact, the White House can call on broadcast and cable news networks to provide simultaneous national coverage of a presidential address. National addresses can give the president a gigantic television audience—although independent television and cable channels increasingly siphon off potential viewers with entertainment programs.

No law requires broadcast or cable networks to defer to presidential requests for time. Indeed, not until President Lyndon Johnson's administration (1963–1969) did these become customary. Perhaps because of the Nixon administration's Watergate troubles, networks began to evaluate such requests more critically after 1974. Presidents since Nixon have all been turned down at least once when they asked for time.

Congressional Television

The Senate began allowing television coverage of its committee hearings (subject to committee chair approval) as early as the 1950s, leading to some notable television public-affairs coverage that had repercussions on the careers of participants. The first televised House committee deliberations came in 1974 when the Judiciary Committee debated the Nixon impeachment resolution.

For decades, however, neither congressional body allowed broadcast access to its legislative sessions. Finally, the House agreed to live television as well as radio coverage in 1979. Even then, it refused to let outsiders run the show, establishing its own closed-circuit television system run by House employees. Broadcasters could carry the signal, live or recorded, at will.

Unenthusiastic about coverage they do not themselves control, few broadcasters used even excerpts of House debates. Cable subscribers could see gavel-to-gavel coverage on C-SPAN, the cable industry's noncommercial public-affairs network.

Not until 1986 did the Senate approve television coverage of all floor sessions. C-SPAN added a second channel to its service to cover the Senate when in session, covering other events at other times with C-SPAN I (Exhibit 11.f). Congress had at last given itself a degree of video parity with the White House.

Both Congress and the White House have developed sophisticated facilities for exploiting television. The House and Senate maintain fully equipped studios for the personal use of members, and the White House can make news feeds and interviews available on short notice from its own studio facilities, as well as handle call-in interviews from local stations.

Court Television

While presidents and members of Congress often use television for political purposes, televising court

Exhibit 11.f

Congressional Television

In 1951 his role in televised hearings on organized crime catapulted Senator Estes Kefauver (D-TN) onto the Democratic ticket in the following year's presidential campaign. The televised Army-McCarthy hearings in 1954 led to the censure of Wisconsin's senator Joseph McCarthy by his colleagues and ultimately to the end of McCarthyism, the far-right communist witch-hunt that he had led. The 1973 Watergate hearings chaired by folksy but astute Senator Sam Erwin (D-NC) made him a star, while the 1974 Nixon impeachment proceedings in the House propelled many members of Congress to fame—and hastened regular television coverage of congressional sessions.

The C-SPAN networks came into being originally to cover the House of Representatives (beginning in 1979) and the Senate (by C-SPAN-II, beginning in 1986). Cameras, controlled by Congress, rarely show that floor debate often takes place in nearly deserted chambers.

In 1991 the nation watched transfixed as the Senate reopened nomination hearings on Judge Clarence Thomas, a prospective member of the U.S. Supreme Court. Thomas was accused of sexual harassment by Anita Hill (pictured at right), an articulate law professor who had once been on his staff. Although the Senate, by a close vote, approved Thomas for the post, the hearings raised consciousness about sexual harassment; television had indeed reset the agenda of millions of Americans. Although many Congressional hearings have been televised since the Thomas hearings—most notably a series of hearings on the seemingly never-ending "Whitewater" scandal, nothing recently seems to have piqued the interest of the American viewer to the extent of the famous hearings noted above.

Source: Photo from UPI/Bettmann Archive.

proceedings primarily serves the function of informing the public about the administration of justice. Nonetheless, the widespread fascination with the O. J. Simpson murder trial and the success of the cable network *Court TV* demonstrate that courtroom proceedings can make compelling TV viewing.

Each state in the United States decides the extent to which trials and other court proceedings can be covered by television. Most states now allow some television coverage, although the extent of coverage varies widely. The United States Judicial Conference, led by the Chief Justice of the United States, has the power to allow television coverage of federal judicial proceedings. The Conference has permitted limited coverage of some civil cases on a trial basis, but opposition by the Chief Justice and other members of the federal judiciary has kept television cameras out of federal courtrooms. Thus, the third branch of our federal government is closed to the electronic media.

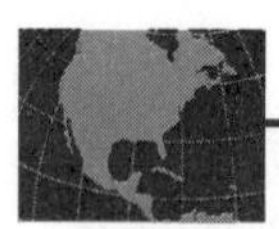

11.6 Entertainment

Any major news story that captures headlines may become grist for comment in late-night talk shows or the subject of a special drama or miniseries; it may also influence future episodes of established series. When President Bush suggested in his acceptance speech at the 1992 GOP convention that the country needed families more like television's old *The Waltons* and less like *The Simpsons*, the Fox program included the comment in an animated episode that appeared just 48 hours later. Similarly, when Vice President Dan Quayle criticized the television series *Murphy Brown* for having the lead character, an unmarried woman, become pregnant and decide to be a single parent, the writers of the show responded to Quayle's comments via the characters in the show.

Docudramas, those controversial blends of fact and fiction, such as the 1996 production on the ill-fated voyage of the *Titanic*, change facts around to suit dramatic needs, thus adding more distortions to already-simplified versions of reality presented to broadcast and cable audiences.

Stereotypes

Fiction influences audience perceptions by reinforcing stereotypes, versions of reality that are deliberately oversimplified to fit preconceived images, such as the stock characters of popular drama: the Italian gangster, the inscrutable Oriental, the mad scientist, the bespectacled librarian, or the befuddled father.

Authors resort to stereotypes when writing for television to save time, both their own and that of the medium:

> Television dramas have little time to develop situations or characters, necessitating the use of widely accepted notions of good and evil. Since the emphasis is on resolving the conflict or the problem at hand, there is little time to project the complexities of a character's thoughts or feelings or for dialogues which explore human relationships. To move the action along rapidly, the characters must be portrayed in ways which quickly identify them. Thus the character's physical appearance, environment, and behavior conform to widely accepted notions of the types of people they represent. (U.S. Commission on Civil Rights, 1977: 27)

Stereotypical images on television clearly help perpetuate those same images in the minds of viewers. As the U.S. Commission on Civil Rights put it, "To the extent that viewers' beliefs, attitudes, and behavior are affected by what they see on television, relations between the races and the sexes may be affected by television's limited and often stereotyped portrayals."

World of Fiction

As researchers examine roles in television series, they invariably find fictional characters to be markedly different from real people in the real world. Compared with life, for example, the world of fiction has far more men than women, most of them young adults, with few very young or elderly persons. Many have no visible means of support, but those who do work have interesting, exciting, action-filled jobs.

Fiction therefore contains an unrealistically high proportion of detectives, criminals, doctors, lawyers, scientists, business executives, and adventurers compared with the real world, where unglamorous, dull, and repetitive jobs dominate. Most people in the real world solve their personal problems undramatically, even anticlimactically or incompletely, using socially approved methods. Fictional characters tend to solve their problems with decisive, highly visible acts, often entailing violence.

Of course, none of this should surprise us. Fact may be stranger than fiction, but fact does not occur in neatly packaged half-hour episodes, with regular commercial breaks.

Children's Socialization

Nevertheless, the make-believe world of electronic media serves as a model of reality for countless people—especially children at the very time when they are eagerly reaching out to learn about the world. Those too young to read, those who never learned the habit of reading, and those who have little access to printed sources of information and entertainment—all depend heavily on electronic media to tell them of the world outside their own immediate surroundings.

Viewers and listeners identify with heroes, participating vicariously in their adventures. Research indicates that young children tend to believe what they see on television, making no distinction among fact, fiction, and advertising.

Electronic media have become major agents of socialization—the all-important process that turns a squalling infant into a functioning member of society. Socialization, though a lifelong process, occurs intensively during the first few years of life, the time when children begin to learn the language, detailed rules of behavior, and value system of their culture.

In the past, socialization was always the jealously guarded prerogative of family and religion, formalized by education and extended by peer-group experiences. Intrusion of a new, external agent of socialization—television—has been a profound change. Of course, all electronic media function as part of national culture, too, but they come from beyond the immediate family circle and its community-linked supports. The media import ideas, language, images, and practices that may be alien to the local culture.

The question of how the intrusion of broadcasting and cable has affected socialization has been and continues to be widely researched and debated. Electronic media can, of course, have both good (prosocial) and bad (antisocial) effects. Producers designed such programs as *Sesame Street* and *Fat Albert and the Cosby Kids* with prosocial effects in mind. Follow-up research indicates that such programs do in fact succeed in achieving prosocial results, though each (especially *Sesame Street*) reinforces a sense that the world comes in short segments with 30-second breaks.

Much more effort, however, has gone into research to prove the *anti*social consequences of television, especially the effects of violence. These effects are studied in Section 11.7.

Influence of Time Spent

"There is no more clearly documented way in which television has altered American life than in the expenditure of time" (Comstock, 1980: 29). At the very least, critics argue, time spent watching or listening could have been spent in some other way—perhaps on some more useful, constructive activity. Some critics take it for granted that *anything* active would be more beneficial than passive absorption. Somehow it seems "wrong" for people to waste their time staring like zombies at the television tube.

Most researchers readily accept that by the time a young person graduates from high school, he or she has watched an average of 22,000 hours of television—more time than any activity other than sleeping. Some futurists predict that, when the 500-plus channel superhighway fully opens for travel, people will exponentially increase the amount of time they spend in front of a video monitor. Others suggest that most will simply divide the available leisure time among the many new options.

In any case, no research has demonstrated that listening and watching necessarily displace more useful and active forms of recreation. In the absence of radio and television, people would do other things with their time, of course, but these would not necessarily be more beneficial. In fact, by the late 1990s, many parents were expressing concern over the amount of time their children were spending in front of another "tube," the personal computer.

Importance of Television

For some groups television is especially important. A team of nine psychologists, who spent five years assessing the medium, concluded:

> Children, the elderly, ethnic minorities and women are the heaviest users of television (the elderly being the highest) because television is a default option used when other activities are not available. These populations are limited by physical restrictions and/or lack of resources. The more time people spend at home or near a television set, the more likely they will watch television. (APA, 1992: 2)

For the elderly, on which more research attention has been focused in recent years as the overall population ages, news and information programs were most popular among those who used television as a "social network to replace the informal network previously supplied by the workplace" (APA, 1992: 3). Clearly, for some groups television is virtually indispensable—nothing else can perform the same function.

Such findings raise questions about "heavy viewers" or "television addicts." Studies suggest that anywhere from 2 percent to 12 percent of the population may watch up to twice the average weekly amount of television, are unhappy about that viewing, yet seem unable to do much about it. Further, they usually watch when they are sad, angry, or lonely (or just plain bored), are subject to "attentional inertia" (simply watching what's on rather than being selective), and become more passive as they watch more television.

"Glow-and-Flow" Principle

The play theory helps to explain the widely acknowledged notion that programs seem of secondary importance as long as *something* fills the screen. Early in television history, one observer suggested that "it is the television set and the watching experience that entertains. Viewers seem to be entertained by the glow and the flow" (Meyersohn, 1957: 347).

A landmark study of television audience attitudes found that most people surveyed said they felt more satisfied with television as a *medium* than they were with specific *programs.* The researcher noted that "a large number of respondents were ready to say television is both relaxing and a waste of time" (Steiner, 1963: 411). Similar studies of attitudes conducted a decade or two later showed that this mixture of reactions persisted (Bower, 1973, 1985).

The glow-and-flow principle becomes especially important when electronic media are the only companions people have. Upon asking respondents to describe satisfactions they derived from watching television, the researcher sometimes received moving testimonials such as this:

> I'm an old man and all alone, and the TV brings people and talk into my life. Maybe without TV, I would be ready to die; but this TV gives me life. It gives me what to look forward to—that tomorrow, if I live, I'll watch this and that program. (Steiner, 1963: 26)

At extreme levels of deprivation—in hospitals and similar institutions—television has a recognized therapeutic function as the most valuable nonchemical sedative available.

In short, electronic media answer a compelling need of the mass audience simply to kill time painlessly, to fill an otherwise unendurable void. The media give people a highly effective way of *performing leisure.* But social critics worry that while we relax with our guard down watching television, violence and other antisocial activities portrayed in many programs may produce antisocial effects.

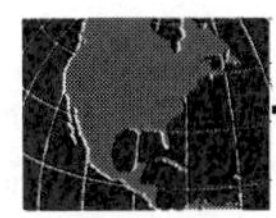

11.7 *Violence*

No aspect of electronic media impact raises more worry than does the amount of violence portrayed on the air and the presumed effect of that cumulative violence. But concern that portrayals of violence and crime might have antisocial effects preceded television by many years. The first systematic research on media violence dates to film studies in the 1930s (Jowett, 1976: 220). Accusations about the potential effects of violence shifted to radio and comic books in the 1950s and, later, to television and rock videos. Along the way, emphasis progressed from merely measuring violent content and deploring its presumed effects toward buttressing conclusions about effects with explanatory theories.

Direct Imitation

We occasionally see news reports of real-life violence apparently modeled on similar actions in films or television programs.

One particularly repellent example of apparent imitation led to an unprecedented lawsuit. In 1974 NBC broadcast a made-for-television film called *Born Innocent*, in which inmates of a detention home for young delinquents "raped" a young girl with a mop handle. Four days after the telecast, older children subjected a nine-year-old California girl to a similar ordeal. Parents of the child sued NBC, asking $11 million in damages for negligence in showing the rape scene, which they alleged had directly incited the attack on their daughter.

Network attorneys persuaded the trial judge to define the issue as a First Amendment question, rather than as one of negligence. The case then collapsed: It was impossible to show that NBC had *deliberately* incited the children to attack their victim and thus lose its First Amendment protection (Cal., 1981: 888).

Recent examples include reenactment of a scene from the movie *The Program*, in which people lie in the middle of the road as automobiles speed by. Such mimicry led to the deaths of several college students. Similarly, a young boy who had watched the characters in *Beavis and Butt-head* play with fire and say it was cool subsequently set his house on fire, killing his baby sister. It seems natural to assume that such imitation proves that televised violence sometimes causes violent behavior. Exhaustive research dem-onstrates, however, that viewing a violent act serves as a *contributing factor* to any subsequent imitation of that act. In other words, the act of viewing takes place within a larger context—one's background, education, predispositions, and the like. Taken together, all these influences *may*, in *some* circumstances, lead to violent behavior.

General Effects

Public concern about media violence arises primarily from fear of its general impact rather than from the risk of possible direct imitation.

This point of view emerged in another much-publicized court case in 1977. The state of Florida charged a 16-year-old boy, Ronnie Zamora, with murdering an elderly neighbor during an attempted robbery. The boy's attorney tried to build his defense on the argument that Zamora could not be held responsible for his violent behavior because he had become a television addict, "intoxicated" by the thousands of murders he had seen enacted on the screen. The trial judge rejected this argument, and the jury convicted the boy of murder. Though ill-considered, his lawyer's attempt to blame television for the crime drew its inspiration from the findings of research on the generalized adverse effects of televised violence that had accumulated during the 1970s (see Exhibit 11.g).

Exhibit 11.g

Government Studies of TV Violence

The Surgeon General's Scientific Advisory Committee on Television and Social Behavior sponsored a group of studies in 1969–1971. Congress allotted a million dollars for the project, which in 1972 resulted in five volumes of findings and a final report. When questioned by a Senate committee, the Surgeon General said flatly:

> [B]roadcasters should be put on notice. The overwhelming consensus and the unanimous Scientific Advisory Committee's report indicates that televised violence, indeed, does have an adverse effect on certain members of our society. . . . [I]t is clear to me that the causal relationship between televised violence and antisocial behavior is sufficient to warrant appropriate and immediate remedial action. (Senate CC, 1972: 26)

A comprehensive analysis of previous research, commissioned by the committee, indicated that television's linkage to aggressive behavior had been intensively analyzed. That every research method available had been employed in these studies made evidence of television's impact even more convincing. As researchers summarized:

> The evidence is that television may increase aggression by teaching viewers previously unfamiliar hostile acts, by generally encouraging in various ways the use of aggression, and by triggering aggressive behavior both imitative and different in kind from what has been viewed. Effects are never certain, because real-life aggression is strongly influenced by situational factors, and this strong role for situational factors means that the absence of an immediate effect does not rule out a delayed impact when the behavior in question may be more propitious. (Comstock et al., 1978:13)

As a result of this work, and of research done since—including a two-volume revisit a decade later by some of the same researchers involved in the 1972 report—Congress and private groups pressured the FCC to limit televised violence. Chiefly for fear of violating the First Amendment and because of the trend toward deregulation, the FCC declined to act.

Still, pressure continued. At the invitation of the Department of Health and Human Services, many researchers involved in earlier studies gathered in Washington late in 1992 to review two decades' worth of research and to assess the impact of changing technology. One study suggested that larger screens and improved picture definition made televised events more real—and thus potentially more dangerous. Others hinted that excessive television watching could impair brain development in young viewers because they would not face a sufficiently wide variety of experiences (Washington Post, 12 October 1992: D5).

Finally, in 1996, Congress passed legislation addressing the violence issue. The bill required the industry to develop a television rating code that would inform parents about the content of programs. Additionally, new television sets must be manufactured with a *V-chip* (V for violence) that will allow parents to program their televisions to exclude shows with a certain rating. For more on this aspect of television regulation see Exhibit 9.d and Section 13.9.

Violence vs. Reality

Three decades ago, during the 1967–1968 television season, George Gerbner and his associates at the University of Pennsylvania began conducting annual analyses of television violence. From these data they constructed a violence "profile" that counted every violent act in a sample week of prime-time and weekend-morning network entertainment programs. In this way they tracked changes in the level of violence from year to year according to network and program type. The

Gerbner data indicated, for example, that animated cartoons had a higher percentage of violent acts than any other program category (the coding sys-tem counted comic as well as serious acts of violence). Still being collected in the late 1990s, these data contribute to what has become a valuable long-term indicator of television content.

Some recent television violence research goes beyond counting violent acts and considers the *context* of the situation in which the violence occurs. The researchers doing this work believe that the context of a violent act may somehow modify the effect of the act on the audience.

Gerbner theorized that violence in programs creates anxiety in viewers because they tend to see the real world in terms of their television experience. Viewers identify with those victims of violence in fiction who resemble themselves. Gerbner found that elderly, poor, and black people have high "risk ratios," or expectations of becoming victims. This anxiety effect, he said, may be a more important by-product of television violence than the imitation effect.

Desensitization

The Gerbner risk-ratio hypothesis or *cultivation theory* seems more persuasive than older hypotheses predicting that people exposed to fictional violence would become *desensitized* to it. According to this view, when television dramatic violence becomes routine, people grow indifferent to real-life violence. Many instances of urban violence and nonresponse by passers-by in recent years lend credence to this hypothesis.

A related notion, the *sanitization hypothesis*, holds that self-imposed program codes cause television to depict violence unrealistically because graphic literalness would encounter viewer backlash. Even news reports, especially in recent years, have been sanitized to some extent (at least during dinner-hour newscasts) to conform with perceived public preferences. As a possible result, televised consequences of fictional violence seem so neat and clean that viewers remain indifferent. They never see or hear the revolting, bloody, painful aftermath of real-life violence.

Violence and Urban Unrest

Some of the most extreme televised violence appears not in entertainment but in news—especially during coverage of live events. When south-central Los Angeles erupted in urban violence in April 1992 in the wake of a controversial trial verdict, two examples of the violence problem emerged.

First, critics argued that the jurors in the case—who acquitted police officers of a vicious assault on a black man—had become desensitized to violent behavior despite the existence of a graphic videotape of the beating taken by a bystander. Second, television news editors again faced the balancing act of how much riot coverage to show without contributing to the very violence happening in front of the cameras. Television had earlier faced this problem in the late 1960s during repeated episodes of urban unrest nationwide brought about by racial and economic inequality, on the one hand, and by protests against the Vietnam War, on the other.

Television's technology improved between the 1960's riots and those of the 1990s. Cameras today are less intrusive and thus less likely to incite trouble by their mere presence. But the basic problem remains—how much to show while the rioting is actually taking place? Ironically, technological improvements have made this problem worse in that stations and networks can now put these dramatic pictures on the air almost instantly with little time to consider their potential audience impact.

In Defense of Violence

Given the steadily increasing range and depth of research evidence showing antisocial effects of television violence, the media business has found it increasingly difficult to defend its widespread use. Yet for television to serve as a serious medium for adult viewers, it cannot ban violence. After all, violence occurs in all forms of literature, even fairy tales for children. Bodies litter the stage when the curtain falls on some of Shakespeare's tragedies.

Writers would face a difficult challenge if they had to meet television's relentless appetite for drama without resorting to violent clashes between opposing forces. A playwright illustrated the problem by explaining that authors have only four basic conflicts around which to build any plot (Baldwin & Lewis, 1972):

- *man against nature:* "This is usually too expensive for television."
- *man against God:* "Too intellectual for television."
- *man against himself:* "Too psychological, and doesn't leave enough room for action."
- *man against man:* "This is what you usually end up with."

Only the last form of conflict can easily be shown on television; it also happens to be the one most likely to involve personal violence.

Under growing pressure from the public, interest groups, and Congress to address—and curb—the amount of violence aired, the broadcast and cable industries took limited action in 1992–1993. The Television Program Improvement Act of 1990 had granted cable and broadcast networks and programmers a temporary antitrust exemption to encourage development of common standards to limit the total amount and some specific types of violence on the air.

The broadcast networks agreed to a generalized common set of standards late in 1992, and a few months later major cable programmers agreed to try to do the same, joining the networks in developing a national research conference to air the issues.

In 1994, under a growing threat of congressional action if they did take more positive steps, the four broadcast television networks agreed, soon after a similar pact among cable networks, to designate UCLA's Center for Communications Policy as an independent monitor to evaluate programs for violent content and summarize findings in an annual public report. Broadcasters also agreed to supply a violence warning—including a *Parental Advisory* symbol (PA)—to newspapers and *TV Guide*.

Cable went further, agreeing to adopt a rating system to indicate the amount of program violence and to make available to viewers a "V-chip" circuit to block out programs containing excessive violence. Despite strong opposition by broadcasters, Congress in 1996 passed legislation requiring manufacturers to install V-chips in all new TV sets (see Section 13.9).

Several television stations, partly in reaction to viewer criticism and partly in an effort to improve ratings, began in the mid-1990s to downplay violence in at least some of their local newscasts. They offered what they variously called "family," "family-sensitive," or "family-safe" programs. These reduced reliance on graphic depictions of crime and violence and often included a specific "good news" segment. Some stations, however, limited this policy to early-evening newscasts—at 11:00 P.M. it was violence as usual.

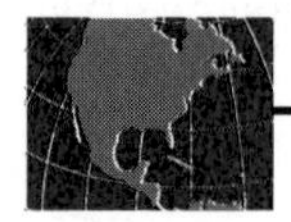

11.8 Advertising

Other than violence, no aspect of electronic media impact has been more measured than advertising. Advertisers demand, and receive, vast amounts of applied research on the reach and effect of their messages. Here, we focus on the broader, long-range social impact of advertising.

Advertising as Subsidy?

At first glance advertising appears to play a useful social role by reducing the direct cost of media to the public. In traditional commercial broadcasting, for example, advertising appears to cover the entire cost of this service, making it "free" to the public. In fact, consumers eventually pay the full cost, because companies include advertising expenses in the final price of their goods and services. Advertiser-supported basic cable combines two viewer charges, adding direct subscriber fees to indirect advertising costs. And for all electronic media, the audience must purchase and maintain receivers, which can be thought of as a kind of consumer subsidy of advertisers.

Advertising's General Impact

Some critics suggest that advertising actually *disserves* consumers by creating a desire for unnecessary purchases—what economist John Kenneth Galbraith termed the *synthesizing of wants.* Electronic media advertising does often stimulate widespread demand for goods and services for which consumers may have little real need. Advertising can build overnight markets for virtually useless products or "new and improved" versions of old products.

Critics often assume from these successes that advertising can overcome most consumer resistance. Advertising practitioners find themselves wishing that were true. Failure of a high proportion of new products every year belies the assumption that advertising is all-powerful.

Not only do many products fail to catch on but leading products also often give way to competitors. Marketers recognize transfer of *brand loyalty* as an ever-present threat. For this reason, much cable, television, and radio advertising aims not at moving merchandise, but at simply keeping brand names visible and viable in the marketplace.

Advertising to Children

The possible impact of commercials in children's programs raises special issues. Children start watching television early in their lives and often find commercials just as fascinating as programs. Action for Children's Television (active until 1992) as well as other consumer organizations have long believed that commercials take unfair advantage of young children—especially preschoolers, who are not yet able to differentiate between advertising and programs.

During the 1970s academic researchers, the industry, and government agencies conducted extensive research on the impact of children's television—both programs and advertising. Industry lobbying fended off effective action by either the Federal Trade Commission or the Federal Communications Commission. The FCC adopted rather ineffectual guidelines but refused to impose hard-and-fast rules. Deregulation in the 1980s further watered down the FCC's guidelines.

Finally, under pressure from Action for Children's Television and others, Congress in 1990 passed a law limiting advertising in children's programs—and calling for more programs that address the educational and informational needs of children (see Section 13.9).

FEDERAL COMMUNICATIONS COMMISSION
U S

Chapter 12

Regulation and Licensing

CHAPTER CONTENTS

The previous chapter surveyed the pervasive social effects of electronic mass media—some real, some probably only imagined. As long as people believe that effects actually occur, they want to have some control over them. They want to maximize beneficial effects (from wholesome entertainment and useful information, for example) while minimizing perceived harmful effects (from pornography and misleading propaganda, for example).

Direct social control comes primarily from government licensing, which determines who is allowed to operate stations, cable systems, and other program outlets. Licensing not only selects operators but also enables monitoring and regulating their performance. Regulation starts with the Constitution, on which all laws and administrative controls are based. These controls, in turn, are embedded in the Communications Act, which defines the role of the Federal Communications Commission.

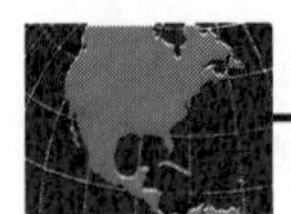

12.1 *Federal Jurisdiction*

Specific constitutional justification for Congress's taking control of electronic media comes from Article 1, Section 8(3), which gives Congress power "to regulate commerce with foreign nations, and among the several states." This *commerce clause* has played a vital role in American history, preventing states from erecting internal trade barriers to national unity. The provision forms a link in a chain of responsibility from Constitution to citizens, as shown in Exhibit 12.a

Interstate vs. Intrastate Commerce

The commerce clause gives Congress jurisdiction over *interstate* and *foreign* commerce but not over commerce within individual states. The law regards radio communications, including all kinds of broadcasting, as *interstate by definition*. Radio (including television) transmissions, even when low in power and designed to cover a limited area within a state, do not simply stop at state boundaries.

Wire and cable communication, however, differs from radio-based services in the level of government that regulates it, because it is usually limited by a specific geographical boundary. State regulation governs telephone and related *common-carrier* services that operate solely within a state (*intrastate*), whereas federal regulation governs those that cross state lines (*interstate*). Common carriers may be broadly defined as those businesses that deliver goods or services from point-to-point, that must make their facilities available to all who want to use them (and can pay for them), whose rates are governmentally regulated, but who have no control over content.

States have their own *public-utility commissions* (PUCs) that approve changes in telephone rates and service for in-state systems. Long-distance carriers that cross state lines, however, have traditionally needed federal approval of rate changes. Even in the 1990s, after much deregulation, such controls still affect AT&T, MCI, Sprint, and others.

Cable (discussed in detail in Sections 12.8 and 12.10) is a *hybrid* service. In some ways it acts as a common carrier; in others it does not. Once controlled primarily at the state and local level, cable is now subject to extensive federal regulation as well.

Delegated Congressional Authority

It is impossible for Congress to dictate regulatory details in dozens of specialized fields. Therefore, it has delegated to a series of *independent regulatory agencies* the authority to oversee such often complex areas as power, transportation, labor, finance—and communication.

Although the president appoints FCC commissioners, with the advice and consent of the Senate, the Commission remains what some call a "creature of Congress." Congress defined the FCC's role in the Communications Act of 1934, and only Congress can change that role by amending or replacing the Act. Though commissioners have considerable leeway in determining regulatory direction, House and Senate subcommittees on

Exhibit 12.a

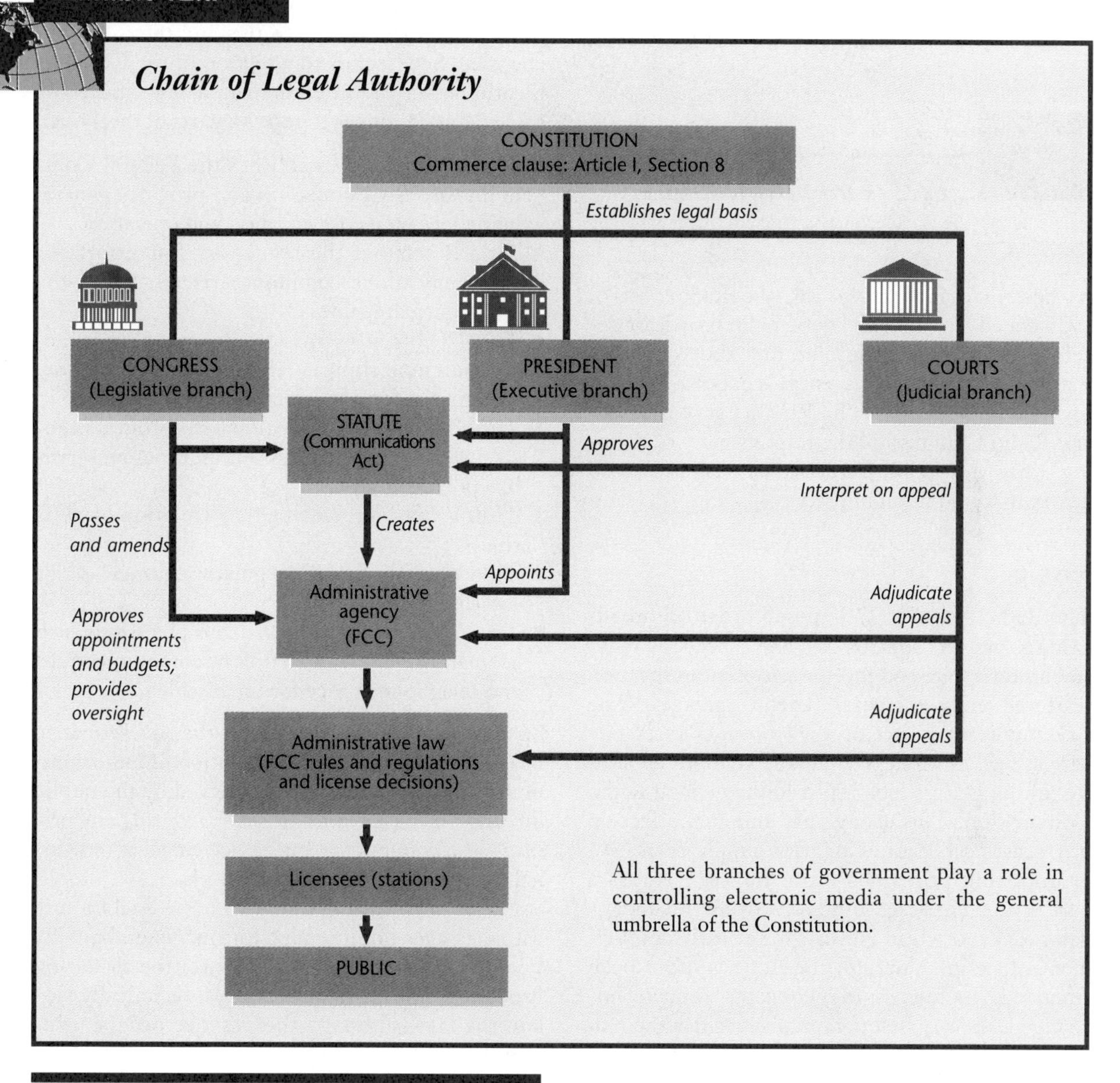

All three branches of government play a role in controlling electronic media under the general umbrella of the Constitution.

communications constantly monitor the FCC, which must come back to Congress annually for budget appropriations. Since 1983 Congress has reauthorized the very existence of the Commission every two years.

Congress gave the FCC power to adopt, modify, and repeal regulations concerning *interstate* electronic media, with the overarching goal of serving the public interest, convenience, and necessity. These regulations carry the force of federal law, deriving their power from Congress through the Communications Act. An understanding of the Act is vital to understanding the FCC's day-to-day role.

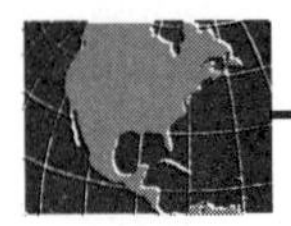

12.2 *Communications Act*

As described in Section 2.6, the Radio Act of 1927 ended a period of chaotic radio broadcast development. Congress for the first time crafted a statute—and created an agency—concerned explicitly with broadcasting. The 1927 act gave the Federal Radio Commission the responsibility of defining what the public interest, convenience, and necessity would mean in practice.

Passage

The Radio Act of 1927 imposed order on broadcasting, but left control of some aspects of radio and all interstate and foreign *wire* communication scattered among several federal agencies. The Communications Act of 1934 brought interstate wire as well as wireless communication under control of the FCC, which replaced the Federal Radio Commission. This change had minimal effect on broadcasting because Congress simply reenacted broadcasting provisions from the 1927 law as a part of the 1934 Act. Thus electronic media law dates back more than 70 years to the early development of radio. Although the Act has often been amended, its underlying concepts remain unchanged. Its very first paragraph specifies the reasons for the FCC's creation: "to make available, so far as possible, to all the people of the United States a rapid, efficient, nationwide and worldwide wire and radio communication service with adequate facilities at reasonable charges."

The Communications Act of 1934, as amended numerous times by Congress, provides the foundation for regulation of electronic media in the United States. The Act is separated into seven units, called *titles*, which focus broadly on different aspects of regulation. These titles are further subdivided into separate *sections*, which address regulations more specifically. (In this and the following chapter, "Sec." refers to a section of the Communications Act or other indicated law. "Section" spelled out continues to refer to parts of this book.)

- Title I of the Act establishes the Federal Communications Commission and provides general guidelines on its organization and operation.
- Title II sets out the regulatory framework for communications common carriers such as the telephone companies.
- Title III contains provisions relating to over-the-air communications, including radio, television, and various forms of microwave transmission.
- Title IV focuses on additional procedural regulations covering Commission inquiries and trial-like proceedings.
- Title V outlines penalties for violation of FCC rules.
- Title VI establishes the regulatory framework for cable television.
- Title VII contains miscellaneous provisions such as unauthorized publication of communications and telephone service for the disabled.

Specific sections in Title III of the Act formalize broadcasting's role as "dissemination of radio communications intended to be received by the public directly." This excludes from *broadcasting* any private radio communication service aimed at individuals or specific groups of individuals.

Nor is broadcasting supposed to be used for private messages not intended for the general public. People who greet their families over the air during broadcast interviews, for example, technically violate the law—although the practice occurs regularly and without consequences.

The Act forbids *disclosure* of nonbroadcast messages to people for whom they were not intended. Congress, the courts, and the FCC have also made it illegal for people even to intercept some nonbroadcast signals, such as pay cable, unless they are subscribers.

Finally, Sec. 3 of the Act defines *radio communication* as "transmission by radio of writing, signs, signals, pictures, and sounds of all kinds, including all instrumentalities, facilities, apparatus, and services . . . incidental to such transmission." By giving the term *radio* such a broad definition, Congress made it possible for the Act to incorporate television when it became a licensed service nearly 15 years after adoption of these words in 1927.

As communication technology changes, Congress often finds it necessary to change the law regulating that technology. In 1984, for example, Congress amended the Communications Act by adding a new Title VI and defined cable television as *neither* a common carrier *nor* a broadcasting service. Congress left the FCC with little responsibility for cable, far less than it had for broadcasting and interstate wire services, and limited the power of states to regulate cable programming and subscriber rates. But Congress amended the Act again in 1992, reinstating many cable regulatory powers to the FCC and local authorities.

Probably the most significant changes to the Communications Act came with the passage of the Telecommunications Act of 1996. This legislation changed many aspects of how the electronic media industries in the United States are regulated. The law allows companies once permitted to operate certain specified businesses to move into other areas not previously open to them. For example, telephone companies are now allowed to provide cable television services in their local service areas. Local telephone services are permitted to enter the long-distance arena, and, *vice versa*, long-distance companies may now offer local services. The 1996 Act also minimized many of the ownership restrictions that limited the number of radio and television stations any one entity could own. The specifics of these changes are addressed throughout the next two chapters.

Provision for FCC

The president appoints the five FCC commissioners to five-year terms, subject to Senate confirmation, and designates which of them shall serve as chair. Congress sought to minimize political bias by allowing no more than three commissioners from the same party.

Throughout much of the FCC's history, Senate confirmation of presidential appointees to the Commission was virtually automatic. This changed in the 1980s, when the Senate began using the confirmation process to chastise the Commission for actions it disapproved of. In some cases, the Senate even refused to schedule confirmation hearings for appointees, effectively blocking their appointments.

Sec. 4 of the Communications Act assigns the Commission broad power to "perform any and all acts, make such rules and regulations, and issue such orders . . . as may be necessary in the execution of its functions." Originally, in only a few instances did Congress tie the Commission's hands with hard-and-fast requirements, such as specific station license terms and restrictions concerning foreign ownership of broadcast stations.

Then, in the 1990s, as Congress involved itself more in the oversight of the electronic communications industries, legislation amending the Communications Act became increasingly specific. Provisions of both the 1992 and 1996 amendments to the Act direct the Commission to regulate in very specific ways, with little room for interpretation or discretion.

PICON Standard

Congress created a highly flexible yet legally recognized standard—*public interest, convenience, or necessity* (PICON)—to limit FCC discretion. PICON occurs regularly in the Act's broadcasting sections. For example, Sec. 303 begins: "Except as otherwise provided in this Act, the Commission from time to time, as *public convenience, interest, or necessity* requires, shall . . ." and goes on to list 19 powers, ranging from classification of radio stations to making regulations necessary to carry out the Act's provisions. The PICON phrase similarly occurs in sections dealing with the grant, renewal, and transfer of broadcast licenses.

In sharp contrast, important 1984 cable amendments to the Communications Act nowhere use the phrase "public interest." Because of its strongly deregulatory nature, the 1984 amendments relied far more on marketplace than on PICON decision making. Cable amendments in 1992 *did* make reference, in a new Sec. 628(a), to promoting "the public interest, convenience, and necessity by increasing competition and diversity in the multichannel video programming market," thus signalling a return to a more regulatory climate.

Similarly, the 1996 amendments to the Act contain several references to the public interest standard, particularly in relation to television broadcasters' use of a second video channel for the delivery of digital broadcast services such as high-definition television.

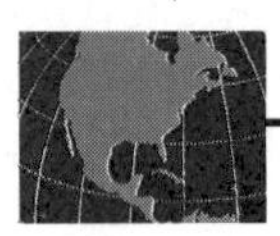

12.3 FCC Basics

While theoretically an independent regulatory agency, the FCC, as a creation of Congress, acts on behalf of the legislative branch. When it makes regulations, the Commission acts in a quasi-legislative capacity. It functions as an executive agency when it puts the will of Congress and its own regulations into effect. And when the FCC interprets the Act, conducts hearings, or decides disputes, it takes on a judicial role.

Budget and Organization

In the 1990s Congress appropriated between $200 and $250 million annually for the FCC. This budget makes the Commission one of the smaller federal agencies, employing about 2,000 persons.

Throughout most of its existence, the FCC's budget came out of general tax revenues. In 1989 Congress began requiring the FCC to collect part of its budget directly from the industries it regulated. In 1996, Congress expected the FCC to collect $126.4 million, roughly half of its budget allocation for that year, from regulated industries. This money comes from annual regulatory fees and application fees. Fees defray the costs of the Commission's enforcement, policy and rule-making, user information, and international activities.

Annual fees vary widely, depending on the service. For example, Class C AM stations pay $280, while VHF television stations in the top 10 markets pay $36,000. DBS operators are assessed a fee of $70,575 per satellite, and cable systems pay 55 cents per subscriber. Many of these fees were scheduled to rise significantly in 1997, because Congress has directed the FCC to collect $151 million for that year. Clearly these fees are passed on to consumers as higher rates for subscription services and higher costs of consumer goods stemming from increased advertising costs.

Exhibit 12.b depicts the FCC's organizational structure. The Mass Media Bureau, one of the units of most interest here, has four divisions:

- *Audio services.* Processes applications for construction permits, licenses, and license renewals for radio stations.
- *Video services.* Processes the same for television stations and MMDS operations.
- *Policy and rules.* Handles FCC proceedings that produce new rules and conducts studies needed for policy-making decisions.
- *Enforcement.* Processes complaints, ensures compliance with statutes and rules, issues interpretations of rules, and represents the Bureau at hearings within the Commission.

Because of greatly increased cable regulatory responsibilities, the FCC in 1994 established a Cable Services Bureau (an earlier cable bureau had operated from 1972 to 1981). Beginning with about 60 employees and scheduled to climb to about 240, it has four divisions:

- *Financial Analysis and Compliance.* Responsible for all cable television rate-related cases before the Bureau.
- *Consumer Protection and Competition.* Handles consumer and competition issues, such as controversies over leased access requirements and rules governing the availability of program networks to competing multichannel providers.
- *Engineering and Technical Services.* Provides support on technology policy, and oversees cable signal leakage regulations and other general operational requirements for cable systems.

Exhibit 12.b

FCC Organization

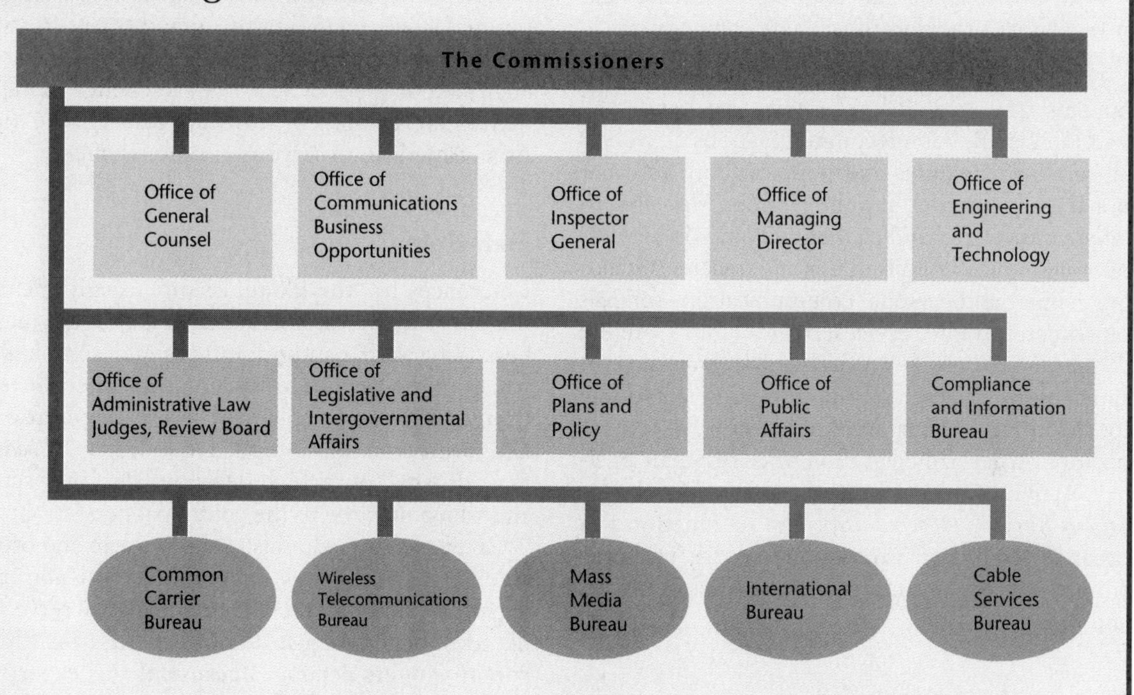

Bureaus shown as ovals authorize services; the other units provide support to commissioners and authorizing bureaus. The Compliance and Information Bureau has the task of informing the public about FCC regulations, policies, and practices plus obtaining compliance with FCC rules. Whereas most FCC employees are located at the national headquarters in Washington, D.C., Compliance and Information Bureau employees are scattered throughout the country in FCC regional offices. The Office of Inspector General conducts audits and investigations of internal programs and operations of the agency, while the managing director handles personnel, budget, and administrative matters.

- *Policy and Rules.* Handles all rule-making activities.

The Common Carrier Bureau remains the unit most closely involved in matters associated with telephony. The Bureau has six operating divisions:

- *Accounting and Audits.* Oversees universal service funding and cost-accounting rules.
- *Network Services.* Oversees telephone numbering issues, network reliability, and telephone hearing aid requirements.
- *Enforcement.* Handles consumer protection issues and complaints as well as mergers and acquisitions.
- *Industry Analysis.* Collects and distributes data on the telecommunications industries.

- *Competitive Pricing Division*. Develops pricing policies for interstate services and administers telecommunications tariffs (agreements concerning service offerings and rates).
- *Policy and Program Planning*. Develops rules and regulatory programs to facilitate competition.

In 1994 the Commission added two additional bureaus: the *International Bureau*, to handle all FCC international telecommunications as well as all satellite programs and policies (both international and domestic), and the *Wireless Telecommunications Bureau*, to handle all FCC domestic wireless telecommunications programs and policies (including cellular and personal communications services plus amateur radio services, but excluding satellite communications). The Wireless Bureau is also responsible for implementing competitive bidding for spectrum auctions. Also under the 1994 reorganization, what had been the Office of Small Business Activities was renamed the *Office of Communications Business Opportunities*, and its function was expanded to include promotion of participation in communication by minorities, small businesses, and women.

Commissioners

Most of the work of the FCC is done by the staff. Only the most important decisions are made by the Commissioners themselves, although they establish overall policy that must be followed by Commission employees.

The five commissioners (seven until 1982) serve five-year terms and may be reappointed. Commissioners must be citizens, may not have a financial interest in any type of communications business, and must devote themselves full-time to the job. In practice, many commissioners never finish their terms of office, because other opportunities entice them into higher paying positions outside the government. There are exceptions, however. James Quello, a Democrat from Michigan, was first appointed to the Commission in 1974 by President Nixon and served until 1997.

The first woman commissioner, Frieda Hennock, served from 1948 to 1955. The next woman was not appointed until 1971, but since 1979 at least one commissioner has been a woman. The first African-American commissioner was Benjamin Hooks (1972–1977), later the head of the NAACP. Hispanics as well as blacks have been appointed since, including the appointment in 1997 of William Kennard as the first African-American chairman of the FCC. In 1994 Rachelle Chong, a San Francisco–based attorney, became the first Asian American to serve as a commissioner.

Staff Role

References to "the Commission" usually include not only the five commissioners but also staff members. The staff handle inquiries and complaints, which seldom come to the commissioners' attention, as well as thousands of letters, applications, and forms from FCC-regulated industries. *Processing rules* spell out which decisions staff may settle and which must go to the commissioners.

Except for top administrators (bureau and office chiefs), whom the chair appoints, Commission staff is part of the federal civil service. Many serve for decades, developing in-depth expertise on which commissioners depend. Because of this expertise, the professional staff often exert considerable influence not only on the day-to-day operations of the Commission but often on long-term policy as well.

Rule-Making Process

The rule-making function generates a large body of FCC administrative law called *rules and regulations* (there is no difference between the two—the phrase is traditional). Whether the Commission acts on a petition to consider a certain action or acts on its own, it often begins with a *notice of inquiry* (NOI) for a subject that needs preliminary comment and research, or a *notice of proposed rule making* (NPRM) when offering specific new rules for public comment, or a combination of both (see Exhibit 12.c). These notices invite comment from interested parties, mostly attorneys representing affected individuals, stations, companies, or industries. On rare occasions, proposed rule changes of

Exhibit 12.c

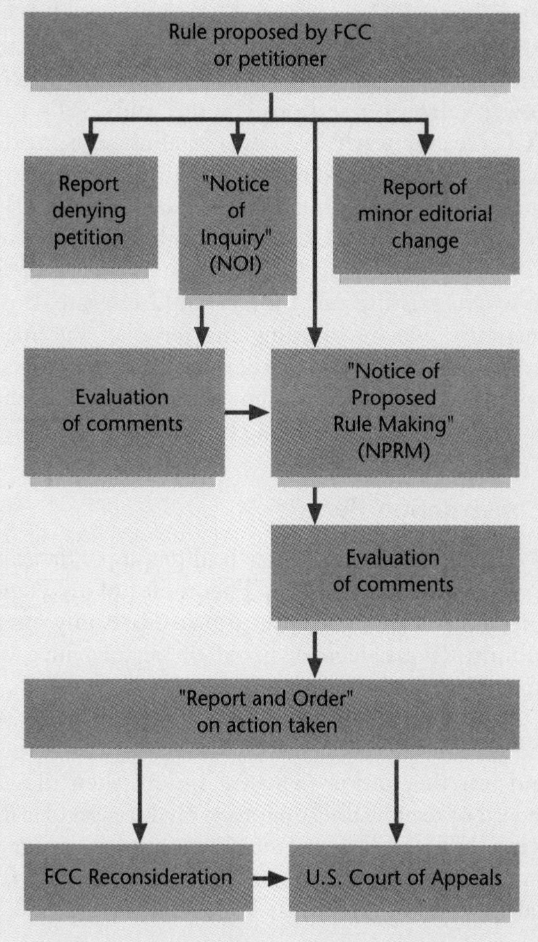

A petition for a new rule or a change of existing rules can come from the public (rare), the regulated industries (common), another part of the executive branch, or a unit within the FCC itself. Parties who are dissatisfied with denials or new rules often appeal for reconsideration by the FCC and, if still not satisfied, to the courts. Each step of the process must be documented in the *Federal Register* so that all interested parties can keep themselves informed of rule-making actions.

special significance or of a controversial nature may be scheduled for oral argument before the five commissioners.

After digesting outside comments, the staff prepare a proposed decision for the commissioners, usually as a recommended *report and order* with background discussion explaining and defending the action. Once a proposed report and order is adopted, that action becomes subject to petitions to the Commission for reconsideration and/or appeal to the U.S. Courts of Appeals. Increasingly in recent years, one party or another appeals nearly every "final" FCC decision, sometimes delaying implementation by months or years.

Adjudication

A second type of FCC decision-making is adjudication, which settles specific disputes—whether between outside parties (for example, rival applicants for a television channel) or between the FCC and an outside party (such as a broadcaster who protests a fine).

The staff may settle simple disputes quickly, but others become the subject of hearings. Electronic media owners avoid hearings if at all possible. Such hearings require expensive legal representation and can take a long time.

FCC Critique

Over the years the FCC has been among the most frequently analyzed and scathingly criticized of all federal regulatory agencies. Official investigations and private studies of the Commission and its methods reach negative conclusions with monotonous regularity. A few criticisms are recurrent:

- The political process for choosing commissioners often fails to select qualified people. FCC appointments do not rank high in the Washington political pecking order. The president often uses appointments to regulatory commissions to pay off minor political debts.
- As a consequence, commissioners usually lack expertise and sometimes the dedication assumed by the Communications Act. Most appointees

have been lawyers. People with experience in engineering, the media, or common carriers—or in relevant academic specialties—rarely receive appointment because they lack the political constituencies required for serious consideration.

- Commissioners' hopes for future employment with regulated industries may underlie their usually narrow reading of "the public interest." Not many stay in office long enough to attain great expertise; the more ambitious and better-qualified appointees soon move on to higher-paying positions, usually in private legal practice, specializing (of course) in communication law.

However, it is somewhat misleading to speak of *the* FCC. As its membership (along with the administration in Washington) changes, the Commission takes on varying complexions—sometimes reform minded, sometimes more interested in the welfare of the businesses it regulates than in the public interest.

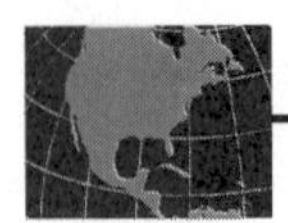

12.4 Broadcast Licensing

The *authorizing of service*—licensing of broadcasting stations—is the FCC's single most important function. (Local authorities that franchise cable systems—see Section 12.8—play a parallel role.) Most other regulation grows out of this licensing process. The FCC acts as gatekeeper, using its licensing power to offer station operation to some and not others. No one can legally "own" any part of the electromagnetic spectrum. Conscious of both real and potential claims to the contrary, the Communications Act repeatedly codifies the ban on channel ownership.

Finding a Channel

A would-be broadcast licensee applies for specific facilities (channel, power, coverage pattern, antenna location, time of operation, and so on). Most desirable commercial channels had been assigned by the 1970s, so that a new licensee nearly always had to buy an existing station. Later, several new opportunities developed.

In 1982 the Commission approved low-power television (LPTV)—whose channel use could fit in among existing outlets, inasmuch as LPTVs are limited in power so as not to interfere with full-power television stations. In the mid-1980s the FCC created nearly 700 new FM allotments, made possible by improved technology limiting interference with stations already on the air. Finally, early in the 1990s, the FCC extended the AM band to 1705 kHz, adding 10 new medium-wave channels to which existing AM outlets could migrate, thus cutting down on existing interference. But in a strong shift away from earlier policy, the Commission restricted access to some of these "new" channels to old players—existing broadcasters.

Construction Permit

Would-be licensees for new facilities apply first for *construction permits* (CPs). The holder of a CP applies for a regular license to broadcast only after submitting satisfactory proof of performance of transmitter and antenna.

A CP gives its holder a limited time (usually 24 months for television and 18 for radio) to construct and test the station. The CP holder then files a record of technical testing, to apply for a regular license. With FCC permission, a permittee may (and most do) begin on-air program testing pending final approval of the license.

Licensee Qualifications

Sec. 310(b) of the Communications Act forbids foreign control of a broadcast license. Some government and business leaders in the mid-1990s began questioning that restriction. These leaders suggest that U.S. licenses could be opened up to foreign ownership in return for greater participation by U.S. companies in foreign markets. Sec. 308(b) limits licensees to those *U.S. citizens*, or corporations primarily owned by U.S. citizens, who qualify as to

- *Character.* Applicants should have personal and business histories free of evidence suggesting defects in character that would cast doubt on their ability to operate in the public interest. Criminal records and violations of antitrust laws may constitute such evidence. Any misrepresentation to the FCC would be an almost-fatal defect.
- *Financial resources.* Applicants must certify that they have "sufficient" financial resources. The FCC has issued varied definitions of "sufficient," but currently it means the ability to construct and operate facilities for 90 days without reliance on station revenue.
- *Technical ability.* Most applicants hire engineering consultants to prepare technical aspects of their applications. They show how a proposed station will get maximum physical coverage without causing objectionable interference to existing stations.

Mutually Exclusive Applications

When two or more qualified applicants seek the same channel, the FCC exercises its right under the Act to specify "other qualifications." The FCC can make a choice among these applicants only after conducting comparative hearings—an often drawn-out and costly process for all concerned.

No comparative hearings have been concluded since 1994, holding up the applications of several hundred prospective licensees since that time. The Court of Appeals in that year declared the FCC's comparative licensing criteria illegal, and, at the time this book went to press, the Commission had not formulated new criteria. Until the Commission acts, all mutually exclusive broadcast applications subject to a hearing are on hold.

License by Lottery

So many thousands of competitive applications piled up for LPTV stations in the early 1980s that, responding to the FCC's plea, Congress amended the Act in 1982 to authorize selection of new (not renewal) LPTV licensees by means of a *lottery* rather than time-consuming comparative hearings (47 USC 309[i]). Using lotteries, the FCC could check an LPTV applicant's qualifications to be a licensee *after* the applicant had already been chosen by the lottery, going back to choose another applicant if the first winner was deficient in some way.

Responding to the demands of Congress to increase the diversity of LPTV ownership among minorities and others, the Commission assigns statistical preferences to minority applicants and those applicants who own no or few other media outlets. This gives these preferred applicants a greater chance of winning licenses through the lottery process.

License by Auction

Although the FCC has authority to grant certain licenses through an auction process (see Section 12.10), that authority did not extend to broadcast licensing at the time this book went to press. As part of its effort to balance the federal budget, Congress will require the auction of much of the existing television spectrum when TV broadcasters give up their analog channels early next century. Some lawmakers estimate the value of this spectrum, primarily in the existing VHF band, to be as high as $30 to $35 billion.

Services Requiring No License

The FCC requires virtually all spectrum-using electronic mass media and common carriers to obtain licenses in order to operate. The few exceptions include

- radio-station auxiliary services transmitted on subcarriers and neither related to broadcasting nor decoded by regular radio receivers;
- teletext and closed-captioned services broadcast by television stations;
- carrier-current radio stations such as those used on some college campuses; and
- very low-power AM radio stations designed to serve small, discrete areas.

Also, the FCC regulates the manufacture of many nonbroadcast devices that use part of the electromagnetic spectrum (for example, remote garage-door openers, microwave ovens, and cordless telephones), but does not require that consumers be licensed to use them.

12.5 Operations

Ideally the FCC should monitor station operations constantly to ensure operation in the public interest. But it simply is not feasible for the Commission to keep watch over all the programs of thousands of stations. In the course of normal operations, broadcasters experience little official supervision.

Employment Practices

Congress explicitly directed the FCC to monitor employment practices. Broadcast licensees and cable franchisees employing five or more persons full-time must set up a "positive, continuing program of practices assuring equal employment opportunities." These *equal employment opportunity* (EEO) and affirmative-action requirements refer to women in all cases, and to minority ethnic groups in cases where they form 5 percent or more of the work force in a station's or cable system's service area.

Stations and cable systems must submit to the FCC annual reports that also must be kept in the station's or cable system's public file (discussed next). The FCC reviews a licensee's EEO record at license midterm and when considering its license renewal, and periodically also examines cable system reports.

The Commission has issued detailed processing guidelines (see Exhibit 12.d). Congress extended broadcast EEO requirements to cable systems in the 1984 cable act. The 1992 Cable Act expanded those requirements further to "any multichannel video programming distributor" and directed the FCC to review its overall EEO program. In 1994 the Commission reported to Congress that, although its existing policies and procedures were effective, continuing monitoring was necessary. The agency also reorganized its Office of Communications Business Opportunities to coordinate and oversee all EEO matters (FCC 94-255).

Reacting to concerns that some affirmative action programs—in electronic media and elsewhere—had resulted in *reverse discrimination*, Congress initiated a broad reexamination of these policies during the mid-1990s. Pending changes in the Communications Act or a court opinion invalidating the regulations, the Commission continues to enforce its EEO requirements strictly, as is evidenced by the numerous fines levied against broadcast stations and cable systems that violate the guidelines.

Public File

The FCC requires stations to keep license-renewal applications and other relevant documents in a file, readily available for public inspection. All stations, commercial and noncommercial, must maintain such a file, retaining some documents for seven years in the case of a radio station and five in the case of television. Cable systems have to maintain a public file containing only employment-related documents. The more inclusive broadcast public file includes the following:

- the latest construction permit or license application, including any for major changes
- the latest license renewal application, as well as ownership reports and annual employment reports
- the EEO model program, if required
- the now-obsolete pamphlet entitled "The Public and Broadcasting—A Procedural Manual," issued in 1974
- a record of the disposition of any political-broadcast time requests for the preceding two years
- a quarterly listing of programs the licensee believes provided the most significant treatment of local community problems of public importance
- records required by the Children's Television Act of 1990 concerning programming for and advertising to children under the age of 16
- any time-brokerage agreements in force

Exhibit 12.d

EEO Mandates and Staff Size

Number of Full-Time Employees	EEO Requirements
1 to 4*	Need not file an EEO plan.
5 to 10	The FCC will review system or station EEO programs unless the percentage of minority and women employees equals half of their representation in the local labor force. In other words, if a market's labor force, as defined by the Census Bureau, is half black, at least one-quarter of a station's or cable system's employees in that market should be from that minority group. In the top job categories (officers and managers, professionals, technicians, and salespersons), a station or system should have a minority-employee ratio of at least one-quarter of that minority's representation in the local labor force. If half of the local labor force is black, at least one-eighth of the top-four station or system jobs should be held by blacks.
11 or more	Should employ at least half as many minorities and women as are represented in the local labor force overall *and* in the top-four job categories (officers and managers, professionals, technicians, and salespersons).
50 or more†	Same as for stations or systems with 11 or more employees, but, in addition, EEO programs are regularly reviewed.

*This category consists mainly of radio stations.
†Most of these are television operations.

- copies of (television) station decisions concerning must-carry and retransmission consent governing its relationship with any local cable systems
- letters received from members of the public (to be kept for three years) and any agreement with citizens groups

Deregulatory decisions such as the abandonment of formal ascertainment of community needs, deletion of programming guidelines, and reduction of renewal applications to postcard form reduced the public file's size, but most media owners still regard it as a waste of time. Indeed, members of the public rarely ask to see it.

Keeping Up with Washington

Electronic media managers must keep current with changes in FCC regulations. Trade organizations and publications and an army of communications attorneys in Washington help licensees with this task. Well over 1,000 attorneys representing electronic media clients are members of the Federal Communications Bar Association (FCBA). Personal contacts with FCC staff members enable Washington lawyers to get things done faster than can distant licensees unfamiliar with the federal bureaucracy. The National Association of Broadcasters, National Cable Television Association, and other trade associations offer legal clinics and regular publications on FCC rules.

From all of these sources, stations and systems receive continuous advice on what is new and what not to forget. For example, licensees often seem unaware of differing federal or state requirements on such things as employee drug use and over-the-air use of live or recorded telephone conversations. The timely filing of regular ownership, EEO, and other reports required by the Commission is a mundane but important aspect of using legal counsel.

Monitoring Performance

The FCC rarely monitors electronic media. Inspectors from the Compliance and Information Bureau check technical aspects of station and cable-system operations, but only occasionally and in random fashion. Questions about programming and commercial practices, if they arise at all, usually come to FCC attention through complaints from the public, consumer groups, or competing media.

The Mass Media Bureau's Enforcement Division receives more public comments than any other federal agency except those dealing with environmental protection and consumer product safety. Few complainers seem to understand the FCC's legal limitations—most complaints are discarded because they ask the Commission to violate the First Amendment by censoring material that the writer personally dislikes. The leading topics vary only slightly from year to year, often influenced by organized letter-writing campaigns as well as by program trends.

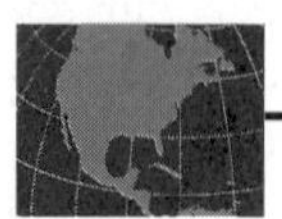

12.6 Renewal

Under the Communications Act, licenses may be awarded for only "limited periods of time," currently a maximum of eight years, and must regularly come up for renewal. Although the FCC renews more than 98 percent of all licenses without asking any searching questions, possible nonrenewal seems always to be lurking in the background. Challenges to renewal applications can come from the FCC but may also be lodged by dissatisfied citizens in the licensee's community. Even if the incumbent wins a contested renewal (and they usually do), defending it can be both expensive and time consuming.

Application Routes

Sec. 309 of the Communications Act stipulates that "licenses shall be renewed if the public interest, convenience, and necessity would be served" by renewal. Before deregulation of renewal procedures, applications often required piles of supporting documents showing how the licensee had served—or would serve—the public interest (see Exhibit 12.e). In 1981 the FCC began using a simple postcard-sized renewal form, later expanded to a still relatively simple five-page form, which most applicants supplement with additional multipage exhibits.

Renewal applications take one of three paths: uncontested, FCC staff contested, and the petition to deny.

- Some 98 percent of all renewal applications fall in the *uncontested* category. The FCC staff uses its delegated authority to renew the station almost automatically. In fact, consumer advocates complain that the Mass Media Bureau merely "rubber-stamps" uncontested applications, no matter how mediocre a station's past performance may have been.
- Occasionally the FCC staff will initiate a proceeding to deny a license renewal if they find serious or frequent violations of U.S. law or FCC regulations. Allegations of criminal law breaches, antitrust activity, or concealing information from the Commission could lead to such action.
- The now-rare *petition-to-deny* renewal comes from a citizen group or other party that opposes an incumbent licensee. Such groups claim that incumbents have failed to meet public-interest

Exhibit 12.e

Renewal Paperwork Before and After Deregulation

(A) In 1971, before deregulation, it took more than 16 pounds of paperwork to file for renewal of four Nebraska stations. (B) In the 1990s this small stack of papers along with a simple five-page form does the same job.

Source: Photos courtesy *Broadcasting & Cable* magazine.

standards (most often allegations about EEO compliance). Although the petition process is important in giving citizens the right to challenge licensees who may not be serving the public interest, in practice the petition-to-deny process rarely results in the denial of renewal.

Until the Telecommunications Act of 1996, another type of renewal scenario known as the *comparative renewal* was possible. This situation arose when a would-be licensee tried to displace the incumbent licensee, claiming that it could do a better job serving the public interest. Comparative renewals put the Commission in the odd position of having to judge the worthiness of two applicants based on entirely different qualifications: The incumbent was judged on its past record while the challenger was judged solely on the paper promises contained in its application. In practice, the incumbent broadcaster nearly always won its renewal, but often after a long and expensive hearing and appeal process. Congress changed the renewal law in 1996 by stating that the Commission, when considering a broadcast license renewal, could not consider a competing application unless it had first determined that the original license should not be renewed.

License Transfers

Other than uncontested renewal, the most frequent licensing process is license transfer, which occurs when an owner sells a station to another party. As noted in Chapter 6, station sales have increased dramatically since the passage of the Telecommunications Act of 1996. A large majority of station owners have acquired their stations by buying them rather than by securing new stations from the FCC.

All station ownership transfers require Commission approval. After the parties to the transfer agree on the terms of the sale, they submit their transfer application to the FCC. The Commission issues a public notice of the application, and interested parties have 30 days to submit comments. In most instances, no one objects and the Commission staff routinely approve the license transfer. Individuals or groups may, however, file petitions to deny the sale. If the FCC finds that a petition raises serious issues, it will order a hearing to determine whether the proposed transfer would serve the public interest. Such findings rarely occur; in most cases, the Commission dismisses petitions to deny.

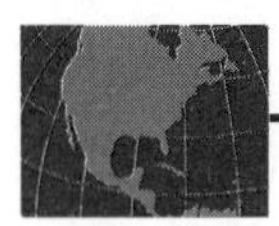

12.7 *Enforcement*

The FCC can impose a variety of penalties on licensees for infractions of its rules. They range from simple letters of admonition, through fines, to short-term license renewal or—the ultimate penalty—no renewal at all. When a licensee proves incorrigible, the FCC can either refuse to renew the license or revoke it outright. The FCC prefers the nonrenewal option because it puts the burden of proof on the licensee, whereas revocation places the burden of proof on the Commission. None of these sanctions can be invoked without legal due process, often beginning with a formal hearing.

Due Process

A fundamental safeguard of individual liberties under the Constitution, the Fifth Amendment *due process clause*, guarantees that government may not deprive a person of "life, liberty, or property without due process of law." Among many other applications, this means that the FCC may not use its powers arbitrarily. Fairness, the goal of due process, requires that applicants and petitioners have ample opportunity to argue their cases under nondiscriminatory conditions and that parties adversely affected by decisions may appeal for review by authorities other than those that made the initial decision. Many due process rights are detailed in the Administrative Procedure Act of 1946, which specifies how agencies such as the FCC must conduct their proceedings to ensure due process for all participants. For example:

- The Commission must advise unsuccessful license applicants of its reasons for rejecting their applications. The applicants may reply, and if the Commission still decides against them, it must then set the matter for a hearing.
- On the other hand, if the Commission grants a license application *without* a hearing, for the ensuing 30 days (after a public notice announcing this action) the grant remains open to protest from "any party in interest." If the FCC finds that some party raises pertinent issues, the Commission must then postpone the effective date of the grant and hold a hearing.
- If the Commission wishes to fine a licensee, it must invite the licensee to "show cause" why such action should *not* be taken.

FCC Hearings

Senior staff attorneys, called *administrative law judges* (ALJs), preside over initial FCC hearings. They conduct the proceedings somewhat like courtroom trials, with sworn witnesses, testimony, evidence, counsel for each side, and so on. Appeals of initial decisions of ALJs are reviewed, first by the FCC's Review Board and then by the commissioners themselves. Procedural rules head off frivolous interventions and intentional delays by carefully defining circumstances that justify hearings and qualifications of parties entitled to *standing*—that is, the right to participate.

Court Appeals

Even after all safeguards in FCC hearings and rehearings have been exhausted, the Communica-

Exhibit 12.f

Citizen Involvement— The Landmark WLBT Case

Until the late 1960s, a broadcast station's audience members had no right to take part in regulatory proceedings concerning that station. That situation changed as a result of a court case that began in 1955 when a group of citizens made the first of a series of complaints to the FCC about the conduct of WLBT, a VHF television station in Jackson, Mississippi. The group accused the station of blatant discrimination against blacks, who formed 45 percent of its audience. The FCC dismissed the citizens' complaints, saying that they had no legal right to participate in a licensing decision. When WLBT's license again came up for renewal in 1964, local groups obtained legal assistance from the Office of Communications of the United Church of Christ (UCC) in New York.

The UCC petitioned the FCC on behalf of the local groups for permission to intervene in the WLBT renewal application proceeding, but the FCC again rejected the petition, saying that citizens had no *legal standing* to intervene. At that time the commission recognized only signal interference or economic injury (to another broadcaster) as reasons to give parties the right to participate in renewal hearings. Thus only other broadcasters had standing to challenge existing licensees.

The UCC went to the Court of Appeals, claiming that the FCC had no right to bar representatives of the viewing public from intervening in renewals, or to award a renewal without a hearing in the face of substantial public opposition. The court agreed, directing the FCC to hold hearings on WLBT's renewal and to give standing to representatives of the public (F, 1966). The FCC held a hearing and grudgingly permitted UCC to participate as ordered. However, it once again renewed WLBT's license.

The UCC returned to court and in 1969 an exasperated appeals court reconsidered the case—14 years after the first complaints had been recorded. In the last opinion written by Warren Burger before he became Chief Justice of the United States, the court rebuked the FCC for "scandalous delay." It ordered the FCC to cancel WLBT's license and to appoint an interim operator pending selection of a new licensee (F, 1969). But 10 *more* years passed before the FCC finally selected a new permanent licensee. Altogether, the case dragged on for more than a quarter of a century.

As the FCC had feared, the WLBT case triggered many petitions to deny renewal of other licenses. This "reign of terror," as a trade magazine put it, resulted in few actual hearings and still fewer denials. Of the 342 challenges filed in 1971–1973, only 16 resulted in denials of license renewal. An exacting standard of evidence established by the FCC and approved by the court ensured this high rate of petition failure. Only after an opponent presented overwhelming evidence would the FCC schedule a license-renewal hearing (F, 1972).

tions Act gives parties adversely affected by FCC actions still further recourse. The Act provides that all appeals concerning station licenses must go before the U.S. Court of Appeals for the District of Columbia Circuit, in Washington, D.C. Appeals from FCC decisions in cases that do not involve licensing may be initiated in any of the 12 other U.S. Courts of Appeals. Each serves a specific region of the country and is known as a *circuit court of appeal.*

The court may confirm or overturn Commission actions, in part or in whole. It may also *remand* a case, sending it back to the FCC for further consideration in keeping with the court's interpretation of the Communications Act and other laws (see Exhibit 12.f).

From any of the federal circuit courts, final appeals may be sought before the Supreme Court of the United States. A request for consideration by the Supreme Court, called a *writ of certiorari*, is most often turned down ("cert. denied"). If that happens, the appeal process is over. Refusal to hear a case does not necessarily mean that the Supreme Court agrees with lower court findings, but the earlier decision holds nonetheless, though without standing as a compelling nationwide legal precedent.

Loss of License

Since the FCC's creation in 1934, only about 150 stations have involuntarily lost their licenses. *Nonrenewal* accounts for two-thirds of these losses, outright *revocation*—not waiting for the current license to expire—about one-third. The average rate of nonrenewals or revocations was less than three stations per year. Though an ever-present background threat, loss of license only rarely becomes a reality among the thousands of stations on the air.

Stations that have lost their licenses have largely been obscure radio outlets whose licensees have accumulated long lists of willful misdemeanors. Transgressions accumulate because the FCC usually treats them leniently as long as the guilty licensee candidly admits error and contritely promises to reform. Eventually, another violation causes the FCC to lose patience, and some specific misdeed results in nonrenewal or revocation.

Lesser Penalties

Not all offenses, of course, warrant the capital punishment of license loss. For lesser offenses, the FCC inflicts milder sanctions:

- *Short-term renewal* (usually a year or two instead of the full renewal term) puts a licensee on a kind of probation pending correction of deficiencies evident during its preceding license period.
- *Conditional renewal* is granted pending correction of some specific fault.
- *Forfeitures*, or *fines*, ranging up to a maximum of $250,000 for repeated violations may be assessed for broadcaster infractions of FCC rules. Fines against common carriers can be as high as $1 million. Most forfeitures come from technical violations, few from program violations—and most are for much less than the maximum amount. Because cable systems have no license that the FCC can threaten, fines are the agency's chief means of enforcing cable rules.
- Relatively minor infractions may result in a *letter* being placed in the station's file for consideration when it applies for license renewal.

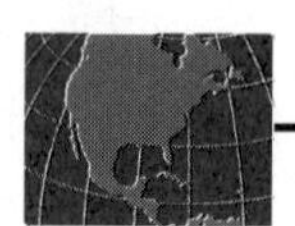

12.8 Cable

Cable "licensing," or *franchising*, follows a pattern totally different from broadcast licensing (see Exhibit 12.g). Local (and in a few instances, state) rather than federal authorities issue franchises because cable systems use streets and other public property that is subject to municipal jurisdiction, rather than federally controlled airways. Federal cable laws contain provisions covering subscriber fees, ownership, equal employment opportunities, programming, customer service, and technical standards. These laws, passed in 1984, 1992, and 1996, amended the 1934 Communications Act.

Cable Acts—General Provisions

The Cable Communications Policy Act of 1984 created a new Title VI of the Communications Act that established a loose federal *de*regulatory framework for cable television. Key parts of that act were reversed in the strongly *re*-regulatory Cable Television Consumer Protection and Competition Act of 1992, which amended Title VI. This 1992 legislation grew out of public and congressional anger over sharp increases in cable subscription rates—in some cases three times the rate of inflation—in the five years before it became law. The 1996 act modified a few provisions and authorized a new provider for the delivery of cable services—the telephone companies. The combined result of the three acts includes the following key provisions:

Exhibit 12.g

Broadcast Licensing and Cable Franchising: A Comparison

Broadcast Licensing	Cable Franchising
Licensed by FCC	Franchised by local or state authority
License term: 8 years	Term set in franchise: 10–15 years standard
"Comparative" renewals not permitted	Comparative renewals allowed, but unusual
Obligations set by Congress and FCC	Obligations set by Congress, FCC, and franchising authority
Station transfers approved by FCC	System transfers approved by franchising authority
Licensing process uniform for all licensees	Franchising process varies from community to community

- The 1984 act defined cable television as a one-way video programming service; neither it nor the 1992 and 1996 revisions regulate two-way (interactive) services.
- The 1992 act transforms the relationship between broadcasters and cable systems. Under this act, each commercial television station has the right every three years (beginning in October 1993) either to charge cable systems within its coverage area a negotiated fee for use of its signal (termed *retransmission consent*) or simply to require that cable systems carry its signal (the *must-carry rule*). If a station and cable system cannot reach agreement on terms, that station will not be carried on that system for at least three years. Stations that fail to choose between must-carry and retransmission consent will, by default, come under the must-carry option. Local noncommercial stations may claim must-carry rights but do not have the option of seeking retransmission consent. MMDS and SMATV systems are also subject to retransmission consent rules.

Cable systems with 12 or fewer channels must devote at least three to local broadcast signals. Systems with more than 12 channels may be required to devote up to a third of their capacity to such signals. None of this applies to the relatively few systems with fewer than 300 subscribers.

The must-carry requirement, once an FCC regulation, had been twice overturned on First Amendment grounds by a federal appeals court in the late 1980s. In 1997 the Supreme Court upheld the requirements, finding that the protection of over-the-air broadcasting was an important government interest.

- A television station may also demand carriage on the same channel number on cable that it uses over the air.
- Reversing a key aspect of the 1984 law, the 1992 act required local authorities or the FCC to regulate subscriber rates for basic tiers of cable service. Congress retreated a bit in 1996 by ending all rate regulation of upper-tier basic services effective 31 March 1999. No rate regulation of pay-cable channels such as *HBO* or *Showtime* is

allowed. Cable systems are exempt from rate regulation if they serve fewer than 30 percent of the households in their area or are subject to "effective competition" from some other multichannel competitor (MMDS, SMATV, DBS, or telephone companies) that is available to at least 50 percent—and actually subscribed to by at least 15 percent—of an area's households.

- In May 1993 the FCC issued a 450-page decision detailing how fair cable subscription rates were to be determined. The plan was sufficiently complicated that, when the dust had settled, about two-thirds of cable subscribers had seen a drop in their rates, but a third had actually seen increases. The anger engendered by use of seeming loopholes in the rules led to a second FCC order to cut prices—this time by about 7 percent—in 1994. Cable rates have increased since then; rates rose around 4 percent in 1995 and 7.8 percent in 1996—the largest increase since 1991.
- Cable program providers (for example, ESPN or HBO) that are owned even in part by cable-system operators must make their services available to multichannel cable competitors (such as SMATV, MMDS, and DBS) at the same prices offered to cable systems. This provision, which expires in 2002, is designed to give competing delivery systems a chance to get an audience foothold.

Cable Acts—Franchising

Whereas the 1984 act severely limited regulatory powers of state and local authorities, the 1992 law reinstated many of those same powers and added new FCC oversight of many aspects of cable franchising.

- Local franchising authorities may require public, educational, and governmental access (PEG) channels over which the cable operator has little editorial control.
- The 1992 act added protection for children against indecent programming on commercially leased and public-access channels. Much of this regulation was subsequently declared unconstitutional by the U.S. Supreme Court.
- The 1984 act called for cable systems with more than 36 channels to set aside 10 to 15 percent of those channels for leased access to outside parties. The cable owner sets rates for leasing access channels but has no editorial control over their content—other than enforcement of obscenity and indecency limitations (as discussed later). The 1992 act requires the FCC to set maximum rates for use of such channels.
- Reinforcing existing practices and FCC guidelines, Congress allowed local franchise authorities to charge annual franchise fees of up to 5 percent of cable-system gross revenues.
- When granting franchise renewals, local authorities may require operators to upgrade cable facilities and channel capacity.
- The FCC may require that specific customer service levels and technical standards be included in new and renewal franchise agreements. Initial detailed FCC rules were issued early in 1993.
- The 1992 act required the FCC to evaluate every franchise-granting authority in the country (some 20,000 cities, counties, and other political units) and to certify that each had authority and capacity to issue and enforce such franchises—and that the franchise provisions agree with all requirements in the amended Communications Act. If any franchise authority is not so certified, then the FCC itself is to act as authority for cable in that market until a local franchise authority is certified.

Franchise Process

When a local franchising authority (city, town, or county) decides it wants cable service, it first develops an *ordinance* or legal codification describing the conditions under which a cable system will be allowed to operate. Drawn up in many cases with the advice of outside experts, ordinances typically stipulate

- the term of the franchise (usually 10 to 15 years, but sometimes longer);
- the quality of service to be provided;
- technical standards, such as the minimum number of channels to be provided, time limits on

construction, and interconnection with other systems;
- the franchise fee; and
- PEG channel requirements, if any.

Bidders base their offers on the design and timetable outlined in the franchise authority's *request for proposals* (RFP). Although franchisers usually grant a franchise to only one bidder, multiple awards (termed *overbuilds* if they cover the same region) are possible. In an attempt to stimulate local competition, the 1992 act states that local authorities may not unreasonably refuse the grant of a competing franchise. Still, the vast majority of cable systems operate as local monopolies.

Renewals

A local authority need not find that renewal will serve the public interest or meet any other standard but, rather, may simply renew a franchise without ceremony. If, however, the local authority wants to deny renewal, the Communications Act requires that it hold a hearing, in effect raising public-interest issues by deciding whether the incumbent operator has (1) complied with the law; (2) provided a quality of service that is "reasonable in light of the community needs"; (3) maintained the financial, legal, and technical ability to operate; and (4) prepared a renewal proposal that is "reasonable to meet the future cable-related community needs and interests" (Sec. 626). The Act does not require consideration of proposals from competing would-be franchisees during the course of renewal grants or renewal hearings, but the franchising authority may consider other proposals if there are other applicants for the franchise.

12.9 Other Electronic Media

While broadcasting and cable television certainly are the largest electronic mass media in terms of the number of people served, other electronic media are also regulated by the FCC. Based on the characteristics of those media, FCC regulation may vary.

Satellite Master Antenna Television Systems

Satellite master antenna television (SMATV) systems are functionally similar to cable television systems except that they serve multi-unit dwelling complexes without crossing public streets or rights of way. As such, they are sometimes termed "private" cable systems, and they are not subject to local regulation. Thus, SMATV systems do not need franchises to operate, nor are they subject to local access channel and equipment requirements. The FCC does, however, require SMATV systems to follow broadcast television must-carry rules.

"Wireless" Cable

Wireless cable—sometimes referred to as multichannel television—is a subscription service much like cable television except that it uses over-the-air microwave frequencies to deliver its signals rather than coaxial cable and fiber. Currently wireless cable utilizes multichannel multipoint distribution service (MMDS) to distribute up to 30 channels of programming to subscribers. The FCC licenses MMDS systems via a lottery in cases where more than one applicant seeks frequencies in the same markets. Equal employment opportunity regulations and broadcast signal carriage rules apply to MMDS systems. Because these systems use over-the-air frequencies, local governments have no jurisdiction over MMDS services.

In 1997 the FCC authorized a new service for the delivery of multichannel television known as the local multipoint distribution service (LMDS). This service, which also uses microwave frequencies but reuses them in a way similar to cellular telephone service, promises to permit delivery of over 100 video channels to subscribers. Although the public interest requirements for LMDS had not been established by mid-1997, it was clear that the FCC intended to license the service via spectrum auctions.

Direct Broadcast Satellite

The FCC licensed the first direct broadcast satellite (DBS) systems in the early 1980s much the same as they granted broadcast licenses. More recently the Commission has granted new licenses through the spectrum auction process. To date DBS systems operate on a subscription basis and deliver much the same programming as cable systems, but without local television signals. DBS operators may deliver selected broadcast stations to individual subscribers, but only to households that cannot receive broadcast signals over the air. Thus, households located in so-called white areas may receive broadcast network affiliates as part of their DBS service, usually at a slight extra fee. FCC EEO rules apply to DBS licensees, and the 1992 amendments to the Communications Act require DBS service providers to set aside up to 7 percent of their channel capacity for public interest programming. Because this requirement was tied up in litigation, the FCC only began to implement the law in 1997.

A new satellite service authorized in 1997, digital audio radio satellite (DARS) service, promises to deliver multichannel, CD-quality audio service directly to homes, automobiles, and portable radios nationwide. Although service requirements had not been set by the Commission when this book went to press, licensing of the two available satellite radio slots was conducted in April, 1997, via spectrum auction.

Telephone Companies

From the mid-1960s to 1996 the FCC did not allow telephone companies to provide cable television services in the same area that they provided telephone service. The Commission enacted this "cross-ownership" restriction so that the powerful telephone companies would not dominate the nascent cable television business. Congress followed the FCC's lead and codified the cross-ownership ban in the 1984 amendments to the Communications Act.

Following several successful court challenges of the cable/telco cross-ownership restrictions in the mid-1990s, Congress removed the ban in the Telecommunications Act of 1996 as part of its effort to stimulate local competition in the multichannel video market. According to the 1996 legislation, telephone companies may now offer video services in their local telephone service areas under four different regulatory regimes: as over-the-air-providers such as MMDS operators, subject to Title III regulation; as common carriers subject to Title II regulation; as cable television operators subject to Title VI regulation; or as a newly created hybrid service called an *open video system* (OVS).

Operation as an OVS allows the telephone company to deliver programming of its own choosing while giving other programmers access to the company's distribution system. Essentially the telephone company acts as a cable system operator and common carrier at the same time. The Act requires that if there is more demand for system capacity than exists, then the telephone company may control no more than one-third of the system's capacity; the other two-thirds of the system capacity must be apportioned among other users. Although OVS operators must satisfy some of the requirements imposed on cable operators—signal carriage and EEO, for example—they are exempt from some aspects of cable regulation. OVS systems are certified by the FCC and are not franchised locally. Also, most rate regulation does not apply to OVS operations. Local authorities are permitted to exact a share of the OVS system revenues, much like the franchise fee levied against cable operators.

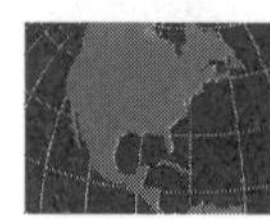

12.10 Deregulation

Many motives underlie efforts to deregulate—in communication and several other economic sectors. Least controversial is the recognized need to discard outdated rules, to simplify unnecessarily complex rules, to ensure that those rules that remain can actually achieve their objectives, and to lighten administrative agency loads. Deregulation based on these motives began in the 1970s with wide political support.

A more controversial drive to deregulate stems from ideological motives growing out of a specific vision of a limited governmental role in national life. This view opposes government intervention, advocating instead reliance on the marketplace as a nongovernmental source of control over private economic behavior. This approach first emerged as a force on the national agenda in the late 1970s under President Jimmy Carter, escalated when Ronald Reagan came to the White House in 1981, and continued during the Bush and Clinton administrations. With respect to the regulation of media content, however, many of President Clinton's goals, and those of former FCC chairman Reed Hundt, were distinctly *re*-regulatory. These efforts to increase government oversight of programming content are addressed in the next chapter.

Theoretical Basis

Deregulatory theory holds that marketplace economic forces can stimulate production of better, more varied, and cheaper consumer goods and services without official guidance from above. Where government regulation may be necessary, on the other hand, it should be tested by a cost-benefit formula to make sure that the costs of such regulation do not outweigh its gains.

The FCC did not go so far as to advocate abandoning all regulation. It divided rules into behavioral and structural categories. *Behavioral* regulation, which deregulators seek to discard or at least minimize, controls what licensees may or may not do in conducting their businesses; *structural* regulation controls the overall shape of the marketplace and terms on which would-be licensees can enter. Rules requiring a licensee to limit advertising in children's programs are one example of behavioral regulation; rules preventing a licensee from owning more than *x* number of stations are an example of structural regulation.

Deregulation, however, tilts toward the structural regulation approach, which enhances competition by making marketplace entry easier, preserves competition by preventing monopoly—and stays away from program-based decisions. Critics argue that deregulation serves mainly to enrich a few companies and individuals, often at the expense of broader public concerns.

Even theorists favoring deregulation admit that *market failure* can occur. Sometimes competition fails to produce expected favorable results. Indeed, uncontrolled competition may produce corporate giants that suppress competition (major airline control of most departure gates at "hub" airports is one example). And some public "goods," as economists call desirable services, may fall outside the realm of marketplace economics and therefore fail to materialize. If, for example, public television is thought desirable but costs too much to produce with limited private support, marketplace economics may have to be supplemented by giving such broadcasters government aid.

Deregulation in Practice

In the mid-1970s early deregulation did away with many incidental bureaucratic rules. During the 1980s a more zealously deregulatory FCC began dismantling or softening more basic rules, first in commercial radio (1981), next in educational broadcasting (1984), and then in commercial television (1985). In the 1980s the FCC also began loosening its structural regulations by allowing station owners to acquire more stations. At the same time Congress also decreased regulatory burdens by lengthening the terms of broadcast licenses and permitting licensing by lottery in some situations.

In the 1990s the deregulatory trend continued in most instances as Congress further lengthened broadcast license terms to a maximum of eight years and eliminated all national ownership restrictions for broadcast stations (with a national audience cap for television owners). (See Section 13.6 for further discussion of the ownership rules.) Congress also permitted licensing via auction in the mid-1990s. In 1995 Congress established a goal of

Exhibit 12.h

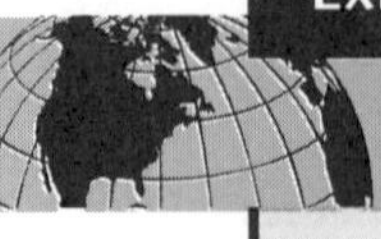

Before and After Deregulation

The FCC modified or eliminated many broadcasting and cable rules during its deregulatory drive in the 1980s. Moving against the deregulatory tide, Congress mandated limits on advertising in children's programming and required television licensees to serve the special needs of children in its programming (1990). The FCC strengthened that requirement in 1996 with a rule requiring television stations to air three hours of children's programming per week. Congress substantially re-regulated cable television in the 1992 cable act, but removed some of it in the 1996 Telecommunications Act. The FCC further reversed its deregulatory course on technical standards when it set new standards for the next generation of television in the United States. The Telecommunications Act of 1996 is both deregulatory and re-regulatory: Many structural regulations were eliminated, thus allowing more competition in telecommunications markets, but several aspects of programming content, most notably violence and sexually explicit material, are now subject to greater regulation.

Before 1980	Changes Since 1980
Licensing	
License runs three years for any broadcast station.	License lasts five years for TV, seven for radio (action by Congress, 1981); all broadcast licenses now last eight years (Congress, 1996)
Comparative hearings are required for choosing among competing applications.	Applicants for some services (LPTV, MMDS, etc.) chosen by lottery (action by Congress, 1982; by FCC, 1983); and other services (e.g., PCS, LMDS) chosen by spectrum auction (Congress and FCC, 1995)
Ownership	
Limit of seven AM, seven FM, and seven TV outlets	Limit raised to 12-12-12 (action by FCC, 1985); radio limits raised again to 18-18; (action by FCC, 1992) and in 1994 to 20-20; all numerical limits on ownership removed in 1996; TV coverage limitation 35 percent of U.S. TV households
Duopoly rule: Only one station of each kind per market	May own more than one radio station in larger markets (action by FCC, 1992); sliding scale now in effect based on market size; may own up to eight radio stations in largest markets (Congress, 1996).
Trafficking rule: Must hold a station at least three years	May sell license (with FCC okay) at any time (action by FCC, 1982)
TV networks may not own cable systems; telephone companies may not own cable systems in their local service areas.	Network-cable and cable-telco cross-ownership provisions eliminated (Congress, 1996)
Programming	
FCC application processing guidelines (not rules) call for minimum amounts of nonentertainment programming.	No qualitative program guidelines (action by FCC for radio, 1982; for TV, 1985)
Specific rules requiring determination of local program interests, needs (ascertainment)	Rules dropped; must file a quarterly "programs/problems" list in public file (action by FCC for radio, 1982; for TV, 1985)

Before 1980	Changes Since 1980
1974 policy (no requirement) to serve the educational needs of children	Requirement to program some educational and cultural programs for children (action by Congress, 1990; by FCC, 1991); rule strengthened to mandate three hours of educational children's programming per week (FCC, 1996)
Controversial issues must be aired and opposing sides treated fairly (fairness doctrine).	Fairness doctrine dropped (action by FCC, 1987)
Network affiliates in the top 50 markets may carry no more than three hours of network or off-network programming during prime time, with exceptions for some program categories.	Prime-time access rule dropped (FCC, 1996)
ABC, CBS, NBC cannot syndicate TV programming and cannot take a financial interest in prime-time television programming.	Syndication and financial interest rules eased (FCC, 1991) and later repealed (FCC, 1995)
No regulation of violent programming	Rating code established for TV programming, TV sets must be equipped with "V-chip" (Congress, 1996).
Advertising	
Guidelines (not rules) on maximum amount of advertising time allowed	Guidelines dropped (action by FCC for radio, 1982; for TV, 1985)
1974 policy (no requirement) suggesting limitations in commercialization of children's programs	Specific limits on advertising in children's programming set (action by Congress, 1990)
Technical Rules	
Specific technical standards selected/mandated for each new service (e.g., color TV, FM stereo)	Starting with AM stereo decision, standards left to marketplace forces (action by FCC, 1982; Congress mandates FCC to select a specific standard, 1992); HDTV standard chosen (action by FCC, 1994/1995/1996).
Engineers ranked by class, and specific classes required for certain jobs	Engineers no longer ranked by class (action by FCC, 1981); certain jobs no longer require a license (FCC)
Rules requiring signal quality and avoiding interference	Stations may use own methods (actions at various times by FCC)
Cable Television	
Cable systems must carry all local TV stations.	Must-carry rule dropped (U.S. Appeals Court, 1985, 1987); rules reinstated (action by Congress, 1992)
Syndicated exclusivity rules allow a station to force local cable systems to black out syndicated programs on superstations that are also carried by the local station.	Rules dropped (action by FCC, 1980); then reinstated (action by FCC, 1988, effective 1990)
Local communities may control rates charged by cable systems.	Local regulatory role largely eliminated (action by Congress, 1984); then reinstated (action by Congress, 1992)

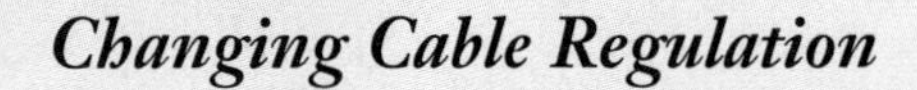

Exhibit 12.i

Changing Cable Regulation

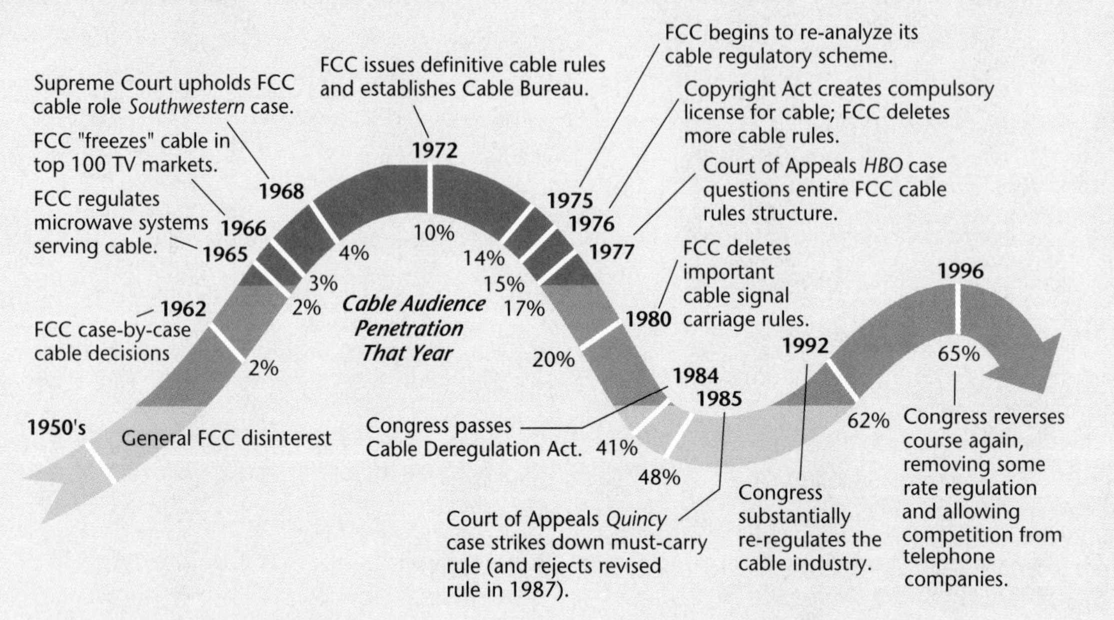

1950's
General FCC disinterest
1962
FCC case-by-case cable decisions
1965
FCC regulates microwave systems serving cable.
1966
FCC "freezes" cable in top 100 TV markets.
1968
Supreme Court upholds FCC cable role *Southwestern* case.
1972
FCC issues definitive cable rules and establishes Cable Bureau.
1975
FCC begins to re-analyze its cable regulatory scheme.
1976
Copyright Act creates compulsory license for cable; FCC deletes more cable rules.
1977
Court of Appeals *HBO* case questions entire FCC cable rules structure.
1980
FCC deletes important cable signal carriage rules.
1984
Congress passes Cable Deregulation Act.
1985
Court of Appeals *Quincy* case strikes down must-carry rule (and rejects revised rule in 1987).
1992
Congress substantially re-regulates the cable industry.
1996
Congress reverses course again, removing some rate regulation and allowing competition from telephone companies.
Cable Audience Penetration That Year
2%
2%
3%
4%
10%
14%
15%
17%
20%
41%
48%
62%
65%

Over nearly four decades, the FCC's regulation of cable television has undergone substantial changes in direction. From virtually no regulation in cable's earliest years to substantial and detailed FCC regulation of the medium in 1972, cable was steadily more controlled. For the next two decades, cable was progressively deregulated, especially with the 1984 cable act that removed most remaining local franchise limitations. In 1992, Congress reversed course again and passed a strongly re-regulatory cable act, only to change course again four years later.

raising more than $14 billion by auctioning parts of the electromagnetic spectrum. By early 1996 the FCC had raised more than $9 billion in spectrum auctions. Much of this money came from auctions for the new personal communications services, a variation on cellular telephone services. But some of the auction proceeds derived from DBS auctions and auctions for interactive digital video service (IDVS), a new service permitting interactive communication with broadcast programming. Exhibit 12.h summarizes many of these changes.

With respect to cable, the FCC had dropped most of its cable regulation by 1980, as shown in Exhibit 12.i. Congress followed suit with the 1984 cable act, limiting both local franchise and FCC authority. The cable television regulatory pendulum swung back fairly quickly as Congress, reacting to subscriber rate hikes and allegations of poor service, passed a highly re-regulatory bill in 1992. By 1996 the mood of Congress had changed again, and much of the rate regulation established in 1992 was rescinded effective in 1999.

Although many industry leaders praise deregulation in theory, in practice they sometimes argue against it because regulation often protects them from increased competition. For example, a 1980 staff report on the regulation of television networks indicated that network regulation had utterly failed to bring about its stated goals of increased program diversity. Instead, the staff indicated that the regulations had stifled competition, and they recommended that the FCC undo much of its existing network regulation. When the FCC attempted to carry out that recommendation, industry lobbyists, fearful of losing their protected positions, brought the efforts to a standstill. Significant network deregulation was not accomplished for almost a decade, and then much of it was attained through litigation. Likewise, leaders in the cable television business argue for deregulation when it is in their best interests, regarding subscriber rates or signal carriage rules, for example. But the industry very much opposed the deregulation of the cable-telco crossownership rules, not wanting the increased competition from the telephone companies.

Technical Standards

In the past the FCC assumed responsibility for setting technical standards for a new technology when it was made commercially available. Universally mandated standards protect the public—at least to a point—from investing in new products that might later prove incompatible with existing products or of less than optimum quality. The deregulatory approach, however, tends to reject official standard setting, leaving choices to the marketplace.

The issue was defined when the Commission declined to select a specific standard while approving AM stereo in 1982, a precedent for a decade of future refusals to impose standards. Similar "nondecisions" followed regarding DBS, teletext, and other standards, indicating that economists had superseded engineers in Commission policy making, even in largely technical matters.

The "marketplace" in each case actually consisted not of consumers buying receivers but of station owners, operators, and manufacturers choosing equipment. Broadcasters would have to decide for themselves which transmission standard to use, hoping that set makers would eventually gear up to supply receivers that consumers would buy.

In AM stereo's case, that did not happen—a decade later few stations were broadcasting in stereo and few consumers had stereo-equipped AM receivers. Both the industry and the Commission recognized that deregulation had failed to achieve its goals. Finally, acting on congressional mandate, the FCC in 1993 selected Motorola's "C-Quam" system, although few observers expected the decade-late decision to make much difference for beleaguered AM licensees.

Industry debate over the FCC's approach to technical standards rose to a fever pitch over high-definition television. At issue were the same questions: Should government help industry select specific standards, and would selection of a standard stifle technical development? But HDTV's financial stakes were far higher—the eventual replacement of all the country's broadcast and consumer television equipment if HDTV succeeded, and concerns about further imbalance of international trade if HDTV devices were largely imported. Perhaps having learned a valuable lesson from its AM-stereo debacle, the Commission in the 1990s began the arduous but vital process of setting standards for this new technology, with billions of dollars riding on the outcome. Finally, after years of testing and compromise, the FCC adopted an industry-developed transmission standard for advanced television services in 1996 (see Section 5.7).

Deregulation Critique

Deregulation has had many positive effects. It greatly speeded FCC actions, eliminated outdated rules, encouraged development of new technology, and gave audiences more program choice.

But deregulation has had negative results as well. Although too much regulation can prove harmful, some areas still need closer governmental oversight. Congress, often cool to deregulation, has shifted fundamentally toward a more traditional oversight role for government—and in several areas has forced the FCC to take a more active regulatory stand. Indeed, the FCC can see limits on some of its own earlier decisions:

- Abandonment by the FCC of many record-keeping requirements as a money-saving tactic in the early 1980s left blanks in knowledge about electronic media. For example, discontinuation of annual broadcasting and cable financial reports limited the FCC's own understanding of the changing financial health of broadcasters and cable systems amid substantial transition.
- Deregulation—no matter what the source—tends to put business interests ahead of those of the public. With its 1984 cable act, for example, Congress curbed the right of cities to control cable subscription fees, allowing systems to raise rates freely, given their monopoly status in most markets. Subsequent cable price increases and poor service pushed Congress to reinstate FCC cable oversight eight years later, though the Commission argued against the renewed powers, fearful of the multimillion-dollar cost and staff time they would require.
- The FCC's 1980s vision of all human activity reduced to simplistic economic terms ("the marketplace") treated programs as the equivalent of manufactured goods with little concern for diversity, quality, or cultural content—what then-chairman Mark Fowler called "a toaster with pictures."
- Sometimes deregulation simply does not deliver the results expected by Congress. Fully a year after passage of the Telecommunications Act of 1996, which was expected to open the way for new competition in the cable television industry as well as local and long-distance telephony markets, many of the expected changes have not happened. The telephone companies' ardor for entering the cable television business waned rapidly, though there were some exceptions (see Section 6.4). And the regulatory requirements for permitting increased competition in telephony effectively delayed new entrants in those markets.

12.11 *Other Regulation*

Many other federal laws, as well as state and municipal statutes, affect electronic media. The following sections briefly survey some of the most relevant.

Treaties

The tendency of radio signals to ignore political boundaries necessitates international treaties to prevent interference. The United States and its neighbors have entered into separate regional treaties governing AM, FM, and television broadcasting. AM agreements cover the widest territory, because long-distance sky-wave propagation affects the scattered islands of the Caribbean as well as the two common-border nations, Canada and Mexico; agreements on simpler FM and television allocations have also long been in place with both countries.

On American international telecommunications policy, the FCC works closely with the U.S. State Department and the National Telecommunications and Information Administration (NTIA)—an agency of the Department of Commerce that acts as the president's chief adviser on telecommunications questions. The FCC provides technical expertise and helps coordinate private-sector participation.

Press Law

Electronic media share with print media a body of laws, precedents, and privileges known as the law of the press. Typical areas include defamation, obscenity, trial coverage, freedom of access to information, right of privacy, labor laws, and copyright (many of which are discussed in Chapter 13).

One other area, *reporters' privilege*, concerns the asserted right of news personnel to withhold the identity of news sources and to refuse to surrender personal notes, including any audio and video "outtakes." This and many other aspects of press law fall under state jurisdiction, with results varying widely from state to state.

Regulation of Advertising

The FCC itself has no authority to punish licensees directly for unfair advertising. All it can do is cite fraudulent advertising as evidence that an applicant or licensee lacks the requisite character qualification to be a licensee and refer cases of deceptive advertising to a sister agency, the Federal Trade Commission (FTC).

The FTC settles most cases of alleged advertising deception by *stipulation*, an informal (and hence time-saving) way of getting advertisers to drop objectionable practices voluntarily. If a formal complaint becomes necessary, the FTC can seek a *consent order*, another nonpunitive measure under which the advertiser agrees to stop the offending practice without admitting guilt. Actual guilt has to be proved before the FTC can obtain a *cease and desist order* forcing compliance with the law. Such orders can be appealed to the courts. The appeal process can mean long delays in bringing the objectionable advertising to an end if the advertiser chooses to use it.

Even this clumsy machinery ground to a halt in the 1980s when the FTC chairperson declared that market forces rather than government regulations should protect consumers from exploitation. However, as the 1990s dawned, the FTC showed signs of renewed vigor. For example, in mid-1990 the FTC Bureau of Consumer Protection obtained its first consent order canceling allegedly fraudulent program-length commercials. The offending company, which sold "cures" for baldness, impotency, and obesity, agreed to pay refunds to consumers and to identify future program-length commercials as paid advertisements.

The only legal product officially banned from broadcast and cable advertising is cigarettes and related tobacco products. A federal statute adopted in 1971 imposed the ban. Many broadcasters claim that they were unfairly singled out because cigarette ads continue to appear—even increase—in other media.

As noted elsewhere, broadcasters and distilled spirits producers traditionally have voluntarily refrained from airing ads for hard liquor products. This total abstinence ended in 1996 as several stations and cable networks began running ads for Seagrams. This led to policy discussions in Congress, the FTC, and the FCC, which many broadcasters fear may result in a ban on all liquor advertising, including wine and beer spots.

The FCC does have jurisdiction over commercialization levels for children's television. A 1990 act of Congress limited the time that could be devoted to advertising in children's programs to 10.5 minutes per hour on weekends and 12 minutes per hour on weekdays. The FCC is in charge of enforcing these rules, and the Commission has assessed thousands of dollars in fines against stations violating these limits.

Lotteries

The advertising of illegal lotteries can subject a licensee to a fine and imprisonment for each day's offense, according to a provision of the U.S. Criminal Code (18 USC 1304). This statute concerns broadcasters because they frequently use contests in advertising and promotional campaigns. They must take care not to let an innocent promotional gimmick turn into an illegal lottery—defined as a *chance* to win a *prize* for a *price.*

A contest becomes a lottery if it requires participants to pay any kind of fee or go out of their way (*price* or *consideration*), chooses the winner by lot (*chance*), and awards the winner something of value (*prize*). Contests involving these elements can get a station into serious trouble. The advertising of legal city and state lotteries is exempted from the antilottery statute.

State and Local Laws

Under the Constitution, federal laws prevail over state laws in subject areas designated as being of federal concern. This means that a state law cannot rise above the Communications Act. Nevertheless, state laws govern many electronic media activities that are not covered by federal statutes. Scores of state laws affect print and electronic media, especially those concerned with

- individual rights (regarding defamation and privacy, as discussed in Section 13.4),
- advertising of specific products and services;
- noncommercial broadcasting (many states have commissions to coordinate statewide public radio and television activities);
- standard business operations (state taxes, for example); and
- aspects of cable television—including franchising or cable right-of-way disputes, theft of service, and the attachment of cable lines to utility poles. Municipalities, too, have enacted many local regulations on cable franchising.

Informal Controls

We have seen how all three branches of government—executive, legislative, and judicial—participate in the formal regulatory process. These forces divide the FCC's allegiance. The Commission finds itself under constant and often conflicting pressures not only from the White House, Congress, and courts but also from industries it regulates, lobbyists representing special interests, and the public.

Although Congress gave the FCC a mandate in the Communications Act of 1934 to act on its behalf, it continually brings the Commission up short if it wanders far afield. In addition to Senate approval of nominations to the FCC and congressional control of federal budgeting, Congress conducts frequent *oversight* hearings on the Commission's performance and plans.

Since the 1920s the White House has found ways, usually indirect, to influence electronic media policy. In the early 1990s, for example, the Bush administration slowed the pace of federal regulatory efforts—including new FCC rules—through the activities of the "Council on Competitiveness," a pro-business group chaired by Vice President Dan Quayle and abolished by the Clinton administration in early 1993. The Clinton administration has taken a much more activist approach to telecommunications policy making. Twice during his first term President Clinton invited electronic media leaders to the White House for policy discussions, once on the topic of violence and again to discuss children's television requirements. Vice President Al Gore has also been a leading administration spokesperson for telecommunication reform. The overall effect of White House pressure varies with each administration and its interest in communications matters.

Consumer Action

The highlight event that crystallized interest in broadcast reform came in the consumerism heyday of the 1960s. For the first time, the WLBT case (see Exhibit 12.f) gave a station's audience the legal right to file complaints against a station and participate in license renewal proceedings. Hundreds did, though with relatively little lasting impact.

In recent years consumer action has increasingly taken the form of single-issue groups with an ax to grind. The threat of a *boycott* has become the strongest action such groups can take to try to influence electronic media behavior. Boycotts involve refusal to buy advertised products. In a pluralistic society, boycotters have difficulty achieving sufficient consensus and discipline to do substantial economic damage.

Concern over abortion, gun control, prayer in the schools, sex and violence in the electronic media, and other "family-values" issues heightened after 1980 (indeed, they were central to the 1992 and 1996 presidential campaigns). Programs and ads debated them all. Protests stimulated widespread news coverage.

Even noncommercial broadcasting, supposedly freer to present more points of view because it does not cater to advertisers, has succumbed to group pressure. Programs on gay lifestyles, avant-garde art, and Third World political movements have all been criticized—and sometimes temporarily removed from the air.

Self-Regulation

Many professions and industries adopt voluntary codes of conduct to cultivate favorable public relations and forestall abuses that might otherwise bring on government regulation. But codes that restrict the freedom to compete run afoul of antitrust laws.

The National Association of Broadcasters (NAB) began developing a code for radio programming and advertising practices in the late 1920s and extended them to television in 1952. These codes used broad generalizations and many "shoulds" and "should nots" but left most decisions to the discretion of station management. Nonetheless, the codes reinforced generally observed standards, especially as to the amount of time broadcasters should devote to advertising.

Despite NAB precautions to avoid any hint of coercion, the Department of Justice brought suit against the Television Code in 1979. The suit alleged that code time standards "artificially curtailed" advertising, repressing price competition and "depriving advertisers of the benefits of free and open competition." In 1982 a federal district court approved a consent decree by which the NAB disbanded its code-making activities.

For a time, networks and many stations continued to follow their own internal standards—some even tougher than the former NAB codes. By the 1990s, the networks had cut back their program standards departments as part of more stringent budget controls. In an effort to restore a sense of industry standards, several members of Congress in 1997 proposed legislation that would exempt electronic media from anti-trust laws so that industry codes of conduct could again be developed.

Much self-regulation results from media worries about externally imposed controls. As explained earlier, broadcasters generally refrained from carrying hard liquor ads because they feared a Congressional reaction (a fear that was looking prophetic in mid-1997). Similarly, broadcast and cable's agreement to monitor violent program content and develop a television rating code resulted from outside pressure, not altruistic motives.

Press Criticism

Coverage of electronic media in both the trade and popular press often affects policy. Congress, the White House, the FCC, and other agencies closely follow reporting about the media, seeking reports of their actions.

Reviewers in major newspapers and magazines seem to have more impact on news and public-affairs programs than on entertainment shows. Further, they influence producers of such programs more than they do the general audience.

The First Amendment includes the Internet
PROTECT OUR CHILDREN
XXX
Porn Poison's Kids

Chapter 13

Constitutional Issues

CHAPTER CONTENTS

The First Amendment to the U.S. Constitution, together with the Fourteenth (Exhibit 13.a), prohibits government regulation of speech and press, yet the Communications Act imposes federal licensing and other limitations on those who own electronic media. This paradox isn't unique, as society often demands a balance between the ideal of *absolute* individual freedom and the practical need to limit speech that might harm others (for instance, one may be punished for yelling "fire" in a crowded theater if there is none). The essence of the constitutional controversies discussed here is how to compromise between these opposing goals.

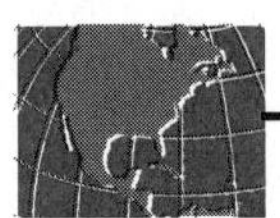

13.1 *First Amendment*

The freedoms of speech and press guaranteed by the First Amendment were intended from the start—and the courts have construed them since—to encourage a wide-open *marketplace of ideas.* First Amendment theory holds that ideas and opinions from different sources should compete in such a marketplace. As the Supreme Court noted in a landmark electronic media decision, "It is the purpose of the First Amendment to preserve an uninhibited marketplace of ideas in which truth will ultimately prevail, rather than to countenance monopolization of the market" (US, 1969: 390).

Although freedom of expression is only part of one of the 10 amendments that make up the Bill of Rights, that freedom has played a pivotal role in the American political system (see Exhibit 13.a). In the words of Supreme Court Justice William O. Douglas, the First Amendment "has been the safeguard of every religious, political, philosophical, economic and racial group amongst us" (US, 1951: 584).

"No Such Thing as a False Idea"

"Under the First Amendment," said the Supreme Court, "there is no such thing as a false idea. However pernicious an opinion may seem, we depend for its correction not on the conscience of judges and juries, but on the competition of other ideas" (US, 1974: 339)—again, the marketplace metaphor.

The First Amendment *encourages* disagreement. "A function of free speech under our system of government is to invite dispute," wrote Justice Douglas. "It may indeed best serve its highest purpose when it induces a condition of unrest, creates dissatisfaction with conditions as they are, or even stirs people to anger" (US, 1949: 4).

Anger certainly arose during the 1980s in Dodge City, Kansas, when station KTTL-FM broadcast daily hour-long sermons by two fundamentalist ministers attacking Jews, blacks, and other groups. They urged listeners to ignore police officers and attack such groups at will. The invective poured out in such abundance that several local groups protested renewal of KTTL's license when it expired in May 1983.

A huge media uproar resulted, with a congressional subcommittee hearing and extensive press coverage of the station owner's extremist conservative views favoring local armed vigilantes. In mid-1985 the FCC designated the license for a comparative hearing with another applicant for the same frequency. In a controversial decision to renew, the FCC said that the First Amendment protected the broadcasts, offensive though they might be to many listeners (FCCR, 1985a). The Commission found that because the material did not present a *clear and present danger* to the public, a test long established by the U.S. Supreme Court, it qualified as protected speech (US, 1919: 52). KTTL's exhortations to take illegal or dangerous actions did not, as the FCC had said in an earlier case, "rise far above public inconvenience, annoyance, or unrest" (FCCR, 1972c: 637).

Government Censorship

Many assume that the First Amendment also provides protection from private parties. But in fact the amendment aims at protecting people only from government (censorship), not from one another (broadcaster or cable editorial decisions).

Exhibit 13.a

"Congress Shall Make No Law . . ."

The First Amendment protects four fundamental rights of citizens that governments throughout history have had the most reason to fear and the greatest inclination to violate—freedom to *believe*, to *speak*, to *gather together*, and to ask rulers to *correct injustices*. The amendment conveys all this in only 45 words, of which just 14 guarantee freedom of expression:

> Congress shall make no law respecting an establishment of religion, or prohibiting the free exercise thereof, or abridging the freedom of speech, or of the press; or the right of the people peaceably to assemble, and to petition the Government for a redress of grievances.

These words limit not only Congress but also state and local governments, thanks to the Fourteenth Amendment, passed in 1868, which says, "No state shall make or enforce any law which shall abridge the privileges or immunities of citizens of the United States. . . ." Section 326 of the Communications Act of 1934 explicitly extends the First Amendment's protection to broadcasting:

> Nothing in this Act shall be understood or construed to give the Commission the power of censorship over the radio communications or signals transmitted by any radio station, and no regulation or condition shall be promulgated or fixed by the Commission which shall interfere with the right of free speech by means of radio communication.

Source: Photo from FPG International.

Station, system, and network officials who edit, cut, bleep, delete, revise, and otherwise mangle programs may be guilty of bad judgment, excessive timidity, and other faults, but editorial control is not censorship, and it does not violate the First Amendment.

Such control becomes a violation only when it results from the kind of government intrusion known as *state action*. In promoting the "family viewing" concept in the 1970s, for example, the FCC attempted to reduce television violence not by rule but by "jawboning"—pressuring the television industry to regulate itself. When the NAB responded to this pressure by amending its Television Code, a court construed this private action as state action in violation of the First Amendment (F, 1979: 355).

Prior Restraint

The principal evil that First Amendment authors intended to address is *prior restraint*—that is, preventing someone in advance from saying, showing, or otherwise publishing something. Only in the most extreme and exceedingly rare case will a court issue an *injunction* barring, for example, a television program from broadcasting an allegedly libelous story (see Section 13.4). If libel does occur, the remedy is punishment after the fact, not censorship before.

Religious Freedom

Another First Amendment clause guarantees religious freedom. The active role of television evangelists in the national elections of 1980 and since (especially the candidacy of Pat Robertson for the GOP nomination in 1988, and some GOP platform planks in 1992) disturbed many people who were sensitive to First Amendment concerns about establishment of any specific religion. Yet the same amendment protects the evangelists' right to have their say.

Doubly protected by the freedoms of speech *and* religion, stations owned by religious groups have often claimed near-immunity from FCC requirements because they regard their right of religious freedom (from any government action) as absolute. Though the FCC has held religious licensees to the same standards of regulation for all broadcasters, it has resisted close monitoring of any licensees. On the other hand, cable operator arguments that rules requiring them to carry religious television stations violated the systems' own First Amendment rights to religious freedom have proved unsuccessful.

13.2 Broadcasting's Limited Rights

Just what do the First Amendment words *speech* and *press* mean in today's context? Speech encompasses not only that amplified by public-address systems but also film, broadcasting, videos, cable, and the Internet. And the press includes pornographic magazines. Can *all* ways of expressing oneself and *all* forms of the press claim equal First Amendment rights?

Broadcasting vs. Print

More specifically, should all regulated electronic media have First Amendment *parity*—equality—with unregulated media, especially the printed press? The answer has generally been "no"—the First Amendment protects some media more completely than others. Broadcasting, it is argued, has unique attributes that justify imposing certain regulations that would violate the First Amendment if imposed on newspapers. The three main arguments favoring this assumption follow.

Channel Scarcity

Not everyone who wants to own a station can do so, because intolerable interference would otherwise result. Because of this channel scarcity, the government (the FCC) has to choose among applicants. Where mutually exclusive applications seek facilities whose activation would cause interference, only one license can be granted—thereby abridging the freedom of other applicants. In contrast, anyone who can afford it can publish a newspaper or magazine without limit or license.

Development of new and improved services *has* decreased scarcity as a justification for regulating broadcasting. The huge increase in the number of stations since Congress wrote the 1927 Radio Act—from about 600 to more than 15,000, plus more than 11,000 cable systems—makes scarcity a relative factor at best. And although demand for channels in major markets continues, cable, MMDS, and DBS provide a multitude of additional channels—converting scarcity into abundance.

In contrast, others point to the many applicants for desirable stations that become available—and the huge prices often paid for them—to suggest that scarcity continues. They argue that availability of cable channels does not fully relieve scarcity because cable reaches about two-thirds of the population while broadcasting reaches virtually everyone. In any event, cable, MMDS, and DBS differ sharply from "free" broadcasting in requiring subscription fees over and above receiver expenses.

Others argue that even if spectrum scarcity does exist, that fact does not justify treating broadcasting differently from the printed press in terms of the First Amendment. Judge Robert Bork, in a case before the U.S. Court of Appeals, addressed the issue this way:

> . . . the line drawn between the print media and the broadcast media, resting as it does on the physical scarcity of the latter, is a distinction without a difference.
>
> It is certainly true that broadcast frequencies are scarce but it is unclear why that fact justifies content regulation of broadcasting in a way that would be intolerable if applied to the editorial process of the print media. All economic goods are scarce, not the least the newsprint, ink, delivery trucks, computers, and other resources that go into the production and dissemination of print journalism. Not everyone who wishes to publish a newspaper, or even a pamphlet, may do so. Since scarcity is a universal fact, it can hardly explain regulation in one context and not another. (F, 1986:508)

According to this argument, economic scarcity is what drives media ownership, not physical scarcity. Anybody *can* speak via a broadcast station—as long as he or she has enough money to buy one.

Conflict in Licensing

When writing the Communications Act, Congress explicitly confirmed "the right of free speech by means of radio communication" (see Exhibit 13.a). Yet the Act also required the FCC to grant or renew licenses only if they serve the public interest. It also told the FCC to ensure that candidates for public office had equal opportunities to use broadcasting.

Any such requirements—by definition—place limits on licensees' freedom of speech. That fundamental contradiction came before the Supreme Court in the 1943 *NBC* case, where broadcasters argued that the FCC had the right to regulate only *technical* factors, and that any further regulation violated the First Amendment. But the Court emphatically rejected this argument, saying:

> We are asked to regard the Commission as a kind of traffic officer, policing the wave lengths to prevent stations from interfering with one another. But the Act does not restrict the Commission merely to supervision of the traffic. It puts upon the Commission the burden of determining the composition of that traffic. (US, 1943: 215)

Intrusiveness

Radio and television enter directly into virtually every U.S. household, readily available to people of all ages and types. Because of the intrusive nature of the medium, people never know exactly what they will see or hear when they first turn on their set, and that initial exposure might be upsetting. The medium also is easily accessible to children; it requires no special training to turn on a television or radio and flip through the channels or stations. Children who are not yet old enough to read can still use a broadcast receiver even when their parents are not around. Because of its intrusiveness and easy accessibility to children, Congress, the courts, and the FCC have often imposed content regulation on broadcasters that would not be allowed in the print media. Some content acceptable in other media—indecency, for example—may be regarded as intolerable in broadcasting, at least when children form a substantial portion of the audience.

First Amendment purists respond that where such distinctions need to be made, the marketplace can make them better than government. If people object to a program, they can—and do—complain to station and system operators, networks, or advertisers, or they can simply hit the "off" button. Regarding protection of children, opponents of content regulation argue that parents should be responsible for the viewing and listening behavior of children, not the government.

13.3 First Amendment Status of Other Electronic Media

Because of the factors noted in the previous section, broadcasting enjoys substantially less First Amendment protection than the printed press.

Generally speaking, as new electronic media emerge, they are not regulated (and thus match the print model) until and unless government takes some specific action—Congressional attempts to control indecency on the Internet, for example. Government must then discern how those media should be treated in terms of the free speech and free press guarantees. The Supreme Court has repeatedly stated that, for First Amendment purposes, each new medium must be judged based on its own unique characteristics. When judging new media, the government tends to analyze them in terms of what is already familiar. Often the analysis comes down to "is the new service more like print or more like broadcasting?"

On a scale of First Amendment protection, cable television systems fall somewhere between the print and broadcast media. Courts consistently find that cable is much different than a broadcast station in terms of its physical characteristics. Cable systems do not use the electromagnetic spectrum in the same way broadcasters do, so spectrum scarcity typically is not an issue. Because cable is a subscription service, cable is not considered as intrusive as broadcasting; subscribers choose to bring the service into their homes every month as they pay their cable bills. For similar reasons, cable is not considered as accessible to children. Plus, cable systems have long provided special lock boxes that can lock out channels that parents may find unsuitable for children.

For these reasons, certain programming restrictions that have been upheld for broadcasters—indecency regulations, for example—have been judged invalid when applied to cable television systems. Critics consider this rather curious. Most consumers do not think much about how their television signals are delivered so long as they get a good picture with good service. But under our First Amendment rulings, television content is regulated differently based on how it is delivered to the home. On the other hand, courts have approved content-related regulations for cable systems—the must-carry rules offer a good example—that would never be approved if applied to newspapers or magazines.

Other electronic media whose First Amendment status has been considered by the courts have been treated similarly. Speech and press protections for telephone systems, DBS, and the Internet so far seem to fall somewhere between the protections afforded newspapers and broadcasters. In all cases, the courts must balance the interests of those seeking the broadest protection of speech and those arguing for protection from certain types of content.

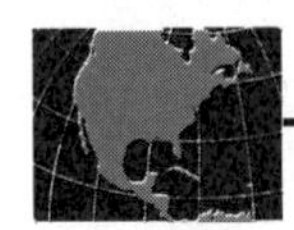

13.4 *Things You Can't Say*

Despite that seemingly strict First Amendment command "Congress shall make no law . . . ," government *does* in fact make laws that punish *unprotected* speech. This includes defamation (libel and slander), invasion of privacy, and obscenity.

Libel—Protecting Reputations

Libel is *defamation* by published untrue words that may expose its subject to public hatred, shame, or disgrace. Spoken defamation is called *slander*; but because electronic media spread spoken words far and wide, and because the words are often preserved on tape, broadcast or cable defamation is treated as libel.

Libel laws constitute another example of conflicting social interests. Although libel should be punishable because society has an interest in protecting its citizens, society also needs to expose official corruption and incompetence. Harassing libel suits can serve as a screen to protect dishonest politicians. In the United States all public figures, not just politicians, must be prepared to face harsh, sometimes even unfair and ill-founded criticism from the media, because it is extremely difficult to prove the *actual malice* standard necessary for their libel suit to succeed (see Exhibit 13.b).

By the early 1990s the primary sources of libel trouble for electronic media were radio "shock jocks" such as Howard Stern and radio call-in

Exhibit 13.b

Times *v.* Sullivan
Libel Landmark

The leading case establishing the relative protection of media from libel suits filed by public figures occurred during the civil rights protests of the 1960s. By chance it involved an instance of "editorial advertising," not investigative reporting. Supporters of the Montgomery, Alabama, bus boycott protesting segregation bought a large display advertisement in the *New York Times* that criticized Montgomery officials. Some of the statements in the advertisement were false, although they apparently were not *deliberate* lies. Sullivan, one of the officials, sued for libel in an Alabama court (all libel suits have to be brought at the state level). The court awarded Sullivan a half-million dollars in damages, and the Alabama Supreme Court affirmed the decision.

On appeal to the U.S. Supreme Court, the *Times* won a reversal. Criticism of public officials, said the Court, had broad First Amendment protection. Even though some of the allegations against the unnamed officials were untrue, they did not constitute libel. Argument over public officials, the Court continued, should be "uninhibited, robust, and wide-open." It may include "vehement, caustic, and sometimes unpleasantly sharp attacks on government and public officials." Such freewheeling debate would be discouraged if, in the heat of controversy, the critic must pause to weigh every unfavorable word:

> The constitutional guarantees require, we think, a federal rule that prohibits a public official from recovering damages for a defamatory falsehood relating to his official conduct unless he proves that the statement was made with "actual malice"—that is, with knowledge that it was false or with reckless disregard of whether it was false or not. (US, 1964: 279)

Subsequent libel cases broadened the term *public officials* to include anyone who, because of notoriety, could be classed as a *public figure*. People so classified have little chance of bringing a successful libel suit against the media. Even when stories about public figures are false, plaintiffs find it exceedingly difficult to prove deliberate malice.

shows (because licensees are liable for any defamation spoken over the air by those calling in).

A 1992 study by the Libel Defense Resource Center traced some 300 libel trials dating back to 1980 and found that the media won only a quarter of the jury trials in those 12 years—though only 58 of 167 initial awards were finally upheld on appeal and judgments paid. The same study found that awards averaged more than $8 million in 1990 and 1991. Libel damage awards came down later in the decade. The Libel Defense Resource Center reported that of the six cases in 1996 in which juries awarded libel damages, the average award was about $3.8 million, but that was more than double the average award of the previous year.

In fact, about 90 percent of all such libel suits are resolved long before they reach trial: they are dropped, or settled, or dismissed by a judge during pretrial proceedings. Even where the media win, they spend a lot of money defending themselves. More important, such suits have a *chilling effect* on investigative reporting.

Preserving Privacy

Privacy as an individual right, though not spelled out in the Constitution, is implied in the Fourth Amendment: "The right of the people to be secure in their persons, houses, papers, and effects . . . shall not be violated." Individuals have several privacy rights:

- the right to physical solitude
- protection from intrusion on private property or publication of the details of one's personal life

- protection from being presented in a "false light" (for example, being photographed walking on campus and then being included in a visual accompanying a television feature about drug use in college)
- protection from unauthorized use of one's name or image for commercial gain

Although the courts have held that public officials, performers, and people involved in news events have a lesser right to privacy because of legitimate public interest in those persons or events, privacy laws still limit the media by generally supporting individual rights.

Restricting Court Coverage

The due process clause of the Fifth Amendment ensures fair play to persons accused of crimes. The Sixth Amendment also spells out some of the elements of due process, among them the right to a fair and public trial. Ordinarily, news media freely cover trials, but that coverage can subject participants to such extensive publicity that a fair trial becomes impossible. In this *free press vs. fair trial* confrontation, the constitutional rights of the media sometimes have to give way to the constitutional rights of defendants.

For decades there was virtually no live or recorded radio or television coverage of trials. After messy coverage of the Lindbergh kidnapping trial (see Exhibit 13.c), the American Bar Association (ABA) recommended in 1935 that judges discourage radio and photo coverage because it tended "to detract from the essential dignity of the proceedings, degrade the court, and create misconceptions with respect thereto in the mind of the public."

By the 1970s, less intrusive equipment and seemingly more mature broadcast journalism judgment, persuaded the ABA to recommend that judges be given wider latitude in allowing photo and video coverage of trials. In 1981 the Supreme Court—although it bans cameras and microphones in its own proceedings—noted the improved technology, holding that "the risk of juror prejudice . . . does not warrant an absolute constitutional ban on all broadcast coverage" (US, 1981: 560). By the early 1990s all but a handful of states either allowed cameras in their courtrooms or had conducted experiments to that end. That the American public has an apparently insatiable appetite for televised real-life trials is demonstrated by the arrival in 1991 of Court TV, a national 24-hour cable courtroom service, and by the unprecedented coverage given to the prosecution, beginning in 1994, of football star turned media personality O. J. Simpson for the murder of his former wife and her male friend. Wanting to avoid a second media circus, the judge in Simpon's subsequent civil trial refused to allow television coverage of the proceedings.

Federal courts have remained off limits to such coverage, despite repeated media attempts to breach that barrier. Late in 1994, after a multiyear experiment with television coverage of civil cases in six federal district and two appeals courts, the Judicial Conference of the United States voted to restore its flat ban on cameras in federal courtrooms. For this reason, cameras and recording equipment were not allowed in the courtroom during even high-profile federal proceedings such as the Timothy McVeigh, Oklahoma City bombing trial.

Broadcast Hoaxes and Undercover Reporting

While the broadcasting of "news" that isn't really news dates back at least to the 1938 Orson Welles' broadcast of *War of the Worlds*, there have been relatively few cases of intentional news hoax broadcasts—until recent years. Indeed, a rash of cases in the early 1990s convinced the FCC that it needed stronger rules and a means of fining stations rather than just sending letters of concern. Several stations broadcast stories of false murders, one station said a nearby trash dump was about to explode because of methane gas build-up, another used the Emergency Broadcast System and said a nuclear attack was underway, and still another reported that a

Exhibit 13.c

Covering Courts—Changing Pictures

(A)

(B)

(C)

Broadcast coverage of court trials began with the 1925 Scopes case and reached the height of overkill with the 1935 trial of Bruno Hauptmann for the kidnap and murder of Charles Lindbergh's son. Reaction to that intrusive coverage by radio and film led to the ABA's "Canon 35," which banned trial coverage for decades. (A) An experiment in television coverage in the 1960s, Billie Sol Estes's trial for fraud, led to a Supreme Court reversal because of television intrusiveness. (B) As restrictions on trial coverage declined in the 1980s, sometimes lurid trials came into people's homes—such as the 1992 rape trial of William Kennedy Smith. The jury found him not guilty, but the trial became known for its often graphic testimony and the media's use of a fuzzy dot over the victim's face to protect her identity. (C) Court coverage reached historic proportions in 1994 and 1995 when live cameras, hour after hour, watched every move by prosecution and defense in the bizarre case of sports and media personality O. J. Simpson, charged with murdering his former wife and her friend.

Source: Photo (A) from UPI/Bettmann Archive; (B) and (C) from AP/Wide World.

volcano had erupted in a suburban Connecticut town. All were presented as real events, and emergency forces responded accordingly.

In 1992 the Commission adopted a rule against such "harmful" hoaxes—those that might divert police or fire or other safety forces from real events. The FCC made clear that it had the First Amendment in mind as the rule was developed and, hence, limited its actions to cases where "broadcast of such information will cause substantial public harm, and broadcast of the information does in fact directly cause such harm." Such broadcasts can result in fines of up to $25,000 for each day on which a harmful hoax occurs—to a maximum FCC fine of $250,000 for repeated violations. Several broadcasters agreed that the industry had not adequately policed itself and thus had brought about this action.

Improved technology has made another type of hoax readily available. As cameras get smaller and operate at lower light levels, options for use of hidden cameras increase. Of growing concern in the 1990s are the number of station and network or syndicated "undercover investigations" in which reporters using hidden cameras impersonate another person (*not* a reporter) to obtain information. This process of omission (not making clear that a reporter is asking questions) raises substantial questions of ethics.

In 1996, ABC News learned firsthand about the dangers of undercover reporting when a North Carolina jury awarded $5.5 million to Food Lion supermarkets, which had sued ABC News over its news gathering techniques. ABC producers had misrepresented themselves to Food Lion in order to get jobs with the company. While on the job, they used hidden cameras to record the repackaging for sale of allegedly spoiled and outdated meat. Food Lion did not sue ABC for libel or invasion of privacy; indeed, the accuracy of the report was not challenged in court. Instead, Food Lion sued on the basis of fraud, trespass, and breach of fiduciary duty by the newsmen in getting their positions with the company under false pretenses. ABC announced that it would appeal the decision, but journalists publicly worried that the award would have a chilling effect on covering important stories.

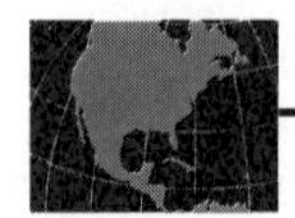

13.5 Obscenity and Indecency

One of the most controversial topics regarding electronic media concerns sexually explicit content. Although the terms "obscene" and "indecent" might be considered synonyms in polite conversation, in law they have different and specific meanings, though they both refer to sexually explicit material. If something is found to be "obscene," it has no First Amendment protection whatsoever. According to the Supreme Court,

> All ideas having even the slightest redeeming social importance—unorthodox ideas, controversial ideas, even ideas hateful to the prevailing climate of opinion—have the full protection of the [First Amendment] guarantees. . . . But implicit in the history of the First Amendment is the rejection of obscenity as utterly without redeeming social importance. . . . We hold that obscenity is not within the area of constitutionally protected speech or press. (US, 1957: 484)

"Indecent" speech, on the other hand, is protected expression under the First Amendment, but it can be regulated under certain conditions to protect children from being exposed to it. The major problem with this area of the law—for both students and the media—is defining and understanding the terms "obscene" and "indecent."

What Is Obscene?

The current definition of obscenity dates to the 1973 *Miller* case in which the Supreme Court upheld a California obscenity law. The decision reaffirmed the basic notion that obscene material does not receive protection under the First Amendment. The Court then emphasized that *community standards* vary from place to place: "It is neither realistic nor constitutionally sound to read the First Amendment as requiring that the people of Maine or Mississippi accept public depiction of conduct found tolerable in Las Vegas or New York City."

Nevertheless, the Court warned, state laws must carefully confine what they classify as *obscene* to "works which, taken as a whole, *appeal to the prurient interest in sex, which portray sexual conduct in a patently offensive way, and which, taken as a whole, do not have serious literary, artistic, political, or scientific value*" (US, 1973b: 24; emphasis added).

The *Miller* case, along with later decisions that added minor modifications, restricted obscenity censorship to "hard-core" materials, leaving to the states the difficult job of defining obscenity. Courts now interpret the First Amendment as preventing censors from taking such past arbitrary actions as

- condemning an entire work because of a few isolated obscene words;
- using outdated standards no longer common to the local community;
- applying as a standard the opinions of hypersensitive persons not typical of the general public; or
- ignoring serious artistic, scientific, literary, or political purpose in judging a work.

Generally speaking, courts must consider the "average person" when applying contemporary community standards to potentially obscene material. Historically those standards have been stricter for material available to children.

Limiting Indecency in Broadcasting

Sec. 1464 of the U.S. Criminal Code reads:

> "Whosoever utters any *obscene*, *indecent*, or *profane* language by means of radio communication shall be fined not more than $10,000 or imprisoned for not more than two years, or both." (18 USC 1464; emphasis added)

Note the use of *indecent* and *profane* in the statute. As with *obscene*, definitions and degrees of control have varied for these terms over the years. In a 1992 policy statement, the Commission defined indecency as "language or material that, in context, depicts or describes, in terms patently offensive as measured by *contemporary community standards for the broadcast medium*, sexual or excretory activities or organs" (FCCR, 1992b: 6464, note 4, emphasis added).

The Commission long remained in doubt as to its power to enforce the Criminal Code provision. Because of ready availability of electronic media to children, material acceptable in other media might be regarded as unacceptable by many in the audience. Furthermore, because broadcast and cable network services have national reach, they confront a great variety of local standards.

Broadcasting's traditional conservatism delayed its response to the tolerant social climate of the 1960s, but the liberalization of standards in other media had its effect. In the 1970s a "topless radio" fad triggered thousands of complaints to the FCC. The format invited women to call in and talk on the air about intimate details of their sex lives. Although such talk shows are commonplace today—on television as well as radio—one Illinois radio broadcast in 1973 triggered a flood of complaints. The FCC imposed what was then a fairly steep $2,000 fine on the offending FM station (FCCR, 1973: 919). The Commission had actually hoped that the station would contest the fine, thus precipitating a test case, but the station dutifully mailed in a check instead (for an amount equal to only a few minutes' advertising revenue in a major market).

Pacifica Decision

That same year, the FCC finally got its test case. A noncommercial station, WBAI-FM in New York, included in a discussion of social attitudes about language a recording of a nightclub act by comedian George Carlin. Called "Filthy Words," the monologue satirized society's hang-ups about seven sexually oriented words not likely to be heard on the air. This time, though, they were heard—no fewer than 106 times in 12 minutes. The single complaint came from a man who, as it later turned out, was associated with a group called Morality in Media. He heard the early afternoon broadcast with his teenage son—a crucial element in the case.

On the basis of that lone complaint, the FCC advised station management that the broadcast appeared to violate the obscenity statute. The licensee, Pacifica, challenged the ruling as a matter of First Amendment principles. The FCC received an initial setback in the appeals court but won Supreme Court approval of its reasoning. Focusing its argument on the Carlin monologue as *indecent* rather than obscene (as the Supreme Court had defined obscenity), the FCC stressed that the broadcast came when children would normally be in the audience.

The FCC argued that children need protection from indecency. Instead of meeting the First Amendment directly by flatly banning such material as the Carlin monologue the FCC said it should be *channeled* to a part of the day when children are least likely to be in the audience.

The channeling concept has precedence in nuisance law, which recognizes that something acceptable in one setting could be an illegal nuisance in others. The Supreme Court agreed with this rationale when, recalling that a judge had once said that a nuisance "may be merely a right thing in a wrong place—like a pig in the parlor instead of the barnyard," the Court added that if the FCC "finds a pig has entered the parlor, the exercise of regulatory power does not depend on proof that the pig is obscene." The Court also accepted the FCC's narrowing of community standards in adding the words *for broadcasting*, saying, "We have long recognized that each medium of expression presents special First Amendment problems. . . . And of all forms of communication, it is broadcasting that has received the most limited First Amendment protection" (US, 1978: 748).

Political Football

In 1987 a rising flood of complaints about perceived obscenity or indecency on the air caused the FCC to issue a statement announcing plans to extend enforcement of Sec. 1464 of the Criminal Code beyond Carlin's seven words at issue in the *Pacifica* case. The Commission reprimanded three radio stations for using overly explicit language, especially at times when children might be in the audience (FCCR, 1987a: 2726). The FCC argued that marginal program material (indecent, perhaps, but not obscene) might be scheduled during the midnight-to-6:00 A.M. period. For some time the FCC had informally allowed such material to begin as early as 10:00 P.M. The narrower definition of allowable time led to a court appeal.

Several broadcasters joined in challenging the FCC decision and in 1988 won reversal of the Commission action on the grounds that limiting indecent language to the midnight-to-6:00 A.M. hours violated the First Amendment. The court remanded to the FCC the question of how best to "promote parental—as distinguished from government—control" of children's listening (F, 1988: 1332).

Thereupon Congress entered this hotly controversial arena. In approving the FCC budget for the 1989 fiscal year, legislators (mindful that they were in the midst of an election campaign) ordered the FCC to ban indecent material at *all hours* of the day.

While the controversy raged in the courts, the FCC took action against a score of radio stations, fining them for various indecency violations. In 1991 the Supreme Court declined to review an appeals court ruling that the 24-hour ban was too broad a limit on freedom of expression (F, 1991: 1504). In 1992 the Commission initiated a rule-making procedure to define and justify a "safe harbor" for such programming and stepped up its prosecution of indecency violations at a number of radio stations.

Meanwhile, the 1992 funding act for public broadcasting included a provision restricting indecent programming on all stations, commercial and public, to the midnight-to-6:00 A.M. period—right back where the FCC was in 1987. Adding to the pressure was an appropriations bill containing just such a definition of safe harbor, signed by the president late in 1992. In early 1993 the FCC rule-making concluded (sagely) that the midnight-to-6:00 A.M. safe harbor seemed the best compromise for all concerned. Yet another appeal, however, knocked the safe harbor back to 8:00 P.M.-to-6:00 A.M. while arguments pro and con raged in court.

The battle continued late in 1993 when a three-judge panel of the U.S. Court of Appeals for the District of Columbia struck down the FCC's 6:00-A.M.-to-midnight prohibition of indecent programming as being unconstitutionally broad. In 1995 the full Court of Appeals apparently settled the issue by ordering the FCC to reinstate its original 10:00 P.M.-to-6:00 A.M. safe harbor. The court also stated that Congress could define the safe harbor as midnight to 6:00 A.M. without running afoul of the Constitution (F, 1995). The Commission appealed the decision but the Supreme Court declined to hear the case.

Enforcement

The Commission has initiated 52 forfeiture actions against stations broadcasting indecent material outside the safe harbor. Forty-nine of those actions were brought since the new enforcement standards were announced in 1987. All of the actions except one have been against radio stations, indicating that the Commission considers the fare of radio "shock jocks" more serious than the titillating discussions of daytime television talk shows or steamy afternoon soap operas (Singleton and Copeland, 1997).

Since 1987, fines have ranged from a low of $2,000 for a single offense to over $100,000 for repeated violations. By 1997, the typical fine against a station for its first offense was $10,000. Stations licensed to Infinity Broadcasting have been fined the most, due to the repeated alleged violation of the indecency standards by its popular shock jock Howard Stern. In 1995 Infinity agreed to pay $1.715 million in fines rather than contest the FCC's actions in court. This did not spell the end of Stern's (or Infinity's) problems with the Commission—complaints about his program have continued into the late 1990s.

Regulating Indecency on Cable

The cable act of 1984 provided for fines or imprisonment for anyone who "transmits over any cable system any matter which is obscene or otherwise unprotected by the Constitution" (47 USC 639).

Although the Supreme Court has held that cable television *does* have First Amendment protection, it has also said that such protection must be balanced against what it called "competing social interests," without specifying what those interests may be or how that might be done (US, 1986: 488).

States had previously tried to go even further by applying the Supreme Court's *Miller* and *Pacifica* decisions to questionable cable content. Four federal court decisions in the mid-1980s concluded that state laws banning cable indecency violated the First Amendment because they were too broad in scope. Because cable does not use open spectrum, the scarcity argument that had supported indecency limits in broadcasting could not be applied. The courts reasoned that cable is not as "uniquely intrusive" as broadcasting, nor as available to children, because subscribers pay for the service—and that *Pacifica* therefore did not apply to cable.

Congress also has attempted to regulate indecency on cable systems, with mixed results. The 1992 cable act for the first time gave system operators the authority to prohibit indecent programming on public and leased access channels and empowered system operators to enforce written policy on such programming. Additionally, the 1992 act required cable operators who did allow indecent programming on these channels to put all such programming on one channel and to provide that channel only to subscribers who wanted it (47 USC 612, 638). In 1996, the Supreme Court upheld the provision allowing cable operators to prohibit indecency on leased access channels, but declared the other requirements of the law unconstitutional infringements on cable programmers' and subscribers' First Amendment rights (U.S., 1996).

In 1996, Congress passed legislation requiring cable operators to block fully the audio and video of channels primarily dedicated to sexually oriented programming. The intent of this provision is to prevent non-subscribers from receiving any part of the channel. The law further required that until the full channel blocking is accomplished, the cable operator must protect the child audience by carrying the channel only at times of the day when children are not likely to be watching (47 USC 561). Several "adult" oriented programmers, including the Playboy Channel have challenged the new law in court. They also sought to stop enforcement of the law while the issues were being litigated, but the court declined their request. Thus, the law must be enforced while the legal challenges are pending.

In 1991, the FCC and the Justice Department agreed to divide up their overlapping concerns about obscenity and indecency. Justice would take primary responsibility for any actions against cable or pay-cable services, whereas the FCC would concentrate on broadcast cases (Ferris *et al.*, 1983–1992: 8–38/9).

Indecency and Other Electronic Media

Although no precedent exists as yet with respect to the indecency standards for other multichannel technologies such as DBS and wireless cable, they will most likely be treated similarly to cable television because they are primarily subscription services. The courts have spoken about the transmission of sexually explicit material on two other electronic media: the telephone system and the Internet.

Since the early 1980s the FCC had been trying to control the transmission of sexually explicit messages via the telephone, commonly known as "dial-a-porn," but several of its attempts at regulation were struck down by the Court of Appeals as being too restrictive. Finally Congress acted in 1988 with legislation that sought to prohibit dial-a-porn by making it a crime to transmit obscene or indecent commercial telephone messages. The Supreme Court unanimously held the statute to be unconstitutional as applied to the transmission of indecent messages. The Court noted that indecent speech was constitutionally protected for adults, and that as long as other techniques were available to block such messages from children, a total ban on telephone indecency violated the First Amendment (US, 1989). Consequently, so long as sexually explicit message providers take steps to make sure that children cannot access their services—through credit card sales, scrambling techniques, or verifiable passwords, for example—their transmissions of indecent material are protected.

The other great pornography battle ground of the 1990s is the Internet. Spurred by press reports of easily accessible sexual material on the Internet, plus anecdotal evidence of children's harmful exposure to this type of content, Congress, as part of its revision to the Communications Act in 1996, passed the Communications Decency Act (CDA). The CDA makes it a crime to make, create, solicit, initiate the transmission of, or display obscene or indecent material over the Internet to any person under the age of 18. Almost immediately after the CDA's passage, a coalition of free speech advocates, Internet users, and computer and telecommunications industry groups went to court, arguing that the CDA was unconstitutional.

In June of 1996 a special three-judge panel of the U.S. District Court in Pennsylvania agreed with the plaintiffs in the case. As a first step, the court noted that access to the Internet is much more complicated than access to broadcasting, implying that a different First Amendment standard should be applied to the Internet. According to the court,

> . . . Internet communication, while unique, is more akin to telephone communication, at issue in *Sable*, than to broadcasting, at issue in *Pacifica*, because, as with the telephone, an Internet user must act affirmatively and deliberately to retrieve specific information online. Even if a broad search will, on occasion, retrieve unwanted materials, the user virtually always receives some warning of its content, significantly reducing the element of surprise or "assault" involved in broadcasting. (FS, 1996: 851–52)

The judges noted that while the CDA only criminalized transmissions to minors, there was no way to know the age of an Internet user. Thus, content providers might stifle their messages, not knowing who might receive them. Since indecent speech is fully protected as to adults, the court concluded that this "chilling effect" on such content violated the First Amendment rights of the speakers. The decision was appealed directly to the Supreme Court. In June, 1997, the Supreme Court, using similar arguments, upheld the lower court's decision, thus making the CDA unenforceable with respect to indecency. Following the Court's decision, both Congress and President Clinton vowed to continue the fight against pornography on the Internet.

Seeking a Balance

Obscenity/indecency law as applied to electronic media will evolve as society's standards evolve. After three decades of increasing liberalization, during which many taboos fell and audiences grew more tolerant of explicit language and scenes, the pendulum for a while swung to a more conservative standard—restrictions aimed at preserving "family values." The country's trend toward political conservatism in the 1980s favored restrictions despite another conservative article of faith—reliance on the marketplace to set any rules.

On the other hand, no indication of a disappearing market for programs that pushed limits was evident in the early 1990s. Syndicated shows (such as *Studs*) and even network shows (such as ABC's *NYPD Blue* with its partial nudity and strong language) suggested that the search for what was acceptable—how far one could go—continued.

The 1990s' trend seemed to be toward more permissive interpretation of laws dealing with sexuality for cable as compared with broadcast programming. This difference arises largely because cable lacks the universal reach of broadcasting and comes into the home only after a conscious decision to subscribe—a decision reviewed monthly when the bill comes in.

Although an all-out invasion of X-rated cable programs has not yet occurred, cable channels such as Playboy and Spice regularly offer only minimally sanitized pornographic videos. Indeed, the owner of one of the nation's largest producers and distributors of X-rated materials has predicted that if the much-touted 500-channel future ever arrives, "there's going to be an enormous new market for adult entertainment" (*The Wall Street Journal*, 11 July 1994: A1).

13.6 Ownership

First Amendment theory stresses the value of many diverse and antagonistic sources of information and opinion. Unregulated competition can lead to monopoly control of media outlets, thus limiting the number of speakers in the market. Accordingly, to promote the goals of the First Amendment and prevent the monopolization of the electronic media, the government limits the number and kinds of media outlets that any one entity can own. This *diversification* of ownership and control has long been a major FCC goal.

Every broadcast station enjoys a limited monopoly—exclusive use of a given channel in a particular market. Similarly, wireless cable operators have monopoly control over specific frequencies in local markets. DBS operators have such control over their orbital slots. Cable, a monopoly service in most markets, has more far-reaching control. Once viewers subscribe, installers usually disconnect rooftop antennas so that home owners receive even broadcast stations only (or at least primarily) by cable. A cable operator thus has control over *all* video signals coming into some homes, other than rented or purchased tapes. As will be explained in this section, government policies on media control and ownership are designed to lessen the impact of these monopoly situations.

Antitrust Law

The Sherman Act of 1890 and the Clayton Act of 1914 together aim to prevent excessive concentration of control by one or a few companies over a particular segment of the economy. They provide the basis for such government actions as the one that led to the 1984 breakup of the giant telephone monopoly, AT&T.

Courts have long held that despite the First Amendment, business law—including antitrust statutes—may be applied to media. Sec. 313 of the Communications Act requires the FCC to consider revoking licenses held by companies found guilty of violating antitrust laws. However, the deregulatory atmosphere of the 1980s weakened enforcement of antitrust. The Justice Department rarely questioned either the huge media mergers and takeovers during the 1980s or the trend toward ever-larger combinations in cable television.

This is changing. Following the passage of the Telecommunications Act of 1996, which relaxed many of the media ownership restrictions, the Department of Justice stepped up its review of media acquisitions. Several radio mergers, which would have given one owner too much control over a local radio market, have been challenged by the Justice Department, and several stations had to be divested. According to the guidelines that seem to be emerging in such cases, if a proposed merger or acquisition would result in one owner controlling over about 40 percent of the radio advertising revenues in a market, the Department of Justice might object to the transaction based on antitrust principles.

Broadcast Ownership Defined

In general, the FCC defines ownership of a broadcast station as any active or participating equity interest in the station of 10 percent or more. Complex *attribution rules* pinpoint exactly what other kinds of interests may be deemed as ownership under the FCC's rules (47 CFR Sec. 3555). In the 1990s, several group owners (including ABC, CBS, Fox, and Tribune) attempted to avoid being considered owners of stations by acquiring passive (noncontrolling) interests, thus staying within the letter of the FCCs ownership limitations. In other situations, broadcasters operate stations without owning them under *local marketing agreements* (LMAs). The FCC planned to address these practices as part of its overall review of the broadcast attribution rules.

Multiple Ownership

Prior to 1996, the FCC limited the total number of broadcast stations anyone could control. By 1994 those limits confined a single owner to no more than 20 AM, 20 FM, and 12 full-power television stations nationwide; for radio, the cap for minority-owned stations was 25 AM and 25 FM; for television there was an additional limit of reaching markets totaling no more than 25 percent of the television homes in the country (UHF and minority-controlled TV stations counted only one-half the homes they actually reached); and minority-owned TV station caps were 14 outlets or 30 percent coverage.

The Telecommunications Act of 1996 changed the national ownership rules substantially. Now there are no limits on the number of radio stations that one entity may own. As discussed in Chapter 6, this has lead to an unprecedented number of radio station sales, with some owners now controlling over 100 radio stations across the country. Congress also removed the limits on the number of television stations that can be controlled by one owner. Congress maintained the national audience cap for television owners, though it raised that limit to 35 percent of television homes in the country. None of these regulations applies to ownership of noncommercial broadcast stations. Note also that while Congress removed these national ownership limits, local ownership restrictions are still in effect. (See Exhibit 13.d.)

Neither DBS nor low-power TV has any ownership limitation, though the FCC has considered restricting the number of "full CONUS" orbital slots DBS operators could control. These slots are very valuable because they provide the ability to cover the entire continental United States (CONUS). There are only three full CONUS slots available, and the FCC has considered allowing DBS owners to control no more than one. MMDS's only limitations are one per market of two available and a 5 percent ownership cap on co-located cable systems.

Duopoly

On the individual market level, a *duopoly* rule the FCC issued in 1940 and enforced for half a century held that no single owner could have more than one station of the same type (for example, more than one AM station) in the same market.

In 1992, while changing its radio ownership rules, the FCC struck down the duopoly limitation, allowing ownership of up to four stations in the largest radio markets. In 1996, as part of the Communications Act revisions, Congress went even further and codified new local ownership standards for radio depending on the size of the market (see Exhibit 13.d). Congress directed the FCC to conduct a rule-making proceeding to determine if the television duopoly rule should be retained, but as of this writing, the TV duopoly rule remains in force.

Cable Systems

Whereas the 1984 Cable Act *allowed* the FCC to establish cable ownership rules if it chose (it didn't), the 1992 Cable Act *mandated* the Commission to do so.

In 1993 the FCC issued rules limiting ownership to systems that pass no more than 30 percent of all U.S. television households (35 percent for minority-controlled systems). However, a U.S. District Court held this restriction unconstitutional and the rules are suspended pending further judicial review.

The Commission also limited so-called *vertical integration* by ruling that a cable system could devote no more than 40 percent of its channels to program services in which the system operator has an ownership interest (45 percent for minority-owned systems). The limit applies to only the first 75 channels.

Cross-Media Ties

Every instance of media *crossownership*, such as common control of newspapers and broadcast stations in the same market, reduces diversification. This reduction in alternative information sources (sometimes called media *voices*, as opposed to actual *outlets* such as stations or systems) is especially undesirable in small communities, in which the only newspaper might own the only broadcast station or cable system.

The FCC issued rules banning new daily-newspaper/broadcasting crossownership in 1975 but allowed all but a very few existing combinations to continue (a process called *grandfathering*). Though both the FCC and Congress have considered limits on newspaper/cable crossownership on several occasions, no such limits have yet been adopted.

In 1970 the Commission prohibited broadcast television network ownership of cable systems as well as crossownership of telephone companies or television stations and cable systems in the same market area (except in rural areas of fewer than 5,000 people). Congress eliminated both the network-cable and the cable-telco crossownership restrictions in the Telecommunications Act of 1996 (broadcast networks had always been allowed to own cable networks). The 1992 Cable Act banned crossownership by a cable system of any MMDS or SMATV service in its franchise area. The FCC, as usual, grandfathered combinations existing when the act was passed (Sec. 613(a)(2)).

Enter the Telcos

At the end of the 1980s, pushed in part by technical and structural changes in the telephone industry, a new crossownership controversy developed that found broadcasters and cable owners on the same side. Both groups feared that the FCC might allow local telephone companies to own cable systems—and possibly even broadcast stations—within their telephone service areas. A ban on cable-telco crossownership in the 1984 cable act had kept the regulated common-carrier and media businesses apart in the same market, but telephone industry representatives increasingly desired a share in the information or content business.

Exhibit 13.d

Local-Market Media Ownership Regulations

Multiple Ownership

	Market Size	Limitation
Radio	45 or more commercial stations	Eight, no more than five in one service (AM or FM)
	30 to 44 commercial stations	Seven, no more than four in one service
	15–29 commercial stations	Six, no more than four in one service
	14 or fewer commercial stations	Five, no more than three in one service
		And, no more than 50 percent of stations in market

Television—Duopoly rule still in effect; cannot own two stations in same market

MMDS—Can control only one of two sets of frequencies available per market

Crossownership

Newspaper/Broadcast—Cannot own a newspaper and a broadcast station in the same market; existing combinations grandfathered

Cable/Broadcast—Cannot own a cable system and a broadcast station in the same market

TV/Radio—Cannot own a TV station and a radio station in the same market; existing combinations grandfathered

Cable/MMDS/SMATV—Cable operators cannot own MMDS or SMATV systems in the same market

In the early 1990s the FCC sought to increase telephone company participation in electronic media ownership for three reasons: to develop competition with largely monopoly cable systems, to hasten installation of fiber-optic transmission lines with their high-quality broadband capacity, and to encourage development of two-way (interactive) video services.

The FCC's video dial tone (VDT) decision of 1992 allowed telephone companies to provide means of audio and video program delivery, though they could control no more than a 5 percent interest in the programs themselves (FCCR, 1992a: 5781). The Commission prohibited telephone companies from buying an existing cable system in their service area because that would simply replace one monopoly provider with another.

In 1994 the FCC revised its VDT rules, saying it had not intended to restrict ownership of programming. Instead, telcos would be limited to 5 percent ownership of *video information providers* (VIPs)—entities that provide video programming directly to subscribers and determine how video programming is presented for sale. The new rules also required telcos to obtain Commission approval of their VDT plans, including rates they would charge their customers, and to show how costs of the new service would be covered by VDT revenues, not by income from telephone service.

Meanwhile, a number of federal courts had ruled that the 1984 telephone-cable crossownership ban violates telcos' First Amendment Rights. Congress ended the debate once and for all by eliminating the cable-telco crossownership restriction in the 1996 Telecommunications Act. As detailed in the previous chapter, telephone companies are now allowed to provide multichannel video services in competition with cable television systems and other video service providers. In order to foster competition at the local level, Congress expressly disallowed telephone companies from acquiring existing cable systems in their service areas and vice versa. Joint ventures between the local telephone company and the cable system also are not allowed.

Not to be outflanked, some large cable MSOs are investigating different kinds of telephone and other voice and data services they might provide. Video compression and other technological developments aid the competing industries as they move onto one another's turf. How regulators will control the shifting alliances and technologies promises to feed major electronic media policy debates throughout the 1990s and into the next century.

Minority Ownership

For years the FCC ignored minority status of owners as an aspect of diversifying media control. A series of court reversals in the 1960s and early 1970s, forced it to reexamine its position, and it slowly began to give an advantage to minority applicants.

Since 1978 the Commission has taken several steps to enhance opportunities for members of ethnic and racial minority groups (African-Americans, Hispanics, Asians, American Indians) to become licensees:

- *Tax certificates* encouraged sales of stations and cable systems to minorities by allowing sellers to avoid or at least defer paying capital gains taxes on their profits.
- *Distress sales:* Normally the Commission will not permit an owner whose license is in serious danger of nonrenewal to sell anything other than the station's physical assets. But to encourage sales to minority applicants, the FCC permits endangered licensees to recover some, though not all, of the market value of "intangible assets" (effectively the station license and the operation's public image).
- *Minority minorities:* In 1982 the Commission sought to further encourage minority ownership by allowing members of a racial minority holding as little as 20 percent of the equity in a licensee to take advantage of both certificates and distress sales rules—provided that the minority owner had voting control. At the same time, the FCC made cable-system sales eligible for tax certificate consideration.
- *Lotteries* used in choosing LPTV and MMDS licensees give preference to minority ownership.
- *Bidding preferences* awarded to companies owned by minorities or women participating in the FCC's auction of personal communication services (PCS) licenses.

Minority ownership of stations has increased from about 30 stations in 1976, to about 300 in the 1990s (of which fewer than 10 percent are television). The National Black Media Coalition reported that to achieve ownership of stations in proportion to their actual numbers in society some 1,250 broadcast stations would have to be owned by African-Americans and about 450 by Hispanics. The industry has a long way to go before minorities achieve such ownership parity.

Despite this record, the deregulatory-minded FCC briefly suspended the distress sale minority preference option in the 1980s because of ideological opposition to "reverse discrimination," as opponents called special breaks for minorities. Pressure from Congress soon forced the Commission to reinstate it. Several legal attacks on this and other minority preference policies were consolidated into one case but rejected by the Supreme Court, which held that such preference policies "do not violate equal protection principles" (US, 1990: 547).

Minority preferences came under renewed attack in 1995. First, a Supreme Court decision striking down minority preferences for federal contracting cast doubt on the continuing constitutionality of the FCC's race-based policies (US, 1995). Next, the United States Court of Appeals barred the FCC from proceeding with a second planned auction of licenses for new wireless telephone and data services, under which small businesses owned by women or minorities would receive a 25 percent discount. At the same time, Congress—angered by reports that Viacom might avoid some $600 million in taxes by selling its cable systems to a black-owned company backed by TCI, the nation's largest cable operator—abolished the tax certificate program as part of its overall review of affirmative action programs (see Section 12.5).

In response to these actions, the FCC initiated two proceedings relating to minority and female ownership of electronic media. The first was a broadly based inquiry seeking insights on its existing minority policies and efforts to extend them (FR, 1995). The second seeks comments on identifying and eliminating market-entry barriers for small businesses, especially small businesses owned by minorities and females (FR, 1996). A final outcome of these and related matters could take years.

Foreign Control

Sec. 310(a) of the Communications Act forbids control of a U.S. broadcast or common-carrier radio station by a foreign entity. Specifically, no more than 25 percent of the station's stock or other means of control can be in the hands of foreign investors. Subscription-based electronic media are not restricted this way: Canadian and other foreign interests, for example, own many U.S. cable systems.

The application of this rule was put in question in 1985 when Rupert Murdoch acquired the Metromedia group of stations that formed the basis of his Fox network. Murdoch, an Australian, became a U.S. citizen, and the FCC approved the ownership transfer. In 1994, however, both NBC and the NAACP raised the ownership question anew, arguing that the Fox O&Os were owned by Murdoch's Australian-based News Corporation.

NBC later withdrew as a contestant (after negotiating carriage of its CNBC and NBC Super Channel on News Corporation's satellite-delivered Asian Star Television), but the underlying question remained. Fox argued that, although News Corporation—a foreign entity—holds 99 percent of the equity (ownership) in the stations, Murdoch—now an American citizen—owns 76 percent of the *voting* (and, hence, controlling) stock, and controls News Corporation through ownership in that entity as well. Fox also pointed out that the FCC may permit foreign ownership if it decides that doing so would be in the public interest, and that creation of a viable fourth network, long encouraged by the Commission, met that test.

None of these arguments addressed the more serious issue of whether Fox had misrepresented to the FCC the ownership/control structure when it sought the Metromedia stations in 1985. A finding of *lack of candor* has historically been the most flagrant violation of Commission rules.

In May 1995 the FCC determined that Fox was in violation of Sec. 310, but that it had not lied about its foreign ownership. Fox filed a petition offering a partial restructuring and reasserting that a waiver of the rule would be in the public interest. In July, by a 5-to-0 vote, the Commission ruled that Fox could keep the company intact, that it had indeed served the public interest by competing with the Big-Three networks, multiplying program choices, and creating thousands of jobs.

In 1997 the World Trade Organization passed a trade agreement that would allow foreign participation and often control of domestic common carrier facilities. It is unclear at this time how this trade agreement will affect the enforcement of Sec. 310, but critics are worried that a similar trade agreement might permit foreign ownership of broadcast stations. Several concerned members of Congress circulated proposals that would require Congressional approval of any trade agreement that changes U.S. law.

13.7 *Political Access*

The Communications Act regulates programs most explicitly when they involve candidates for public office. Congress correctly foresaw in 1927 that broadcasting would one day exert a major influence on voters. If the party in power could monopolize electronic media, opposing candidates would stand little chance of winning elections.

"Equal Time"

The commonly but imprecisely used term *equal time* does not appear in the Communications Act. Indeed, the correct term, *equal opportunities*, requires more than equal time. A literal interpretation of equal time would, for example, permit a broadcaster to run a 30-second commercial for a favored candidate during prime time and run another for the opponent at 4:00 A.M.—a ploy not permitted under the equal opportunities rule.

Sec. 315 of the Act requires that

> If any licensee shall permit any person who is a legally qualified candidate for any public office to use a broadcasting station, he shall afford equal opportunities to all other such candidates for that office in the use of such broadcasting station; *Provided*, That such licensee shall have no power of censorship over the material broadcast under the provisions of this section. No obligation is imposed under this subsection upon any licensee to allow the use of its station by any such candidate. (47 USC 315a)

A 1971 amendment to the Act mandates allowing time for *federal* candidates by adding a new basis for license revocation: the "willful or repeated failure to allow reasonable access to" federal candidates (47 USC 312a(7)).

Buying Political Time

One of the trickiest problems in political broadcasting is defining what constitutes a station's "lowest unit charge." This phrase in Sec. 315(b) of the Communications Act defines the maximum rate that licensees may charge candidates who buy time for political purposes.

In effect, stations may charge a political candidate no more for a given commercial than would be charged to the lowest-paying advertiser for the same spot. For example, a commercial advertiser might have to buy several hundred spots to qualify for the maximum quantity discount, but a political candidate benefits from that discount even when buying only a single spot.

Even with the lowest unit rate, candidates spend tens of millions of dollars on broadcast advertising during political campaigns. This inevitably leads to calls for campaign reform. Following the 1996 elections, several members of Congress proposed legislation that would give federal candidates up to 30 minutes of free air time during election campaigns. Under this proposal, candidates could purchase additional time at a rate of 50 percent of the lowest unit cost. Many politicians suggest that this would be a small price to pay for the broadcasters' use of additional spectrum for digital television. President Clinton has voiced his support for these proposals, as has former FCC Chairman Reed Hundt. Broadcasters for the most part oppose these plans, arguing that they should not be required to foot the bill for campaign finance reform. As of mid-1997, none of the campaign reform proposals had passed Congress, and most critics agree that little will happen unless and until the public gets sufficiently upset over the issue.

Candidates in the News

What about the presence of candidates in news stories? Do such "uses" also trigger Sec. 315's equal opportunities obligation? At first the FCC said no; later it said yes. Congress thereupon amended Sec. 315 to exempt news coverage of all candidates:

> Appearance by a legally qualified candidate on any
> (1) bona fide newscast,
> (2) bona fide news interview,
> (3) bona fide news documentary (if the appearance of the candidate is incidental to the presentation of

the subject or subjects covered by the news documentary), or

(4) on-the-spot news coverage of bona fide news events (including but not limited to political conventions and activities incidental thereto),

shall not be deemed to be use of a broadcasting station within the meaning of this subsection. (47 USC 315(a); emphasis added)

Although the amendment liberated news coverage from political equal-time harassments, it also left the FCC with many knotty problems of interpretation. Questions regarding political candidates' rights are among those most frequently asked by licensees seeking interpretation of FCC rules. Some examples of the issues confronting stations are detailed in Exhibit 13.e. In addition to providing paid-for access, electronic media have a responsibility to inform the electorate about political issues in a nonpartisan way. This tricky combination of enforced cooperation (which does not exist for newspapers or other print media), profit-making, and the obligations of journalistic objectivity continues to create problems for media managers.

13.8 Public Access

Not everyone with an idea to express can own a station or cable system, nor can everyone expect access to stations or systems owned by others. The FCC long struggled with this problem by trying to provide access to broadcasting for *ideas* rather than for specific *people.* But even access for ideas has to be qualified. It would be impractical to force stations to give time for literally every idea that might be put forward. The FCC mandated access only for ideas about *controversial issues of public importance.*

Rise of the Fairness Doctrine

The FCC slowly elaborated its access ideas into a formalized set of procedures called the *fairness doctrine.* This doctrine obligated stations (a) to schedule time for programs on controversial issues of public importance, and (b) to ensure expression of opposing views on those issues. Both stations and the FCC largely ignored (a), focusing their attention on (b). In practice, therefore, most fairness doctrine complaints came as reactions to ideas that had already been discussed on the air, rather than as complaints about the failure to initiate discussion of issues.

Licensees had great latitude in deciding whether a subject qualified as both a controversial issue and one of public importance, how much time should be devoted to replies, when replies should be scheduled, and who should speak for opposing viewpoints.

In 1969 the Supreme Court upheld the FCC's fairness doctrine, as well as its related personal attack and political editorializing rules, in its landmark *Red Lion* decision (see Exhibit 13.f). By unanimously supporting the FCC in this case, the Court strongly affirmed the fairness doctrine concept in an opinion that emphasized four key principles relevant to broadcasting's First Amendment status (US, 1969: 367):

- *On the uniqueness of broadcasting:* "It is idle to posit an unabridgeable First Amendment right to broadcast comparable to the right of every individual to speak, write, or publish."
- *On the fiduciary principle:* "There is nothing in the First Amendment which prevents the Government from requiring a licensee to share his frequency with others and to conduct himself as a proxy or fiduciary."
- *On the public interest:* "It is the right of the viewers and listeners, not the right of the broadcasters, which is paramount."
- *On the scarcity factor:* "Nothing in this record, or in our own researches, convinces us that the [spectrum] resource is no longer one for which there are more immediate and potential uses than can be accommodated, and for which wise planning is essential."

Although the fairness doctrine is no longer in effect, these underlying constitutional principles remain intact.

End of the Doctrine

By 1985 the FCC, by then largely made up of Reagan administration appointees dedicated to

Exhibit 13.e

What Does "Equal Opportunities" Mean?

- *Who gets equal opportunities?* Candidates for nomination in primary elections and nominees in general elections get equal opportunities. But equal opportunities can be claimed only by candidates for the same specific office; purchase of time by a candidate for Congress, for example, entitles all other candidates for that post in that district to equivalent opportunities, but not candidates for other districts or other offices.
- *Do presidential news conferences count as bona fide news programs?* Yes—they are exempt from equal opportunities claims.
- *Do presidential candidate debates count as news?* Though the rules have changed many times, the answer is "yes." However, third-party candidates are less likely to be heard (networks or other sponsors exclude them as being of little interest—Ross Perot in 1992 being an exception, but even he was excluded in 1996)—a news exception that flies in the face of Sec. 315's intent to let voters hear all the candidates.
- *Are regularly scheduled interview and talk programs exempt from equal opportunity requests?* Yes—including "infotainment" programs such as *Oprah*.
- *How much time constitutes "reasonable access"?* The FCC refuses to provide any set amount, but stations are prohibited from limiting candidates for federal office to set numbers of spots or amounts of program time, even when their schedules do not easily allow extended political speeches.
- *May live news appearances be recorded for later playback and still be exempt?* Yes.
- *Is a candidate entitled to whatever time of day he or she wants?* No. Candidates cannot decree when they appear. Stations may keep political spots out of particular periods—such as the times scheduled for news programs.
- *Must a station broadcast a candidate's use of obscene material?* No. When in 1984 *Hustler* publisher Larry Flynt threatened to use arguably obscene material in a possible campaign for the presidency, licensees asked the FCC for policy guidance. The Commission held that the ban on obscenity overrode the "no censorship" provision of Sec. 315.
- *Are photos of aborted fetuses in political ads obscene, and therefore not allowed?* No. During the 1992 and 1996 elections, several candidates ran on antiabortion platforms and some used graphic pictures of aborted fetuses in their advertising. Yielding to both viewer and licensee concerns, the Commission held that such depictions, although well within the rights of candidates' uncensored access, might be indecent and thus could be restricted to the "safe harbor" hours of 10 P.M. to 6 A.M. A federal appeals court disagreed and ruled that broadcasters could not restrict such ads only to the safe harbor hours. The court argued that this would deprive many voters from hearing the candidates' message (F, 1996). Stations can run a disclaimer that federal law required the ad to be carried.
- *Must a station carry potentially libelous material by a candidate?* Yes. *Is the station legally liable?* No.
- *May electronic media endorse political candidates in editorials?* Yes; but if they do, they must notify opposing candidates for the same office and offer reply time, even if the editorial occurs within a newscast.
- *Are cable systems held to the same regulations?* Regulations apply only to channels on which the cable operator *originates* programming. Cable systems are not responsible for political content of stations or services whose programs they carry but over which they have no content control.
- *May a licensee evade political broadcasting problems by banning all political advertising?* No—at least not for federal candidates (who must receive "reasonable access").
- *Is any appearance of a candidate a "use" of broadcast facilities?* Probably yes. Until 1992, any time a candidate appeared on the air—even in an entertainment or other nonpolitical situation—that time constituted a use. In 1992 the FCC narrowed the definition to cover only time that is authorized or otherwise controlled by a candidate or his or her committee—in other words, overtly political use of time. In 1994 the Commission again reversed course and seemed to define "use" in the traditional way—*any* appearance by a candidate, whether purposely political or not. Stay tuned for further flip-flops.

Exhibit 13.f

A Place and a Case Called Red Lion

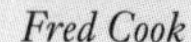

Fred Cook

Billy James Hargis

An unlikely small-town station formed the setting for one of the leading Supreme Court decisions on electronic media. During the 1960s right-wing preachers inundated radio with paid syndicated political commentary, backed by ultraconservative supporters such as Texas multimillionaire H. L. Hunt through tax-exempt foundations. Purchased time for these religious/political programs provided much-needed radio income in small markets.

The landmark case got its name from WGCB, a southeastern Pennsylvania AM/FM outlet licensed to John M. Norris, a conservative minister, under the name Red Lion Broadcasting. In 1964 one of the Reverend Billy James Hargis's syndicated broadcasts carried by the station attacked author Fred Cook, who had criticized defeated Republican presidential candidate Barry Goldwater and had written an article on what he termed the "hate clubs of the air," referring to the Hargis series *Christian Crusade*, among others. Hargis attacked Cook on the air, charging him with communist affiliations and with criticizing the FBI and the CIA—the standard litany of accusations Hargis routinely made against liberals.

Cook then accused the station of violating FCC rules by failing to inform him of a personal attack. When he wrote asking for time to reply, the station responded with a rate card, inviting him to buy time like anyone else. Cook appealed to the FCC, which agreed that he had a right to free airtime for a reply. It ordered WGCB to comply.

It would have been easy for the Reverend Mr. Norris to grant Cook a few minutes of time on the Red Lion station, but he refused on First Amendment grounds, appealing the Commission decision. The court of appeals upheld the FCC, but Norris took the case to the Supreme Court, which, in 1969, also upheld the FCC, issuing an opinion strongly defensive of the FCC's right to demand program fairness.

Several years later, Fred Friendly, a former head of CBS News but by then a Columbia University journalism professor, began looking into the background of this well-known case for a book about the fairness doctrine (Friendly, 1976). He discovered that Cook had been a subsidized writer for the Democratic National Committee and that his fairness complaint had been linked to a systematic campaign mounted by the Democrats to discredit right-wing extremists such as Hargis. According to Friendly, the Democrats set out to exploit the fairness doctrine as a means of harassing stations that sold time for the airing of ultraconservative political programs. Cook and the Democratic National Committee claimed that Friendly had misinterpreted their activities, maintaining that Cook acted as a private individual, not as an agent of the Democratic party.

Source: Photos from George Tames/New York Times Pictures.

deregulation, had joined a mounting chorus of opposition to the fairness doctrine. After a lengthy proceeding, the Commission concluded:

> [Based on] our experience in administering the doctrine and our general expertise in broadcast regulation, we no longer believe that the fairness doctrine, as a matter of policy, serves the public interest. . . . Furthermore, we find that the fairness doctrine, in operation, actually inhibits the presentation of controversial issues of public importance to the detriment of the public and in degradation of the editorial prerogatives of broadcast journalists. (FCCR, 1985b: 143)

Acting on a court remand requiring the FCC to reconsider an earlier fairness decision, the Commission instead abolished the doctrine entirely (FCCR, 1987b: 5043). The personal attack and political editorializing rules, however, remained in place, as discussed below.

That should have ended the matter, but the FCC's defiance in ending a practice that Congress had gone on record as favoring enraged key legislators. They vowed to reinstate the fairness mandate, in the meantime making their dissatisfaction known by holding up approval of appointments to vacant FCC seats. Bills reinstating the doctrine were introduced and passed twice but were vetoed both times by President Reagan. When President Clinton stated early in 1993 that he would sign a fairness doctrine reinstatement bill if it reached his desk, many predicted renewed congressional action. As of mid-1997, Congress had not acted further to reinstate the fairness doctrine and appeared highly unlikely to do so.

A *Right* of Reply?

Two specific fairness-related requirements, the personal attack and political editorializing rules (both adopted in 1967 and retained even after the fairness doctrine's demise in 1987), continue to cause concern to opponents of FCC program interference.

The *personal attack rule* requires stations to inform individuals or groups of personal attacks on their "honesty, character, integrity or like personal qualities" that occur in the course of discussions of controversial public issues. Within a week after the offending broadcast, licensees must advise those attacked, explaining both the nature of the attack and how replies can be made. Specifically exempted from the right of reply are on-the-air attacks made against foreigners, those made by political candidates and their spokespersons during campaigns, and those occurring in news interviews, on-the-spot news coverage, and news commentaries (47 CFR 73.1920).

The *political editorializing rule* requires that all candidates or their representatives be given a chance to respond if a licensee endorses any of their opponents. If a station editorially opposes a candidate, that candidate must, likewise, be given an opportunity to respond. A station must inform such candidate(s) of their rights to respond within 24 hours of such editorials. The rule does not apply to use of a station's facilities by opposing candidates, a situation covered by the equal opportunities procedures. The Radio-Television News Directors Association (RTNDA) has long urged the FCC to abolish the personal attack and political editorial rules, but the FCC consistently has refused to act. RTNDA finally went to court and in 1996 obtained an order forcing the FCC to consider the issue. Early in 1997 the FCC invited public comment on the prospect of modifying or repealing the rules.

Rights for Advertisers

Access by advertisers to broadcasting has been governed more by considerations of taste and public acceptance—and income potential—than by government regulation. However, cigarette advertising became the subject of a famous fairness doctrine complaint. In 1968 the FCC decided that the Surgeon General's first report on the dangers of smoking, as well as Congress's 1965 act requiring a health warning on cigarette packages, justified treating cigarette advertising as a unique fairness doctrine issue. Therefore, stations would have to carry antismoking spots if they carried cigarette commercials. These counter-ads ended, of course, after Congress banned all broadcast advertising of cigarettes in 1971, though occasional public-service announcements concerning smoking still appear.

Editorial advertising or "advertorials," posed a different kind of problem. Traditionally, electronic media have declined to let advertisers use commercials as vehicles for comment on controversial issues, arguing that

- serious issues cannot be adequately discussed in short announcements;
- selling larger blocks of time for editorializing by outsiders involves surrender of editorial responsibility; and
- not everyone can afford to buy time, so selling to those with funds is inherently unfair.

The Supreme Court upheld the principle of licensee *journalistic discretion* in a case dealing with a fairness doctrine demand for access to editorial advertising:

> Since it is physically impossible to provide time for all viewpoints . . . the right to exercise editorial judgment was granted to the broadcaster. The broadcaster, therefore, is allowed significant journalistic discretion in deciding how best to fulfill the Fairness Doctrine obligations, although that discretion is bounded by rules designed to assure that the public interest in fairness is furthered. (US, 1973a: 111)

Despite deletion of the fairness doctrine, the question of editorial advertising continually reappears. Controversies arise periodically between advertisers and broadcasters. In several cases, ads that were turned down for television have appeared on cable or in print—with pointed commentary about the refusal of broadcasters to sell time for such advertising. The controversy over advertising of condoms is but one example: Few broadcasters accepted condom advertising or editorial statements calling for their use to prevent AIDS, for fear that substantial parts of their audiences would be offended by such messages. Moreover, the refusal of many broadcasters to air the messages of anti-abortion advocates led abortion foes to run for Congress. As detailed in Section 13.7, broadcasters cannot censor the ads of political candidates; this gives anti-abortion candidates an opportunity to put their views on the air.

Editorial Discretion

By their nature, news and public-affairs programs necessarily involve controversial issues, often leading partisans to bring charges of unfairness. It becomes difficult for broadcasting and cable journalists to deal with serious issues if their corporate bosses prefer to avoid controversy. Yet electronic media cannot win full public respect and First Amendment status without taking risks similar to those the printed press has always faced.

The FCC and courts generally assume that reporters and editors use editorial discretion, which calls for fair and considered news treatment of events, people, and controversies. No one believes that journalists always use the best judgment or that they totally lack bias or prejudice. First Amendment philosophy holds, however, that it is better to tolerate journalists' mistakes—and even their prejudice and incompetence—than to set up a government agency as an arbiter of truth. The Supreme Court reaffirmed this notion in confirming yet another FCC decision rejecting a fairness complaint:

> For better or worse, editing is what editors are for, and editing is selection and choice of material. That editors—newspapers or broadcast—can and do abuse this power is beyond doubt, but that is not reason to deny the discretion Congress provided. Calculated risks of abuse are taken in order to preserve higher values. (US, 1973a: 124)

News Bias

Critics often accuse electronic media, especially television network news departments, of news bias. CNN discovered in an ironic way that it had "arrived" as a widely used news source when, in the wake of its 1991 Gulf War coverage, many complained of its alleged bias by keeping reporters in Baghdad.

Such charges usually come from political conservatives, many of whom regard all news media as too liberal in outlook. They argue that the cumulative effect of alleged liberal bias over time tends to build up one-sided perceptions of issues. Professional gadflies, such as the misnamed Accuracy in Media (which seems always to find *in*accuracy and bias in every media report it investigates), regularly take up conservative causes against alleged television bias. Reagan and Bush administration elements lent support to such complaints, particularly after Bush's reelection defeat in 1992. More recently, conservative critics point to the disparity in media coverage of the sexual harassment allegations against Supreme Court nominee Clarence Thomas on the one hand and President Clinton on the other. Content-analysis research is used to support both sides of the bias questions, depending on who finances the studies. The debate continues.

Localism

The public needs access to diverse voices—but they also need voices of their own. The notion of a broad right of public access has two aspects—access to ideas of others and to the means of expressing those ideas. The widespread availability of stations and cable systems can facilitate such two-way access—especially within local market communities.

From its inception well into the 1960s, the FCC encouraged *localism*—reflection of the local community's needs and interests—by distributing as many local outlets to as many localities as possible, and by issuing guidelines that encouraged airing of programs reflecting community needs and interests. The Communications Act gives the Commission its own localism guidelines when it acts on applications for new stations:

> The Commission shall make such distribution of licenses, frequencies, hours of operation, and of power among the several States *and communities as to provide a fair, efficient, and equitable distribution of radio service to each of the same.* (47 USC 307(b); emphasis added)

This policy of localizing stations has provided listeners in major markets 40 or more radio stations from which to choose, whereas people in rural areas have far fewer or, in some cases, none at all. As cable expanded, it equalized and increased television channel choices in most urban and suburban areas; but it, too, underserves rural areas.

Yet cable does take pressure off commercial broadcasters, inasmuch as coverage of meetings and local events has largely transferred to local cable government and access channels. Most cable systems have "PEG" (public, educational, and governmental) channels as encouraged by federal law and often required by local franchise. Though budget problems facing most local governments and school systems have restricted their use, the potential of cable's growing number of channels to serve a variety of local concerns has increased greatly in the past decade.

Still, when threatened by changing technology—direct broadcast satellites or telephone industry takeover, for example—broadcasters immediately cite their unique public interest in providing locally relevant programs. They argued long and hard to the FCC and Congress, though to no avail, that cable and DBS are almost entirely national services, whereas broadcasters still hold to the ideal of station-based localism. However, as any discerning listener can attest, while broadcasting and cable do have local *outlets*, the *voices* they carry are increasingly the same nationwide.

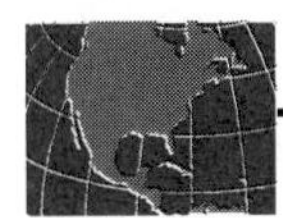

13.9 Serving Children

Reacting to constant pressure from Action for Children's Television and others, Congress passed legislation in 1990 regulating the amount of advertising in children's programs as well as establishing a general obligation to serve child audiences. Its major provisions include the following:

- Advertising per hour of children's programs is limited to no more than 10½ minutes on weekends and 12 minutes on weekdays. This "limitation" applies to broadcast licensees *and* cable systems—for all programming produced and broadcast primarily for children.

- At license renewal, the FCC must consider whether stations have served the "educational and information needs of children" through their overall programming, which must include "some" (otherwise undefined) programming specifically designed to meet children's needs.
- Establishment of a National Endowment for Children's Educational Television was proposed, to be funded at $2 million for fiscal year 1991 and $4 million for fiscal 1992. The endowment was eventually funded at about half this rate and administered by NTIA.

Licensee Duties

With the basic provisions of the Children's Television Act having set the stage, the FCC conducted rule makings in 1990–1991 to put more specific regulations in place. After considerable input from the broadcast and cable industries and others, the Commission arrived at the following decisions:

- It took a very wide view of *programming* by defining "children" as those 16 and younger, thus allowing stations to include music and dance programs aimed at teens—not what Congress had in mind (the congressional limitation on *commercials*, for example, affects programs aimed at children 12 and under).
- It prohibited program-length commercials for children. A program built around a fantasy hero may no longer include advertising for products related to that hero—they must run in another, unrelated children's program. (Since 1974 the FCC had policies against "host selling" and the close intermixing of program and advertising elements.)
- It defined the "educational and cultural" programming called for in the Children's Television Act to mean "programming that furthers the positive development of the child in any respect."
- It declared that stations need not program for all children—if desired, they may select one or more age groups.
- It encouraged (but did not require) licensees to "assess" the needs of children in their communities.
- It required keeping of records, in a form chosen by the licensee, that must be made a part of the station's public file.
- It decided that although public broadcasters must meet the programming requirement, they are not held to the specific record-keeping regulations. Cable systems are not obligated to meet either the programming or record-keeping requirement, but they are bound by all advertising limits.
- Finally, it established a fine of $10,000 for each violation of any of the rules growing out of the Children's Television Act (47 CFR 73.660).

The Commission began to review compliance with this list of requirements in 1992 and found that 95 percent of monitored program hours met the commercial limits within weeks of their being put in place; a year later compliance stood at 98 percent. But several television stations and cable systems *did* run afoul of these limits and became widely publicized examples of the FCC's intent to enforce the new rules rigidly (*Broadcasting & Cable*, 25 April 1994: 10).

As for the programs themselves, one consumer group reported in late 1992 that some broadcasters were meeting the letter of the law's requirements (and those of the loose FCC reporting rules) simply by claiming positive aspects of existing cartoon or rerun comedy shows, or by placing what few new programs were being produced at very early morning hours when few would watch (*New York Times*, 30 September 1992: 1, B8).

That study, and rising consumer group and congressional displeasure with broadcaster performance, led to 1993 hearings in Congress and inception of an FCC inquiry into whether its children's program requirements should be made more specific.

In 1996 President Clinton took the unusual step of becoming directly involved in this issue when he invited television industry and government leaders to the White House for a discussion of the children's television requirements. At this meeting a consensus arose among the participants that television broadcasters should, at a minimum, air three hours of educational programming for children per week. The FCC subsequently adopted this standard as a rule that all television broadcasters must satisfy.

Television Violence

As detailed in Section 11.7, television violence and its effects on children have elicited more concern and study than any other entertainment program issue. The first congressional hearings on television violence were held in the 1950s, and there have been many more since then. Until relatively recently, however, the government has declined to regulate violent programming based primarily on First Amendment grounds. Throughout the controversy broadcasters have resisted regulation, arguing that parents and not the government should control children's television viewing. Not satisfied with broadcasters' efforts at self-regulation, Congress finally acted on this issue in 1996.

As part of the Telecommunications Act of 1996, Congress set in motion a process that is intended to help parents control the programming their children watch. The legislation envisioned the establishment of a television ratings system that would rate every television program. That rating would then be listed in television programming guides and displayed at the beginning of every program. The legislation also requires that new television sets be equipped with a device—known as a "V-chip" ("V" is for violence)—that would allow parents to program their televisions to block out all programs of a particular rating. Each program would carry its rating in the signal's vertical blanking interval (Section 4.8) so that the V-chip could identify which programs could be shown and which should be blocked.

The act allows broadcasters to establish a ratings system which would then be submitted to the FCC for approval. If the broadcasters' system is not approved, the Commission is empowered by the legislation to appoint a committee to establish a system more to its liking. After many months of study, an industry committee headed by Motion Picture Association of America (MPAA) President Jack Valenti unveiled its ratings system at the end of 1996. Similar to the MPAA movie ratings scheme, the initial television system rated programs according to age categories.

Almost immediately critics of the proposal attacked the age-based system, arguing that the legislation intended a content-based ratings system instead. Parents rights groups and government leaders alike charged that the age-based system did not provide enough information to help parents decide what was appropriate for their children. They demanded that the system be changed in such a way that the ratings reflect objectionable content: V for violence, S for sexual situation, and L for strong language. Broadcasters generally argued that a content-based rating system was too confusing and that advertisers would shy away from programs rated for objectionable content.

By mid-1997 Valenti and his committee succumbed to the pressure and added content descriptions to the ratings (listed in Exhibit 9.c). NBC, believing that the content ratings were too intrusive, rejected the newer system and stated it would stay with the earlier age-based system. At the time of this writing, the FCC was considering whether to accept either or both of the ratings plans.

13.10 Copyright

The Constitution recognizes the fundamental importance of encouraging national creativity. Article I, Section 8, calls on Congress to "promote the progress of science and the useful arts, by securing for limited times to authors and inventors the exclusive right to their respective writings and discoveries."

When broadcasting began, authors and composers had to rely on the Copyright Law of 1909, which dealt primarily with printed works and live performances. Despite amendments, the old act never caught up with the times. After long study and debate, Congress finally passed a new copyright law in 1976, effective for works published in 1978 and later.

Basics

Key provisions of the 1976 Copyright Act, administered by the Copyright Office (part of the Library of Congress), include the following:

- *Purpose.* Copyright holders license others to use their works in exchange for payment of *royalties.* "Use" consists of making public by publishing, performing, displaying—or broadcasting.
- *Copyrightable works.* In addition to traditionally copyrightable works—books, musical compositions, motion pictures, and broadcast programs—such works as sculptures, choreographic notations, and computer programs can be copyrighted. Things *not* copyrightable include ideas, slogans, brand names, news events, and titles. (However, brand names, logos, and slogans can be protected under trademark regulations.)
- *Length of copyright.* In general, a copyright lasts for the life of the work's creator plus 50 years. Copyrights initially granted to corporations last 75 years from the date the copyright was registered or 100 years from the date the work was created, whichever comes first. After that, a work enters the *public domain* and can be used without securing permission or payment of royalties.
- *Compulsory licensing.* Owners of copyrighted material who license television stations to use their work must grant *retransmission consent* to cable systems that lawfully pick up such programs off the air and deliver them to subscribers. Cable systems, in turn, must pay a preset small proportion of their revenue for these rights.
- *Fair use.* The new act retained the traditional copyright concept of fair use, which permits limited uses of copyrighted works without payment or permission for certain educational and critical or creative purposes (such as reviews of books and musical works).

The advent of new media forms (electronic delivery of newspapers, digital recording, and the Internet, for example) presents novel and difficult copyright questions. Royalty obligations to authors and others, whose works were prepared for single use by one medium but are then reused by others, will be the subject of negotiation, litigation, and legislation well into the next century.

Music Licensing

Broadcast and cable programmers obtain rights to use recorded music by reaching agreements with the copyright licensing organizations ASCAP, BMI, and SESAC. Most stations hold *blanket licenses* from these agencies, which allow unlimited air play of any music in their catalogs in return for payment of an annual percentage of the station's gross income. Some radio stations, especially those with news and talk formats, pay on a per-use basis. These licenses, whether blanket or per use, permit the *broadcast* of the music by the stations. They do not cover other uses of the music, such as adding the music to a local commercial or public service announcement. Such uses require additional permissions from the copyright owner.

Cable, Copyright, and the CRT

The Copyright Act of 1976 established a Copyright Royalty Tribunal (CRT). For cable television, the CRT was to establish rates that cable had to pay for use of imported distant television broadcast signals, pool the resulting revenue, and divide pooled royalty money among copyright holders.

The process of apportioning payments became snarled in legal proceedings from the day the CRT began work. Throughout its short history, the

CRT lacked sufficient staff and expertise to carry out its contentious assignments. In 1982 the agency's chairperson called the whole process unworkable and unfair. Finally, in 1993, Congress abolished the CRT, turning its duties over to a series of arbitration panels to be supervised by the Library of Congress and its copyright office.

Cable Piracy

Until the mid-1980s *signal piracy* chiefly involved illicit hookups to cable television feeder lines, enabling cable reception without payment of monthly subscriber fees. As pay cable developed, illegal "black boxes" (decoders) made possible the reception of even scrambled signals. The cable industry estimated that in the 1990s an equivalent of a quarter of cable revenue was lost to pirate activities.

Pirated cable reception violates Sec. 705 of the Communications Act, which defines penalties for unauthorized "publication" of communications intended for reception only by subscribers paying to use special unscrambling devices. Several cable systems prosecuted violators and publicized the resulting felony convictions, which entailed jail terms and fines of up to $25,000 for the first violation and $50,000 for subsequent violations. More than half the states have passed antipiracy laws prohibiting the manufacture, sale, or use of unauthorized decoders or antennas.

TVRO Piracy

Piracy increased when the prices of television receive-only (TVRO) Earth stations began to fall in the 1980s, making them affordable by ordinary households. Both pay- and basic-cable operators began to scramble satellite signals to protect their investments from pirates.

As a result, Congress modified Sec. 705 to allow individuals to pick up any *nonscrambled* satellite programming if they have obtained authorization, usually for a fee. By the early 1990s nearly all satellite channels were scrambled. Since the cable industry had long refrained from marketing its services to such scattered audiences, a number of third-party brokers had already begun to assemble packages of cable services for sale to TVRO users, although the use of illegal descramblers continued.

VCR Piracy

In 1976 a number of program producers brought suit against Sony, the pioneering manufacturer of home video recorders, for "indirect" copyright infringement ("indirect" because Sony provided the *means* of infringement—the machines that made recording possible—but did not itself do the copying). A district court found in favor of Sony, concluding that home recordings of broadcast programs not sold for a profit fell within the fair-use provision of the new copyright law. The Supreme Court affirmed the decision by a 5-to-4 vote (US, 1984a: 417). The Court cited audience research showing that people record broadcast signals primarily for time-shifting purposes, which it regarded as a fair use.

The decision left unresolved the legality of recording cable or pay programming; the *Sony* case covered only recording of over-the-air or "free" broadcast material. The legality of showing a taped copy of a copyrighted film or program to a group outside the home remained in question, a problem intensified by increasing numbers of video rentals. Hollywood producers pressured Congress—thus far unsuccessfully—to modify the Copyright Act once more, this time to extract from those who did home recording such indirect royalty payments as a surcharge on recorders or blank tapes at the time of purchase, or on rental fees customers paid to video rental stores.

Digital Audio Piracy

Late in 1992 Congress enacted into law an industry compromise on royalties for recordings produced by the coming generation of digital audio home recording devices. Digital audio tape (DAT) and digital compact cassette (DCC) as well as future hardware systems would include a 2 percent royalty fee (from $1 to $8 per machine), whereas blank tapes and discs would include a 3 percent fee on their wholesale price. The funds raised would go to the Copyright Office and be disseminated to copyright holders.

The law also supported the industry's serial copy management system (SCMS) technical standard, which prevents a second-generation digital dub from being made on any of these machines. In other words, an owner can make a copy but not a copy of a copy (*Television Digest*, 28 September 1992: 10).

Copyright and the Internet

The ease and speed with which information travels on the Internet poses unique and thorny copyright problems. Anyone with an image scanner and a computer can easily put copyrighted material into the hands of thousands of Internet users in a matter of seconds. Technically, any time a copyrighted work is transmitted from one computer to another and copied in the receiving computer's memory a violation of the copyright law has occurred. Moreover, digitized information is easily manipulated, so copyright notices can be removed without much effort. A further question involves the copyright liability of Internet service providers (ISPs) such as Compuserve and Prodigy. Because these services store information provided by their subscribers and make that information available to others, are the ISPs likewise liable for copyright infringement when the information is protected by copyright?

Internet user groups argue that the copyright laws should be interpreted as allowing the freest exchange of information possible. They contend that restrictive copyright enforcement will stifle the freewheeling and robust interchange of ideas that occurs on the Internet. Commercial interests and the Clinton administration, on the other hand, counter that copyright enforcement on the Internet should be increased in order to protect the economic interests of copyright owners. The very few court cases that have addressed these issues thus far have reached mixed conclusions on how the copyright laws should apply to Internet users.

These and other issues were addressed both nationally and internationally in 1996. In the United States, a task force looking into copyright and new technology produced a report generally concluding that uses of copyrighted material on the Internet should come in line with the use of such material in more traditional media. On the international front, the World Intellectual Property Organization drafted treaties in 1996 dealing with copyright and the Internet that would attempt to slow the flow of copyrighted information on the Internet and make copyright violations easier to identify and prosecute. To be enforceable in the United States, the treaties must first be ratified by Congress. No action had been taken on the treaties by the time this book went to press.

13.11 Changing Perspectives

In the late 20th century, numerous media compete for attention and consumer dollars, but more can too often become less in a marketplace dominated by giant corporations. These well-financed communicators blanket the nation, reducing the diversity of competing voices.

Two totally different approaches to this problem are actively debated. Here stands the *deregulator*, confident that the marketplace will regulate itself, given maximum competition and unconstrained consumer choice. There stands the *regulator*, demanding that government intervene once more to oppose monopolistic media tendencies and protect the public interest from the effects of unrestrained competition.

Many practical examples of this fundamental clash of views exist:

- Should indecency in electronic media have the same First Amendment protections it has on newsstands and in movie theaters? If so, how best can children be protected from such content?

- Should electronic media have special responsibilities with regard to airing controversial issues of social importance? If so, why—and to what degree—should they be treated differently than print media that bear no such responsibilities?
- Does violence in electronic media have such potentially damaging social effects as to justify tilting the regulatory playing field against it?
- As cable systems retransmit broadcast signals, as programming is delivered by satellite directly to the home, as cable systems move into telephony and telephone companies into cable, should and how can a level regulatory field be ensured?
- What controls should—and realistically can—be imposed (and by whom) on the unlimited-channel universe on the superhighway and in cyberspace, with media that for some are mass and for others personal, with interactive, multiplayer computer video games and "private" bulletin boards?

In sum, as the 20th century winds down, the continued conservatism of the FCC and growing conservatism of the courts suggest intensifying media competition. How well the public will be served by this overflow of technological options cannot be predicted. Certainly traditional broadcasting's long-time primacy in American lives continues, though with a lesser public responsibility role than it once played.

Chapter 14

Global View

CHAPTER CONTENTS

Few Americans realize that most other countries operate electronic media quite differently than we do. Yet because radio waves ignore national boundaries (especially when sent from satellites) and telephone and computer networks reach almost all of the world, electronic media are truly international. Countless organizations here and abroad participate in global exchanges of equipment, programs, and training. Fully understanding electronic media in America requires at least a sense of how they operate elsewhere.

Although countries have the same theoretical potential for using electronic media, each adopts a system uniquely suited to serve its own conditions, economic abilities, and needs. Three attributes of these media promote a mirrorlike relationship between a nation's character and its electronic media:

- Broadcasting invites *government regulation* because it requires use of electromagnetic spectrum, which all governments view as public property. Each government defines its responsibility to control spectrum in the context of its own history, geography, economics, culture, and most important, political philosophy.
- Transmissions using spectrum can cause *interference*, creating a need for controlling the number and power of transmitters. Political realities affect how, and how much, a country chooses to regulate based on interference control.
- Electronic media have *political and cultural impact* because they can bypass government entities, going directly and instantly to the public. Whatever its politics, no nation can afford to leave such persuasive media totally unregulated.

14.1 Controlling Philosophies

Controls that a nation imposes on electronic media reflect its government's attitudes toward its own people. Despite dramatic political changes over the past decade, worldwide electronic media structures still generally fall into three broad categories: permissive, paternalistic, and authoritarian.

Permissivism

Electronic media in America provide the classic example of a predominantly *permissive* or *laissez-faire* system. Our Constitution encourages free enterprise and makes freedom of communication a central article of faith. The industry argues that resulting commercialism creates more lively, popular, and expertly produced programs than are usually found elsewhere.

Generally speaking, countries now or once within American influence (such as the Philippines and many Latin American nations) have adopted similar permissive, profit-driven systems. Many other countries deplore American commercialism because it focuses almost exclusively on what people *want* rather than on what critics, experts, and government and religious leaders think they *need*.

Paternalism

Most countries perceive electronic media as playing a positive social role: preservation and enhancement of predominant national cultures. They take a *paternalistic* approach, putting special emphasis on preserving national languages, religions, and social norms by ensuring a "balanced" program diet—meaning not too much light entertainment at the expense of information and culture. And many nations take special care with children's programs, ensuring positive and culturally relevant examples.

The British Broadcasting Corporation (BBC) is the classic example of paternalism. Growing out of a brief period of commercial operation, the BBC converted to a public-service role in 1926, seeking to avoid America's "mistake" of radio commercialization. BBC funds today come primarily from government-imposed license fees on television receivers, removing dependence on advertising. A well-articulated ideal emerged from the BBC experience. Adopted by many other countries, *public-service broadcasting* includes such principles as

Exhibit 14.a

Broadcasting House, London

In pretelevision days, broadcasters from all over the world journeyed to this famous art deco building in the heart of London as a kind of broadcaster's mecca. The BBC moved here from its original quarters on the bank of the Thames in 1932. Though Broadcasting House tripled the corporation's previous space, the new building proved too small for BBC activities even before its completion. The giant BBC television center is located in a London suburb—and BBC radio and television own or lease many other buildings around the city. The British sometimes refer to the BBC as "Auntie."

Source: Photo courtesy BBC.

- balanced programming, representing all principal genres;
- control by a public body, independent of both political and commercial pressures;
- relative financial autonomy, usually secured by partial or complete dependence on receiver license fees, income from which is dedicated to support of broadcasting;
- program services that can be received by, and that hold interest for everyone, including rural dwellers and minorities;
- strict impartiality in political broadcasts; and
- respect for the artistic integrity of program makers.

This public-service broadcasting philosophy, adapted to varying national circumstances, soon spread worldwide. Thousands of foreign broadcasters have visited BBC's famous Broadcasting House in London (Exhibit 14.a). Many take BBC training courses in all aspects of production, engineering, and management. Other national services modeled to some extent on BBC traditions include Australia's ABC, Canada's CBC, and Japan's NHK.

Authoritarianism

Traditionally, dictatorships of whatever political flavor take an authoritarian approach to electronic media. The state itself often finances and operates media systems, along with other telecommunication services. Indeed, governments once owned and operated most broadcasting systems, although the number of privately owned services began to increase during the 1980s and at an accelerating pace in the 1990s.

Authoritarian approaches generally ignore popular taste as programs have a propaganda or educational goal. Broadcasters pay little attention to marketing techniques that permissive and even paternalistic broadcasters employ to ensure attractive and cost-effective programs. Indeed, authoritarians look on broadcasting as a one-way medium—all give and no take. Either they do not understand or they actively suppress its potentially *democratic nature*—a medium that can offer access to many different ideas and that depends on free-will listening by audiences.

Third World Authoritarianism

Lack of purchasing power limits ownership of even simple radio sets in many Third World countries. Absence of rural electrical power as well as of relay facilities for networking further restricts growth—especially for line-of-sight television service. Neither receiver license fees nor advertising (alone or in combination) can bring in enough revenue to support broadcasting adequately. Therefore, most Third World governments own and operate their broadcasting systems. Dictators heading often shaky Third World regimes dare not allow broadcasters to incite illiterate masses. Prudence—and ideology—dictate tight control. (Note that in political upheavals, government-operated broadcast stations, usually in the capital city, are initial targets for rebel takeover.)

14.2 *Pluralistic Trend*

As we approach the new millennium none of these three controlling philosophies—permissive, paternalistic, or authoritarian—exists anywhere in pure form. Compromise leads to combinations of features to create a fourth approach: *pluralism.*

Role of Competition

Almost 80 years' experience has proved that pluralistic electronic media systems work best in most national circumstances. Pluralism means more than simple competition among rival services. Given that similar motives drive all services, media would merely imitate one another in the absence of pluralistic competition.

Pluralism means putting a variety of motives to work, each with approximately equal status—usually a mix of commercial and public-service motives. Healthy competition between differently motivated broadcasting organizations can stimulate creativity, encourage innovation, and ensure variety. The result is usually a wider range of genuine program choices than any single-motive system could produce.

British Pluralism in Transition

The British eventually developed a widely admired pluralistic system based on control by two noncommercial public authorities: the BBC and the Independent Broadcasting Authority (IBA), the latter set up in 1955. A nonprofit, government-chartered corporation, IBA selected and supervised regional commercial television companies known collectively as Independent Television (ITV), as well as specialized national television networks and, eventually, local radio stations. Like the BBC, IBA owned and operated its own transmitters, so commercial motives could not distort geographical coverage by placing outlets only in major cities. Unlike the BBC, however, privately owned commercial companies provided programming and sold advertising time.

BBC's national and regional radio services initially had little competition. Its networks feature pop music (Radio 1), middle-of-the-road programs (Radio 2), serious music and talk (Radio 3), and news/current affairs (Radio 4). In 1991 it added Radio 5 to bring instructional/educational programs to schools as well as to provide such programming as African music to underserved audiences. Today, over 40 BBC local radio stations provide regional programming of interest to audiences in England, Scotland, Wales, and Northern Ireland, while competing with an increasing number of local commercial radio stations.

This "comfortable duopoly" operated for 35 years, ending in 1990. A new broadcasting law, reflecting something of an American deregulatory approach, included the following features:

- The BBC continues as the "cornerstone of public-service broadcasting," deriving revenue from receiver license fees. The BBC's royal charter was extended in 1996, maintaining the BBC's special status in the U.K. until the end of 2006.
- The commercial network became Channel 3 at the beginning of 1993, with the BBC's two networks renamed Channels 1 and 2. Channel 4 (begun in 1982 as an "electronic publisher" to carry programs produced by others) continued and now sold its own commercial time. The Channel 3 companies are responsible for Channel 4's financial health.
- A new commercial Channel 5, able to reach about 70 percent of the population, was authorized in 1994 with a 10-year franchise and program requirements similar to Channel 3. Regional franchises for Channel 3 and the new Channel 5 were auctioned to the highest bidders, subject only to their ability to meet program-quality requirements. Channel 5 began broadcasting in 1997.
- The IBA and a separate Cable Authority merged into a new "light touch" regulatory authority called the Independent Television Commission (ITC). A separate Broadcasting Standards Council was created to develop a program code controlling program sex and violence. And a new Radio Authority is to auction 200 to 300 new local commercial radio station franchises.

Britain's 1990 law reflected changes taking place to varying degrees worldwide: increases in competition, more local (in addition to national) services, greater private ownership—and deregulation.

14.3 Deregulation

Starting around 1980 U.S.-inspired deregulatory practices swept the telecommunications world. Although American influence alone did not create this trend, foreign experts studied American deregulation intensively, while U.S. government and media officials promoted both deregulation and private ownership abroad. Deregulation appealed overseas as it did here for another reason—no regulator could keep up with fast-changing technology.

The Clinton administration has negotiated aggressively in international discussions to further the deregulatory trend, seeking to stimulate global competition, avoid anticompetitive conduct, and open international markets to foreign participation. Talks began in 1995 and were concluded in 1997 with several international trade agreements that make it easier for companies from one country to participate in the telecommunications markets of other countries. These agreements are all subject to ratification by the individual nations involved, and not all countries agree with these goals. Similar agreements have been reached among the members of the European Union; On January 1, 1998, all telecommunications services and infrastructure in the EU member nations will be opened to full competition. Worldwide, we have come a long way from the monopolist, nationally controlled media of the past.

Impact on Public-Service Broadcasting

With deregulation, traditional public-service operators faced new competition. Cable television, satellite-distributed programs for both cable and direct-to-home viewing, videocassettes, and new commercial broadcast channels fought for audiences, advertisers, and programs. Some critics argue that excessive competition may force public-service broadcasters to lower program standards as they struggle to maintain viable audience shares. Others push for private ownership of government broadcasters.

Pressures on the Canadian Broadcasting Corporation (CBC) illustrate the problem. Created in the 1930s to overcome American radio's dominance among Canadian listeners, CBC operates French and English services on both radio and television, the latter in part commercially supported. The French services are very popular—but most

English-speaking Canadians tune to American programs, either directly or as rebroadcast on commercial Canadian networks competing with CBC. In the mid-1990s, CBC's huge federal subsidy (some $800 million, nearly three times what Congress gives to the Corporation for Public Broadcasting) came under concerted attack as unneeded in an era of multiple channels, commercial diversity, and government deficit. An early 1995 subsidy cut of 25 percent brought resignation of CBC's president and questions of how CBC services could be maintained (parallel again to similar events in American public broadcasting). The pressure to boost revenues and maintain its audience share continues to intensify as the service adjusts to competition from cable television and DBS.

Privatization

Converting state-owned facilities to private ownership (or creating new private outlets), known as *privatization*, rapidly took hold in France and Italy. Other European countries, as well as Australia, Canada, Japan, and New Zealand, plus some Latin American nations have taken similar, though usually less drastic, deregulatory steps. Two European examples illustrate the process.

Government once operated all of France's broadcasting stations and networks. Private French stations and networks were first authorized in 1984. Two years later conservatives won control and took the unprecedented step of selling off France's leading public-service television broadcasting network, TF1, and of authorizing competition from new privately owned television networks. By 1994 TF1 controlled 40 percent of the audience and 55 percent of television advertising. A French educational over-the-air channel began daytime operation in 1994 on a fifth television channel, already used in the evening by Arte, a Franco-German cultural network.

Privatization came earlier in Italy and by a different route. In the 1970s the official Italian broadcasting monopoly, RAI (*Radio Televisione Italia*—the initials are traditional and date back to radio-only days) went to court to suppress unauthorized cable operations. The Italian Constitutional Court ruled in 1975 that RAI's legal monopoly covered only *national network* broadcasting, not local cable or local broadcast operations controlled by private owners. This decision opened the floodgates to thousands of private stations. In the 1990s these stations were still operating without benefit of formal regulation because the Italian Parliament, though always seemingly about to pass a new law, could not agree on the form it should take.

Former Communist States

In the Soviet Union and (to an even greater degree) in its East European client-states, democratization brought dramatic change. In 1985 new Soviet leader Mikhail Gorbachev inaugurated a policy of *glasnost*—roughly translated as "candor" or "openness." In addition, electronic media spillover signals from neighboring countries, insistent penetration by the Voice of America and other external services, satellite-borne programs from Western Europe, and videocassettes all helped push communist regimes toward greater openness.

Each of these media played an active part in the revolutions of 1989–1991. Eastern European regimes in 1989 toppled domino-like in the wake of Gorbachev's reforms. Nothing in history matches the startling suddenness with which four decades of rigid communist control, exemplified by the 30-year-old Berlin Wall, crumbled almost overnight in country after country—culminating with the collapse of the Soviet Union itself in late 1991.

Russian broadcasters soon scheduled more live shows, spontaneous interviews (in place of obviously rehearsed recitals formerly used), and telephone call-in programs. They extended schedules into early-morning and late-night dayparts. For the first time, broadcasting began to please and inform Russians instead of lecturing them.

The first privately owned and commercially sponsored television station appeared in 1990, to be joined by two others in 1993. All three were working toward developing national coverage by 1995. Their competition for—and success in gaining—audiences encouraged more advertising revenue and greater use of investigative reporting and often critical analyses (of the late-1994 Russian invasion of Chechnya, for example) with vivid pictures contributing to sharp political debate. Resulting pressure to change corrupt old ways at one of the three state-run television stations led to the widely reported 1995 killing of a highly popular on-air personality who had just been named manager. But such setbacks underlined the growing pluralism in a system that for years had harbored no dissent.

By 1994 private broadcasting and cable was making headway in several formerly communist nations. TV NOVA took over a government station in Prague, providing 19 hours of daily programming to Czech listeners. Supported by advertising, the station expected to break even in about five years. In Poland, a quarter of the nation's viewers could receive Western-style satellite-delivered cable services, including a new one based in Warsaw. Home Box Office marketed itself actively in Hungary. Developing fiber-optic cable networks promised more broadband capacity. On the other hand, some old habits die hard. In several countries, including Hungary and Slovakia, new legislation placed strong controls on media content, including limits on antigovernment news or statements.

International Telecommunication Union

As with changes in individual countries, many international controls were revised during the 1980s. Most nations belong to the International Telecommunication Union (ITU), a United Nations affiliate headquartered in Geneva, Switzerland. As sovereign nations, members cannot be forced to obey ITU regulations—though they usually find it in their best interests to accept ITU recommendations and to adopt ITU standardized terms and procedures.

Acting together, ITU member-states allocate frequency bands to specific services to avoid interference. Throughout the world, AM and FM radio, television, and broadcast satellite services have similar spectrum allocations. (The ITU also issues initial letters for station-identification call signs—hence the *K* or *W* used by American stations, whereas Canadian station calls begin with *C* and Mexican outlets with *X*.)

In the 1980s the ITU expanded its aid to help developing nations improve their telecommunication facilities. At the same time, with many newly independent Third World members, ITU conferences became more politicized. Third World countries argue that ITU spectrum and orbital allocations should be planned for the long term. In other words, they want channels and slots *reserved* for their future use, even though decades might pass before most developing nations could activate them. The United States and most other heavy users of satellite communications, however, need immediate access to these resources to meet current needs and to foster technological development. Clashes on this controversy occur at each ITU meeting.

In the early 1990s the ITU underwent its most dramatic reorganization in a half century—evidence that its dozens of Third World members were having more impact on ITU decisions and processes alike. The ITU elevated assistance to Third World countries to equal status with its more traditional technical standard setting and frequency allocation roles.

Competing Technical Standards

Thanks to ITU uniform spectrum allocations and standards, taking a portable radio to listen abroad works in most places. But don't try travelling internationally with a television set. For although television has similar channel allocations throughout the world, its technical standards are *not* uniform. No fewer than 14 monochrome and 3 principal color technical standards require converters to interchange programs (as in Olympic or World Cup sports coverage) or to use one country's television equipment elsewhere.

Today's three basic color television systems—NTSC, PAL, and SECAM—reflect American, German, and French government promotion in the

1960s. Each country sought to persuade other governments to adopt its standard. Adoptions meant not only national prestige but also huge revenues from international sales by manufacturers of equipment using the favored system. PAL is most widely used; the American NTSC system is nearly universal in North and South America and Japan, whereas France, the former USSR, and their close allies or former colonies adopted SECAM.

A second start on uniform world television standards could result from ongoing ITU study of direct-broadcast satellite (DBS) and high-definition television (HDTV) system technologies. Certainly satellite-to-cable and DBS services designed to cover many countries simultaneously would benefit from such standardization.

By 1990 Europe, Japan, and the United States each had different and incompatible analog HDTV standards on the drawing boards. The American breakthrough on digital HDTV (see Section 5.7) led early in 1993 to the end of Europe's attempt to develop its own analog standard. Consequently, a consortium of European countries began developing a blanket video delivery standard that will apply to cable, satellite, and terrestrial distribution systems. In Japan, an analog version of satellite-delivered HDTV has been in service since 1994 and serves approximately 330,000 households. In 1997 the Ministry of Posts and Telecommunications announced that it will promote a new digital system that will closely resemble the U.S. standard.

PTT Deregulation

Deregulation has also had widespread impact on common-carrier services, loosening control by highly centralized national post, telephone, and telegraph (PTT) monopolies. In addition to their telephone operations, some PTTs have long held exclusive rights to install and operate broadcast transmitters and relays, as well as cable facilities. However, in many countries, transmission and programming were regulated by separate government authorities.

For cable television, marketing and promotion play key roles: Consumers need to be convinced that this new service will benefit them. Lack of experience in facing competition generally means monopoly PTTs have little marketing expertise. Progress in cable installations has often languished under PTT leadership. Rising public and advertiser demand for cable and other services has helped to break down PTT dominance in many countries since the mid-1980s.

Privatization of the telephone and cable networks in Central and Eastern Europe has led to some interesting business collaborations that would not have been thought possible a decade earlier. For example, Matav, the Hungarian telephone company, in 1993 sold a 30 percent stake in its operations—worth $875 million—to a consortium of Ameritech International and Deutsche Telekom. The company hoped to increase the number of telephone lines in the country at least 15 percent per year for six years. The Slovak Republic and Romania also are planning to privatize their telephone networks. In the Czech Republic, a foreign consortium purchased a portion of the telephone system, SPT Telecom, and U.S. West acquired 28.6 percent of the largest Czech cable television company, Kabel Plus. U.S. West and Bell Atlantic hold 49 percent of the only Czech mobile telephone company, Eurotel.

14.4 Access

That all electronic media can inform, persuade, and cultivate values makes access to them a jealously guarded prerogative. Traditionally most countries limit access to professional broadcasters, experts on topics of public interest, people currently in the news, entertainers, and politicians.

Politicians

Democracy requires fairness in political use of electronic media—without at the same time crippling their role as a means of informing voters. In America, weak parties and candidates have the same theoretical access rights as major parties—

provided they can afford to buy advertising time. No other industrialized democracy permits such broad access for candidates—and such commercialization of elections. In most European countries, for example, despite strict fairness regulations on paper, ruling political parties often evade rules by controlling appointments to state broadcasting services and regulatory agencies.

Britain severely limits political broadcasts while providing them free of charge. Campaign broadcasts focus on parties rather than on individual candidates (in keeping with the parliamentary system in which party membership is more important than in American elections). Given that British national campaigns last only 30 days, voters are spared endless merchandising of candidates. Nor do candidates have to beg for donations and accept money from lobbyists to pay for expensive broadcast advertising. Britain is also one of the few countries that allow live televising of Parliament debates—similar to C-SPAN's coverage of Congress here. Such a ready means of watching can bring the governing process closer to voters.

Citizens

During the 1960s people in many countries sought airway access. They argued that if spectrum is really a shared natural resource, then everybody should get a chance to use it. This access movement paralleled a widespread rise in ethnic and regional awareness.

The access movement at first had little impact on the centralized systems of Europe. Access seekers petitioned authorities to create regional and community stations. Many did. France, for example, legitimized more than a thousand small, privately owned FM radio stations following passage of a new broadcasting law in 1982. Many had started as pirate stations, which the French government had suppressed rigorously. The 1990 British broadcasting law placed only limited regulation on some 300 new local radio stations, as well as on new commercial networks first permitted under the new law. In Scandinavia, governments finance low-power FM *när radio* (neighborhood stations), inviting local groups to help program them, free of virtually all regulation.

Groups

Another way of dealing with access demands is to shift emphasis from individuals to groups to which those individuals belong. The now-defunct FCC fairness doctrine applied such a strategy (as detailed in Section 13.8).

The uniquely structured access system of the Netherlands has gone farthest in assuring different social groups their own programs on national broadcast facilities. Most program time on government-run networks is turned over to citizen broadcasting associations. Some represent religious faiths; some have a nonsectarian outlook. Even very small constituencies, such as immigrant workers or people from Dutch colonies, can regularly obtain airtime. An umbrella organization, Netherlands Broadcasting Foundation (NOS), coordinates Dutch time-sharing and produces programs of broad interest, such as national and international news and major sporting events. Another organization handles all sales of advertising time, revenue from which goes to a central program fund.

Access can also work in a negative fashion. The once white-controlled *apartheid* government of South Africa for years limited black access to outside sources of news. Radio news directed at tribal groups in their own languages was channeled onto FM radio. As FM and short-wave tuning were not ordinarily available in the same receiver, the use of FM for controlled domestic news limited black listeners' access to external news sources.

14.5 Economics and Geography

Economics is second only to politics in shaping a country's electronic media. National systems vary widely in audience size, facilities, revenue sources, and the ability to produce homegrown programs. Economic constraints account for these differences, though geography also plays an important role.

Audience Size

Some 200 countries and dependencies have their own radio broadcasting systems. Fewer than 40 countries (mostly small islands) lacked television stations by the mid-1990s. Television set penetration per household varies from nearly 100 percent in countries like the United States, the United Kingdom, and Switzerland, to less than 10 percent in parts of Asia and Africa. High penetration levels depend only partly on economics—important as well are programs popular enough to motivate set purchase, and a policy of licensing local stations. But set penetration does not tell the whole story of viewing and listening in other countries. Cultural differences can influence the use of electronic media even when set penetration levels are comparable. Television viewing time per household in Asia, for example, is much less than in the United States.

Former communist governments invested heavily in transmitters, relays, and production facilities, yet set penetration remains disproportionately low. The People's Republic of China did not begin television broadcasts until 1958, and set penetration lagged far behind most other Asian countries for many years. Recent reforms and the growth of a consumer culture have spurred the growth of television in China.

In tropical Third World countries, radio sets do not work because of humidity and a shortage of batteries. Government investment in transmitters and production facilities can therefore be a waste of money. It costs just as much in program and transmission expense to reach a few scattered individuals as to reach everyone within a transmitter's coverage range. Lack of communications infrastructure—electric power, telephones, and relay facilities for networks—further impedes Third World electronic media development. However, a few of the oil-rich Middle East states have achieved high radio set penetrations—for example, 69 per 100 people in Kuwait and 54 per 100 in Bahrain, compared with an average of fewer than 20 in most African nations. By contrast, the average person in the United States owns two or more radio sets.

Revenue Sources

Electronic media financial support comes from three main sources: government appropriations, receiver license fees, and advertising. Most countries' electronic media still depend at least partially on government funds.

In industrialized democracies, substantial support depends on user license fees—an annual payment per household or receiver (about $150 a year in most European countries by the mid-1990s). This somewhat insulates broadcasting organizations from inevitable biases caused by dependence on either direct government funding or advertising. But the protection is only partial, because governments create fee payment rules and set fee levels. As operational costs rise faster than audience fees increase, many formerly fee-supported systems increasingly turn to advertising, though theoretically advertising contributes too small a part to give advertisers any real influence over programs.

Program Economics

Television consumes expensive programs at such a rate that even developed nations with strong economies cannot afford to program several different television networks exclusively with homegrown productions. Britain is unusual in having five full-scale broadcast television networks, but even it imports some entertainment programs, though with a voluntary limit of 14 percent of total airtime for such imports.

In smaller countries, the lack of programs from domestic sources stimulated cable's growth even before satellite-to-cable program networks came into being. Community antennas in small countries can often pick up a half-dozen services from neighboring countries.

Most imported programs come from the United States (as discussed in Section 14.6), though many come from Britain. A few Third World countries (especially Brazil, Mexico, and India) are active in supplying international program markets.

International syndication and various forms of cost and talent sharing have helped increase the amount of programming available. For example, the European Broadcasting Union (EBU), an association of official broadcasting services in Europe and nearby countries, shares programs through Eurovision, which arranges regular exchanges (mainly sports and news) among its members. Similar associations exist in Asia, the Middle East, and the Caribbean, although they are not as active as the EBU in handling program exchanges.

Co-production is increasingly used to cover high program costs. Producers from two or more countries combine financial and other resources to co-produce television series or movies. They divide capital expenses and profits from ensured distribution of programming in participating countries.

Coverage Problems

Cost-effective media coverage of a country depends on both its shape and size. Alaska and Hawaii, for example, had to await satellite transmission to enjoy same-time coverage with the U.S. mainland.

Indonesia's 6,000 or so widely scattered islands with diverse populations speaking many different languages present even more formidable coverage problems. And Russian territory extends so far east and west that national broadcast schedules have to be adapted to serve 10 different time zones (contrasted with only four in the continental United States). Program distribution needs prompted Russia and Indonesia to become early users of domestic satellites.

Signal Spillover

Geography (and a lack of short-wave radios) insulates most American listeners and viewers from programs spilling over from foreign countries. But spillover *from* the United States has strongly influenced not only broadcasting but also satellite and cable development in neighboring Canada. Most Canadians live within 100 miles of the American border and can easily receive American broadcast signals off-air and/or by cable (and vice versa, of course). As a result, Canada became the first heavily cabled country in the world.

To avoid being overwhelmed by American programs—and to ensure work for its own creative community—Canada imposes quotas limiting the amount of syndicated programming that electronic media may import. To help fill the gap, Canada's government subsidizes Canadian production, leading to a vibrant Canadian popular music industry, for example. In addition, some American films and TV programs have been co-produced in Canada to lower their cost.

14.6 *Programs*

Interchangeability and sharing of programming simply reflects that similar program formats exist throughout the world. News, commentary, public affairs, music, drama, variety, studio games, sports events—such program types appeal everywhere. National differences are evident only in the way these genres are treated.

News and Public Affairs

Prime-time daily newscasts are universally popular, but their content and style differ among countries. Parochialism, chauvinism, and ideology affect the choice, treatment, and timing of news stories. Each country emphasizes its own national events, few of which hold interest elsewhere.

In authoritarian countries such as Iraq, Cuba, and North Korea, news focuses heavily on the national leader to the virtual exclusion of all else. Some broadcasting systems devote their external pickup facilities entirely to reporting on the head of state's every public move, and developing a cult of personality.

Program Balance

Audiences everywhere prefer entertaining comedy and drama to more serious informational content. Accordingly, where audience appeal controls programs—as through audience ratings—entertainment dominates. Most industrialized democracies (including the United States, thanks to public broadcasting and news/culture-oriented cable channels) try to strike a balance among light entertainment and news, information, culture, and educational programs.

Wealthy members of any society who buy most consumer products and services, and enjoy access to political and social groups, see and hear themselves through electronic media far more than do the poor and powerless. Accordingly, electronic media depict a largely urban, fairly well-educated, affluent people, whether fictional or real.

Third World nations can afford few local productions with popular appeal. Therefore, they import foreign entertainment that attracts audiences for a fraction of what it would cost to produce programs locally. But imports throw schedules out of balance (as discussed next) by overemphasizing entertainment and playing up foreign cultures often quite different from viewer surroundings.

Schedules

Though common in America, broadcast days of 18 to 24 hours were rare even in other developed countries until quite recently. Traditionally most foreign radio services would broadcast a short morning segment, take a break before a midday segment, then another break before evening programs, closing down for the night relatively early. Even in Britain, the BBC began 24-hour radio only in 1979, when Radio 2 filled in the previously silent hours of 2:00 to 5:00 A.M. Television in much of the world still begins only late in the afternoon, signing off by about 11:00 P.M. Extending programming into early-morning and late-night hours in the 1980s was one sign that new networks, stations, and cable systems were meaningful competition for long-time broadcasters.

International Syndication

Cultural differences often impact programming. One major difference between American and foreign television (especially in Europe) is the tolerance of nudity in programs and advertising. Full frontal nudes are not uncommon on some European television channels. What is acceptable in one country often is not in another, creating problems for program exchange or syndication. American programs, for example, are often criticized abroad for being excessively violent.

U.S. programs, long dominant in world syndication, came to new prominence with added demand from new satellite-distributed cable networks, direct-broadcast satellite channels, and VCRs (discussed in Sections 14.8 and 14.9). These services vacuum up programs from whatever sources they can find. Their enhanced demand has intensified old fears of American cultural domination. The low price for American programs (because their original cost has already been repaid through domestic sales) and their almost sure-fire mass appeal keep them in demand. Exhibit 14.b shows typical prices paid overseas for syndicated American programs.

In the 1990s, American international syndicators' sales slowed down. In response, many American firms invested in overseas media, though such ventures may encounter legal and cultural opposition to American cost-efficient operating methods. Nor could investors count on quick or easy profits from newly introduced, advertiser-supported, privately owned media services. Rundown economies of former communist countries, for example, require cash for the most basic economic necessities, leaving audiences with little ability to purchase media-advertised goods. Despite that pressure, those countries also adopted local program requirements—the Czech Republic, for example, required 60 percent indigenous programming on a new commercial television station in Prague.

Cultural Imperialism

American dominance of international syndication has long fed Third World concern about cultural im-

Exhibit 14.b

Program Bargains from U.S. Syndicators

Prices for U.S. syndicated material vary widely according to potential audience and ability to pay. Prices for feature films also fluctuate widely based on the quality of the movie.

Cost per Program (U.S. Dollars)

Purchasing Country	Music/Arts (per hour)	Series: Half hour	Series: Hour	Movies: TV	Movies: Theatrical
Australia	$20,000–500,000	$ 10,000	$ 19,000	$ 72,000	$ 7,900
Austria	10,000	6,000	12,000	20,000	17,500
Belgium	4,000	1,400	5,000	7,000	56,000
Canada	25,000–50,000	15,000–50,000	30,000–70,000	100,000–160,000	200,000–300,000
Czech Republic	1,000	700	1,200	3,000	2,500–15,000
France	60,000	15,000–40,000	75,000–250,000	40,000–120,000	145,000–2,000,000
Germany	20,000–50,000	40,000–65,000	65,000–130,000	80,000–250,000	100,000–450,000
Holland	4,500	6,000	8,000–12,000	15,000	25,000–100,000
Japan	20,000–400,000	7,000–8,000	15,000	N/A	100,000–1,500,000
Russia	N/A	2,000–3,000	4,000–6,000	5,000–7,000	12,000–17,000
Spain	N/A	10,000–15,000	40,000–60,000	35,000–45,000	40,000–300,000
United Kingdom	15,000	5,000–30,000	75,000–250,000	50,000	10,000–3,000,000

Source: Based on data in "Global TV Price Guide," *Variety* (September 30, 1996): M6. Reprinted with permission of VARIETY, Inc. © 1997.

perialism. Critics argue that images and values shown in imported television undermine local cultures. For example, American shows are criticized for encouraging excessive consumption, materialism, and disregard for other traditions. Further, every imported program denies an opportunity for locals to showcase their own talents. Cheaper imported programs can thus perpetuate dependence on foreigners.

Nor do such complaints come only from the Third World. Even developed nations with extensive production of their own limit the amount of entertainment their national systems may import. In 1992 the 12-nation European Community (representing a market larger than America) imposed program-import restrictions despite strong American objections. A year later France led Europe to protect its media market from American product often sold at bargain-basement prices (about a tenth of the cost to produce the same thing locally). There was interest—but no specific requirement—in increasing the amount of time given to French music on French stations.

American producers understandably strongly resist attempts by other countries to limit program imports. At the newspaper and broadcast industries' behest, U.S. diplomats successfully advocated a *free-flow* international communication policy when the United Nations was created in 1945. But more than 70 new nations, most extremely conscious of their prior histories as colonies of developed countries, have since joined the UN. They ask what value free flow has for them when it runs almost entirely in one direction—*from* the United States and other industrialized countries to the Third World. Instead, they call for *balanced* flow—defined as news and entertainment that treats Third World countries fairly and in proportion to their population significance. The debate between these polar positions continues in many international organizations today.

On the other hand, such American cultural creations as MTV attracted some 200 million viewers in other countries by the late 1990s just as the cable music service had done here 15 years earlier. MTV International now operates several regional services: MTV Europe, MTV Asia, and MTV Latin America. The company plans to target its programming even more by offering country-specific channels such as MTV India, MTV in the U.K., and MTV Japan. These channels, produced in Singapore, Taipei, and London, will feature national talent and playlists designed to appeal to each channel's target market.

NBC Super Channel, also based in London and transmitting to much of Europe, offers a selection of the network's American domestic product to new audiences—apparently with considerable success. With similar aims, ABC has purchased part interest in several European production firms.

UNESCO's Role

Much of this debate came to a head in the 1970s and 1980s at the Paris-based UN Educational, Scientific and Cultural Organization (UNESCO), which led in defining and calling for establishment of a *new world information and communication order* (NWICO). NWICO called for, among other things, balanced media flow, training (and perhaps licensing) of journalists to operate in other cultures and languages, and support from rich nations to assist development of poor nations' media infrastructure. American media leaders saw this development as an attack on journalistic independence and free-enterprise advertising. Claiming that UNESCO wasted money (a quarter of its budget came from the United States) and had become hopelessly politicized, the U.S. government withdrew as a member in 1984, leaving an agency that it had helped to establish four decades earlier. By the mid-1990s, America said it would rejoin UNESCO as much of the agency's political and budget turmoil had been resolved. Ironically, that reentry has been delayed further—not by politics, but by tight budget conditions in Washington.

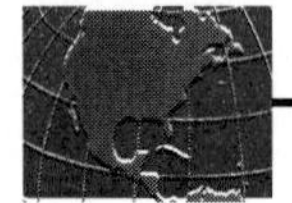

14.7 *Transborder Broadcasting*

The ability to use radio to penetrate political boundaries added a potent element to diplomatic relations. Never before had nations been able to talk directly to masses of foreigners, crossing even the most heavily defended national borders. By the 1990s more than 80 countries were operating official external services—programs aimed at foreign countries.

BBC World Service

Colonial commitments first prompted nations to broadcast radio internationally. The Dutch and Germans started in 1929, the French in 1931. After experimenting for several years, Britain's BBC began what was first called the Empire Service in 1932. It initially broadcast only in English, primarily seeking to maintain home-country ties with expatriates in colonies around the world. During World War II foreign listeners came to regard the BBC as having the most credibility among external broadcasters, and it has retained that reputation ever since. Today approximately 140 million listeners worldwide tune to what is now called BBC World Service, especially when they doubt reports from other sources. It has more than a million American radio listeners; a few listen via short wave, but many tune in via rebroadcasts on Public Radio International. Europeans can also hear it as a direct-to-home satellite service. The service is now broadcast in 45 languages.

BBC World Service Television (WSTV) began in 1991 as a cooperative effort of domestic television news and World Service radio. The TV services came under the control of BBC Worldwide, the commercial arm of BBC, in 1995. Presently, the World Service provides news programming to the satellite-delivered 24-hour news and information channel, BBC World.

Voice of Russia (Radio Moscow)

Though the Soviet Union initially controlled no overseas colonial empire, it used radio to promote its revolutionary ideas. With Radio Moscow's inception in 1929, the Soviets recognized the importance of foreign-language radio as a means of gaining and influencing friends abroad.

Radio Moscow developed fewer overseas relay stations than major Western external broadcasters, though in the 1960s it built one such station in Cuba aimed at the Americas. Like radio services in all communist states, Radio Moscow was relentlessly propagandistic. It lightened its tone to gain wider appeal in the 1980s, even initiating a 24-hour English-language service. Renamed The Voice of Russia in 1995, the service was forced by budget problems to cut its staff by 30 percent and discontinue broadcasting in 16 foreign languages. Today the service broadcasts in 32 languages, and the English language service broadcasts to all continents 24 hours per day.

Voice of America

The United States joined the growing battle of words with Voice of America (VOA) early in 1942. Wary of creating a domestic propaganda agency, Congress forbids broadcast of VOA programs in the United States. However, anyone with a short-wave radio can pick up VOA programs aimed at overseas listeners, and VOA scripts and recordings (as well as those from Radio/TV Martí and Radio Free Europe) can now be obtained on the Internet.

In 1997 VOA used 52 languages in addition to English, broadcasting 700 hours a week, about 85 percent devoted to American events and news. VOA programs originate in Washington, going overseas via leased satellite channels and 114 VOA short-wave transmitters located in Greenville, North Carolina, and several secondary U.S. sites. The VOA also leases sites in a dozen foreign countries, where it maintains transmitters for rebroadcasting programs to listeners in nearby areas.

VOA news and public-affairs programs reflect official American policies. News commentaries are explicitly labeled as coming from the U.S. government. For the sake of credibility, VOA tries to observe the spirit of its original 1942 policy: "Daily at this time we shall speak to you about America. . . . The news may be good or bad. We shall tell you the truth." Truth-telling continues to be VOA policy, despite occasional lapses when partisan officials bend facts to suit momentary political objectives.

Worldnet

The short range of broadcast television makes it useless for external services aimed at distant targets in the manner of short-wave radio. The United States Information Agency (USIA), VOA's parent organization, used to rely entirely on persuading foreign broadcasters to carry American television programs on their own domestic television services.

In 1983 USIA created Worldnet as a daily television service distributed abroad by satellite. To encourage foreign broadcasters to use its material, Worldnet offers interactive teleconferences with U.S. government officials responding to live questions from foreign news personnel. American interviewees speak from a Washington studio to questioners in other countries. Participation by another country's own broadcasters makes the American presence more acceptable on that country's television service. Additionally, Worldnet televises live call-in and public-affairs programs in Arabic, French, Mandarin, Russian, Polish, Serbian, Spanish, and Ukrainian.

Worldnet fills most of its schedule with "passive" programs—news and general information minus the interactive feature. American diplomatic posts throughout the world pick up Worldnet on some 230 TVROs, as do foreign cable systems and broadcast stations in about 200 cities in more than 125 countries. Worldnet is available via INTELSAT (discussed in Section 14.8) to any foreign broadcasting or cable television operation that wishes to use it.

Service costs, as well as research showing very small audiences, led to congressional budget cuts for Worldnet in the early 1990s, thus limiting its ability to provide original programming.

Radio Free Europe and Radio Liberty

In addition to conventional external broadcasting, the United States has long provided *surrogate* domestic services. Such broadcasts simulate domestic networks in target countries, bypassing censored domestic media.

Beginning in the early 1950s, Radio Liberty (RL) targeted the Soviet Union while Radio Free Europe (RFE) aimed at Eastern European states then under Soviet control. CIA support of these services was revealed in the early 1970s, leading to their reorganization under the Board for International Broadcasting. RFE and RL initially built studios and transmitters in Munich, Germany, and additional transmitters in Portugal, and in Spain at sites favorable for sky-wave transmission to their target countries. Broadcasting by radio, entirely in the languages of the target countries, RFE and RL provided domestic and foreign news from the listeners' perspective, often with strong messages encouraging resistance to communist thinking and governments. The United States spent more on operating these two surrogate services than on VOA.

The 1989–91 collapse of communism in Russia and Eastern Europe led to major adjustments in RFE and RL. Both stepped up their activity now that listening to them was no longer illegal, and they soon added transmitters and news bureaus right in their target countries—something undreamed of in Cold War days. Driven by budget concerns and Clinton administration proposals, Congress decided to merge RFE/RL into a new broadcasting bureau (including VOA) within USIA. As an indication of dramatic change among their former target countries, in 1994 RFE began relocating its broadcast functions from Munich to the less expensive conditions of Prague—at the invitation of the Czech government. From its new offices in Prague, RFE/RL broadcasts 700 hours of programming in 23 different languages to a regular audience of over 25 million listeners.

Radio/TV Martí

In May 1985 the United States introduced Radio Martí, a new surrogate service aimed at giving Cuba news and information free of Castro-regime bias. Supporters claim that Radio Martí has had a powerful effect, heightening dissatisfaction with the Castro regime by revealing facts it conceals from its own people. The expatriate Cuban community in south Florida, with solid political ties in Washington, strongly supports the operation. Critics suggest that some Cuban expatriots exert too much control over the station, thus damaging its credibility as a news source. When Cuban military jets shot down two civilian airplanes operated by a Cuban exile group in 1996, President Clinton responded in part by ordering Radio Martí to increase its power to provide an even stronger signal in Cuba.

Congress authorized funds in 1988 to start experiments with a television version of Radio Martí, using a transmitter hung from a balloon tethered 10,000 feet above Cudjoe Key in Florida. This height enabled television line-of-sight signals to reach Havana 110 miles away. Tests conducted in 1990 incited Cuba to *jam* TV Martí transmissions and to interfere with American AM radio stations. The Cuban government appealed to the ITU, claiming TV Martí violated ITU allocations that designate television channels for domestic use only. Many American broadcasters also objected to the scheme, which was characterized by one *Broadcasting* editorial as a "huge, disastrous silliness" (9 April 1990: 98). By 1997, due to technical problems with the transmitter and signal jamming by the Cuban government, it was unclear whether anyone in Cuba had ever viewed a complete program from TV Martí.

Radio Free Asia

The idea of bringing news reports to places otherwise unlikely to hear them took hold on the other side of the world. In 1993 Congress authorized a Radio Free Asia (RFA) service aimed at potential audiences in the People's Republic of China and other Asian countries. Radio Free Asia began broadcasting in 1996 with an hour-long newscast in Mandarin. By mid-1997, RFA was broadcasting news and cultural programming to China, Tibet, Burma (Myanmar), Laos, Cambodia, Vietnam, and Korea. Several of these countries, most notably China and Vietnam, have strongly criticized the RFA operations.

Religious Broadcasters

A different ideological motivation led to transborder religious broadcasting. International radio gave evangelicals their first opportunity to deliver messages directly to potential converts in closed societies dominated by state religions. Official hostility—in some Moslem countries toward Evangelical Christians, for example—can prevent the setting up of on-site missions but cannot easily bar radio messages. Well-funded conservative religious broadcasters have so saturated short-wave bands that listeners can pick them up almost anywhere in the world, 24 hours a day.

Peripherals

Commercial motives account for transborder broadcasters known as *peripherals.* Several European ministates bordering large countries have long operated such stations, capitalizing on unfulfilled demand for popular music and broadcast advertising. Both audiences and advertisers, frustrated by severely regulated domestic services, have welcomed these alternatives.

Peripherals beam commercial radio services in appropriate languages to neighboring countries. They specialize in popular music formats, sometimes supplemented by objective news programs, which are welcome where ruling political parties dominate broadcast news.

The Grand Duchy of Luxembourg, ideally located for peripheral transmitters at the intersection of Belgium, France, and Germany, receives much of its national income from international commercial television as well as radio broadcasting. Exhibit 14.c offers further details. Other notable transborder commercial radio stations operate in the German Saar (Europe No. 1), Monaco (Radio Monte Carlo), Cyprus (Radio Monte Carlo Middle East), Morocco (Radio Mediterranean International), and Gabon (Africa No. 1). Peripherals tend to be rather staid operations, tolerated by their target countries, some of which even invest in them. They still leave some commercial and program demands unsatisfied, creating a vacuum filled by pirate radio outlets.

Pirates

The first offshore *pirate* stations began broadcasting from ships anchored between Denmark and Sweden in 1958. Lacking authorization from their target countries, they were often financed by American interests and copied American popmusic formats, advertising techniques, and promotional gimmicks. They quickly captured large and

Exhibit 14.c

Luxembourg: Home of Peripherals

The tiny Grand Duchy of Luxembourg granted a broadcasting monopoly to a hybrid government/private corporation, now known as RTL (Radio-Télé-Luxembourg), back in 1930. A year later, it began operating as what the French called a *radio periphérique* (peripheral radio). In those days, when official European radio services tended to be rather highbrow and stuffy, listeners far and wide avidly tuned in to its pop-music programs. Legend has it that Radio Luxembourg, received in Liverpool, England, gave the Beatles their first taste of pop music.

Today RTL has high-power long-wave (2,000 kw), medium-wave (1,200 kw), and short-wave (500 kw) radio transmitters radiating across the borders, carrying programs in Dutch, French, and German. After 60 years of service, RTL ended its English-language programming in 1992. It broadcasts television in both the PAL and SECAM systems in order to reach both French and German viewers. RTL holds shares in a number of European privately owned broadcasting services and owns extensive production facilities in Luxembourg.

Television's short range made it impossible for the Grand Duchy of Luxembourg to repeat its radio coverage with the newer medium, but it overcame this problem in part by setting up jointly owned broadcasting services in neighboring countries, notably RTL Plus in Germany (both a broadcast and a satellite-to-cable service), TV (French-speaking Belgium's first commercial broadcast channel), and RTL-Veronique (Holland's first privately owned commercial broadcasting channel).

The ITU's 1977 allocation of DBS orbital slots to European countries offered Luxembourg a chance to extend its television coverage to the whole of Europe. SES (Societé Européenne des Satellites), founded in

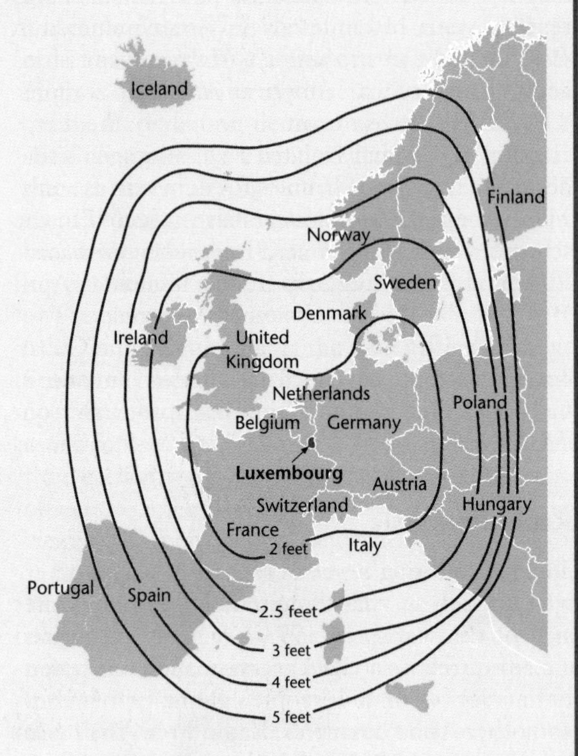

1985, launched ASTRA in 1988—the first privately owned European communication satellite. The numbers indicate the diameter of home TVROs needed to receive ASTRA in the zone defined by the contour lines.

ASTRA contracted with a variety of direct-to-home and satellite-to-cable services to occupy its 16 channels, including Disney and MTV-Europe. It also downlinked the first British DBS service, Sky Television, later known as BSkyB.

devoted youthful audiences—their very illegality adding spice to their attractiveness.

Some pirates made a lot of money, but at considerable risk. They suffered from storms, raids by rival pirates, and stringent laws that penalized land-based firms for supplying or doing other business with them. The appetite for pop music whetted by pirate stations forced national systems to take notice of formerly ignored musical tastes. The BBC, for example, reorganized its national radio network offerings, adding a pop-music network (Radio 1) imitative of the pirates. Some offshore DJs ended up working for the BBC and other established broadcasters.

Controversy over pirate broadcasting caused the fall of one Dutch government years ago, and the country reorganized its broadcasting system as a result. It permitted two former pirate organizations to come ashore and develop into leading legitimate broadcasters. As national systems become more pluralistic in program appeal, the rationale for pirate stations declines—few operate today.

14.8 International Satellites

In recent years commercial services have more readily crossed borders by use of satellite-to-cable networks and DBS facilities.

International Telecommunications Satellite Organization

The United States led the 1964 founding of the International Telecommunications Satellite Organization (INTELSAT). With relay satellites stationed above the Atlantic, Indian, and Pacific oceans, INTELSAT eventually made possible instant worldwide distribution not only of television programs but also of telephone conversations, news agency stories, and business data.

More than 130 countries have part ownership of INTELSAT, but the United States owns the largest block of its shares (about 25 percent), houses its headquarters in Washington, D.C., and has provided its last three director-generals. Communications Satellite Corporation (Comsat) initially operated INTELSAT's network under contract on behalf of the rest of the consortium until INTELSAT itself took over in 1979. In the early 1990s the former Soviet Union and its onetime client-states of Eastern Europe abandoned their Intersputnik rival network and became members of INTELSAT. Exhibit 14.d offers more details on INTELSAT and its recently emerged competition.

Though primarily an international carrier, INTELSAT also leases satellite capacity at reasonable rates to Third World countries for their *domestic* use. Thus INTELSAT enables many such nations to vault directly into the satellite era, avoiding costly construction of microwave and coaxial-cable circuits throughout their territories.

Satellite Launching

The U.S. National Aeronautics and Space Administration (NASA) long monopolized Western capacity to launch communication satellites. In 1984, a consortium of European countries began to challenge NASA's monopoly with their own launch facility, Arianespace (named for *Ariane*, the rocket used). Ariane rockets operate from a site in French Guiana on the northern coast of South America, near the equator. That location gives Arianespace better conditions for attaining equatorial orbit than does NASA's more northerly location at Cape Canaveral, Florida.

After 1986 NASA launches were confined to government projects. This decision opened the launch market to private U.S. and foreign rocket makers. China and Russia made their government launch facilities available to foreign commercial users—although American national security concerns limited U.S. participation (we did not want to give away expensive technology information to potentially hostile countries).

Domsats

The Soviet Union initiated domestic satellite (domsat) development. In 1965 it began launching

Exhibit 14.d

INTELSAT and "Separate Systems"

The globe shows an example of INTELSAT coverage. INTELSAT V (F-15), located above the Atlantic Ocean, beams both to the east and to the west. Note that such satellites can project beams to the entire region or to a specific zone or spot within a region.

American satellite firms, strongly backed by the American government, argue that INTELSAT's monopoly is inefficient, that its rates do not reflect actual costs, and that its sheer size makes it inflexible. Competition from smaller, nimbler satellite firms, they contend, would lower prices for all, enhance services, and encourage innovation. The first private American "separate system" to launch a satellite, Pan American Satellite Corporation (PANAMSAT), offers both domestic and international satellite services. It links the United States and countries of the Hispanic world—Central and South America and the Caribbean—with Europe, and serves Hispanic domestic needs as well.

Peru became the first PANAMSAT customer, followed by the Dominican Republic and Costa Rica on this side of the Atlantic, and by Britain, Ireland, Luxembourg, Sweden, and the former West Germany on the other, with more yet to come. CNN and other U.S. satellite-to-cable networks contracted with PANAMSAT to relay their programs to countries to the south. In 1997, PANAMSAT announced that it was merging with another international satellite operator, Hughes Galaxy.

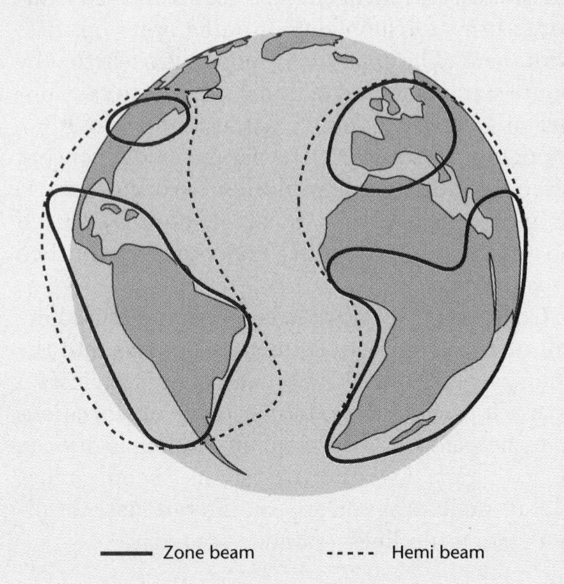

Source: Satellite footprints used by permission of INTELSAT.

the *Molniya* satellite series that enabled relaying of Soviet television throughout its vast territory. Canada's *Anik* domsat series followed, starting in 1972, preceding the first U.S. domsat, Westar, by two years. The 1970s also saw the launching of domestic and regional satellite services elsewhere: Indonesia's *Palapa*, the first Third World satellite; Europe's *Eutelsat*, the first of a regional satellite series; and Japan's first satellite. France, first among the European countries with its own domsat, launched *Telecom* in 1985, in part to enable relays to French overseas territories.

Direct Broadcast Satellite Service

Direct Broadcast Satellite services have shown promise in countries, such as Australia and Japan, where they do not have to compete with high cable penetration. Japan forestalled cable expansion by concentrating on DBS experiments. It led the world with the first full-time DBS service in 1987. By the end of 1997, three commercial DBS operators were competing for Japanese customers,—PerfeTV, Direct TV Japan, and JSkyB, a joint venture among Rupert Murdoch's News Corp., Sony, and Fuji TV Network. And Australia's AUSSAT, though a general-purpose satellite rather than a specialized DBS vehicle, transformed Australian broadcasting in the late 1980s. It brought television for the first time to remote areas too thinly populated to support either terrestrial relay or cable systems.

ASTRA and BSkyB

In 1988 a Luxembourg corporation launched ASTRA, Europe's first privately owned satellite suitable for DBS services, described in Exhibit 14.c. It is used by Sky Television (called BSkyB on the air), a DBS service resulting from a 1990 merger of two competing British firms. The first direct-to-home satellite venture in Britain, Rupert Murdoch's Sky Television, began in 1989; a second service launched a satellite in 1990. The two could not sustain the costly competition for home subscribers—a competition costing Murdoch $3.2 million a week. The merged Sky service provides sports, general programming, films, and an all-news channel. By mid-1997, over 25 million Europeans received some form of satellite-delivered television service.

Asia's StarTV

In 1991 StarTV, more formally known as Satellite Television Asian Region Ltd., began satellite delivery of television programming to all of Asia. In July 1993 Rupert Murdoch's News Corp. purchased nearly two-thirds of the operation for $525 million. Based in Hong Kong, the service offers seven channels featuring news, music, sports, and entertainment programming. Two of the channels are entirely in the Mandarin language; the others feature programs in English and several indigenous languages. The service is viewed in 38 countries by some 260 million people, who welcome a wider choice of viewing options. Large advertisers support the programming, which is free to viewers. StarTV has a potential audience in the billions as receiving dishes become more widely available.

Radio by Satellite

Americans tend to think of DBS services exclusively in television terms, but Europeans can receive some 60 radio services by satellite. In addition to such major public-service broadcasters as the BBC, a number of colorfully named private radio stations can be received by satellite—Kiss FM, Radio Radio, Fun FM, and Skyrock, for example. Europeans receive satellite radio services directly by means of home TVROs or indirectly via cable systems and FM radio stations.

14.9 Cable and Newer Media

Many developed countries had primitive community antenna television (CATV) for years before modern cable emerged. These early systems merely extended domestic broadcast station coverage, sometimes adding a few channels for the purpose of carrying neighboring foreign broadcast networks. Most operated noncommercially, often owned by municipalities. They offered few channels, no local origination, and no pay television.

Cable

Modern cable has developed most extensively in Canada, the United States, and industrialized smaller countries (see Exhibit 14.e). Although cable penetration continues to increase, public demand in many larger countries remains sluggish. If basic cable systems do not offer a wide variety of popular programs, subscribership remains low. High-appeal cable services delivered by satellite did not begin in Europe until 1982 with the launch of Sky Channel, the London-based English-language service noted earlier. After that, satellite networks grew rapidly. More than 100 were serving European cable systems by 1997. Some targeted only a single country; some, several nations that shared common languages.

VCRs

Both as an alternative and as a supplement to cable and DBS, videocassette recorders (VCRs) have

Exhibit 14.e

Television Around the World

Cable penetration (the percentage of television homes subscribing to cable service) varies greatly around the world, as does VCR ownership.

Country	TV Sets per 100 people	Cable Penetration (% TV homes)	VCR players per 100 people
Belgium	45	95%	37
Brazil	26	na	3
Egypt	16	na	3
France	54	6	23
Germany	53	45	21
Ghana	5	na	1
India	5	na	1
Israel	33	58	14
Japan	68	na	24
Poland	41	na	14
Russia	41	na	1
South Africa	11	na	3
Spain	38	1	15
Switzerland	78	73	40
United Kingdom	71	5	30
USA	**77**	**68**	**31**

Source: Cable information from Telenet 1997. TV set and VCR ownership data from BBC International Broadcasting Audience Research, *World Radio & Television Receivers 1995*.

proved a boon to viewers in countries where broadcast services failed to satisfy demand. They offer a relatively inexpensive short cut to programs banned from, or otherwise not available on, national broadcast or cable channels.

In formerly communist countries, rigid political censorship created appetites only VCRs could help satisfy. Clandestine VCR tapes hastened the demise of those repressive regimes. VCRs are common in Britain, used primarily to time-shift desired programs to more convenient viewing hours and to view feature films. In 1995 the BBC reported nearly 93 million VCRs in all of Europe. But penetration varies from a high in Switzerland of 40 per 100 people to the former USSR with one per 100 (see Exhibit 14.e).

VCRs help compensate for often inadequate domestic television schedules and poor program quality. Few individuals in Third World countries can afford to buy a VCR outright, but rentals, club purchases, and group viewing in bars, coffeehouses, and even on buses resolve cost problems. In some cases heavy censorship encourages VCR growth—as in Saudi Arabia, where conservative Moslem standards severely limit broadcast television. A worldwide underground market in VCRs and tapes defeats most government attempts to limit sales and rentals. It also undermines profitability of video sales.

Teletext and Videotex

The BBC and the IBA invented teletext, which has proved far more successful overseas than in the United States (see Section 3.3). Though not a runaway success, it has nevertheless found greater acceptance in Britain and Europe than in this country, where too many alternative media choices exist.

France leads the world in videotex development with its *Minitel* system. The government telecommunications monopoly promotes Minitel, renting terminals to consumers for a few dollars a month. Telephone subscribers forgo printed telephone directories, instead paying about six cents a minute for Minitel's many services, which include, in addition to telephone directory assistance, thousands of independent electronic services such as transportation schedules, banking, and personal "chat lines." By 1997 Minitel served almost 7 million terminals and carried content from more than 17,000 different services, many of which touted Minitel use in their print and television advertising.

The Internet

With millions of users around the world and thousands more logging on every month, the Internet is the fastest-growing electronic medium in the world. The popularity of the Internet and the ability of net users to communicate effortlessly across national boundaries make the medium at once a very liberating and a dangerous technology, depending on one's point of view. Totalitarian governments in particular worry about the uncontrolled access of their citizens to ideas and opinions contrary to the party line. Some countries have even outlawed the private ownership of personal computers in an effort to quell the use of the Internet inside their borders. As with any new medium, the international repercussions of the Internet will not be known entirely for many years. It is definitely a story worth watching.

Conclusion

And so, as we count down the few years remaining to the turn of the century, the electronic media in America—now a huge industry of many competing technologies and owners, all sharing the common goal of providing entertainment and news to American homes—continue to thrive and, in many cases, to set benchmarks to which other nations aspire. Even the newest electronic media services owe much to traditional broadcasting because their programming draws directly from genre developed and still provided by radio and television networks and stations. And audiences rarely care about how program content reaches their home screen or radio.

Bibliography

Abramson, Albert. 1995. *Zworykin: Pioneer of Television.* U. of Illinois Press, Urbana.

Adams, R. C. 1989. *Social Survey Methods for Mass Media Research.* Erlbaum, Hillsdale, NJ.

Advertising Age. Weekly. Crain, Chicago.

Alten, Stanley R. 1994. *Audio in Media,* 4th ed. Wadsworth, Belmont, CA.

Anderson, James. 1987. *Communication Research: Issues and Methods.* McGraw-Hill, New York.

———, and Meyer, Timothy. 1988. *Mediated Communication: A Social Action Perspective.* Sage, Beverly Hills, CA.

APA. 1992. "APA Task Force Explores Television's Positive and Negative Influences on Society." News Release. American Psychological Association, Washington, DC.

Arlen, Michael J. 1969. *Living-Room War.* Viking, New York.

Auletta, Ken. 1991. *Three Blind Mice: How the TV Networks Lost Their Way.* Random House, New York.

Avery, Robert K., and Pepper, Robert. 1979. *The Politics of Interconnection: A History of Public Television at the National Level.* National Association of Educational Broadcasters, Washington, DC.

Baker, W. J. 1970. *A History of the Marconi Company.* Methuen, London.

Baldwin, Thomas F., and Lewis, Colby. 1972. "Violence in Television: The Industry Looks at Itself," in Comstock and Rubenstein, 290–365.

———; McVoy, D. Stevens; and Steinfield, Charles. 1996. *Convergence Integrating Media, Information & Communication.* Sage, Thousand Oaks, CA.

Banning, William P. 1946. *Commercial Broadcasting Pioneer: The WEAF Experiment, 1922–1926.* Harvard U. Press, Cambridge, MA.

Barnouw, Erik. 1978. *The Sponsor: Notes on a Modern Potentate.* Oxford U. Press, New York.

Baudino, Joseph E., and Kittross, John M. 1977. "Broadcasting's Oldest Station: An Examination of Four Claimants," *Journal of Broadcasting* 21 (Winter): 61.

Berger, Arthur Asa. 1991. *Media Research Techniques.* Sage, Newbury Park, CA.

BIB (Board for International Broadcasting). Annual. *Annual Report.* GPO, Washington, DC.

Bilby, Kenneth. 1986. *The General: David Sarnoff and the Rise of the Communications Industry.* Harper and Row, New York.

Birch Scarborough. 1991. "How America Found Out About the Gulf War: A Birch Scarborough Study of Media Behavior," Birch Scarborough Research, Coral Springs, FL.

Bishop, John. 1988. *Making It in Video: An Insider's Guide to Careers in the Fastest-Growing Industry of the Decade.* McGraw-Hill, New York.

Bliss, Edward, Jr. 1991. *Now the News: The Story of Broadcast Journalism.* Columbia U. Press, New York.

Block, Alex Ben. 1990. *Outfoxed: Marvin Davis, Barry Diller, Rupert Murdoch, Joan Rivers, and the Inside Story of America's Fourth Television Network.* St. Martin's, New York.

Block, Eleanor S., and Bracken, James K. 1991. *Communication and the Mass Media: A Guide to the Reference Literature.* Libraries Unlimited, Englewood, CO.

Blum, Eleanor, and Wilhoit, Frances Goins. 1990. *Mass Media Bibliography: An Annotated Guide to Books and Journals for Research and Reference.* U. of Illinois Press, Urbana.

Boorstin, Daniel J. 1964. *The Image: A Guide to Pseudo-Events in America.* Harper and Row, New York.

———. 1978. "The Significance of Broadcasting in Human History," in Hoso-Bunka Foundation, *Symposium on the Cultural Role of Broadcasting,* The Foundation, Tokyo.

Bower, Robert T. 1973. *Television and the Public.* Holt, Rinehart and Winston, New York.

———. 1985. *The Changing Television Audience in America.* Columbia U. Press, New York.

Broadcast Engineering. Monthly. Intertec, Overland Park, KS.

Broadcasting (after 1992, *Broadcasting & Cable*). Weekly. Cahners, Washington, DC.

Broadcasting & Cable Yearbook. (Title varies.) Annual. Broadcasting, Washington, DC (to 1991); Bowker, New Providence, NJ (1992–present).

Brooks, John. 1976. *Telephone: The First Hundred Years.* Harper and Row, New York.

Brooks, Tim, and Marsh, Earle. 1992. *The Complete Directory to Prime Time Network TV Shows: 1946–Present,* 5th ed. Ballantine, New York.

Brotman, Stuart N. 1990. *Telephone Company and Cable Television Competition: Key Technical, Economic, Legal and Policy Issues.* Artech, Norwood, MA.

Brown, Les. 1971. *Televi$ion: The Business Behind the Box.* Harcourt Brace Jovanovich, New York.

———. 1992. *Les Brown's Encyclopedia of Television,* 3d ed. Facts on File, New York.

Browne, Steven E. 1992. *Film Video Terms and Concepts.* Focal, Stoneham, MA.

Buckley, Tom. 1978. "Popularity of '60 Minutes' Based on Wide-Ranging Reports," *New York Times* (December 17): 99.

Buxton, Frank, and Owen, Bill. 1972. *The Big Broadcast: 1920–1950.* Viking, New York.

Cablevision. Weekly. International Thomsom Communications, Denver.

Cal. (California Reporter). 1981. *Olivia N. v. National Broadcasting Co.,* 178 Cal. Rptr. 888 (California Court of Appeal, First District).

Cantril, Hadley. 1940. *The Invasion from Mars: A Study of the Psychology of Panic.* Princeton U. Press, Princeton, NJ.

Carothers, Diane Foxhill. 1991. *Radio Broadcasting from 1920 to 1990: An Annotated Bibliography.* Garland, New York.

Carroll, Raymond L., and Davis, Donald M. 1993. *Electronic Media Programming: Strategies and Decision Making.* McGraw-Hill, New York.

Carter, T. Barton, et al. 1991. *The First Amendment and the Fourth Estate: The Law of Mass Media,* 5th ed. Foundation, Westbury, NY.

———. 1996. *The First Amendment and the Fifth Estate: Regulation of Electronic Mass Media,* 4th ed. Foundation, Westbury, NY.

CCET (Carnegie Commission on Educational Television). 1967. *Public Television: A Program for Action.* Harper and Row, New York.

CCFPB (Carnegie Commission on the Future of Public Broadcasting). 1979. *A Public Trust.* Bantam, New York.

CEMA. 1997. http://www.cemacity.org

CFR (Code of Federal Regulations). Annual. *Title 47: Telecommunications,* 5 vols. GPO, Washington, DC.

Christians, Clifford, et al. 1991. *Media Ethics: Cases and Moral Reasoning,* 3d ed. Longman, White Plains, NY.

Clifford, Martin. 1992. *Modern Audio Technology: A Handbook for Technicians and Engineers.* Prentice Hall, Englewood Cliffs, NJ.

Comstock, George, et al. 1978. *Television and Human Behavior.* Columbia U. Press, New York.

———, and Rubenstein, F., eds. 1972. *Television and Social Behavior: Media Content and Control I.* GPO, Washington, DC.

Coolidge, Calvin. 1926. "Message to Congress," 68 *Congressional Record* 32.

CPB (Corporation for Public Broadcasting, Washington, DC).
Annual. *Annual Report.*
Annual. *Public Broadcasting Revenue* (title varies).

Doyle, Marc. 1992. *The Future of Television: A Global Overview of Programming, Advertising, Technology and Growth.* NTC Business Books, Lincolnwood, IL.

Drake-Chenault Enterprises Inc. 1978. "History of Rock and Roll," Drake-Chenault, Canoga Park, CA.

Duncan, James. Five times a year. *American Radio.* Duncan Media, Indianapolis, IN.

Eastman, Susan Tyler, ed. 1993. *Broadcast/Cable Programming: Strategies and Practices,* 4th ed. Wadsworth, Belmont, CA.

———, and Newton, Gregory D. 1995. "Delineating Grazing: Observations of Remote Control Use," *Journal of Communication* 45 (Winter): 77.

EBUR. 1987. "Morning Has Broken: An Idea Whose Time Has Come," *European Broadcasting Union Review* (Sept.): 12.

EIA (Electronic Industries Association). Annual. *The U.S. Consumer Electronics Industry in Review* (title varies). EIA, Washington, DC (after 1994: Arlington, VA).

Ellerbee, Linda. 1986. *"And So It Goes": Adventures in Television.* Putnam, New York.

Ennes, Harold E. 1953. *Principles and Practices of Telecasting Operations.* Sams, Indianapolis.

F (*Federal Reporter, 2d and 3d Series*). West, St. Paul, MN.
1926. *U.S. v. Zenith Radio Corp.,* 12 F2d 614.
1969. *Office of Communication v. FCC,* 425 F2d 543.
1972. *Brandywine-Main Line Radio v. FCC,* 473 F2d 16.
1975. *Natl. Assn. of Independent TV Distributors v. FCC,* 516 F2d 526.
1977. *Home Box Office v. FCC,* 567 F2d 9.
1979. *Writers Guild v. ABC,* 609 F2d 355.
1986. *Telecommunications Research and Action Center v. FCC,* 801 F2d 501, *rehearing denied,* 806 F2d 1115 (D.C. Cir. 1986).
1987. *Century Communications Corp. v. FCC,* 835 F2d 292.
1988. *Action for Children's Television et al. v. FCC,* 852 F2d 1332.
1991. *Action for Children's Television v. FCC,* 932 F2d 1504.

1995. *Action for Children's Television v. FCC,* 58 F3d 654.

1996. *Becker v. FCC,* 95 F3d 75, 4 CR 882.

FCC (Federal Communications Commission). GPO, Washington, DC.

Annual Report. Annual.

Broadcast and Cable Employment Report. Annual.

1980a. (Network Inquiry Special Staff). *New Television Networks: Entry, Jurisdiction, Ownership and Regulation.* Final Report, Vol. 1. *Background Reports,* Vol. 2.

1980b. *Staff Report and Recommendations in the Low-Power Television Inquiry.* FCC, Washington, DC.

FCCR (*FCC Reports,* 1st and 2d Series; and *FCC Record*). GPO, Washington, DC.

1952. *Amendment of Sec. 3.606* [adopting new television rules] . . . Sixth Report and Order. 41 FCC 148.

1965. *Comparative Broadcast Hearings.* Policy Statement. 1 FCC2d 393.

1972a. *DOMSAT.* Report and Order. 35 FCC2d 844.

1972b. *Cable Television.* Report and Order. 36 FCC2d 143.

1972c. *Complaint by Atlanta NAACP.* Letter. 36 FCC2d 635.

1973. *Apparent Liability of Stations WGLD-FM* [Sonderling Broadcasting]. News Release. 41 FCC2d 919.

1985a. *Cattle Country Broadcasting [KTTL-FM].* Hearing Designation Order and Notice of Apparent Liability. 58 RR 1109.

1985b. *Inquiry into . . . the General Fairness Doctrine Obligations of Broadcast Licensees.* Report. 102 FCC2d 143.

1985c. *Application of Simon Geller.* Memorandum Opinion and Order. 102 FCC2d 1443.

1987a. *New Indecency Enforcement Standards to Be Applied to All Broadcast and Amateur Radio Licenses.* Public Notice. 2 FCC Rcd 2726.

1987b. *Complaint of Syracuse Peace Council Against Television Station WTVH.* Memorandum Opinion and Order. 2 FCC Rcd 5043.

1992a. *Telephone Company-Cable Television Cross-Ownership Rules.* Second Report and Order, Recommendation to Congress and Second Further Notice of Proposed Rulemaking. 7 FCC Rcd 5781. ("Video Dial Tone")

1992b. *Enforcement of Prohibitions Against Broadcast Indecency in 18 USC 1464.* 7 FCC Rcd 6464.

1995. *Minority and Female Ownership,* 10 FCC Rcd 2788.

1996. *Section 257 Proceeding to Identify and Eliminate Market Entry Barriers for Small Businesses,* 11 FCC Rcd 6280.

Ferris, Charles, et al. 1983–present. *Cable Television Law: A Video Communications Guide,* 3 vols. Matthew Bender, New York.

Friendly, Fred. 1976. *The Good Guys, the Bad Guys, and the First Amendment: Free Speech vs. Fairness in Broadcasting.* Random House, New York.

FS (*Federal Supplement*). West, St. Paul, MN.

1996. *American Civil Liberties Union v. Reno,* 929 F. Supp. 824.

Giovannoni, David G., et al. 1992. *Public Radio Programming Strategies: A Report on the Programs Stations Broadcast and the People They Seek to Serve.* Corporation for Public Broadcasting, Washington, DC.

Goldenson, Leonard H., with Wolf, Marvin J. 1991. *Beating the Odds: The Untold Story Behind the Rise of ABC—The Stars, Struggles and Egos That Transformed Network Television.* Scribner's, New York.

Great Britain, Command (Cmnd.) Papers. 1986. *Report of the Committee on Financing the BBC* ["Peacock Report"], Cmnd. 9824. Her Majesty's Stationery Office, London.

Greenfield, Jeff. 1977. *Television: The First Fifty Years.* Abrams, New York.

Greenfield, Thomas A. 1989. *Radio: A Reference Guide.* Greenwood, Westport, CT.

Halberstam, David. 1979. *The Powers That Be.* Knopf, New York.

Head, Sydney W. 1985. *World Broadcasting Systems: A Comparative Analysis.* Wadsworth, Belmont, CA.

Intellectual Property and the National Information Infrastructure: The Report of the Working Group on Intellectual Property Rights. 1996. GPO, Washington, DC.

Isailovic, Jordan. 1985. *Videodisc and Optical Memory Systems.* Prentice-Hall, Englewood Cliffs, NJ.

Johnson, Leland L. 1992. *Telephone Company Entry into Cable Television: Competition, Regulation, and Public Policy.* Rand, Santa Monica, CA.

Jowett, Garth. 1976. *Film: The Democratic Art.* Little, Brown, Boston.

Kaltenborn, H. V. 1938. *I Broadcast the Crisis.* Random House, New York.

Katz Television Group. 1997. *Katz Syndicated Programming Guide, 1997.* Katz Television Group, New York.

Kendrick, Alexander. 1969. *Prime Time: The Life of Edward R. Murrow.* Little, Brown, Boston.

Klapper, Joseph T. 1960. *The Effects of Mass Communication.* Free Press, Glencoe, IL.

Klein, Paul. 1971. "The Men Who Run TV Aren't That Stupid . . . They Know Us Better Than You Think," *New York* (January 25): 20.

Koch, Tom. 1991. *Journalism for the 21st Century: Online Information, Electronic Databases and the News.* Praeger, New York.

Mallinson, John C. 1993. *The Foundations of Magnetic Recording.* Academic, San Diego, CA.

Markey, Edward J. 1988. "Statement Accompanying Congressional Research Service Letter to House Subcommittee on Telecommunications and Finance. Cases Involving the Federal Communications Commission That Were Reversed . . ." (March 21). Library of Congress, Washington, DC.

Mayo, John S. 1992. "The Promise of Networked Multimedia Communications." Speech delivered at Bear Sterns Sixth Annual Media & Communications Conference, Coronado, CA (October 28).

Metz, Robert. 1975. *CBS: Reflections in a Bloodshot Eye.* Playboy Press, Chicago.

———. 1977. *The Today Show: An Inside Look . . .* Playboy Press, Chicago.

Meyersohn, Rolf B. 1957. "Social Research in Television," in Rosenberg and White, 345–357.

Minow, Newton N. 1964. *Equal Time: The Private Broadcaster and the Public Interest.* Atheneum, New York.

———. 1991. "How Vast the Wasteland Now?" Gannett Foundation Media Center, Columbia University, New York (May 9).

Montgomery, Kathryn C. 1989. *Target: Prime Time: Advocacy Groups and the Struggle Over Entertainment Television.* Oxford U. Press, New York.

NAB (National Association of Broadcasters), Washington, DC. Annual. *Radio Financial Report* (to 1992).
Annual. *Television Financial Report.*
Semiannual. *Broadcast Regulation* (title varies).
1978. *The Television Code.*

National Cable Television Association. Quarterly. *Cable Television Developments.* Washington, DC.

New Century Net. 1997. http://www.newcentury.net

NewMedia. Monthly. HyperMedia, San Mateo, CA.

Newseum. 1997. http://www.newseum.org

Newspage. 1997. http://www.newspage.com

Nielsen. 1996. *Television Audience.*

Nielsen Media Research. 1993. *1992–1993 Report on Television.* Nielsen Media Research, New York.

Ohio State University. School of Journalism. Annual. *Journalism and Mass Communications Graduate Survey: Summary Report.* Ohio State University, Columbus, OH.

Paley, William S. 1979. *As It Happened: A Memoir.* Doubleday, New York.

Pareles, Jon. 1989. "After Music Videos, All the World Has Become a Screen," *New York Times* (10 December): E6.

Paterson, Richard. Annual. *TV and Video International Guide.* Tantivy, London.

Paul Kagan Associates. Annual. *The Cable TV Financial Databook.* Carmel, CA.

Pichaske, David. 1979. *A Generation in Motion: Popular Music and Culture of the Sixties.* Schirmer Books, New York.

Pike and Fischer. *Radio Regulation* (after 1995 *Communications Regulation*). Bethesda, MD.

Pohlmann, Ken C. 1989. *The Compact Disc: A Handbook of Theory and Use.* A-R Editions, Madison, WI.

Radio World. Bi-weekly. IMAS, Falls Church, VA.

Rather, Dan. 1987. "From Murrow to Mediocrity?" *New York Times* (March 10): A27.

Roper Organization. 1993. *America's Watching: Public Attitudes Toward Television.* National Association of Broadcasters, Washington, DC, and Network Television Association, New York, NY.

Rosenberg, Bernard, and White, David Manning, eds. 1957. *Mass Culture: The Popular Arts in America.* Free Press, Glencoe, IL.

Rubin, Bernard. 1967. *Political Television.* Wadsworth, Belmont, CA.

Rumsey, Francis. 1990. *Tapeless Sound Recording.* London.

Senate CC (U.S. Congress, Senate Committee on Commerce). 1972. *Surgeon General's Report by Scientific Advisory Committee on Television and Social Behavior.* Hearings. 92nd Cong., 2d Sess. GPO, Washington, DC.

SGAC (Senate Government Affairs Committee). 1991. *Pentagon Rules on Media Access to the Persian Gulf War.* Hearing. 102nd Cong., 1st Sess. GPO, Washington, DC.

Settel, Irving. 1967. *A Pictorial History of Radio,* 2d ed. Grosset and Dunlap, New York.

———. 1983. *A Pictorial History of Television,* 2d ed. Ungar, New York.

Setzer, Florence, and Levy, Jonathan. 1991. *Broadcast Television in a Multichannel Marketplace.* FCC Office of Plans and Policy (Staff Study No. 26, June).

Shawcross, William. 1992. *Murdoch.* Simon and Schuster, New York.

Sieber, Robert. 1988. "Industry Views on the People Meter: Cable Networks," *Gannett Center Journal* 2 (Summer): 70.

Singleton, Loy A., and Copeland, Gary A. 1997. "FCC Broadcast Indecency Enforcement Actions: Changing Patterns Since *Pacifica*," Law and Policy Division, BEA Annual Conference.

Smith, Bruce Lannes, et al. 1946. *Propaganda, Communication and Public Opinion.* Princeton U. Press, Princeton, NJ.

Smith, Richard A. 1985. "TV: The Light That Failed," *Fortune* (December): 78.

Smith, Sally Bedell. 1990. *In All His Glory: The Life of William S. Paley—The Legendary Tycoon and His Brilliant Circle.* Simon and Schuster, New York.

Steiner, Gary A. 1963. *The People Look at Television: A Study of Audience Attitudes.* Knopf, New York.

Stephenson, William. 1967. *The Play Theory of Mass Communication.* U. of Chicago Press, Chicago.

Sterling, Christopher H. 1984. *Electronic Media: A Guide to Trends in Broadcasting and Newer Technologies, 1920–1983.* Praeger, New York.

———, and Kittross, John M. 1990. *Stay Tuned: A Concise History of American Broadcasting*, 2d ed. Wadsworth, Belmont, CA.

Strong, William S. 1993. *The Copyright Book: A Practical Guide*, 4th ed. MIT Press, Cambridge, MA.

TCAF (Temporary Commission on Alternative Financing for Public Telecommunications). 1982–1983. *Alternative Financing Options for Public Broadcasting* (Vol. 1); *Final Report* (Vol. 2). FCC, Washington, DC.

Toffler, Alvin. 1970. *Future Shock.* Random House, New York.

———. 1980. *The Third Wave.* Morrow, New York.

TV Technology. Semi-monthly. IMAS, Falls Church, VA.

US (*United States Reports*). GPO, Washington, DC.

1919. *Schenk v. U.S.*, 249 US 47.

1942. *Marconi Wireless Telegraph Co. of America v. U.S.* 320 US 1.

1943. *NBC v. U.S.*, 319 US 190.

1949. *Terminiello v. Chicago*, 337 US 1.

1951. *Dennis v. U.S.*, 341 US 494.

1957. *Roth v. United States*, 354 US 476.

1964. *New York Times v. Sullivan*, 376 US 254.

1968. *U.S. v. Southwestern Cable*, 392 US 157.

1969. *Red Lion v. FCC*, 395 US 367.

1973. *Miller v. California*, 413 US 15.

1974. *Miami Herald v. Tornillo*, 418 US 241.

1978. *FCC v. Pacifica Foundation*, 438 US 726.

1981. *Chandler v. Florida*, 449 US 560.

1984. *Universal Studios v. Sony*, 464 US 417.

1986. *City of Los Angeles and Department of Water and Power v. Preferred Communications, Inc.*, 476 US 488.

1989. *Sable Communications of California, Inc. v. FCC*, 492 US 115.

1990. *Metro Broadcasting v. FCC*, 497 US 547.

1995. *Adarand Constructors, Inc. v. Pena*, 515 US 200, 115 SCt 2097.

1996. *Denver Area Educational Telecommunications Consortium v. FCC*, 116 SCt 2374.

USC (*United States Code*). Regularly revised. GPO, Washington, DC.

U.S. Commission on Civil Rights. 1977, 1979. *Window Dressing on the Set: Women and Minorities in Television*, 2 vols. GPO, Washington, DC.

USDOS (U.S. Department of State). 1990. *Eastern Europe: Please Stand By—Report of the Task Force on Telecommunications and Broadcasting in Eastern Europe.* GPO, Washington, DC.

Variety. Weekly. Variety, New York.

The Veronis, Suhler & Associates Communications Industry Forecast: Historical and Projected Expenditures for 9 Industry Segments. Annual. Veronis, Suhler, New York.

Webster, James, and Lichty, Lawrence W. 1991. *Ratings Analysis: Theory and Practice.* Erlbaum, Hillsdale, NJ.

Williams, Christian. 1981. *Lead, Follow, or Get Out of the Way: The Story of Ted Turner.* Times Books, New York.

Williams, Margorie. 1989. "MTV as Pathfinder for Entertainment," *The Washington Post* (December 13): A1.

Wood, James. 1992a. *History of International Broadcasting.* Peter Peregrinus/Science Museum, London.

———. 1992b. *Satellite Communications and DBS Systems.* Focal, Stoneham, MA.

World Radio-TV Handbook. Annual. Billboard, New York.

Wurtzel, Alan, and Acker, Stephen R. 1989. *Television Production*, 3d ed. McGraw-Hill, New York.

Zettl, Herbert. 1996. *Television Production Handbook*, 6th ed. Wadsworth, Belmont, CA.

Index